An Integrated Approach to Computational Mathematics

An Integrated Approach to Computational Mathematics

Edited by
Lucas Lincoln

Willford Press
www.willfordpress.com

Published by Willford Press,
118-35 Queens Blvd., Suite 400,
Forest Hills, NY 11375, USA

ISBN: 978-1-68285-564-5

Cataloging-in-Publication Data

An integrated approach to computational mathematics / edited by Lucas Lincoln.
p. cm.
Includes bibliographical references and index.
ISBN 978-1-68285-564-5
1. Mathematics--Data processing. 2. Mathematics. I. Lincoln, Lucas.
QA76.95 .I58 2019
510--dc23

For information on all Willford Press publications
visit our website at www.willfordpress.com

Willford Press

Contents

Preface

Every book is a source of knowledge and this one is no exception. The idea that led to the conceptualization of this book was the fact that the world is advancing rapidly; which makes it crucial to document the progress in every field. I am aware that a lot of data is already available, yet, there is a lot more to learn. Hence, I accepted the responsibility of editing this book and contributing my knowledge to the community.

Computational mathematics is the mathematics that is used in scientific areas where computing is applied for the analysis and regulation of data. The focus of this area of study is computing for varied purposes by creating diverse numerical methods, algorithms or symbolic computations. This book brings forth some of the most innovative concepts and elucidates the unexplored aspects of computational mathematics. The various studies that are constantly contributing towards advancing technologies and evolution of this field are also examined in detail. The book studies, analyses and upholds the pillars of this field and its utmost significance in modern times. Coherent flow of topics, student-friendly language and extensive use of examples make this book an invaluable source of knowledge. Those who want to gain deeper insights into this area of study will find it immensely helpful.

While editing this book, I had multiple visions for it. Then I finally narrowed down to make every chapter a sole standing text explaining a particular topic, so that they can be used independently. However, the umbrella subject sinews them into a common theme. This makes the book a unique platform of knowledge.

I would like to give the major credit of this book to the experts from every corner of the world, who took the time to share their expertise with us. Also, I owe the completion of this book to the never-ending support of my family, who supported me throughout the project.

Editor

Convergence Control Parameter Region for an Unsteady Three Dimensional Navier-Stokes Equations of Flow between Two Parallel Disks by using Homotopy Analysis Method

Selvarani S* and Beulah RD

Department of Mathematics, VLB Janakiammal College of Arts and Science, Coimbatore-641 042, Tamilnadu, India

Abstract

Purpose: The paper aims to find the convergence control parameter region for an unsteady three dimensional Navier-Stokes equations of flow between two parallel disks by using Homotopy Analysis Method.

Findings: The region and value of the convergence control parameter has been found.

Keywords: 3D Navier-Stokes equation; Homotopy analysis method; Convergence; System of nonlinear differential eqiations

Introduction

In mathematics and physics, nonlinear partial differential equations are partial differential equations with nonlinear terms. A few nonlinear differential equations have known exact solutions, but many which are important in applications do not. Sometimes these equations may be linearized by an expansion process in which nonlinear terms are discarded. When nonlinear terms make vital contributions to the solution this cannot be done, but sometimes it is enough to retain a few small ones. Then a perturbation theory may be used to obtain the solution. The differential equations may sometimes be approximated by an equation with small nonlinearities in more than one way, giving rise to different solutions valid over different range of its parameters.

Most scientific and engineering problems are modeled by ordinary differential equations or partial differential equations, Some of them are solved using the analytic methods of perturbation by Nayfeh [1]. In the numerical methods, stability and convergence should be considered so as to avoid divergence or inappropriate results. In the analytic perturbation methods, we should excert the small parameter in the equation. In numerical methods the advantage is that we have to use the small parameter a lot since most problems do not have known analytic solutions, or that if they are known it is too complex to deal with them. The main advantage in analytic method is that it is exact and gives us more context. One of the semi-exact methods which do not need small or large parameters is the Homotopy Analysis Method (HAM), first proposed by Liao in his Ph.D thesis. Liao [2] employed the basic ideas of homotopy in topology to propose a general analytic method for nonlinear problems, namely HAM, which is a powerful analytical method for solving linear and nonlinear differential equations. The HAM also avoids discretization and provides an efficient solution with high accuracy, minimal calculations and avoidance of physically unrealistic assumption. Furthermore, the HAM always provides us with a family of solution expressions with the auxiliary parameter $\hbar$, the convergence region and the rate of each solution might be determined conveniently by the auxiliary parameter $\hbar$. HAM contains the homotopy perturbation method (HPM) discussed by He [3], the Adomian decomposition method (ADM) examined by Allan [4], and the d-expansion method.

The main goal of the present study is to find the value of convergence control parameter for the problem of flow between two disks by the HAM.

Mathematical Formulation

Consider the axis-symmetric flow between two infinite disks with a distance d between them. Both disks are placed in the radial direction with a velocity proportional to the radii. The bottom disk is located in the z = 0 plane. The velocity ratio of the upper disk to the lower one is γ and ε is the amplitude of the disk. For an incompressible fluid without body forces and based on axis symmetric reads from the papers discussed by Dinarvand [5] and Munnavar [6].

$$\frac{1}{r}\frac{\partial}{\partial r}(ru_r)+\frac{\partial u_z}{\partial z}=0, \tag{1}$$

$$\frac{\partial u_r}{\partial t}+u_r\frac{\partial u_r}{\partial r}+u_z\frac{\partial u_r}{\partial z}=\nu(\frac{\partial^2 u_r}{\partial r^2}+\frac{1}{r}\frac{\partial u_r}{\partial r}+\frac{\partial^2 u_r}{\partial z^2}-\frac{u_r}{r^2}), \tag{2}$$

$$\frac{\partial u_z}{\partial t}+u_r\frac{\partial u_z}{\partial r}+u_z\frac{\partial u_z}{\partial z}=\nu\left(\frac{\partial^2 u_z}{\partial r^2}+\frac{1}{r}\frac{\partial u_z}{\partial r}+\frac{\partial^2 u_z}{\partial z^2}\right). \tag{3}$$

Where the velocity vector $\overline{V}=(u_r,u_z),\nu$ is the kinematic viscosity. By using von Karman type similarity transformations, similarity functions can be sought as follows,

$$u_r = rF(\eta),\ and\ \ u_z = dH(\eta),$$

where $\eta=\frac{z}{d}=\frac{z}{\gamma t}$ is the similarity variable. Substituting the similarity functions into the equations (1), (2) and (3). Therefore, the governing equations yields a similarity equation group

$$\begin{cases} F'' &= Re(F^2+ReF+2ReHF+ReH^2F), \\ H' &= -2F. \end{cases} \tag{4}$$

with boundary conditions

***Corresponding author:** Department of Mathematics, VLB Janakiammal College of Arts and Science, Coimbatore - 641 042, Tamilnadu, India
E-mail: s.selvarani91@gmail.com

$$F(0)=1+\varepsilon cost,\quad H(0)=H(1)=0 \quad and \quad F(1)=\gamma. \tag{5}$$

Where $Re=\frac{d^2}{\nu}$ is the Reynolds number of the wall and γ is the parameter of the upper disk showing the velocity ratio of the upper disk to the bottom disk. Without loss of generality, we assumed that $0 \le \gamma \le 1$.

Analytical solution with HAM

Due to basic idea of HAM, as described in detail by Liao [7,8], according to the boundary conditions (5), we choose

$$F_0(\eta)=1-(1+\varepsilon cost-\gamma)\eta+\varepsilon cost, \tag{6}$$

$$H_0(\eta)=0, \tag{7}$$

as initial guesses of F(η), and H(η) which satisfy the boundary conditions (5). Besides, we select the auxiliary linear operators $L_1(F)$, and $L_2(H)$ as

$$\mathcal{L}_1(F)=F'', \tag{8}$$

$$\mathcal{L}_2(H)=H', \tag{9}$$

satisfying the follwing properties

$$\mathcal{L}_1(c_1\eta+c_2)=0, \tag{10}$$

$$\mathcal{L}_2(c_3)=0. \tag{11}$$

Where c_i, I=1, 2, 3 are arbitrary constants. If q ∈ [0, 1] is an embedding parameter and ħ is an auxiliary nonzero parameter, then the zeroth-order deformation equations are of the following form,

$$(1-q)\mathcal{L}_1\left[\hat{F}(\eta;q)-F_0(\eta)\right]=q\hbar\mathcal{N}_1\left[\hat{H}(\eta;q),\hat{F}(\eta;q)\right], \tag{12}$$

$$(1-q)\mathcal{L}_2\left[\hat{H}(\eta;q)-H_0(\eta)\right]=q\hbar\mathcal{N}_2\left[\hat{H}(\eta;q),\hat{F}(\eta;q)\right]. \tag{13}$$

subject to the boundary conditions

$$\hat{F}(0;q)=1+\varepsilon cost,\quad \hat{F}(1;q)=\gamma,$$

$$\hat{H}(0;q)=\hat{H}(1;q)=0,$$

in which we define the nonlinear operators $\mathcal{N}_1$ and $\mathcal{N}_2$ as

$$\mathcal{N}_1\left[\hat{H}(\eta;q),\hat{F}(\eta;q)\right]=\frac{\partial^2\hat{F}(\eta;q)}{\partial\eta^2}-Re(2Re(\hat{H}(\eta;q))(\hat{F}(\eta;q))+Re(\hat{F}(\eta;q))$$

$$+(\hat{F}(\eta;q))^2+Re(\hat{H}(\eta;q))^2(\hat{F}(\eta;q)),$$

$$\mathcal{N}_2\left[\hat{H}(\eta;q),\hat{F}(\eta;q)\right]=\frac{\partial\hat{H}(\eta;q)}{\partial\eta}+2\hat{F}(\eta;q).$$

Clearly, when q = 0 the zero-order deformation equations (12) and (13) give rise to:

$$\hat{F}(\eta;0)=F_0(\eta),\quad \hat{H}(\eta;0)=H_0(\eta),\quad \hat{P}(\eta;0)=P_0(\eta). \tag{14}$$

when q=1, they become:

$$\hat{F}(\eta;1)=F(\eta),\quad \hat{H}(\eta;1)=H(\eta),\quad \hat{P}(\eta;1)=P(\eta). \tag{15}$$

As q increases from 0 to 1, $\hat{F}(\eta;q)$ and $\hat{H}(\eta;q)$ vary from F_0(η) and H_0(η) to F(η) and H(η).

Expanding $\hat{F}(\eta)$ and $\hat{H}(\eta)$ in Maclaurin series with respect to the embedding parameter q and equations (14) and (15), we obtain

$$\hat{F}(\eta;q)=F_0(\eta)+\sum_{m=1}^{+\infty}F_m(\eta)q^m, \tag{16}$$

$$\hat{H}(\eta;q)=H_0(\eta)+\sum_{m=1}^{+\infty}H_m(\eta)q^m, \tag{17}$$

where

$$F_m(\eta)=\frac{1}{m!}\frac{\partial^m\hat{F}(\eta,q)}{\partial q^m}|_{q=0},$$

$$H_m(\eta)=\frac{1}{m!}\frac{\partial^m\hat{H}(\eta,q)}{\partial q^m}|_{q=0}.$$

As pointed by Liao [9], the convergence of the series (16) - (17) strongly depend upon auxiliary parameter ħ. Assume that ħ is selected such that the series (16) - (17) are convergent at q = 1 then due to equations (14) and (15) we have

$$F(\eta)=F_0(\eta)+\sum_{m=1}^{+\infty}F_m(\eta), \tag{18}$$

$$H(\eta)=H_0(\eta)+\sum_{m=1}^{+\infty}H_m(\eta). \tag{19}$$

Differentiating the zero-order deformation equations (12) and (13) m times with respect to q, then setting q = 0 and finally dividing by m! we have the mth - order deformation equations [10-18].

$$\mathcal{L}_1\left[F_m(\eta)-\chi_m F_{m-1}(\eta)\right]=\hbar R_{1,m}(\eta), \tag{20}$$

$$\mathcal{L}_2\left[H_m(\eta)-\chi_m H_{m-1}(\eta)\right]=\hbar R_{2,m}(\eta). \tag{21}$$

with the following boundary conditions

$$F_m(0)=F_m(1)=0 \text{ and } H_m(0)=H_m(1)=0, \tag{22}$$

where

$$R_{1,m}(\eta)=\frac{\partial^2F_{m-1}(\eta)}{\partial\eta^2}-Re\sum_{i=0}^{m-1}(F_i(\eta)F_{m-1-i}(\eta)+ReF_i(\eta)$$

$$+2ReH_i(\eta)F_{m-1-i}(\eta)+\sum_{l=0}^{i}ReH_i(\eta)H_{i-1}(\eta)F_{m-1-i}(\eta),$$

$$R_{2,m}(\eta)=\frac{\partial H_{m-1}(\eta)}{\partial\eta}+2F_{m-1}(\eta),$$

and

$$\chi_m=\begin{cases}0 & m\le 1,\\ 1 & m>1.\end{cases}$$

Then the solutions for equations (20) and (21) can be expressed by:

$$F_m(\eta)=\chi_m F_{m-1}+\hbar\mathcal{L}_1^{-1}[R_{1,m}]+c1\eta+c2,$$

$$H_m(\eta)=\chi_m H_{m-1}+\hbar\mathcal{L}_2^{-1}[R_{2,m}]+c3.$$

Where c1, c2, c3 are integral constants can be found by boundary conditions (22). For example, we can obtain the following result for solving the first-order deformation equation by using symbolic software *MATHEMATICA*, and successively obtain [19-25],

$$F_1(\eta)=\frac{-1}{24}\hbar(-1+n)nRe(9-9n+3n^2+8Re-4nRe+4v+4nv-4n^2v$$

$$+4Rev+4nRev+2V^2+2nv^2+2n^2v^2-4(-3-2Re+n(3+Re-v)$$

$$+n^2(-1+v)-v)Cos[t]+(3-3n+n^2)Cos[2t]),$$

$$H_1(\eta)=\hbar(-3+2n+n^2(-1+v)+v-(3-2n+n^2)Cos[t].$$

Convergence of HAM solution

The totally analytic series solutions of the functions F(η) and H(η) are given in equations (18) - (1819). The convergence of these series and the rate of approximation for the HAM strongly depends upon the value of the auxiliary parameter ħ, as pointed out by Liao

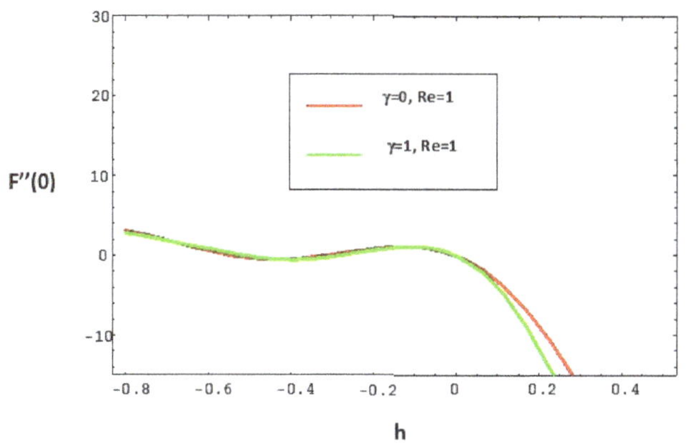

Figure 1: The h – curve of F″(0), obtained by 11th order approximation of the HAM for γ=0, γ=1, ε=1 and t=1 with Re=1.

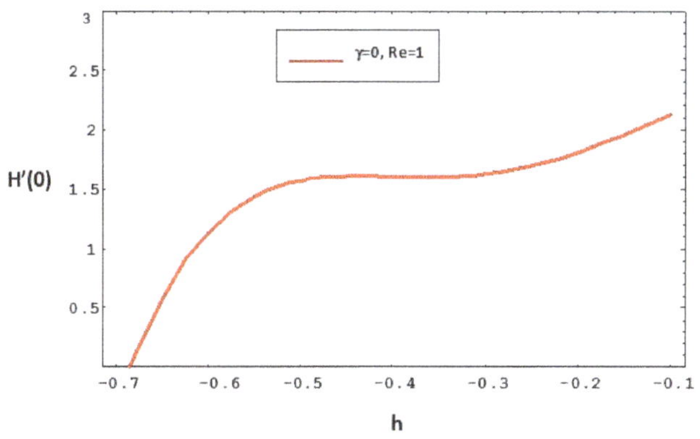

Figure 2: The h – curve of H″(0), obtained by 11th order approximation of the HAM for γ=0, γ=1, ε=1 and t=1 with Re=1.

[8]. In general, by means of the $\hbar$-curve, it is straightforward to choose a proper value of ħ to control the convergence of the approximation series. To find the range of the admissible values of ħ, ħ - curves of $F''(0)$ and $H'(0)$ obtained by the 11th order approximation of the HAM for $\gamma = 0$ and $\gamma = 1$ at $Re = 1$, $t = 1$ and $\varepsilon = 1$ are plotted in Figures 1 and 2, respectively. From these figures, the valid regions of ħ correspond to the line segments nearly parallel to the horizontal axis. Sometimes this region is not perfectly flat to the slowly convergence rate of the series solution which was discussed by Liao in his book Homotopy Analysis Method. However a value of ħ can be picked up. Therefore in this problem we can choose $\hbar = -0.4$ [26-28].

Conclusion

In this paper, the HAM was used for finding the convergence control parameter of the system of nonlinear ODE derived from von Karman type similarity transform for the unsteady state three dimensional Navier-Stokes equations of flow between two parallel disks. Unlike perturbation methods, the HAM does not depend on any small physical parameters. Thus homotopy analysis method is valid for both weakly and strongly nonlinear problems. Different from all other analytic methods, the homotopy analysis method provides us a simple way to adjust and control the convergence region of the series solution by means of auxiliary parameter ħ. Thus, the auxiliary parameter ħ plays a vital role within the frame of HAM which can be determined by the ħ curves.

References

1. Nayfeh AH (1979) Introduction to perturbation techniques. Wiley, New York.
2. Liao SJ (2005) Comparison between the homotopy analysis method and Homotopy Perturbation Method. Journal of Computer and Applied Mathematics 169: 1186-1194.
3. He JH (2006) Homotopy Perturbation Method for solving boundary value problems. Physics Letters A 350: 87-88.
4. Allan FM (2007) Derivation of the Adomain Decomposition Method using Homotopy Analysis Method. Applied Mathematics and Computation 190: 6-14.
5. Dinarvand S, Rashidi MM, Shahmohamadi H (2009) Analytic approximate solution of three dimentional Navier-Stokes equations of flow between two streatchable disks. Wiley online library, Wiley periodicals.
6. Munawar S, Ali A, Saleem N, Naqeep A (2014) Swirling flow over an oscillatory stretchable disk. Journal of Mechanics 30: 339-347.
7. Liao SJ (2003) Beyond Perturbation: Introduction to homotopy analysis method. Chapman and Hall, CRC Press, Boca Raton.
8. Liao SJ (2003) On the analytic solution of magnetohydrodynamic flows of non-Newtonian fluids over a stretching sheet. Journal of Fluid Mechanics 488: 189-212.
9. Liao SJ (2004) On the homotopy analysis for nonlinear problems. Applied Mathematics and Computation 147: 499-513.
10. Abbasbandy S (2006) The application of homotopy analysis method to nonlinear equations arising in heat transfer. Physics Letters A 360: 109-113.
11. Allan FM, Syam MI (2005) On the analytic solution of non-homogeneous Blasius problem. Journal of Computer and Applied Mathematics 182: 362-371.
12. Allan FM (2009) Constructions of analytic solution to chaotic dynamical systems using the homotopy analysis method. Chaos Solitons Fractals 39: 1744-1752.
13. Bouremel Y (2007) Explicit series solution for the Glauert-jet problems by means of homotopy analysis method. Communications in Nonlinear Science Numerical Simulation 12: 714-724.
14. Ganji DD, Hosseini MJ, Shayegh J (2007) Some non linear heat transfer equations solved by three approximate methods. International Journal of Heat Mass Transfer 34: 1003-1016.
15. Hayat T, Ahmad N, Sajid M, Asghar S (2007) On the MHD flow of a second grade fliud in a porous channel. Journal Of Computer and Applied Mathematics 54: 407-414.
16. Hayat T, Sajid M, Ayub M (2007) A note on series solution for generalized Couette flow. Communications in Nonlinear Science Numerical Simulation 12: 1481-1487.
17. He JH (2000) A coupling method for homotopy technique and perturbation technique for nonlinear problems. International Journal of Mechanics 35: 37-43.
18. Ibrahim MO, Egbetade SA (2013) On the homotopy analysis method for an seir tuberculosis model. American Journal of Applied Mathematics and Statistics 1: 71-75.
19. Nave O, Lehavi Y, Dshtein VG (2012) Application of the HPM and HAM to the problem of the thermal explosion in a radiation gas with polydisperse fuel spray. Journel of Applied and Computational Mathematics.
20. Rafei M, Ganji DD, Daniali H (2007) Solution of the epidemic model by homotopy perturbation method. Applied Mathematics and Computation 187: 1056-1062.
21. Rafei M, Daniali H, Ganji DD (2007) Variational iteration method for solving the epidemic model and the prey and predator problem. Journal of Computer and Applied Mathematics 186: 1701-1709.
22. Ran XJ, Zhu QY, Li Y (2009) An explicit series solution of the squeezing flow between two infinite plates by means of the homotopy analysis method. Communications in Nonlinear Science Numerical Simulation 14: 119-132.
23. Rand RH, Armbruster D (1987) Perturbation methods, bifurcation theory and computer algebraic. Applied Mathematical Sciences, Springer-Verlag, New York.

24. Rashidi MM, Domairry G, Dinarvand S (2009) Approximate solutions for the Burger and regularized long wave equations by means of the homotopy analysis method. Communications in Nonlinear Science Numerical Simulation 14: 708-717.

25. Tsai CC (2012) Homotopy method of fundamental solutions for solving certain nonlinear partial differential equations. Engineering Analysis with Boundary Elements 36: 1226-1234.

26. White FM (1991) Viscous fluid flow (2nd edn). McGraw-Hill, New York.

27. Ziabakhsh Z, Domairry G (2009) Solution of the laminar viscous flow in a semi-porous channel in presence of a uniform magnetic field by using the homotopy analysis method. Communications in Nonlinear Science Numerical Simulation 14: 1284-1294.

28. Zurigat M, Momani S, Alawneh A (2013) The multistage HAM: Application to a biochemical reaction model of fractional order 91: 1030-1040.

2

Sensitivity Analysis for General Nonlinear Nonconvex Variational Inequalities

Salahuddin*

Department of Mathematics, Jazan University, Jazan, Kingdom of Saudi Arabia

Abstract

In this communication, we proved that the parametric general nonlinear nonconvex variational inequalities are equivalent to the parametric general Wiener-Hopf equations. We use this alternative equivalence formulation to studied the sensitivity analysis for general nonlinear nonconvex variational inequalities without assuming the differentiability of the given data.

Keywords: Sensitivity analysis; Parametric general nonlinear nonconvex variational inequalities; Fixed point; Parametric general Wiener-Hopf equations; (φ,ψ)-relaxed cocoercive mapping; Lipschitz continuous mappings; uniformly r-prox regular sets; Hilbert spaces

AMS Mathematics Subject Classification: 49J40, 47H06

Historical background

The variational inequality theory was introduced by Stampacchia [1] has become a rich source of inspiration and motivation for the study of a large number of problems arising in economics, finance, transportation, networks, structural analysis and optimizations [2-5]. It should be pointed that almost all the results regarding the existence and iterative scheme for solving variational inequalities and related optimization problems are being considered in the convex setting. Consequently all the techniques are based on the properties of the projection operators are convex sets which may not hold in general when the sets are nonconvex. It is known that the uniformly r-prox regular sets are nonconvex and included the convex sets as a special cases [6-9].

Over the last decade there has been increasing interest in studying the sensitivity analysis of variational inequalities and variational inclusions. Sensitivity analysis for variational inclusions and inequalities have been studied extensively [2,3,10-13].

The techniques suggested so far vary with the problems being studied. Dafermos used the fixed point formulation to considered the sensitivity analysis of the classical variational inequalities. These techniques have been modified and extended by many authors for studying the sensitivity analysis of the other classes of variational inequalities and variational inclusions. It is known that the variational inequalities are equivalent to Wiener-Hopf equations [14]. This alternative equivalence formulation has been used by Noor [15-17] to developed the sensitivity analysis frame work for various classes of (quasi) variational inequalities.

In this paper we develop the general frame work of sensitivity analysis for general non-linear nonconvex variational inequalities. First we establish the equivalence between the parametric general nonlinear nonconvex variational inequalities and the parametric general Wiener-Hopf equations by using the projection techniques. By using the fixed point formulation, we obtain an approximate rearrangement of the Wiener-Hopf equations. We use this equivalence to developed the sensitivity analysis for general nonlinear nonconvex variational inequalities without assuming the differentiability of the given data.

Preliminaries

Let H be a real Hilbert space whose inner product and norm are denoted by $\langle .,. \rangle$ and $\|.\|$ respectively. Let K be a nonempty closed subset of H.

Definition 2.1 The proximal normal cone of K at a point u $\in$ H with u $\notin$ K is given by $N_k^p(\mathrm{u}) = \{\xi \in \mathrm{H} : \mathrm{u} \in \mathrm{P_k}(\mathrm{u}+\alpha\xi)$ for some $\alpha > 0\}$

Where $\alpha > 0$ is a constant and

$P_k(\mathrm{u}) = \{\mathrm{v} \in \mathrm{k} : \mathrm{d_k}(\mathrm{u}) = \| \mathrm{u}-\mathrm{v} \| \}$

Where dK(.) or d(.;K) is the usual distance function to the subset of K, that is

$d_k(\mathrm{u}) = \inf \| \mathrm{u}-\mathrm{v} \|$.

The proximal normal cone $N_k^p(\mathrm{u})$ has the following characterizations:

Lemma 2.2 Let K be a nonempty closed subset in H. Then $\varsigma \in N_k^p(\mathrm{u})$ if and only if there exists a constant $\alpha = \alpha(\varsigma, \mathrm{u}) > 0$ such that

$\langle \varsigma, v-u \rangle \leq \alpha \| \mathrm{v}-\mathrm{u} \|^2, \forall v \in k$

Lemma 2.3 Let K be a nonempty closed and convex subset in H. Then $\varsigma \in N_k^p(\mathrm{u})$

$\langle \varsigma, v-u \rangle \leq 0, \forall v \in k$

The Clarke normal cone denoted by $N_k^c(\mathrm{u})$ is defined by

$N_k^c(\mathrm{u}) = \mathrm{co}[\mathrm{N}_k^p(\mathrm{u})]$

where co mean the closure of the convex hull.

Clearly $\mathrm{N}_k^p(\mathrm{u}) \subseteq \mathrm{N}_k^c$ but the converse is not true in general. Note that $N_k^c(\mathrm{u})$ is always closed and convex cone where as $N_k^c(\mathrm{u})$ is convex

***Corresponding author:** Salahuddin, Department of Mathematics, Jazan University, Jazan, Kingdom of Saudi Arabia
E-mail: salahuddin12@mailcity.com

but may not be closed, see [5, 12].

Definition 2.4 For any given $r \in (0, +\infty]$ a subset Kr of H is said to be normalized uniformly r-prox regular (or uniformly r-prox regular) if and only if every nonzero proximal normal to Kr can be realized by an r-ball that is for all $u \in k_r$ and $0 \neq \varsigma \in N_{kr}^{P}(u)$ with $\|\varsigma\| = 1$

$$\langle \varsigma, v-u \rangle \leq \frac{1}{2r} \| v-u \|^2, v \in k_r$$

Lemma 2.5 A closed set K $\subseteq$ H is convex if and only if it is proximally smooth of radius r for every r>0:

If r=1 then uniformly r-prox regularity of Kr is equivalent to the convexity of K: If Kr is uniformly r-prox regular set, then the proximal normal cone N_{kr}^{P} (u) is closed as a set valued mapping. If we take $n = \frac{1}{2r}$ it is clear that r $\to \infty$ then n=0:

Proposition 2.6 [12] For r>0, let Kr be a nonempty closed and uniformly r-prox regular subset of H. Set

$$u = \{u \in h : 0 \leq d_{kr}(u) < r\}$$

Then the following statements are holds:

for all $u \in u$, $p_{kr}(u) \neq \varnothing$

for all $r' \in (0, r)$, p_{kr} is a Lipschitz continuous mapping with constant $\delta = \frac{r}{r-r'}$ on

$$u = \{u \in h : 0 \leq d_{kr}(u) < r'\}$$

(i) the proximal normal cone is closed as a set valued mapping.

Assume that F; T : H ! 2H are set valued mappings, g; h : H ! H the nonlinear single valued mappings such that $k_r \subseteq g(h)$ and N : H X H $\longrightarrow$ H the mapping. For any constants n>0 and p>0, we consider the problem of finding $u \in h, x \in T(u), y \in F(u)$ such that $h(u) \in k_r$ and

$$\langle pN(x,y) + h(u) - g(u), v - h(u) \rangle + n \| v - h(u) \|^2 \geq 0, \forall v \in k_r$$

The equation (2.1) is called general nonlinear nonconvex variational inequalities. Now we consider the problem of solving general Wiener-Hopf equations. To be more precise, let QKr = I - h^{-1}PKr where PKr is the projection operator, h^{-1} is the inverse of nonlinear mapping h and I is an identity mapping. For given nonlinear mappings T; F; h; g; consider the problem of finding $z, u \in h, x \in T(u), y \in F(u)$ such that $N(x,y) + p^{-1}Qk_r z = 0$ is called general Wiener-Hopf equations.

Lemma 2.7 $u \in H, x \in T(u), y \in F(u), h(u) \in k_r$ is a solution of (2.1) if and only if $u \in H, x \in T(u), y \in F(u), h(u) \in k_r$ satisfies the relation h(u) = PKr [g(u) _N(x; y)] where PKr is a projection of H onto the uniformly r-prox regular set Kr:

Lemma 2.7 implies that the general nonlinear nonconvex variational inequality (2.1) is equivalent to the fixed point problem (2.3).

Now we consider the parametric version of equations (2.1), (2.2) and (2.3). To for- mulate the problem, let Ω be an open subset of H in which parameter λ takes values. Let $T, F : \Omega XH \to 2^h$ be the set valued mappings, N : H X H $\longrightarrow$ and g; h : Ω _H $\longrightarrow$ H the nonlinear single valued mappings such that $K_r \subseteq g(h)$ and N : H X H $\longrightarrow$ H the mapping. For any constants n>0 and p>0, we consider the problem of finding $u \in H, x \in T(u), y \in F(u)$ such that $h(u) \in K_r$ and

$$\langle pN(x,y) + h(u) - g(u), v - h(u) \rangle + n \| v - h(u) \|^2 \geq 0, \ \forall v \in K_r$$

The equation (2.1) is called general nonlinear nonconvex variational inequalities. Now we consider the problem of solving general Wiener-Hopf equations. To be more precise, let $Q_{Kr} = I - h^{-1}P_{Kr}$ is the projection operator, h^{-1} is the inverse of nonlinear mapping h and I is an identity mapping. For given nonlinear mappings T; F; h; g; consider the problem of finding $z, u \in H, x \in T(u), y \in F(u)$ such that $N(x,y) + P^{-1}Q_{kr}Z = 0$ is called general Wiener-Hopf equations.

Lemma 2.7 $u \in H, x \in T(u), y \in F(u), h(u) \in K_r$ is a solution of (2.1) if and only if $u \in H, x \in T(u), y \in F(u), h(u) \in K_r$ satisfies the relation

$$h(u) = P_{kr}[g(u) - pN(x,y)] \tag{2.3}$$

where PKr is a projection of H onto the uniformly r-prox regular set Kr: the single valued mappings. We define $g\lambda(u) = g(u,\lambda), h\lambda(u) = h(u,\lambda), x\lambda(u) = x(u,\lambda) \in T_\lambda(u), y_\lambda(u) = y(u,\lambda) \in F_\lambda(u)$ unless otherwise specified. The parametric general non-linear nonconvex variational inequality is to find $(u,\lambda) \in HX\Omega, x\lambda(u) \in T_\lambda(u), y_\lambda(u) \in F_\lambda(u), h_\lambda(u) \in K_r$ such that

$$\langle PN(x\lambda(u), y\lambda(u)) + h\lambda(u) - g_\lambda(u), v - h_\lambda(u)) \rangle \geq 0, \forall v \in K_r \tag{2.4}$$

We also assume that for some $\overline{\lambda} \in \Omega$ problem has a unique solution $\overline{u}$ Related to the parametric general nonlinear nonconvex variational inequality (2.4), we consider the parametric general Wiener-Hopf equation. We consider the problem of finding $(z,u,\lambda) \in HXHC\Omega, x\lambda(u) \in T_\lambda(u) \in F_\lambda(u)$ such that

$$N(x_\lambda(u), y_\lambda(u)) + p^{-1}Q_{kr}Z = 0 \tag{2.5}$$

where p > 0 is a constant and QKr (z) is define on the set (z, λ) with $\lambda \in \Omega$ and takes values in H. The equation (2.5) is called parametric general Wiener-Hopf equation.

Lemma 2.8 If H is a real Hilbert space. Than the following two statements are equivalent: an element $u \in H, x_\lambda(u) \in T_\lambda(u), y_\lambda(u) \in F_\lambda(u), h_\lambda(u) \in K_r$ is a solution of (2.4), the mapping $E\lambda(u) = u - h\lambda(u) + P_{kr}[g\lambda(u) - pN(x\lambda(u), y\lambda(u))]$ has a fixed point.

One can established the equivalence between (2.4) and (2.5), by using the projection techniques, see Noor [10,11].

Lemma 2.9 Parametric general nonlinear nonconvex variational inequality (2.4) has a Solution $(u,\lambda) \in Hx\Omega, x\lambda(u), y\lambda(u) \in F\lambda(u)$ if and only if parametric general Wiener-Hopf equation (2.5) has a solution $(z,u,\lambda) \in HxHx\Omega, x_\lambda(u) \in T\lambda(u), y\lambda(u) \in F\lambda(u)$

$$h_\lambda(u) = P_{krz} \tag{2.6}$$

$$z = g\lambda(u) - pN(x\lambda(u), y\lambda(u)) \tag{2.7}$$

From Lemma 2.9, we see that Parametric general nonlinear nonconvex variational inequalities (2.4) and parametric general Wiener-Hopf equations (2.5) are equivalent. We use these equivalence to study the sensitivity analysis of general nonlinear non-convex variational inequalities. We assume that for some $\overline{\lambda} \in \Omega$ problem (2.5) has a solution $\overline{Z}$ and X is a closure of a ball in H centered at $\overline{Z}$. We want to investigate those conditions under which for each λ is a neighbourhood of $\overline{\lambda}$ then (2.5) has a unique solution $z(\lambda)$ near $\overline{z}$ and the function $z(\lambda)$ is (Lipschitz) continuous and differentiable.

Definition 2.10 For u; v $\in H, \lambda \in \Omega$ the mapping $N : HXH \to H$

is said to be (φ,ψ) relaxed cocoercive with respect to first argument and $g:\Omega XH \rightarrow H$ with constants $\varphi>0,\psi>0$, and Lipschitz continuous with respect to first and second argument if there exists a constants $\alpha>0,\beta>0$ such that

$$\langle N(x_\lambda(u),y\lambda(u))-N(x_\lambda(v)),g_\lambda(u)-g\lambda(v)\rangle \geq -\varphi \| N(x_\lambda(u),y_\lambda(u))-N(x_\lambda(v),y\lambda(v))\|^2$$

$$+\psi \| g\lambda(u)-g\lambda(v)\|^2$$

And

$$\| N(x\lambda(u),y\lambda(u))-N(x\lambda(v),y\lambda(v))\| \leq \alpha \| x\lambda(u)-x\lambda(v)\| + \beta \| y\lambda(u)-y\lambda(v)\|$$

$$\forall x\lambda(u)\in T\lambda(u), x\lambda(v)\in T\lambda(v), y\lambda(u)\in F\lambda(u), y\lambda(v)\in F\lambda(v)$$

Definition 2.11 A single valued mapping $g:HX\Omega \rightarrow H$ is said to be Lipschitz continuous if there exists a constant $\gamma>0$ such that

$$\| g\lambda(u)-g\lambda(v)\| \leq \gamma \| u-v\|, \forall u,v\in H$$

Definition 2.12 The set valued mapping $T:HX\Omega \rightarrow 2^H$ is said to be D-Lipschitz continuous if there exists a constant v > 0 such that

$$D(T_\lambda(u),T_\lambda(v)) \leq v \| u-v\|, \forall u,v\in H, \lambda\in\Omega$$

where D is the Hausdorff metric.

Definition 2.13 Let $h:\Omega XH \rightarrow H$ be a single valued mapping. Then $h\lambda$ is said to be ξ-relaxed cocoercive if there exists a constant $\xi<0$ such that

$$\langle h\lambda(u)-h\lambda(v),u-v\rangle \geq \xi \| h\lambda(u)-h\lambda(v)\|^2$$

and Lipschitz continuous if there exists a constant $\mu>0$ such that

$$\| h_\lambda(u)-h\lambda(v)\| \leq \mu \| u-v\|, \forall u,v\in H, \lambda\in\Omega$$

Main Results

In this section, we consider the case when the solution of the parametric general Wiener-Hopf equations (2.5) lies in the interior of X. We consider the map

$$E\lambda(z)=Pkrz-pN(x\lambda(u),y\lambda(u))=g\lambda(u)-pN(x\lambda(u),y\lambda(u)), \forall(z,\lambda)\in X\Omega \quad (3.1)$$

Where

$$h_\lambda(u)=Pk_r z \quad (3.2)$$

We have to show that the map $E\lambda(z)$ has a fixed point which is a solution of parametric general Wiener-Hopf equations (2.5). First of all we prove the map $E\lambda(z)$ defined by (3.1) is a contraction map with respect to z uniformly in $\lambda\in\Omega$ by using the techniques of Noor [10].

Lemma 3.1 Let P_{kr} be a Lipschitz continuous operator with constant $\delta=\frac{r}{r-r^1}$ Let $N:H\times H\rightarrow H$ be the Lipschitz continuous with first argument and second argument with Constants $\alpha>0,\beta>0$ respectively. Let h; g : $H\times\Omega\rightarrow H$ be the Lipschitz continuous with constants $\mu>0,\gamma>0$ respectively and $h\lambda$ be the ξ relaxed cocoercive with respect to the constant ξ < 0: Let T; F :$\Omega\times$ H⟶ 2H be the D-Lipschitz continuous with constants v; x > 0; respectively. Let N be the (φ,ψ) relaxed cocoercive with respect to first argument and $g\lambda$ with constants $\varphi,\psi>0$ respectively. We have

$$\| E_\lambda(z_1)-E_\lambda(z_2)\| \leq \theta \| z_1-z_2\|$$

Where

$$\theta=\delta\frac{\sqrt{\gamma^2-2p(-\varphi(\alpha v+\beta x)^2+p^2(\alpha v+\beta x)^2}}{1-k} \quad (3.3)$$

$$k=\sqrt{1-2\xi\mu^2+\mu^2}, \delta>0 \quad (3.4)$$

For

$$| p-\frac{\psi\gamma^2-\varphi(\alpha v+\beta x)^2}{(\alpha v+\beta x)^2} |<\frac{\sqrt{\delta^2(\psi\gamma^2-\varphi(\alpha v+\beta x)^2)^2-(\alpha v+\beta x)^2(\delta^2\gamma^2-(1-k)^2)}}{\delta(\alpha v+\beta x)^2}$$

$$\delta(\psi\gamma^2-\varphi(\alpha v+\beta x)^2)>(\alpha v+\beta x)\sqrt{(\delta\gamma-1+k)(\delta\gamma+1-k)}$$

$$\delta(\psi\gamma^2-\varphi(\alpha v+\beta x)^2)>(\alpha v+\beta x)$$

$$\psi\gamma^2>\varphi(\alpha v+\beta x)^2 \quad (3.5)$$

Proof. For all z1; z2 $\in x,\lambda\in\Omega$ from (3.1) we have

$$\| E_\lambda(z_1)-E_\lambda(z_2)\| = \| g_\lambda(u)-g_\lambda(u)-g_\lambda(v)-p(x_\lambda(u),y_\lambda(u))-N(x_\lambda(v),y_\lambda(v)))\| \quad (3.6)$$

Now

$$\begin{aligned} &\| g\lambda(u)-g_\lambda(v)-p(N(x_\lambda(u),y_\lambda(u))-N(x_\lambda(v),y_\lambda(v)))\|^2 \\ &\leq \| g_\lambda(u)-g_\lambda(v)\|^2 -2p\langle N(x_\lambda(u),y_\lambda(u))-N(x_\lambda(v),y_\lambda(v)),g_\lambda(u)-g_\lambda(v)\rangle \\ &+p^2 \| N(x_\lambda(u),y_\lambda(u))-N(x_\lambda(v),y_\lambda(v),y_\lambda(v))\|^2 \end{aligned} \quad (3.7)$$

Since N is Lipschitz continuous with respect to _rst and second argument and T; F are D-Lipschitz continuous with constants v; x > 0 respectively, we have

$$\begin{aligned} \| N(x_\lambda(u),y_\lambda(u))-N(x_\lambda(v),y_\lambda(v))\| &\leq \alpha \| x_\lambda(u)-x_\lambda(v)\| + \beta \| y_\lambda(u)-y_\lambda(v)\| \\ &\leq \alpha D(T_\lambda(v))+\beta D(F_\lambda(u)-y_\lambda(u),F_\lambda(v)) \\ &\leq \alpha v\| u-v\| + \beta x \| u-v\| \\ &\leq (\alpha v+\beta x)\| u-v\| \end{aligned} \quad (3.8)$$

And

$$\| g_\lambda(u)-g_\lambda(v)\| \leq \gamma \| u-v\| \quad (3.9)$$

From the (φ,ψ))-relaxed cocoercive mapping of N with respect to first argument and g_λ we have

$$\begin{aligned} \langle N(x_\lambda(u),y_\lambda(u))-N(x_\lambda(v),y_\lambda(v)),g_\lambda(v)\rangle &\geq -\varphi \| N(x_\lambda(u),y_\lambda(u))-N(x_\lambda(v),y_\lambda(v))\|^2 \\ &+\psi \| g_\lambda(u)-g_\lambda(v)\|^2 \\ &>-\varphi(\alpha v+\beta x)^2 \| u-v\|^2 + \psi\gamma^2 \| u-v\|^2 \\ &\geq (-\varphi(\alpha v+\beta x)^2+\psi\gamma^2)\| u-v\|^2 \end{aligned} \quad (3.10)$$

Hence from (3.7)-(3.10), we have

$$\begin{aligned} &\| g_\lambda(u)-g_\lambda(v)-p(N(x_\lambda(u),y_\lambda(u))-N(x_\lambda(v)))\|^2 \leq \gamma^2 \| \gamma^2 \| u-v\|^2 \\ &-2p(-\varphi(\alpha v+\psi\gamma^2))\| u-v\|^2 + p^2 + p^2(\alpha v+\beta x)^2 \| u-v\|^2 \\ &\leq (\gamma^2-2p(-\varphi(\alpha v+\beta x)^2+p^2(\alpha v+\beta x)^2 \| u-v\|^2 \end{aligned} \quad (3.11)$$

Therefore from (3.6) and (3.11), we have

$$\| E_\lambda(z_1)-E_\lambda(z_2)\| \leq \sqrt{\gamma^2-2p(-\varphi(\alpha v+\beta x)^2+\psi\gamma^2)+p^2(\alpha v+\beta x)^2} \| u-v \quad (3.12)$$

Also from (3.2) and Lipschitz continuity of projection mapping PKr with constant δ; we

Have

$$\begin{aligned} \| u-v\| &\leq \| u-v-(h_\lambda(u)-h_\lambda(v))\| + \| P_{kr}(z_1)-P_{kr}(z_2)\| \\ &\leq \| u-v-(h_\lambda(u)-h_\lambda(v))\| + \delta \| z_1-z_2\| \end{aligned} \quad (3.13)$$

Since h is Lipschitz continuous with constant $\mu>0$ and ξ relaxed cocoercive with constant ξ <0 we have

$$\begin{aligned}&\| u-v-(h_\lambda(u)-h_\lambda(v))\|^2 \leq \| u-v\|^2 -2\langle h_\lambda(v), u-v\rangle + \| h_\lambda(u)-h_\lambda(v)\|^2\\ &\leq \| u-v\|^2 -2\xi \| h_\lambda(u)-h_\lambda(v)\|^2 + \| h_\lambda(u)-h_\lambda(v)\|^2\\ &\leq \| u-v\|^2 -2\xi\mu^2 \| u-v\|^2 +\mu^2 \| u-v\|^2\\ &\leq \| u-v-(h_\lambda(u)-h_\lambda(v))\| \leq k\| u-v\|\end{aligned} \tag{3.14}$$

Where $k=\sqrt{1-2\xi\mu^2+\mu^2}$ From (3.13) and (3.14) we have

$$\| u-v\| \leq \frac{\delta}{1-k}\| z_1-z_2\| \tag{3.15}$$

Combining (3.12),(3.15) and using (3.3) we have

$$\begin{aligned}&\| E_\lambda(z_1)-E_\lambda(z_2)\| \leq (1-\sigma_n)\| z_1-z_2\|\\ &+\sigma_n\delta\frac{\sqrt{\gamma^2-2p(-\varphi(\alpha v+\beta x)^2+p^2(\alpha v+\beta x)^2}}{1-k}\| z_1-z_2\\ &=(1-\sigma_n)\| z_1-z_2\| +\sigma_n\theta\| z_1-z_2\|\end{aligned} \tag{3.16}$$

Where

$$\theta=\delta\frac{\sqrt{\gamma^2-2p(-\varphi(\alpha v+\beta x)^2+\psi\gamma^2)+p^2(\alpha v+\beta x)^2}}{1-k}$$

From (3.5) it follows that $\theta<1$ and consequently the map $E_\lambda(z)$ de_ne by (3.12) is a contraction map and has a fixed point $z(\lambda)$ which is the solution of parametric general Wiener-Hopf equations (2.5).

Remark 3.2 From Lemma 3.1, we see that the map $E_\lambda(z)$ define by (2.1) has a unique fixed point $z(\lambda)$ that is $z(\lambda) = E_\lambda(z)$ Also by assumption the function $\bar{z}$ for $\lambda=\bar{\lambda}$ is a solution of parametric general Wiener-Hopf equations (2.5). Again by Lemma 3.1 we see that $\bar{z}$ for $\lambda=\bar{\lambda}$ is a fixed point of $E_\lambda(z)$ and it is also a fixed point of $E_{\bar{\lambda}}(z)$ Consequently, we conclude that $z(\bar{\lambda})=\bar{z}=E_{\bar{\lambda}}(z(\bar{\lambda}))$.

Using Lemma 3.1 we can prove the continuity of the solution $z(\lambda)$ of parametric general Wiener-Hopf equations (2.5). However for the sake of completeness and to convey the idea of the techniques involved, we give the proof.

Lemma 3.3 Assume that the mappings $T_\lambda(.), F_\lambda(.)$ are D-Lipschitz continuous and $g_\lambda(.), h_\lambda(.)$ are Lipschitz continuous with respect to the parameter λ If the mapping N is Lipschitz continuous with first and second argument respectively, and the map $\lambda\to p_{kr}(z), \lambda\to T_\lambda(u), \lambda\to F_\lambda(u),_\lambda\to g_\lambda(u), \lambda\to h_\lambda(u)$ are continuous (or Lipschitz continuous), the function $z(\lambda)$ satisfying the (2.3) is Lipschitz continuous at $\lambda=\bar{\lambda}$

Proof. For all $\lambda\in\Omega$ invoking Lemma 3.1 and the triangle inequality, we have

$$\begin{aligned}\| z(\lambda)-z(\bar{\lambda})\| &\leq \| E_\lambda(z(\lambda))-E_\lambda(z(\bar{\lambda}))\| + \| E_\lambda(z(\bar{\lambda}))-E_{\bar{\lambda}}(z(\bar{\lambda}))\|\\ &\leq\theta\| z(\lambda)-z(\bar{\lambda})\| + \| E_\lambda(z(\bar{\lambda}))-E_{\bar{\lambda}}(z(\bar{\lambda}))\|\end{aligned} \tag{3.17}$$

From (3.1) and the fact that the mapping $N, T_\lambda, F_\lambda, h_\lambda$ and g_λ are Lipschitz continuous with respect to the parameter λ we have

$$\begin{aligned}&\| E_\lambda(z(\bar{\lambda}))-E_{\bar{\lambda}}(z(\bar{\lambda}))\| = \| g_{\bar{\lambda}}(u(\bar{\lambda}))-p(N(T_\lambda(u(\bar{\lambda})), F_\lambda(u(\bar{\lambda})))-N(T_{\bar{\lambda}}(u(\bar{\lambda})), F_{\bar{\lambda}}(u(\bar{\lambda}))))\|\\ &\leq\gamma\|\lambda-\bar{\lambda}\| + p(\alpha\| T_{\bar{\lambda}}(u(\bar{\lambda}))\| +\beta\| F_{\bar{\lambda}}(u(\bar{\lambda}))\|)\\ &\leq\gamma\|\lambda-\bar{\lambda}\| + p(\alpha v\|\lambda-\bar{\lambda}\| +\beta x\|\lambda-\bar{\lambda}\|)\\ &\leq(\gamma+p(\alpha v+\beta x))\|\lambda-\bar{\lambda}\|\end{aligned} \tag{3.18}$$

Combining (3.17) and (3.18), we obtain

$$\| z(\lambda)-z(\bar{\lambda})\| \leq \frac{(\gamma+p(\alpha v+\beta x))}{1-\theta}\|\lambda-\bar{\lambda}\|, \forall\lambda,\bar{\lambda}\in\Omega$$

from which the required results follows.

We now state and prove the main result of this paper which is motivation of the next result.

Theorem 3.4 Let $\bar{u}$ be a solution of parametric general nonlinear nonconvex variational inequalities (2.4) and $\bar{n}$ be the solution of parametric general Wiener-Hopf equations (2.5) for $\lambda=\bar{\lambda}$. Let $h_\lambda(u)$ be ξ _-relaxed cocoercive mapping and Lipschitz continuous mapping, and $T_\lambda(u), F_\lambda(u)$ be the D-Lipschitz continuous mappings and N be (φ,ψ))-relaxed coco-ercive mapping with respect to first argument and g_λ and g_λ be the Lipschitz continuous for all $u, v\in x$ If the map $\lambda\to p_{kr}(z), \lambda\to T_\lambda(u), \lambda\to F_\lambda(u), \lambda\to g_\lambda(u), \lambda\to h_\lambda(u)$ are Lipschitz (continuous) mappings at $\lambda=\bar{\lambda}$ then there exists a neighbourhood M of Ω of $\bar{\lambda}$ such that for $\lambda\in M$ parametric general Wiener-Hopf equation (2.5) has a unique solution $z(\lambda)$ in the interior of $x, z(\bar{\lambda})=z(\lambda)$ and $z(\lambda)$ is (Lipschitz) continuous at $\lambda=\bar{\lambda}$.

Proof. Its proof follows from Lemma 3.1, 3.3 and Remark 3.2.

References

1. Stampacchia G (1964) An Formes bilineaires coercitives sur les ensembles convexes, C. R. Acad. Sci. Paris, 258: 4413-4416.
2. Agarwal RP, Cho YJ, Huang NJ (2000) Sensitivity analysis for strongly nonlinear quasi variational inclusions. Appl Math Lett 13: 19-24.
3. Anastassiou GA, Salahuddin, Ahmad MK (2013) Sensitivity analysis for generalized set valued variational inclusions. J Concrete Appl Math 11: 292-302.
4. Baiocchia C, Capelo A (1984) Variational Quasi Variational Inequalities, Wiley London.
5. Clarke FH, Ledyaw YS, Stern RJ, Wolenski PR (1998) Nonsmooth Analysis and Control Theory, Springer-Verlag, New York.
6. Dafermos S (1988) Sensitivity analysis in variational inequalities. Math Oper Res 13: 421-434.
7. Giannessi F, Maugeri A (1995) Variational Inequalities and Network Equilibrium Problems, Plenum Press, New York, NY, USA.
8. Kyparisis J (1987) Sensitivity analysis frame work for variational inequalities. Math Prog 38: 203-213.
9. Liu J (1995) Sensitivity analysis in nonlinear programs and variational inequalities via continuous selection. SIAM J Control Optim 33: 1040-1068.
10. Noor MA (1993) Wiener Hopf equations and variational inequalities. JOTA 79: 197-206.
11. Noor MA, Noor KI (2013) Sensitivity analysis of general nonconvex variational inequalities, JIA, 302.
12. Poliquin RA, Rockafellar RT, Thibault L (2000) Local differentiability of distance functions. Trans Amer Math Soc 352: 5231-5249.
13. Qiu Y, Magnanti TL (1989) Sensitivity analysis for variational inequalities defined on polyhedral sets. Math Oper Res 14: 410-432.
14. Shi P (1991) Equivalence of Wiener-Hopf equations with variational inequalities. Proc Amer Math Soc 111: 339-346.
15. Aubin JP, Ekeland I (1984) Applied Nonlinear Analysis, Wiley, New York.
16. Verma RU (2006) Generalized -resolvent operator technique and sensitivity analysis for relaxed cocoercive variational inclusions 7: 70-83.
17. Wen DJ (2010) Projection methods for a generalized system of nonconvex variational inequalities with di_erent nonlinear operators. Nonlinear Anal 73: 2292-2297.

Boundary Element Method and Tangent Operator Technique applied in the Multi Crack Propagation Modelling of Concrete Structures

Ferreira Cordeiro SG and Leonel ED*

Department of Structural Engineering, School of Engineering of Sao Carlos, University of Sao Paulo, Brazil

Abstract

The cohesive crack model has demonstrated its accuracy in the simulation of crack growth phenomena in concrete structures. Moreover, the Boundary Element Method (BEM) is recognised as a robust and efficient numerical technique for addressing fracture problems. In this context, the present study aims at coupling the cohesive crack model to a BEM formulation to simulate the mechanical behaviour of cracked concrete structure. The cohesive cracks are modelled mechanically through the sub-region BEM approach. Because the tractions values along the fractured interfaces depend on the crack opening displacements, this problem is solved in the context of nonlinear solutions. The nonlinear system of equations is solved using the Tangent Operator (TO) technique, in which tangent prevision and tangent correction steps are required. Such scheme assures better convergence and accuracy than the classical Newton approach. The determination of the TO terms is the main contribution of this study. To validate the proposed formulation, it was applied in the simulation of crack growth in concrete structures. The results achieved were compared to numerical and experimental responses available in the literature. Apart from the strong agreement among the results obtained, faster convergence was verified using TO instead of the classical scheme.

Keywords: Cohesive crack growth; Concrete structures; Tangent operator technique

Introduction

Among the usual engineering materials, concrete is widely utilized around world due to its low production cost, geometry adaptability, chemical properties and strength capacities. Moreover, the collapse of engineering structures is caused by the growth of internal discontinuities, such as cracks, which leads to the mechanical failure by fracture. Therefore, the scientific community has applied the fracture mechanics to predict accurately the collapse conditions of concrete. In such material, the mechanical degradation processes, especially due to the crack growth, introduce nonlinear mechanical effects that cannot be disregarded. Consequently, nonlinear models were proposed to represent the concrete fracture, in which the cohesive crack model must be highlighted [1]. This approach was introduced by Barenblatt and Dugdale [2,3] and was extensively used in numerical applications of fracture problems [4-8]. The crack propagation modelling in concrete by the cohesive crack approach requires the solution of a nonlinear system of equations, generally relating the Crack Opening Displacements (COD) to the tractions values along the crack surfaces. The traction values represent the residual material strength at the cohesive zone, which depend on the COD intensity. The residual material strength decrease as the COD values grow.

In this context, this study addresses the mechanical analysis of multi crack growth in concrete structures using a nonlinear Boundary Element Method (BEM) formulation and an efficient algorithm to solve the nonlinear system of equations based on the Tangent Operator (TO) scheme. The BEM is a robust numerical technique for modelling fracture and crack growth problems as only the boundary surfaces discretization is required [9]. Therefore, this numerical technique is well adapted to solve fracture mechanics problems, as the remeshing procedure becomes a less complex task. Moreover, the stresses singularities present at the crack tip are represented accurately due to the non-requirement of a domain mesh [10]. The numerical formulation applied in this study involves the sub-region BEM technique and the cohesive crack model. The first technique divides the entire domain into sub-regions over which, for elastic problems, compatibility of displacements and equilibrium of forces are enforced along its interfaces. In the cohesive crack problems, the displacement compatibility is no longer verified. Therefore, cohesive laws are required to relate the COD to the tractions values along the crack surfaces. Moreover, cracks are assumed to growth along the sub-region interfaces. One advantage of this approach concerns the possibility to analyse nonlinear fracture problems in nonhomogeneous systems, which is a challenge problem in structural engineering [11].

The nonlinear problem is solved using the TO approach. This procedure assures better convergence and accuracy than the classical Newton approach. Such a technique achieves the solution through initial and corrections steps that account the tangent direction of the global equilibrium trajectory. The tangent search is performed by incorporating the derivative set of cohesive laws into the algebraic BEM equations. This type of operator has been utilized successfully in the literature for dealing many different nonlinear engineering problems. For instance, Botta et al. [12], Oliveira et al. [13] and Leonel et al. [14] applied this operator on the modelling of localization phenomenon, fracture analyses, and contact problems, respectively. The TO is derived for the crack growth analysis in concrete materials, which is the main contribution of this study.

Two applications were performed with the developed nonlinear BEM formulation. The results obtained were compared to numerical

***Corresponding author:** Leonel ED, Department of Structural Engineering, School of Engineering of Sao Carlos, University of Sao Paulo, Av. Trabalhador Sao Carlense, 400, Centro, 13566-590, Sao Carlos-SP, Brazil
E-mail: edleonel@sc.usp.br

and experimental responses available in the literature to illustrate its robustness and accuracy. It was also verified that the TO provides faster convergence and lower computational time consuming for the nonlinear solution when compared to classical Newton method.

The Cohesive Crack Model

The cohesive crack model is an appropriate approach to represent the mechanical degradation processes in quasi-brittle materials, such as the concrete. In this model, the degradation phenomena occur along a fictitious crack positioned ahead the real crack tip. Thus, the material degradation zone is reduced by one dimension. The first studies in which the dissipation zone was reduced either to a curve for 2D problems or to a surface for 3D problems [1-3].

In the present study, simple softening laws approximate the residual material strength along the fictitious crack. These laws relate the fictitious crack opening displacement, COD, to the cohesive stresses values. Several cohesive crack laws relating cohesive stresses to the COD have been proposed in the literature. Three of them are often adopted to handle the crack growth analyses in quasi-brittle materials. The simplest law is given by a linear function relating the cohesive stresses to the fictitious COD smaller than the threshold value, COD_c. For fictitious crack openings larger than COD_c, cohesive stresses are assumed as null. The relations that represent the linear cohesive law are the following:

$$\begin{aligned} &\sigma = E\varepsilon && \text{if} \quad \text{elastic conditions} \\ &\sigma_{cohesive}(COD) = f_t\left(1-\frac{COD}{COD_c}\right) && \text{if} \quad 0 \le COD \le COD_c \\ &\sigma_{cohesive}(COD) = 0 && \text{if} \quad COD \ge COD_c \end{aligned} \tag{1}$$

in which E is the Young's modulus and f_t the tensile material strength.

Alternatively, the COD and cohesive stresses may be related through the composition of linear functions. Particularly, such law is defined by two linear functions, which is named bi-linear model. This cohesive model is defined by the following equations:

$$\begin{aligned} &\sigma = E\varepsilon && \text{if} \quad \text{elastic conditions} \\ &\sigma_{cohesive}(COD) = f_t - \left(\frac{f_t - f_t^{"}}{COD^{"}}\right)COD && \text{if} \quad 0 \le COD \le COD^{"} \\ &\sigma_{cohesive}(COD) = \frac{f_t^{"}COD}{COD^{"} - COD_c} + f_t^{"}\left(1 - \frac{COD^{"}}{COD^{"} - COD_c}\right) && \text{if} \quad COD^{"} \le COD \le COD_c \\ &\sigma_{cohesive}(COD) = 0 && \text{if} \quad COD \ge COD_c \end{aligned} \tag{2}$$

For the bi-linear model, the variables f_t,$COD^{"}$ and COD_c are defined as follows:

$$\begin{aligned} &f_t^{"} = \frac{f_t}{3} \\ &COD^{"} = \frac{0.8G_f}{f_t} \\ &COD_c = \frac{3.6G_f}{f_t} \end{aligned} \tag{3}$$

in which G_f represents the material fracture energy.

Finally, an exponential expression represents the third cohesive law utilized in this study. Equation (4) provides the analytical expressions for this cohesive model:

$$\begin{aligned} &\sigma = E\varepsilon && \text{if} \quad \text{elastic conditions} \\ &\sigma_{cohesive}(COD) = f_t e^{\frac{-f_t}{G_f}COD} && \text{if} \quad COD > 0 \end{aligned} \tag{4}$$

Boundary Integral Equations

The BEM has been widely applied to solve engineering problems such as contact, fatigue and crack propagation due to its high accuracy and robustness in modelling strong stress concentration. Considering a two-dimensional homogeneous elastic domain Ω with boundary Γ, the equilibrium equation, written in terms of displacements, is given by:

$$u_{i,jj} + \frac{u_{j,ji}}{1-2\nu} + \frac{b_i}{\mu} = 0 \tag{5}$$

Where μ is the shear modulus, v is the Poisson's ratio, u_i are the components of displacements and b_i are the body forces. The singular integral representation, written in terms of displacements, is obtained applying the Betti's theorem and the Eq.(5). This integral equation is written, disregarding body forces, as follows:

$$c_{ij}(s,f)u_j(s) + \int_\Gamma P_{ij}^*(s,f)u_j(f)d\Gamma = \int_\Gamma U_{ij}^*(s,f)p_j(f)d\Gamma \tag{6}$$

in which p_i and u_i are the tractions and displacements at the boundary, respectively. The free term c_{ij} is equal to $\delta_{ij}/2$ for smooth boundaries. δ_{ij} is the Kroenecker operator and P^*_{ij} and U^*_{ij} are the fundamental solutions for tractions and displacements, respectively, written for the source point s and related to the field point f [15].

Equation (6) leads to another important integral equation, known in literature as the hyper singular integral representation. To obtain it, the Eq.(6) has to be differentiated with respect to the directions *x* and *y*. Then, the relation among displacements and strains is used in order to determine an integral equation written in terms of strains. Then, a constitutive relation, the generalized Hooke's law for instance, is utilized to obtain an integral equation written in terms of stresses, for a boundary source point. Finally, the Cauchy formula is applied and the hyper singular integral equation is obtained, which is written in terms of tractions as follows:

$$\frac{1}{2}p_j(s) + \eta_k \int_\Gamma S_{kij}^*(s,f)u_j(f)d\Gamma = \eta_k \int_\Gamma D_{kij}^*(s,f)p_j(f)d\Gamma \tag{7}$$

in which S^*_{kij} and D^*_{kij} contains the new kernels computed from P^*_{ij} and U^*_{ij} respectively [16,17] and η_k is the outward normal vector at the source point.

Equations (6) and (7) can be applied separately to construct the system of algebraic equations for two-dimensional non-cracked domains. However, the application of only one integral equation, either Eq.(6) or Eq.(7), for modelling cracked structures in a single domain leads to the singularities in the system of algebraic BEM equations. Such singularities appear due to the presence of coincident source points at the crack surfaces. Therefore, to analyse structures containing cracks, some special BEM formulations were proposed in the literature. Among them, it is worth mentioning the dual BEM [16,17], the multi-region approach [10] and the dipole BEM formulation [13]. In the present work, the sub-region BEM technique is adopted, in which compatibility of displacements and equilibrium of forces are enforced along the interfaces of the sub-domains. The coupling procedure of sub-region formulation is presented in the next section.

To assemble the system of BEM equations, as usual, Eq.(6) and/or Eq.(7) are transformed into algebraic relations by discretizing the boundaries and interfaces into elements over which displacements and tractions are approximated by polynomial functions. Besides that, one has to select a convenient number of collocation points to obtain the algebraic representations. The algebraic equations for boundary nodes are calculated considering these boundary collocation points.

After determining the displacement and traction fields at the boundaries, internal values for displacements, stresses and strains can be achieved. Internal displacements are determined using the integral Eq.(6) with the source point located inner the boundary. In this case, the free term c_{ij} becomes δ_{ij}. On the hand, the stress field at internal nodes is obtained through the integral stress representation [15].

Algebraic BEM Equations

One interesting approach to address crack growth problems through BEM is the sub-region BEM technique. In this approach, the entire domain is divided into sub-domains over which Eq.(6) and/or Eq.(7) are applied. Then, the compatibility of displacements and equilibrium of forces are enforced along the sub-domain interfaces. As a result, the cracks are assumed to growth along such interfaces.

The crack surfaces, i.e. the interfaces, are discretized by oppositely oriented boundary elements. To analyse the crack growth in concrete, the cohesive crack model is incorporated into the interface elements. Thus, the cohesive law governs the COD and stresses values along the cohesive crack and, consequently, the nonlinear mechanical behaviour is introduced.

The classical set of algebraic BEM equations is written for a given multi-region domain taking into account the location of the collocation points, i.e. the source points. These points may be located at the left (L) or right (R) opposite boundary elements of a given interface k, or even at the external boundary (B). Hence, for a domain composed by NI interfaces, the BEM algebraic equations is presented as follows:

$$H^B u^B + \sum_{k=1}^{NI}\left[H^{L(k)}u^{L(k)} + H^{R(k)}u^{R(k)}\right] = G^B p^B + \sum_{k=1}^{NI}\left[G^{L(k)}p^{L(k)} + G^{R(k)}p^{R(k)}\right] \quad (8)$$

in which H and G matrices contain the values integrated from the kernels S^*_{kij}, P^*_{ij} and D^*_{kij}, U^*_{kij}, U^*_{ij}, respectively.

Afterwards, displacements and tractions must be rotated from the *x,y* global orientation to the normal η and tangent t directions of each crack side. Thus, one defines $u_\eta^{s(k)}$, $u_t^{s(k)}$, $p_\eta^{s(k)}$ and $p_t^{s(k)}$ as the normal and tangent components of displacements and tractions at a given side S, either *L* or *R*, of an interface k, respectively. Over the algebraic equations presented in Eq.(8), one imposes the boundary conditions at the external boundary, as usual in BEM, which leads to the following:

$$A^B x^B + \sum_{k=1}^{NI}\begin{bmatrix} H_\eta^{L(k)}u_\eta^{L(k)} + H_t^{L(k)}u_t^{L(k)} \\ H_\eta^{R(k)}u_\eta^{R(k)} + H_t^{R(k)}u_t^{R(k)} \end{bmatrix} = F^B + \sum_{k=1}^{NI}\begin{bmatrix} G_\eta^{L(k)}p_\eta^{L(k)} + G_t^{L(k)}p_t^{L(k)} \\ G_\eta^{R(k)}p_\eta^{R(k)} + G_t^{R(k)}p_t^{R(k)} \end{bmatrix} \quad (9)$$

in which, H_η, H_t and G_η, G_t are the sub-matrices of H and G that multiply the normal and tangent displacements and tractions after the rotation, respectively. x^B are the boundary unknown values, A^B is composed by the columns from H^B and G^B, and F^B contains the contribution of prescribed values at the boundary. To solve such problem, the equilibrium condition must be enforced along the interface boundary elements. Then, over such elements, the following conditions are imposed:

$$p_\eta^L = -p_\eta^R = p_\eta \qquad and \qquad p_t^L = -p_t^R = p_t \quad (10)$$

Therefore, based on Eq.(10), one rewrites Eq.(9) as follows:

$$A^B x^B + \sum_{k=1}^{NI}\begin{bmatrix} H_\eta^{L(k)}u_\eta^{L(k)} + H_t^{L(k)}u_t^{L(k)} \\ H_\eta^{R(k)}u_\eta^{R(k)} + H_t^{R(k)}u_t^{R(k)} \end{bmatrix} = F^B + \sum_{k=1}^{NI}\begin{bmatrix} \left(G_\eta^{L(k)} - G_\eta^{R(k)}\right)p_\eta^{(k)} + \\ \left(G_t^{L(k)} - G_t^{R(k)}\right)p_t^{(k)} \end{bmatrix} \quad (11)$$

The displacements calculated at the interfaces are used to define the crack opening displacement (COD) and the crack sliding displacement (CSD). Then, these variables are defined as follows:

$$-u_\eta^{L(k)} - u_\eta^{R(k)} = COD^{(k)} \qquad and \qquad u_t^{L(k)} + u_t^{R(k)} = CSD^{(k)} \quad (12)$$

Where $COD^{(k)}$ is the vector of Crack Opening Displacements whereas $CSD^{(k)}$ is the vector of Crack Sliding Displacements. Each component of these vectors is defined for each pair of coincident nodes of an interface k.

Obviously, when $COD^{(k)}=CSD^{(k)}=0$, elastic condition is assumed for the interface. Otherwise, the cohesive crack model is incorporated. The cohesive crack propagation starts when at least one component of the vector $p_\eta^{(k)}$ overcomes the material tensile strength f_t. In such a case, cohesive behaviour for these tractions follow the stresses values predicted by the cohesive law, which depends on the COD values. Therefore, the problem becomes nonlinear and the tractions at the cohesive regions of the interfaces must satisfy the following constrains:

$$p_\eta = p^C\left(COD\right) \qquad and \qquad p_t = 0 \quad (13)$$

Equation (13) indicates that the normal tractions p_η must represent the residual material resistance, which is governed by the cohesive tractions $p^c(COD^{(k)})$. On the other hand, the second constrain indicates brittle shear behaviour. When a component of the vector $COD^{(k)}$ overcomes the threshold value, COD_c, the respective normal traction is assumed as null and the material has no longer a residual resistance to avoid the opening of the crack surfaces. Hence, to consider cohesive mechanical behaviour, the discontinuities defined in Eq. (12) and the constrains presented in Eq. (13) must be incorporated into the system of BEM equations, Eq.(11). Thus:

$$A^B x^B + \sum_{k=1}^{NI}\begin{bmatrix} \left(H_\eta^{L(k)} - H_\eta^{R(k)}\right)u_\eta^{L(k)} - H_\eta^{R(k)}COD^{(k)} + \\ \left(H_t^{L(k)} - H_t^{R(k)}\right)u_t^{L(k)} + H_t^{R(k)}CSD^{(k)} \end{bmatrix} = F^B + \sum_{k=1}^{NI}\left[\left(G_\eta^{L(k)} - G_\eta^{R(k)}\right)p^C\left(COD^{(k)}\right)\right] \quad (14)$$

Because the cohesive law was included, Eq.(14) becomes nonlinear due to the dependency of the variables p^c and COD.

Nonlinear solution technique using the Tangent Operator approach

To solve the nonlinear problem provided by the cohesive crack model, Eq.(14) has to be rewritten as a function of its variables as follows:

$$\begin{aligned} Y\left(x^B, u_\eta^{L(k)}, COD^{(k)}, u_t^{L(k)}, CSD^{(k)}\right) &= A^B x^B + \sum_{k=1}^{NI}\left(H_\eta^{L(k)} - H_\eta^{R(k)}\right)u_\eta^{L(k)} \\ &- \sum_{k=1}^{NI} H_\eta^{R(k)}COD^{(k)} + \sum_{k=1}^{NI}\left(H_t^{L(k)} - H_t^{R(k)}\right)u_t^{L(k)} + \sum_{k=1}^{NI} H_t^{R(k)}CSD^{(k)} \\ &- \sum_{k=1}^{NI}\left[\left(G_\eta^{L(k)} - G_\eta^{R(k)}\right)p^C\left(COD^{(k)}\right)\right] - F^B = 0 \end{aligned} \quad (15)$$

Where x^B, $u_\eta^{L(k)}$, $COD^{(k)}$, $u_t^{L(k)}$, $CSD^{(k)}$ are the variables of the problem, i.e., v.

An increment Δv must be obtained to solve the problem by the TO scheme. Such an increment has to lead to an equilibrium condition and the cohesive constraints have to be attended. Hence, one writes the function Y(v+Δv) as a linear Taylor expansion at the vicinity of a previous equilibrium configuration of the variables v as follows:

$$Y(v+\Delta v) \approx Y(v) + \frac{\partial Y}{\partial v}(v)\Delta v \tag{16}$$

When the elastic condition is violated, the elastic prediction, i.e. $COD^{(k)}=CSD^{(k)}=0$, provides normal tractions higher than the tensile material strength, $p_n > f_t = p^c(COD=0)$. Therefore, the solution with null discontinuities is no longer acceptable. The difference between p_η and $p^c(COD)$ is the normal traction exceeding vector Δp_η^{exc}, which must be reapplied along the cohesive interface regions to satisfy the constrains in Eq. (13). However, after reapplying it, the equilibrium condition will be violated, i.e, $Y(v)=\Delta F \neq 0$. The vector ΔF represents the non-equilibrated force vector, which appears after the imposition of the cohesive constrains.

The solution of the problem requires $Y(v+\Delta v)=0$. Therefore, from Eq. (16), $\frac{\partial Y}{\partial v}\Delta v \approx -\Delta F$, where $\frac{\partial Y}{\partial v}\Delta v$ is computed by the following differentiation:

$$\frac{\partial Y}{\partial v}\Delta v = -\Delta F \leftrightarrow \frac{\partial Y\left(x^B\ldots\right)}{\partial x^B}\Delta x^B + \frac{\partial Y\left(\ldots u_\eta^{L(k)}\ldots\right)}{\partial u_\eta^{L(k)}}\Delta u_\eta^{L(k)} + \frac{\partial Y\left(\ldots COD^{(k)}\ldots\right)}{\partial COD^{(k)}}\Delta COD^{(k)} + \frac{\partial Y\left(\ldots u_t^{L(k)}\ldots\right)}{\partial u_t^{L(k)}}\Delta u_t^{L(k)} + \frac{\partial Y\left(\ldots CSD^{(k)}\ldots\right)}{\partial CSD^{(k)}}\Delta CSD^{(k)} \approx -\Delta F \tag{17}$$

From Eq.(17), an approximated solution for the increment Δv, which leads to the solution of the problem, can be calculated as follows:

$$\begin{Bmatrix} \Delta x^B \\ \Delta u_\eta^{L(k)} \\ \Delta COD^{(k)} \\ \Delta u_t^{L(k)} \\ \Delta CSD^{(k)} \end{Bmatrix} = \begin{bmatrix} A^B, \\ \sum_{k=1}^{NI}\left(H_\eta^{L(k)} - H_\eta^{R(k)}\right), \\ \sum_{K=1}^{NI}\left(-H_\eta^{R(k)} - \left(G_\eta^{L(k)} - G_\eta^{R(k)}\right)\frac{\partial p^C}{\partial COD^{(k)}}\right), \\ \sum_{k=1}^{NI}\left(H_t^{L(k)} - H_t^{R(k)}\right), \\ \sum_{k=1}^{NI} H_t^{R(k)} \end{bmatrix}^{-1} \{-\Delta F\} \tag{18}$$

In the system presented in Eq. (18), the matrix that multiplies $\{-\Delta F\}$ is the so called the Tangent Operator(TO). This matrix includes information about the derivative of the cohesive laws. Therefore, when incorporated into the BEM equations, it makes the search for the nonlinear solution along the tangent direction of the global equilibrium trajectory.

When linear or partially cohesive laws are adopted, the term $\partial p^c/\partial COD^{(k)}$ becomes constant. Consequently, the TO also becomes constant as well. Therefore, the solution Δv for a given load step can be obtained with only one iteration. On the other hand, when nonlinear cohesive laws are involved, such as an exponential one, the solution Δv must be obtained iteratively as the derivative terms $\partial p^C/\partial COD$ depends on the actual openings $COD^{(k)}$. At the first iteration, the increment $\Delta v^{(1)}$ must be computed form Eq. (18). Therefore, the initial traction exceeding vector $\Delta p_\eta^{exc(1)}$ must be accounted in ΔF and the tangent terms $\partial p^C/\partial COD$ are computed assuming COD=0. After the first increment, non-null COD are obtained. However, because the cohesive law is no longer linear, a new traction exceeding vector $\Delta p_\eta^{exc(2)}$ is observed and must be reapplied at the cohesive regions. Therefore, the second iterative increment $\Delta v^{(2)}$ is computed from Eq. (18) in which the tangent terms $\partial p^C/\partial COD^{(k)}$ were updated with the new values of COD. The stop criterion for such iterative procedure was based on the norm of the traction exceeding vector, which must be smaller than a prescribed tolerance for achieving the converged solution.

As the boundary loads are prescribed incrementally, the TO procedure must be performed at each incremental step.

Applications

Two applications of fracture in concrete specimens are presented. The first concerns the three-point bended beam whereas the second addresses the four-point mixed mode multiple cohesive crack growth.

Three-point bended-notched beam

The first application of this study concerns a mode I fracture test, which was analysed experimentally by Saleh [18] and numerically by Oliveira et al. [13]. The specimen consists of a three point bended concrete beam with an initial notch. To define the three cohesive laws, two parameters were considered: the tensile material strength f_t=3 MPa and the material fracture energy G_t=75 N/m. Figure 1 illustrates the structure analysed, in which the geometric data and the material elastic proprieties are also presented.

The structure was divided into two sub-regions, with a vertical interface, in which the main crack propagates in mode I. The boundary mesh adopted is composed by 55 cubic and 29 quadratic discontinuous boundary elements with 307 source points. The displacements prescribed were imposed into 100 load steps and the tolerance for convergence was assumed as 10^{-6} MPa. Figure 2 shows the initial and the final deformed structural configuration, with displacements magnified 100 times.

The displacements fields in the last load step are presented in Figure 3. It is worth mentioning the physical discontinuity introduced by the crack.

During the numerical analysis, the interface tractions were monitored and its cohesive behaviour is illustrated in Figure 4 (left) for linear, bilinear and exponential cohesive laws. As illustrated in this figure, the cohesive crack becomes a real crack in the end of the incremental load process. The nonlinear structural behaviour was also monitored through the load-displacement curves. The load was calculated as the equivalent force at the upper beam surface.

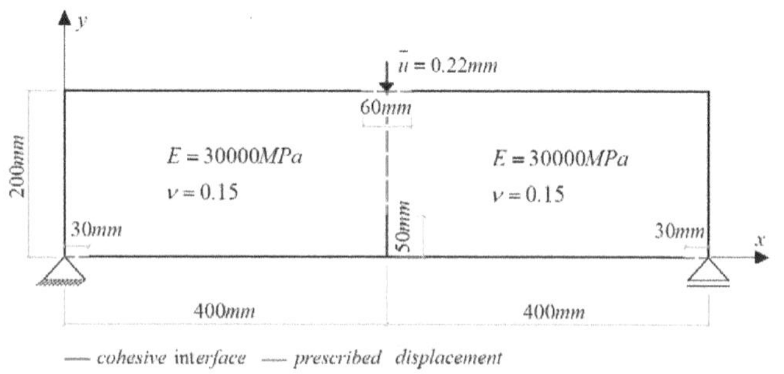

Figure 1: Three-point bended-notched beam.

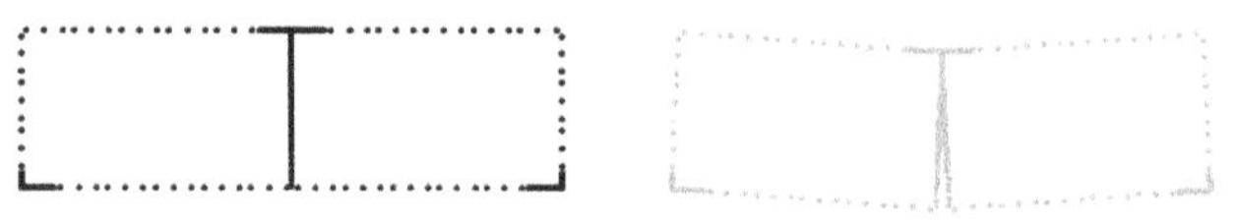

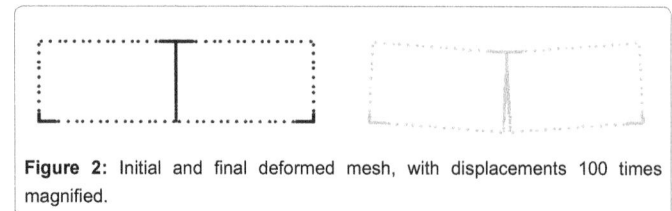

Figure 2: Initial and final deformed mesh, with displacements 100 times magnified.

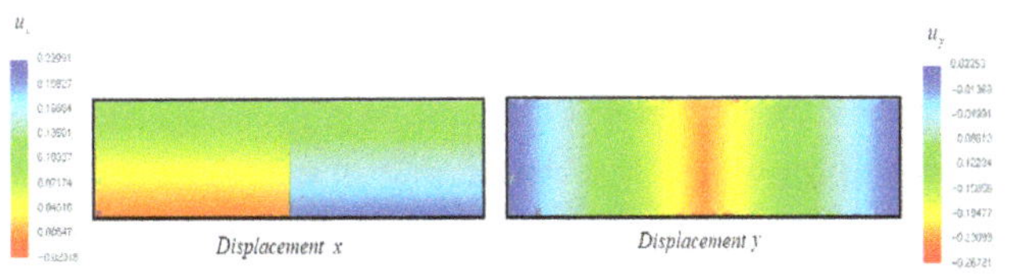

Figure 3: Colour scale displacements fields, Dimensions in cm.

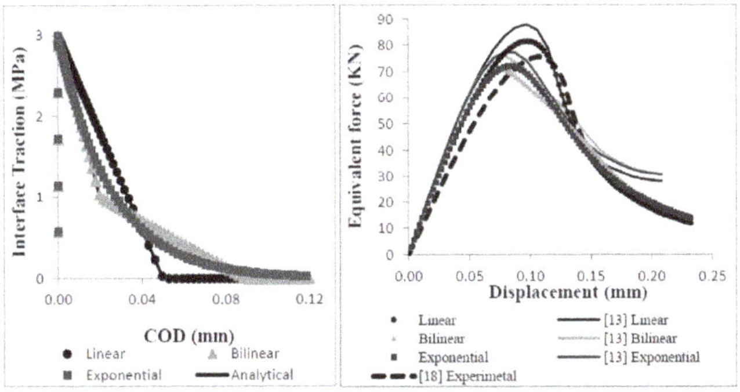

Figure 4: Cohesive tractions at the interface (left) and nonlinear structural response (right).

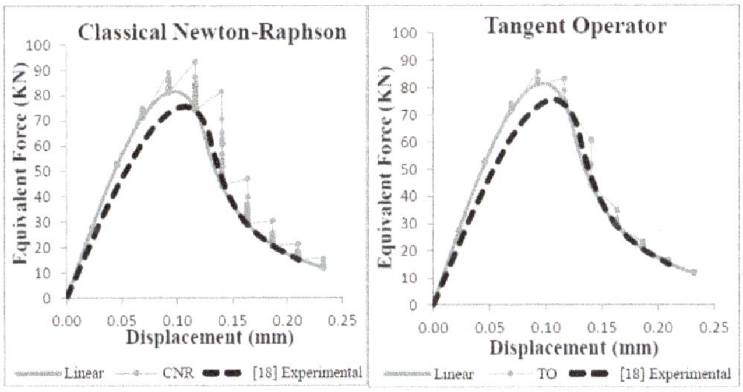

Figure 5: Nonlinear solution search: Classical Newton-Raphson versus the Tangent Operator.

The displacement values were determined by averaging the vertical displacements at the crack mouth. Such nonlinear responses are illustrated in Figure 4 (right), in which experimental and numerical curves obtained by previous references are also illustrated.

The developed BEM formulation was capable to represent accurately the nonlinear mechanical behaviour of the cracked specimen, as illustrated in Figure 4. Moreover, the BEM model with the three cohesive laws achieved the resistant structural load with considerable accuracy if compared to the experimental value. The softening part of the numerical curves reproduce quite well the experimental results.

The convergence path obtained by the TO and the classical Newton approaches are presented in Figure 5. As presented in this figure, the TO approach makes the search for the equilibrium configuration by the tangent direction of the global equilibrium trajectory. Then, it is quite efficient in comparison with the classical approach.

To quantify the efficiency of the TO approch and the classical Newton scheme, the number of iteration required during the analyses were monitorated and presented in Table 1. The improvement in terms of the required iterations to achieve the convergence is remarkable with the TO. Such reduction reaches, for the best case, 92% and 77%, for the worst case. Then, the TO approach is recommended.

Four-point mixed mode multiple crack growth

The second application addresses the mechanical analysis of a mixed mode multiple cohesive crack growth case in a concrete specimen. The specimen is subjected to a four-point shear beam test, which as analysed numerically by Carpinteri [19] with the Finite Element Method (FEM) and by Saleh et al. [6] with the BEM. Both authors adopted the linear cohesive law to describe the residual material strength along the cohesive crack. The crack paths determined in Saleh et al. and Carpinteri [6,19] were adopted as interfaces in the model proposed in this study. Therefore, the entire domain was split into three sub-regions as presented in Figure 6, which also shows geometric data and the material elastic properties considered. In the analyses performed by the proposed BEM model, the cohesive laws linear, bilinear and exponential were adopted. The tensile material strength was defined as f_t=2 MPa and the material fracture energy was assumed equal to G_f=100 N/m.

The adopted boundary element mesh was composed by 72 cubic and 30 quadratic discontinuous boundary elements with 378 source points. The displacements prescribed and the stop tolerance criterions were assumed as the same of the previous application. Figure 7 illustrates the initial and the final deformed configuration of the structure, with displacements magnified 500 times.

The displacements fields in the last load step are presented in Figure 8. The colour scale graphs show the physical discontinuity introduced by the cracks.

The load was considered as the equivalent force at the beam upper

Cohesive law	Classical Scheme	Tangent Operator	Reduction
Linear	2150	180	- 92%
Bilinear	2157	199	- 91%
Exponential	2067	468	- 77%

Table 1: Comparison of iterations 1.

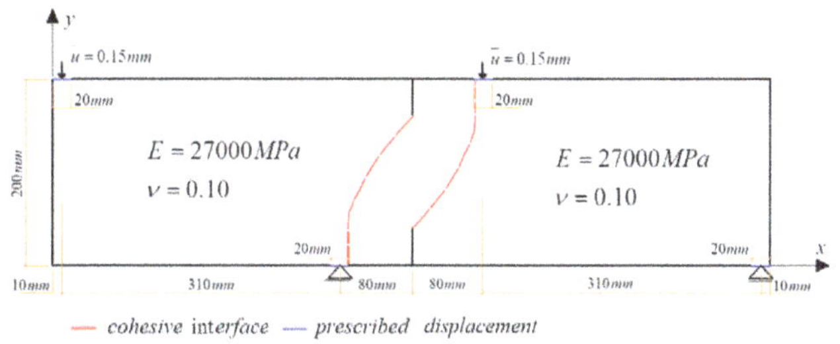

Figure 6: Four point shear beam test.

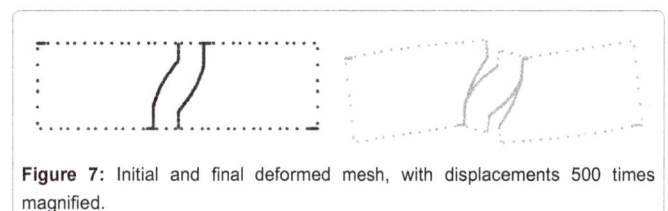

Figure 7: Initial and final deformed mesh, with displacements 500 times magnified.

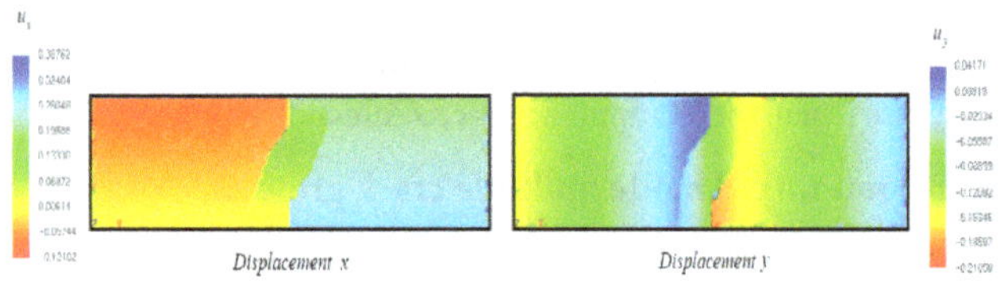

Figure 8: Color scale displacement field. Dimensions in cm.

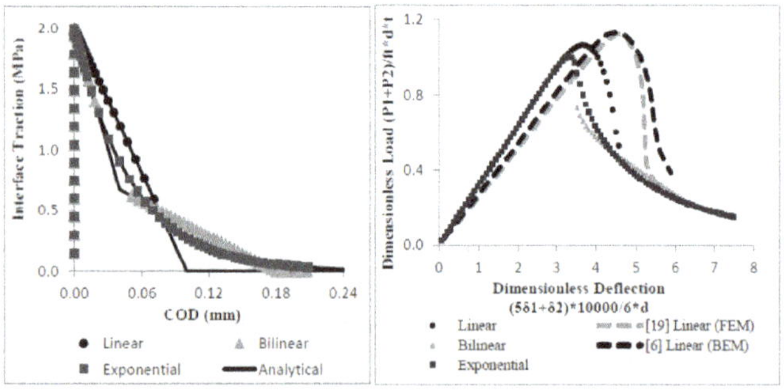

Figure 9: Cohesive tractions at the interface (left) and nonlinear structural answer (right).

Cohesive law	Classical Scheme	Tangent Operator	Reduction
Linear	1281	100	~ 92%
Bilinear	3346	220	~ 93%
Exponential	3527	579	~ 84%

Table 2: Comparison of iterations 2.

face. The displacements values were considered as the prescribed ones. Figure 9 (left) shows the cohesive traction responses and Figure 9 (right) shows the load-displacement curves. Good agreement is observed among the numerical results considered. The developed BEM formulation was capable to represent the nonlinear structural behaviour for this application. As authors Saleh et al. and Carpinteri [6,19] used the linear cohesive law, the results obtained with such a law presented the best agreement. However, the three cohesive laws applied given good accuracy in terms of maximum resistant load.

The numerical efficiency was compared between the TO approach and the classical Newton method. The efficiency in terms of the required amount of iterations is presented in Table 2. Strong improvement is achieved with the TO, 93% for the best case and 84% for the worst case.

Conclusions

A BEM formulation has been presented for modelling the fracture of concrete structures using the cohesive crack model. For the considered applications, the results obtained showed strong agreement with numerical and experimental responses available in the literature. Furthermore, the TO scheme has demonstrated its efficiency as it required an amount of iterations considerable lesser than the classical Newton approach. Therefore, such an operator may lead to remarkable gains in terms of computational time consuming when applied in most complex engineering problems involving multi-fractured structures.

Acknowledgements

Sponsorship of this research project by the Sao Paulo Research Foundation (FAPESP), grant number 2014/18928-2, is greatly appreciated. This research is a part of the activities of research project USP/COFECUB 2012.1.672.1.0.

References

1. Hillerborg A, Modeer M, Peterson PE (1976) Analysis of crack formation and crack growth in concrete by mean of failure mechanics and finite elements. Cement and Concrete Research 6: 773-782.
2. Barenblatt GI (1962) The mathematical theory of equilibrium cracks in brittle fracture. Advances in Applied Mechanics 7: 55-129.
3. Dugdale DS (1960) Yelding of steel sheets containing slits. Journal of Mechanics and Physics of Solids 8: 100-104.
4. Chen T, Wang B, Cen Z, Wu Z (1999) A symmetric galerkin multi-zone boundary element method for cohesive crack growth. Engineering Fracture Mechanics 63: 591-609.
5. Cendon DA, Galvez JC, Elices M, Planas J (2000) Modelling the fracture of concrete under mixed loading. International Journal of Fracture 103: 293-310.
6. Saleh AL, Aliabadi MH (1995) Crack-growth analysis in concrete using boundary element method. Engineering Fracture Mechanics 51: 533-545.
7. Comi C, Mariani S (2007) Extended finite element simulation of quasi-brittle fracture in functionally graded materials. Computer Methods Applied Mechanics Engineering 196: 4013-4026.
8. Abbas S, Alizada A, Fries TP (2010) Model-independent approaches for the XFEM in fracture mechanics. International Journal of Numerical Methods in Engineering 1: 1-10.
9. Oliveira HL, Leonel ED (2013) Dual BEM formulation applied to analysis of multiple crack propagation. Key Engineering Materials 560: 99-106.
10. Cordeiro SGF, Leonel ED (2014) Cohesive Discontinuities Growth Analysis using a Nonlinear Boundary Element Formulation. Journal of Applied & Computational Mathematics 3: 172-181.
11. Benedetti I, Aliabadi MH (2013) A Cohesive Boundary Element Approach to Material Degradation in Three-dimensional Polycrystalline Aggregates. Advances in Boundary Element & Meshless Techniques 14: 51-55.
12. Botta AS, Venturini WS, Benallal A (2005) BEM applied to damage models emphasizing localization and associated regularization techniques. Engineering Analysis with Boundary Elements 29: 814-827.
13. Oliveira HL, Leonel ED (2014) An alternative BEM formulation, based on dipoles of stresses and tangent operator technique, applied to cohesive crack growth modelling. Engineering Analysis with Boundary Elements 41: 74-82.
14. Leonel ED, Venturini WS (2011) Non-linear boundary element formulation applied to contact analysis using tangent operator. Engineering Analysis with Boundary Elements 35: 1237-1247.
15. Brebbia CA, Dominguez J (1989) Boundary Elements: An introductory. McGraw Hill, New York.
16. Hong HK, Chen JT (1988) Derivations of integral equations of elasticity. Journal of Engineering Mechanics 114: 1028-1044.
17. Chen JT, Hong HK (1999) Review of dual boundary element methods with emphasis on hypersingular integrals and divergent series. Applied Mechanics Reviews ASME 52: 17-33.
18. Saleh AL (1997) Crack growth in concrete using boundary elements. WIT Press/Computational Mechanics publisher.
19. Carpinteri A (1994) Cracking of strain-softening materials. In: Aliabadi MH, Brebbia CA, Parton VZ (eds.) Static and Dynamic Fracture Mechanics. Computational Mechanics Publisher 311-365.

Semi Analytical-Solution of Nonlinear Two Points Fuzzy Boundary Value Problems by Adomian Decomposition Method

Jameel AF*

School of Mathematical Sciences, 11800 USM, University Science Malaysia, Penang, Malaysia

Abstract

In this paper the Adomian Decomposition Method (ADM) is employed to solve n^{th} order (n>2) non linear two point fuzzy boundary value problems (TPFBVP). The Adomian decomposition method can be used for solving nth order fuzzy differential equations directly without reduction to first order system. We illustrate the method in numerical experiment including fourth order nonlinear TPFBVP to show the capabilities of ADM.

Keywords: Fuzzy numbers; Fuzzy differential equations; Two point fuzzy boundary value problems; Adomian decomposition method

Introduction

Many dynamical real life problems may be formulated as a mathematical model. Many of them can be formulated either as a system of ordinary or partial differential equations. Fuzzy differential equations (FDEs) are a useful tool to model a dynamical system when information about its behavior is inadequate. FDE appears when the modeling of these problems was imperfect and its nature is under uncertainty. FDEs are suitable mathematical models to model dynamical systems in which there exist uncertainties or vagueness. These models are used in various applications including, population models [1-3], mathematical physics [4], and medicine [5,6]. In recent year's semi -analytical methods such as the Adomian Decomposition Method (ADM), Homotopy Perturbation Method (HPM), Variational Iteration Method (VIM), Optimal Homotopy asymptotic method (OHAM) and Homotopy Analysis Method (HAM) have been used to solve fuzzy first and n^{th} order ordinary differential equations. For n^{th} order fuzzy initial value problems, The ADM was employed in [7] to solve second order linear fuzzy initial value problems. Abbasbandy et al. [8] used the VIM to solve linear system of first order fuzzy initial value problems. Moreover, some of these methods have been also used to obtain a semi-analytical solution of TPFBVP. VIM has been used in [9] to solve linear TPFBVP. Other method like undetermined fuzzy coefficients method has been introduced in [10] in order to obtain an approximate solution of second order linear TPFBVP.

The ADM have been introduced in [11,12] and has been applied to a wide class of deterministic and stochastic problems of mathematical and physical sciences [13-15]. This method provides the solution as a rapidly convergent series with components that are elegantly computed. This method can be used to solve all types of linear and nonlinear equations such as differential and integral equations, so it is known as a powerful method. Another important advantage of this method is that it can reduce the size of computations, while increases the accuracy of the approximate solutions so it is known as a powerful method

In this paper, our aim is to formulate ADM from crisp into fuzzy case in order to solve nonlinear n^{th} order TPFBVP directly. To the best of our knowledge, this is the first attempt at solving the nth order TPFBVP using the ADM. The structure of this paper is as follows: In section 2, some basic definitions and notations are given about fuzzy numbers that will be used in other sections we discussed. In section 3, the structure of ADM is formulated for solving high order TPFBVP. In section 4, we present a numerical example and finally, in section 5, we give the conclusion of this study Figure 1.

Fuzzy Numbers

Fuzzy numbers are a subset of the real numbers set, and represent uncertain values. Fuzzy numbers are linked to degrees of membership which state how true it is to say if something belongs or not to a determined set Figure 2. A fuzzy number [16] μ is called a triangular fuzzy number if defined by three numbers $\alpha < \beta < \gamma$ where the graph of $\mu(x)$ is a triangle with the base on the interval $[\alpha, \beta]$ and vertex at $x = \beta$, and its membership function has the following form:

$$\mu(x;\alpha,\beta,\gamma) = \begin{cases} 0, \ if\, x < \alpha \\ \dfrac{x-\alpha}{\beta-\alpha}, \ if\, \alpha \le x \le \beta \\ \dfrac{\gamma - x}{\gamma-\beta}, \ if\, \beta \le x \le y \\ 1, \ if\, x > \gamma \end{cases}$$

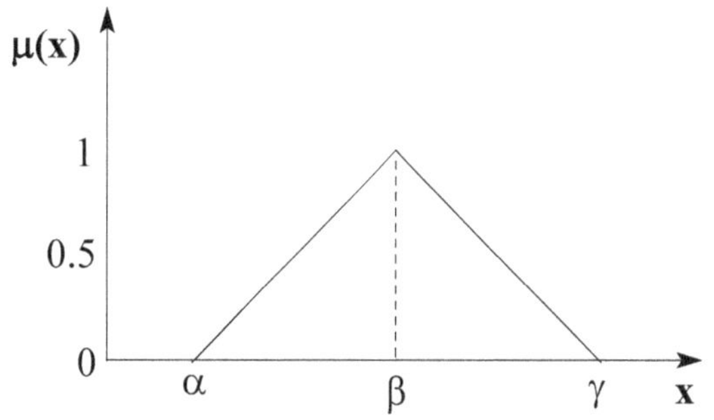

Figure 1: Triangular Fuzzy Number.

***Corresponding author:** Jameel AF, School of Mathematical Sciences, 11800 USM, University Science Malaysia, E-mail: kakarotte79@gmail.com

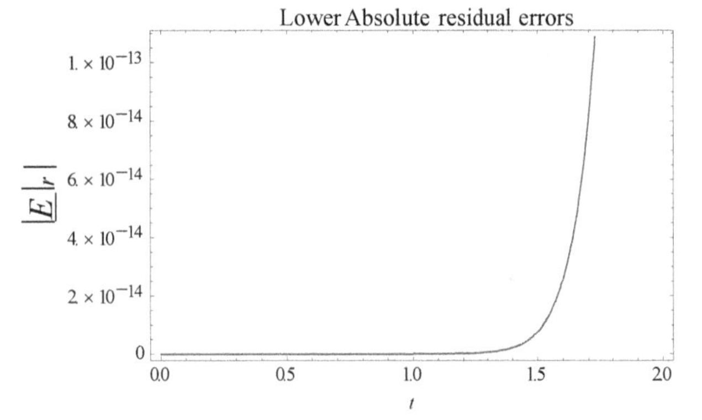

Figure 2: Absolute residual errors of the lower solution of Eq. (18) at r=0.8.

and its r-level is $[\mu]_r = [\alpha + r(\beta-\alpha),\ \gamma - r(\gamma-\beta)],\ r \in [0,1]$ In this paper the class of all fuzzy subsets of R will be denoted by RF and satisfy the following properties [16,17]:

1. $\mu(t)$ is normal, i.e $\exists t_0 \in R$ with $\mu(t_0)=1$,
2. $\mu(t)$ is convex fuzzy set, i.e. $\mu(\lambda t+(1-\lambda)s) \ge \min\{\mu(t),\mu(s)\}\ \forall t,s \in \mathbb{R}, \lambda \in [0,1]$,
3. μ upper semi-continuous on R, and $\overline{\{t \in \mathbb{R} : \mu(t) > 0\}}$ is compact

RF is called the space of fuzzy numbers and R is a proper subset of RF.

Define the r-level set $x \in R$, $[\mu]_r = \{x \setminus \mu(x) \ge r\}, 0 \le r \le 1$ where $[\mu]_0 = \{x \setminus \mu(x) > 0\}$ is compact which is a closed bounded interval and denoted by $[\mu]_r = (\underline{\mu}(t), \overline{\mu}(t))$. In the parametric form, a fuzzy number is represented by an ordered pair of functions $(\underline{\mu}(t), \overline{\mu}(t))$, $r \in [0,1]$ which satisfies [18]:

1. $\underline{\mu}(t)$ is a bounded left continuous non-decreasing function over [0.1].
2. $\overline{\mu}(t)$ is a bounded left continuous non-increasing function over [0.1].
3. $\underline{\mu}(t) \le \overline{\mu}(t)$, $r \in [0,1]$.

A crisp number r is simply represented by $\underline{\mu}(r) = \overline{\mu}(r) = r$, $r \in [0,1]$.

Fuzzification and Defuzzification of ADM

The general structure of ADM for solving crisp n[th] order two point boundary value problems involving ordinary differential equations are mentioned in [18-20]. To solve n[th] order TPFBVP, we need to fuzzify ADM and then defuzzify it Figure 3.

Consider the following general n[th] order TPFBVP

$$\tilde{y}^{(n)}(t) = f\left(t, \tilde{y}(t), \tilde{y}'(t), \tilde{y}''(t), \ldots \tilde{y}^{(n-1)}(t)\right) + \tilde{w}(t),\ t \in [t_0, T] \quad (1)$$

subject to two point boundary conditions

$$\begin{cases} \tilde{y}(t_0) = \tilde{\alpha}^{(0)}, \tilde{y}'(t_0) = \tilde{\alpha}^{(1)}, \ldots \tilde{y}^{(k)}(t_0) = \tilde{\alpha}^{(k)}, \\ \tilde{y}(T) = \tilde{\beta}^{(0)}, \tilde{y}'(T) = \tilde{\beta}^{(1)}, \ldots, \tilde{y}^{(n-k-r)}(T) = \tilde{\beta}^{(n-k-s)} \end{cases} \quad (2)$$

where $0 \le k \le n-2$ is an integer(For cases when n=4 and n=6, [21, 22]), $\tilde{y}(t)$ is a fuzzy function of the crisp variable t. Also, f is fuzzy function of the crisp variable t and the fuzzy variable $\tilde{y}$. Here $\tilde{y}^{(n)}$ is the nth order fuzzy H-derivative [23] of $\tilde{y}(t)$ and the H-derivatives $\tilde{y}'(t), \tilde{y}''(t), \ldots \tilde{y}^{(n-1)}(t)$. Moreover the fuzzy boundary conditions $\tilde{y}(t_0), \tilde{y}'(t_0), \ldots \tilde{y}^{(k)}(t_0), \tilde{y}(T), y', \ldots \tilde{y}^{(n-k-s)}(T)$ are convex fuzzy numbers as in section 2. We denote the fuzzy function y by $\tilde{y} = [\underline{y}, \overline{y}]$, for $t \in [t_0, T]$ and $r \in [0,1]$ it means that the r-level set of $y(t)$ can be defined as:

$$[\tilde{y}(t)]_r = [\underline{y}(t;r), \overline{y}(t;r)],$$

$$[\tilde{y}'(t)]_r = [\underline{y}'(t;r), \overline{y}'(t_0;r)], \ldots [\tilde{y}^{(n-1)}(t)]_r = [\underline{y}^{(n-1)}(t;r), \overline{y}^{(n-1)}(t;r)]$$

$$[\tilde{y}(t_0)]_r = [\underline{y}(t_0;r), \overline{y}(t_0;r)],$$

$$[\tilde{y}'(t_0)]_r = [\underline{y}'(t_0;r), \overline{y}'(t_0;r)], \ldots [\tilde{y}^{(k)}(t_0)]_r = [\underline{y}^{(k)}(t_0;r), \overline{y}^{(k)}(t_0;r)],$$

$$[\tilde{y}'(T)]_r = [\underline{y}'(T;r), \overline{y}'(T;r)], \ldots [\tilde{y}^{(k)}(T)]_r = [\underline{y}^{(k)}(T;r), \overline{y}^{(k)}(T;r)]$$

Where W(t) is crisp or fuzzy inhomogeneous term such that $[\tilde{w}(t)]_r = [\underline{w}(t;r), \overline{w}(t;r)]$.

Since $y^{(n)}(t) = f\left(t, y(t), y'(t), y''(t), \ldots y^{(n-1)}(t)\right) + w(t)$

If we let $(t) = y(t), y'(t), y''(t), \ldots y^{(n-1)}(t)$, such that

$$\tilde{\mathcal{Y}}(t;r) = [\underline{\mathcal{Y}}(t;r), \overline{\mathcal{Y}}(t;r)] = [\underline{y}(t;r), \underline{y}'(t;r), \ldots, \underline{y}^{(n-1)}(t;r), \overline{y}(t;r), \overline{y}'(t;r), \ldots, \overline{y}^{(n-1)}(t;r)]$$

Also, we can write

$$[\tilde{f}(t, \tilde{\mathcal{Y}})]_r = [\underline{f}(t, \tilde{\mathcal{Y}};r), \overline{f}(t, \tilde{\mathcal{Y}};r)] \quad (3)$$

by using Zadeh extension principles [24,25] we have

$\tilde{f}(t, \tilde{\mathcal{Y}}(t;r)) = [\underline{f}(t, \tilde{\mathcal{Y}}(t;r)), \underline{f}(t, \tilde{\mathcal{Y}}(t;r)),]$, such that

$$\underline{f}(t, \tilde{\mathcal{Y}}(t;r)) = F(t, \underline{\mathcal{Y}}(t;r), \overline{\mathcal{Y}}(t;r)) = F(t, \tilde{\mathcal{Y}}(t;r))$$

$$\overline{f}(t, \tilde{\mathcal{Y}}(t;r)) = G(t, \underline{\mathcal{Y}}(t;r), \overline{\mathcal{Y}}(t;r)) = G(t, \tilde{\mathcal{Y}}(t;r))$$

Then we have

$$\underline{y}^{(n)}(t;r) = \mathcal{F}(t, \tilde{\mathcal{Y}}(t;r)) + \underline{w}(t;r) \quad (4)$$

$$\overline{y}^{(n)}(t;r) = \mathcal{G}(t, \tilde{\mathcal{Y}}(t;r)) + \overline{w}(t;r) \quad (5)$$

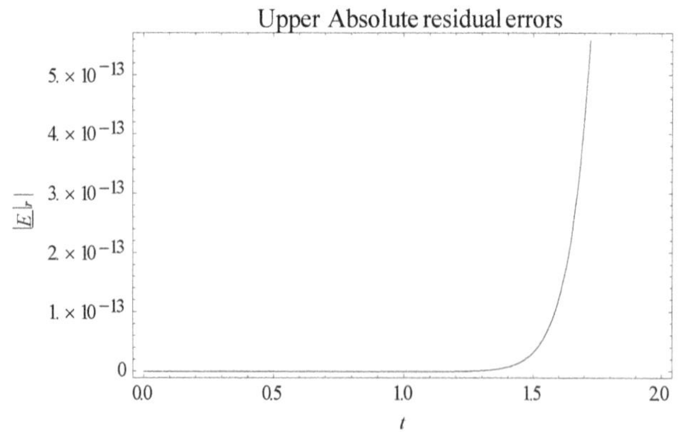

Figure 3: Absolute residual errors of the upper solution of Eq. (18) at r=0.8.

where the membership function of $\mathcal{F}(t,\tilde{\mathcal{Y}}(t;r))+\underline{w}(t;r)$ and $\mathcal{G}(t,\tilde{\mathcal{Y}}(t;r))+\overline{w}(t;r)$ can be defined as

$$\mathcal{F}(t,\tilde{\mathcal{Y}}(t;r))+\underline{w}(t;r)=\min\{\tilde{y}^{(n)}(t,\tilde{\mu}(r)):\mu|\mu\in[\tilde{\mathcal{Y}}(t;r)]_r\},$$

$$\mathcal{G}(t,\tilde{\mathcal{Y}}(t;r))+\overline{w}(t;r)=\max\{\tilde{y}^{(n)}(t,\tilde{\mu}(r)):\mu|\mu\in[\tilde{\mathcal{Y}}(t;r)]_r\},$$

for all r < [0,1], Esq.(4) -(5) can be written as follows

$$L_n\underline{y}(t;r)=\mathcal{F}(t,\tilde{\mathcal{Y}}(t;r))+\underline{w}(t;r),\ t\in[t_0,T] \qquad (6)$$

$$L_n\overline{y}(t;r)=\mathcal{G}(t,\tilde{\mathcal{Y}}(t;r))+\overline{w}(t;r),\ t\in[t_0,T] \qquad (7)$$

where $\tilde{\mathcal{L}}_n=[\underline{\mathcal{L}}_n,\overline{\mathcal{L}}_n]$ are the linear operators with $\tilde{\mathcal{L}}_n=\frac{d^{(n)}}{dt^{(n)}}$ and, F,G are nonlinear operators .define the inverse operators $\tilde{\mathcal{L}}_n^{-1}=\int_0^t\int_0^t\int_0^t\ldots[\cdot]_r d\tau d\tau\ldots d\tau$ and applying it on Eqs . (6) and (7) we have

$$\underline{y}(t;r;\underline{a}_i(r))=\underline{y}_0(t;r;\underline{a}_i(r))+L_n^{-1}\underline{w}(t;r)+L_n^{-1}\mathcal{F}(t,\tilde{\mathcal{Y}}(t;r)) \qquad (8)$$

$$\overline{y}(t;r;\overline{a}_i(r))=\overline{y}_0(t;r;\overline{a}_i(r))+L_n^{-1}\overline{w}(t;r)+L_n^{-1}\mathcal{G}(t,\tilde{\mathcal{Y}}(t;r)) \qquad (9)$$

where $\underline{y}_0(t;r;\underline{a}_i(r))$, and $\overline{y}_0(t;r;\overline{a}_i(r))$ are initials guessing and can be defining as follows:

$$\underline{y}_0(t;r;\underline{a}_i(r))=\underline{a}_1(r)+\underline{a}_2(r)t+\frac{\underline{a}_3(r)}{2!}t^2+\ldots+\frac{\underline{a}_n(r)}{(n-1)!}t^{(n-1)} \qquad (10)$$

$$\overline{y}_0(t;r;\overline{a}_i(r))=\overline{a}_1(r)+\overline{a}_2(r)t+\frac{\overline{a}_3(r)}{2!}t^2+\ldots+\frac{\overline{a}_n(r)}{(n-1)!}t^{(n-1)} \qquad (11)$$

where $\underline{a}_i(r)$ and $\overline{a}_i(r)$ are fuzzy constants to be determined i=1,2…n. The ADM introduces the solution $\tilde{y}(t;r;\tilde{a}_i(r))$ and the nonlinear fuzzy functions F and G by infinite series

$$\tilde{y}(t;r;\tilde{a}_i(r))=\sum_{j=0}^{\infty}\tilde{y}_i(t;r;\tilde{a}_i(r)) \qquad (12)$$

and

$$\begin{cases}F(t,\tilde{\mathcal{Y}}(t;r))=\sum_{j=0}^{\infty}\underline{A}_j(t;r)\\ G(t,\tilde{\mathcal{Y}}(t;r))=\sum_{j=0}^{\infty}\overline{A}_j(t;r)\end{cases} \qquad (13)$$

where $\tilde{A}_i(t;r)$ are the so-called Adomian polynomials [19,20]. Now let

$$\begin{cases}\underline{\varnothing}(t;r;\underline{a}_i(r))=\underline{y}_0(t;r;\underline{a}_i(r))+\underline{\mathcal{L}}_n^{-1}\underline{w}(t;r)\\ \overline{\varnothing}(t;r;\overline{a}_i(r))=\overline{y}_0(t;r;\overline{a}_i(r))+\overline{\mathcal{L}}_n^{-1}\overline{w}(t;r)\end{cases} \qquad (14)$$

then substituting (11) and (12) into (8) and (9) respectively we obtain

$$\sum_{i=1}^{\infty}\underline{y}_i(t;r;\tilde{a}_i(r))=\underline{\Psi}(t;r;\underline{a}_i(r))+\underline{\mathcal{L}}_n^{-1}\sum_{j=0}^{\infty}\underline{A}_j(t;r)\frac{1}{2} \qquad (15)$$

$$\sum_{i=1}^{\infty}\overline{y}_i(t;r;\tilde{a}_i(r))=\underline{\Psi}(t;r;\underline{a}_i(r))+\overline{\mathcal{L}}_n^{-1}\sum_{j=0}^{\infty}\overline{A}_j(t;r) \qquad (16)$$

According to the ADM, the components $\tilde{y}_i(t;r;\tilde{a}_i(r))$ can be determined as

$$\begin{cases}\underline{y}_0(t;r;\underline{a}_i(r))=\underline{\Psi}(t;r;\overline{a}_i(r))]\\ \underline{y}_1(t;r;\underline{a}_i(r))=2\underline{L}_4^{-1}\underline{A}_0(\underline{y}_0(t;r;\underline{a}_i(r)))\\ \underline{y}_{n+1}(t;r;\overline{a}_i(r))=\underline{L}_4^{-1}\underline{A}_n(\sum_{i=0}^{n}\underline{y}_i(t;r;\underline{a}_{i+1}(r)))\end{cases} \qquad (17)$$

and for the upper bound

$$\begin{cases}\tilde{y}_0(t;r;\tilde{a}_i(r))=[\overline{\Psi}(t;r;\overline{a}_i(r))]^2\\ \tilde{y}_1(t;r;\tilde{a}_i(r))=2\overline{L}_4^{-1}\overline{A}_0(\underline{y}_0(t;r;\tilde{a}_i(r)))\\ \overline{y}_{n+1}(t;r;\overline{a}_i(r))=\overline{L}_4^{-1}\overline{A}_n(\sum_{i=0}^{n}\tilde{y}_i(t;r;\overline{a}_i(r)))\end{cases} \qquad (18)$$

Since the approximate solution series $\tilde{y}(t;r;\tilde{a}_i(r))$ contain the constants $\tilde{a}_i(r)$ which can determine easily by using the boundary conditions (2) for all $r\in[0,1]$ and for i=1,2,…,n.

Numerical Example

In this section employ OHAM on two examples of a linear and non-linear fourth order TPFBVP and present their approximate results with absolute errors. These examples were presented in [26] for the crisp case (non- fuzzy).

Consider the fourth order nonlinear TPFBVP

$$y^{(4)}(t)=y(t)^2+1,\ 0\le t\le 2 \qquad (18)$$

$$y(0)=[0.1r-0.1,0.1-0.1r],\ y(2)=[0.1r-0.1,0.1-0.1r]$$

$$y'(0)=[0.1r-0.1,0.1-0.1r],\ y'(2)=[0.1r-0.1,0.1-0.1r]$$

$$\forall\ r\in[0,1].$$

Define the linear operator of Eq. (18) $\tilde{\mathcal{L}}_4$ with the inverse operator $\tilde{\mathcal{L}}_4^{-1}$, then the initial approximation guesses

$$\tilde{y}(0;r;\tilde{a}_i(r))=\tilde{a}_1(r)+\tilde{a}_2(r)t+\frac{1}{2!}\tilde{a}_3(r)t^2+\frac{1}{3!}\tilde{a}_4(r)t^3 \qquad (19)$$

such that

$$\tilde{\Psi}(t;r;\tilde{a}_i(r))=\tilde{y}_0(t;r;\tilde{a}_i(r))+\tilde{L}_4^{-1}(1) \qquad (20)$$

for i=1,2,3,4 and $r\in[0,1]$.

According to section 3 the ADM approximate series solution of Eq. (18) is given by

$$\sum_{j=1}^{\infty}\tilde{y}_j(t;r;\tilde{a}_i(r))=\tilde{\Psi}(t;r;\tilde{a}_i(r))+\tilde{L}_4^{-1}\sum_{j=0}^{\infty}\tilde{A}_j(t;r) \qquad (21)$$

Here $\tilde{A}_j(t;r)=\sum_{j=0}^{n-1}\tilde{y}_j(t;r;\tilde{a}_i(r))\tilde{y}_{n-1-j}(t;r;\tilde{a}_i(r))$ for i=1,2,3,4>1.

According to the ADM in section 3, the n components of $\tilde{y}_j(t;r;\tilde{a}_i(r))$ can be determined as

$$\begin{cases}\tilde{y}_0(t;r;\tilde{a}_i(r))=[\tilde{\Psi}(t;r;\tilde{a}_i(r))]^2\\ \tilde{y}_1(t;r;\tilde{a}_i(r))=2\tilde{L}_4^{-1}[\tilde{y}_0(t;r;\tilde{a}_i(r))\tilde{y}_1(t;r;\tilde{a}_i(r))]\\ \tilde{y}_n(t;r;\tilde{a}_i(r))=\tilde{L}_4^{-1}\tilde{A}_n(\sum_{i=0}^{n}\tilde{y}_i(t;r;\tilde{a}_i(r)))\end{cases} \qquad (22)$$

After ten terms of the ADM approximate solution series we have

$$\tilde{y}(t;r;\tilde{a}_i(r))=\sum_{j=0}^{10}\tilde{y}_j(t;r;\tilde{a}_i(r)) \qquad (23)$$

for i=1,2,3,4 and $r \in [0,1]$. Now to determine the values of $\tilde{a}_i(r)$, we substitute the boundary conditions of Eq.(4.1) with the ADM solution series (23) in (18), we obtained

Substitute these values in Eq. (12) to get 10- order ADM approximate solution series. The following tables show ADM approximate solution series $\sum_{j=0}^{5}\tilde{y}_j(t;r;\tilde{a}_i(r))$ at t=0.8 and for all $\in [0,1]$. Since Eq. (18) is without exact analytical solution, so to show the accuracy of 10-order ADM absolute errors $[\underline{E}]_r$ and $[\overline{E}]_r$ of solutions $\underline{y}(0.8;r)$ and $\overline{y}(0.8;r)$, we define residual absolute error [27] as follows

r	$\underline{a}_1(r)$	$\underline{a}_2(r)$	$\underline{a}_3(r)$	$\underline{a}_4(r)$
0	−0.5	−0.5	1.916150	−2.76970
0.25	−0.375	−0.375	1.503530	−2.27339
0.5	−0.25	−0.25	1.102100	−1.81276
0.75	−0.125	−0.125	0.712063	−1.38835
1	0	0	0.333655	−1.00071

Table 1: Lower bound $\sum_{i=1}^{4}\underline{a}_i(r)$ values.

r	$\overline{a}_1(r)$	$\overline{a}_2(r)$	$\overline{a}_3(r)$	$\overline{a}_4(r)$
0	0.5	0.5	−1.058770	0.1702920
0.25	0.375	0.375	−0.7293630	−0.0642764
0.5	0.25	0.25	−0.3873120	−0.3380610
0.75	0.125	0.125	−0.0328861	−0.6504120
1	0	0	0.3336550	−1.000710

Table 2: Upper bound $\sum_{i=1}^{4}\overline{a}_i(r)$ values.

r	$\underline{y}\left(0.8;r;\sum_{i=1}^{4}\underline{a}_i(r)\right)$	$[\underline{E}]_r$	$\overline{y}\left(0.8;r;\sum_{i=1}^{4}\overline{a}_i(r)\right)$	$[\overline{E}]_r$
0	−0.501024	3.77476×10^{-15}	0.5981310	4.88498×10^{-16}
0.25	−0.367949	6.66134×10^{-16}	0.4562040	4.44089×10^{-16}
0.5	−0.233698	1.11022×10^{-16}	0.3156360	0
0.75	−0.0439827	0	0.1763930	2.22045×10^{-16}
1	0.0384436	0	0.0384436	2.22045×10^{-16}

Table 3: Numerical results by 10-order ADM for the lower bound of Eq. (18) at *t*=0.8 for all $r \in [0,1]$.

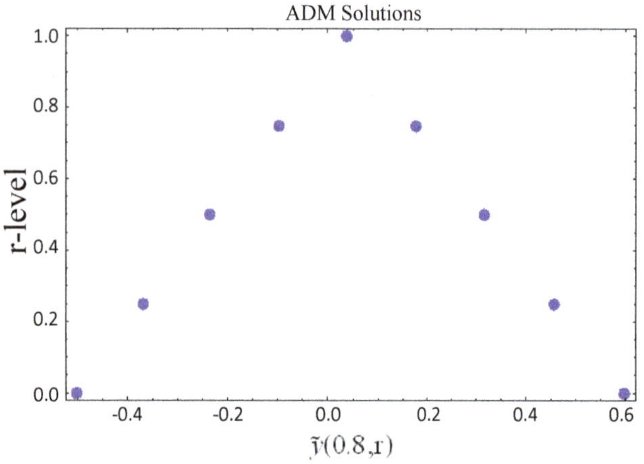

Figure 4: ADM solution of Eq. (18) when t = 0.8 for all $0 \leq r \leq 1$.

$$[\tilde{E}]_r = \left| \left(\sum_{j=0}^{10}\tilde{y}_j(t;r;\tilde{a}_i(r))\right)^{(4)} + e^{-t}\left[\sum_{j=0}^{10}\tilde{y}_j(t;r;\tilde{a}_i(r))\right]^2 - 1 \right| \quad (24)$$

From Tables 1-3 and Figure 4 one can see that the numerical results are satisfies the convex symmetric triangular fuzzy number and we have the same results for all $0 \leq t \leq 2$.

Conclusions

In this study, a semi- analytical method ADM was applied to obtain an approximate solution n^th^ order nonlinear TPFBVP. A scheme based ADM approximate the solution of nth order TPFBVP has been formulated to obtain the approximate solution directly with our reduced in to first order system. From the nonlinear TPFIVP the accuracy of ADM can be determined even these equations without exact analytical solution. ADM give more accurate solution when the number of terms of the series solution was increased. Numerical examples which included nonlinear TPFBVP show the efficiency of the implemented of this semi- analytical method. The problem results satisfied the fuzzy numbers properties by taking the triangular fuzzy numbers shape.

References

1. Omer A, Omer O, Pray A (2013) Pretdour Model with Fuzzy Intial Values. Hacettepe Journal of Mathematics and Statistics 41: 387-395.
2. Tapaswini S, Chakraverty S (2013) Numerical Solution of Fuzzy Arbitrary Order Predator-Prey Equations. Applications and Applied Mathematics 8: 647-673.
3. Zaini M, Ahmad BDB (2009) A Predator-Prey Model with Fuzzy Initial Populations. *IFSA-EUSFLAT*.
4. Naschie MSE (2005) From Experimental Quantum Optics to Quantum Gravity Via a Fuzzy Kahler Manifold, Chaos Solution and Fractals 25: 969-977.
5. Abbod MF, Keyserlingk DGV, Linkens DA, Mahfouf M (2001) Survey of Utilization of Fuzzy Technology in Medicine and Healthcare, Fuzzy sets and system 120: 331-3491.
6. Barro S, Marin (2002) Fuzzy Logic in Medicine, Heidelberg: Physica-Verlag.
7. Wang L, Guo S (2011) Adomian Method For Second-Order Fuzzy Differential Equation, World Academy of Science, Engineering and Technology 5: 783-786.
8. Abbasbandy S, Allahviranloo T, Darabi P, Sedaghatfar O (2011) Variational Iteration Method for Solving N-th order Fuzzy Differential Equations. Mathematical and Computational Applications 16: 819-829.
9. Armand A, Gouyandeh Z (2013) Solving Two-Point Fuzzy Boundary Value Problem Using Variational Iteration Method, Communications on Advanced Computational Science with Applications 1-10.
10. Guo X, Shang D, Lu X (2013) Fuzzy Approximate Solutions Of Second-Order Fuzzy Linear Boundary Value Problems. Journal of Boundary Value Problems 1-17.
11. Adomian G (1989) Nonlinear Stochastic Systems Theory and Applications to Physics, Kluwer, Dordrecht.
12. Adomian G (1994) Solving Frontier Problems of Physics: The Decomposition Method, Kluwer, Dordrecht.
13. Yee E (1993) Application of the Decomposition Method to the Solution of the Reaction Convection-Diffusion Equation. App Math Computation 56: 1-27.
14. Babolian E, Biazar J (2002) Solving the Problem of Biological Species Living Together by Adomian Decomposition Method. App Math Computation 129: 339 -343.
15. Babolian E, Biazar J (2003) Solving Concrete Examples by Adomian Method. App Math Computation 135: 161-167.
16. Dubois D, Prade H (1982) Towards fuzzy differential calculus, Part 3: Differentiation. Fuzzy Sets and Systems 8: 225-233.
17. Mansouri S, Ahmady SN (2012) A Numerical Method For Solving Nth-Order Fuzzy Differential Equation by using Characterization Theorem. Communication in Numerical Analysis 1-12.

18. Kaleva O (1987) Fuzzy Differential Equations. Fuzzy Sets Syst 24: 301-317.

19. Babolian E, Aahidi V, Asadi G (2005) Solving Differential Equations by Decomposition Method, App Math Computation 167: 1150-1155.

20. El-Sayed S (2002) The Modified Decomposition Method for Solving Nonlinear Algebraic Equations. App Math Computation 132: 589-597.

21. Liang S, Jeffrey D (2010) Approximate solutions to a parameterized sixth order boundary value problem. Comput Math Appl 59: 247-253.

22. Liang S, Jeffrey D (2009) An efficient analytical approach for solving fourth order boundary value problems. Comput Phys Commun 180: 2034-2040.

23. Salahshour S (2011) Nth-order Fuzzy Differential Equations under Generalized Differentiability, Journal of fuzzy set value analysis 1-14.

24. Zadeh LA (1965) Fuzzy Sets. Information and Control 8: 338-353.

25. Zadeh LA (2005) Toward A Generalized Theory of Uncertainty. Information Sciences 172: 1-40.

26. Haq S, Idrees M, Islam S (2010) Application of Optimal Homotopy Asymptotic Method to Eighth Order Initial and Boundary Value Problems. International Journal of Applied Mathematics and Computation 2: 73-80.

27. Liang S, David JJ (2003) Comparison of Homotopy Analysis Method and Technique. Applied Mathematics and Computation 135: 73-79.

Effect of Solvents on the Electronic Properties of Fullerene Based Systems: Molecular Modelling

Abdel-Baset H Mekky[1]*, Hanan G Elhaes[2], Mohamed M. El-Okr[3], Abdulaziz S Al-Aboodi[1] and Medhat A. Ibrahim[4]

[1]Buraydah Colleges, Al-Qassim, King of Saudi Arabia
[2]Faculty of Women for Arts, Science, and Education, Ain Shams University, Egypt
[3]Faculty Science, El Azhar University, Egypt
[4]Spectroscopy Department, National Research Centre, Egypt

Abstract

Density functional theory (DFT) was utilized as a backup for the experimental study. Solvent effects on the molecular structure and electronic properties of fullerene C60 and its derivatives were estimated by performing the DFT with B3LYP/3-21G** calculations using the polarized continuum model (PCM) developed by Tomasi. Overall, different forms of quantum-chemistry were used based on performed calculations, including dipole moments (total dipole moment, X, Y, and Z components) and finally, orbital energies, EHOMO, ELUMO; HL gap (gap between EHOMO and ELUMO).

Keywords: Computational methods; Fullerene; Solubility; DFT; polarized continuum model; Molecular modeling; Electronic properties

Introduction

Fullerenes are allotropes of carbon and possess various novel properties that make them useful in the area of nanotechnology and pharmaceuticals. Fullerenes are currently being widely investigated and have potential for various technical applications [1]. In particular, for biomedical testing, water-soluble forms of fullerenes are of outstanding interest.

Numerous efforts have been formed to explain the trends in room-temperature solubility of the fullerenes [2-4]. There are many different approaches to calculate and predict C60 solubility in organic solvents. Some of them are fully mechanistic [5-9], developed from the thermodynamical point of view; others are statistically based, with good correlation coefficients, but not transparent and complicated in interpretation [10-16].

In this work, we aspired to discover simple, transparent relationship and computationally fast approach to predict the solubility of C60 in various solvents. The exact description of solvation phenomena presents a substantial challenge to theoretical chemistry. Two typical approaches are employed to contain the solvent effects. One is the so-called supermolecular approach or the discrete model, where explicit solvent molecules are added and treated at the same level of theory as that employed for the solute. Nevertheless, this supermolecular approach does come with a significant increase in computational expense. Some other approach is the polarizable continuum model (PCM).

Previously, the structures of Fullerenes c60 had been studied in the gas phase by means of the B3LYP/3-21G** level of calculation [17,18]. We made geometry optimization in vacuo, and the calculation was based along the gas-phase optimized geometries. The solvent effects were identified through the polarizable continuum model (PCM) of Tomasi et al. [19] because the PCM model offers a balanced and theoretically sound treatment of all solute–solvent interactions at a very reasonable computational cost.

Computational Methods

All the calculations use the polarizable continuum model (PCM) were utilized to describe the thermodynamics of solvation of the fullerenes C_{60} and its derivatives in some solvents. Solvent effects were evaluated using the (Gaussian 98 w) [20] implementation of the thermodynamically based polarized continuum model (PCM)· the level of calculations is b3lyp/3-21g** for the ground state.

Results and Discussions

Solubility of fullerene C_{60}

The bond lengths and bond angles of the studied compound, C_{60}, were optimized in the gas phase, water, Methanol, Heptane, Acetonitrile, Toluene and Pyrrolidine. Table 1 establishes the important changes of bond length and bond angle for C60 from gas to solve.

a) Minimal value of bond length between carbon atoms;

b) Maximal value of bond length between carbon atoms

c) Minimal value of bond angle between carbon atoms;

d) Maximal value of bond angle between carbon atoms

From Table 1, we can see little changes in the bond lengths of C60 when C60 was optimized with water, Methanol, Heptane, Acetonitrile and Toluene and compared within the gas phase. The great change of bond length and angle value of C_{60} in the Pyrrolidine compared within the gas phase, water, Methanol, Heptane, Acetonitrile and Toluene. The same bond length and angle in water, Heptane, Acetonitrile and Toluene (slight modifications in the attachment angle of C60 in Methanol).

***Corresponding author:** Abdel-Baset H Mekky, Buraydah Colleges, Saudi Arabia
E-mail: hofny_a@yahoo.com

Solvents	Geometry Optimization			
	C – C [a)]	C – C [b)]	C - C – C [c)]	C - C – C [d)]
Gas phase	1.385	1.464	108	120
Water	1.389	1.461	108	120
Methanol	1.393	1.457	107.9	120.2
Heptane	1.389	1.461	108	120
Acetonitrile	1.389	1.461	108	120
Toluene	1.389	1.461	108	120
Pyrrolidine	1.011	1.571	100.7	124.6

Table 1: Selected structural parameters of the optimized C_{60} in gas and different solvents, bond distance (A°) and bond angles (°). Obtained by the B3LYP methods with 3-21G** basis set method which described through the polarizable continuum model (PCM).

Solvent	HOMO	LUMO	Energy Gap (ev)
Gas phase	-6.531	-3.523	3.007
Water	-6.531	-3.523	3.007
Methanol	-6.531	-3.523	3.007
Heptane	-6.531	-3.523	3.007
Acetonitrile	-6.531	-3.523	3.007
Toluene	-6.531	-3.523	3.007
Pyrrolidine	-6.06	-3.359	2.7

Table 2: Calculated HOMO/LUMO energy gap as (ev) for C_{60}, which described through the polarizable continuum model (PCM) and calculated by B3LYP methods with 3-21G** basis set.

Also, Table 1 shows a slight lengthening of the maximum bond length and angle of C_{60} in Pyrrolidine (a widening of the maximum bond length and angle), and a narrowing of minimum bond length and angle. Our simulation predicts a lengthening of the max C-C bond and the max. C-C-C angle (≈+0.11Å and 4.6 degrees, respectively) and a narrowing of the min. bond length and angle (≈-0.873 Å and 7.3 degrees, respectively), whereas the Methanol-solvent results predict smaller changes for the bond lengths and angles.

The solvents (water, Methanol, Heptane, Acetonitrile and Toluene) effect has a minor influence on these geometrical parameters, the calculations do not evidence any significative effect on the lengths and angles, the only observable change is of C_{60} in the Pyrrolidine.

The max. C-C bond length in solvent (Pyrrolidine) is longer than that in a vacuum, and the solvents (water, Methanol, Heptane, Acetonitrile and Toluene), these results indicate that the solvent (Pyrrolidine) can reduce the intensity of the max. C-C bond, which might imply that the solvent (Pyrrolidine) is in favor of the C-C activation and improve the reaction activity.

Table 2, summarizes the highest occupied molecular orbital (HOMO), the lowest unoccupied molecular orbital (LUMO) and HOMO and LUMO energy gaps for C_{60} calculated at the DFT level in the 3-21G** basis set. The values of LUMO and HOMO and their energy gap reflect the chemical activity of the molecule. LUMO as an electron acceptor represents the ability to obtain an electron, while HOMO as an electron donor represents the ability to donate an electron. The smaller the LUMO and HOMO energy gaps, the easier it is for the HOMO electrons to be excited; the higher the HOMO energies, the easier it is for HOMO to donate electrons; the lower the LUMO energies, the easier it is for LUMO to accept electrons. From the resulting data shown in Table 2, it is obvious that the LUMO energy of C_{60} in the Pyrrolidine is lower than those of C_{60} in the gas phase and also, the energy gap of C_{60} in the Pyrrolidine is smaller than that of C_{60} in the gas phase. Consequently, the electrons transfer from HOMO to LUMO in C_{60} in the gas phase is relatively easier than that in C_{60} in the Pyrrolidine. With the decrease of the LUMO energies, LUMO in C_{60} in the gas phase accepts electrons easily. Soft molecules have a small energy gap.

From Table 2, one can see that the solvent sequence with low to high HOMO/LUMO gap (energy gap of C_{60} in Pyrrolidine<energy gap of C_{60} in the gas phase, water, Methanol, Heptane, Acetonitrile and Toluene solutions).

The HOMO–LUMO gaps had been studied in the gas phase, water, Methanol, Heptane, Acetonitrile and Toluene solutions, show no effect and the energy gap remains the same. Based on the above description, it is easy to understand why the water, Methanol, Heptane, Acetonitrile and Toluene solutions are suspension of C_{60}.

The results reported in Table 3, show that C_{60} in the gas phase, water, Methanol, Heptane, Acetonitrile and Toluene solvents possess the lowest dipole moment compared with C_{60} in Pyrrolidine solvent.

By knowing the value of dipole moment we can conclude that the clusters having a higher dipole moment can be treated as a good solvent.

As expected, the molecule becomes more polarized in the presence of the Pyrrolidine solvent (Table 3), and the molecular dipole moment increases with solvent polarity.

Solubility of specific fullerene derivatives

The optimized structures of the $C_{59}X$ (X=C, B, Al, Ga, In, N, P, As, Sb) cages in pyrrolidine solvent, the bond lengths and bond angles in pyrrolidine solvent are listed in Table 4. From Table 4, we can see that all the doped cages in pyrrolidine solvent undergo some distortions due to the dopant atoms, though they still preserve closed cage structures.

a) Minimal value of bond length between carbon atoms;

b) Maximal value of bond length between carbon atoms

c) Minimal value of bond angle between carbon atoms;

d) Maximal value of bond angle between carbon atoms

It is well-known that there are two kinds of C—C bond in C_{60} cage, the [6] bond and the [5,6] bond. In pyrrolidine solvent, The bond lengths are 1.011 and 1.571 Å for [6] and [5,6] bonds, respectively based on our DFT calculations. When the carbon cage is doped by the dopant atom, the C—X bonds are presented. From Table 4, it can be determined that the C—X bond lengths are in the orbit of 1.011 to 2.143Å. The bond lengths increased obviously for X= B, Al, Ga and In, ranging from 1.087Å to 2.143Å and for X= P, As, and Sb, ranging from 1.087Å to 2.013Å. Nevertheless, the C—N bonds in C59N are 1.084 and 1.554Å, and therefore the original [5,6] bond (max. bond length) is even decreased by 0.017 Å compared with that in the pristine cage. It is also found that the C—X bond lengths increase more significantly for the larger dopant atoms.

For example, the C—B bonds are 1.087 and 1.603 Å, while the C—X bonds (X =P, As, Sb) are estimated to be within 1.087-2.013Å. Hence it is clearly that the cage with larger dopant atom gives more obvious distortion.

The bond angles are also listed in Table 4, for fullerene and substituted fullerene in pyrrolidine solvent. From the table it is found that the bond angles are about unchanged for doping with the atoms compared with that of the pristine cage.

Dipole moment contribution as Debye	Gas phase	Water	Methanol	Heptane	Acetonitrile	Toluene	Pyrrolidine
X	0	0	-0.0002	0	0	0	1.9692
Y	0	0	0.0004	0	0	0	0
Z	0	0	-0.0004	0	0	0	-1.2089
Total dipole moment	0.0001	0	0.0006	0	0	0	2.3106

Table 3: Calculated dipole moment as Debye for C_{60}, which described through the polarizable continuum model (PCM) and calculated by B3LYP methods with 3-21G** basis set.

Clusters	Geometry Optimization			
	C – C [a)]	C – C [b)]	C – C – C [c)]	C – C – C [d)]
C60	1.011	1.571	100.7	124.6
C59-B	1.087	1.603	100.7	126.8
C59-Al	1.089	1.828	100.8	127.9
C59-Ga	1.089	1.924	100.8	124.6
C59-In	1.09	2.143	100.8	124.6
C59-N	1.084	1.554	100.4	127.7
C59-P	1.087	1.64	100.7	128.1
C59-As	1.084	1.795	100.8	127.6
C59-Sb	1.085	2.013	100.8	127.6

Table 4: Calculated bond lengths (Å) and bond angles (°) for C_{60} and some substituted fullerenes, which described through the polarizable continuum model (PCM) by using Pyrrolidine and calculated by B3LYP methods with 3-21G** basis set.

Clusters	X	Y	Z	Total dipole moment
C_{60}	1.9692	0	-1.2089	2.3106
$C_{59}B$	-0.7839	-0.0001	-2.5694	2.6863
$C_{59}Al$	-2.3374	0.0004	-1.3346	2.6916
$C_{59}Ga$	2.6356	-1.0881	0.0002	2.8514
$C_{59}In$	3.0786	0.7356	0	3.1653
$C_{59}N$	0.3106	0	-1.6342	1.6635
$C_{59}P$	0.4219	0	-1.8398	1.8875
$C_{59}As$	1.0084	0.0011	-1.6833	1.9622
$C_{59}Sb$	-1.5629	-1.538	0.0011	2.1928

Table 6: Calculated dipole moment as Debye for C_{60} and some substituted fullerenes, which described through the polarizable continuum model (PCM) by using Pyrrolidine and calculated by B3LYP methods with 3-21G** basis set.

Clusters	HOMO	LUMO	Energy Gap (ev)
C60	-6.06	-3.359	2.7
C59-B	-6.261	-3.54	2.72
C59-Al	-6.079	-3.376	2.702
C59-Ga	-6.077	-3.372	2.705
C59-In	-6.127	-3.41	2.717
C59-N	-5.258	-3.557	1.7
C59-P	-4.68	-3.544	1.135
C59- As	-4.765	-3.489	1.275
C59- Sb	-4.651	-3.586	1.064

Table 5: Calculated HOMO/LUMO energy gap as (ev) for C60 and some substituted fullerenes, which described through the polarizable continuum model (PCM) by using Pyrrolidine and calculated by B3LYP methods with 3-21G** basis set.

It is well-known that the frontier orbitals, the highest occupied molecular orbital (HOMO) and the lowest unoccupied molecular orbital (LUMO), act as an important role in chemical reaction for the reactant molecule, so the frontier orbital analysis of the doped cages is necessary. In Table 5, the HOMO and LUMO energy levels of the substituted fullerene were summarized. It can be seen that HOMO levels of the doped (for X=B, Al, Ga and In), are all decreased and for doped by N, P, As and Sb are increased compared with that of $C_{60.}$ As for LUMO levels of the doped cages are all decreased compared with that of C_{60}. As for LUMO levels are increased when doping with N, P, As and Sb atoms, compared with that of the pristine cage.

It is recognized that both the thermodynamic stability and kinetic stability have crucial influence on the relative abundances of different fullerene structures. It has been pointed out that higher kinetic stability is usually associated with a larger HOMO-LUMO energy gap, because exciting electrons from a low HOMO to a high LUMO is energetically unfavorable, which would be necessary to activate a reaction. The calculated Eg of the doped cages is listed in Table 5. It can be found that all the doped cages by N, P, As and Sb present smaller Eg than that of C_{60} cage. So the kinetic stability of the cage is decreased by substitution from viewpoint of HOMO-LUMO gap. And also, It can be found that all the doped cages by B, Al, Ga and In present greater Eg than that of C_{60} cage. So the kinetic stability of the cage is increased by substitution from viewpoint of HOMO-LUMO gap.

Lastly, we have also worked out the dipole moments of the C60 and some substituted fullerenes, which described through the polarizable continuum model (PCM) by using Pyrrolidine and calculated by B3LYP methods with 3-21G** basis set (Table 6).

The dipole moment is defined as the first derivative of the energy with respect to an electric field. The calculated dipole moment can be conveyed in terms of vector in three directions X, Y and Z as indicated in Table 6. The negative mark of both X, Y and Z-axis indicates that the dipole moment is pointing away from X, Y and Z direction, while the plus sign of X, Y and Z-axis indicates that it points in X, Y and Z direction. Regarding Table 5, one can conclude that, the total dipole moment and the net contribution of charge in the directions X, Y and Z are completely exchanged.

As listed in Table 5 the dipole moment is computed. Regarding C_{60} the total dipole moment is 2.3106 Debye distributed as 1.9692, -0.0000 and -1.2089 contributions in the x, y, and z-directions, respectively. As the effect of doping the change in the dipole moment of C_{60} and $C_{59}X$ was being discussed. The value of dipole moment is increased for $C_{59}X$ (X= B, Al, Ga, In) and it is increased as compared with that of C_{60}. And also, The value of dipole moment is increased for $C_{59}X$ (X= N, P, As, Sb) and it is decreased as compared with that of C_{60}.

The C60 and doped-C60 indicate that doping has an outcome on the dipole moment, which can cause subsequent changes in the electrical properties of the molecular species.

Conclusions

The solubility of C60 in the solvents is important to enable purification and chemical change. In universal, the solubility in the majority of solvents is very depressed, because C60 exhibit a high tendency for aggregation. On the other hand the interaction between the solvent molecules and C60 is usually very light, since the fullerene is a nonpolar molecule, which is hardly polarizable due to the large HOMO-LUMO gap. In summary, the solubility of C60 in polar solvents such as methanol and water is about zero. This low solubility can also be considered in the case of, Heptane, Acetonitrile and Toluene as solvents. The best solubilities are obtained in Pyrrolidine, which makes these solvents to the standard solvents for preparative use.

References

1. Dresselhaus MS, Dresselhaus G, Eklund PC (1996) Science of Fullerenes and Carbon Nanotubes.
2. Beck MT, Mandi G (1997) Fullerene Sci. Technol. 5: 291-310.
3. Marcus Y, Smith A, Korobov M, Mirakyan A, Avramenko NA (2001) Solubility of C60 Fullerene J Phys Chem B 105: 2499-2506.
4. Heymann D (1996) Carbon 3: 627-631.
5. Marcus Y (1997) Solubilities of buckminsterfullerene and sulfur hexafluoride in various solvents. J Phys Chem 101:8617–8623.
6. Abraham MH, Green CE, Acree WE (2000) Correlation and prediction of the solubility of Buckminsterfullerene in organic solvents; estimation of some physicochemical properties. J Chem Soc Perkin Trans 2: 281-286.
7. Hansen CM, Smith AL (2004) Using Hansen solubility parameters to correlate solubility of C60 fullerene in organic solvents and in polymers. Carbon 42: 1591-1597.
8. Stukalin EB, Korobov MV, Avramenko NV (2003) Solvation free energies of the fullerenes C60 and C70 in the framework of polarizable continuum model. J Phys Chem B 107: 9692-9700.
9. Kiss IZ, Mandi G, Beck MT (2000) Artificial neural network approach to predict the solubility of C60 in various solvents. J Phys Chem A 104: 8081-8088.
10. Liu H, Yao X, Zhang R, Liu M, Hu Z (2005) Accurate quantitative structure-property relationship model to predict the solubility of C60 in various solvents based on a novel approach using a least-squares support vector machine. J Phys Chem B 109: 20565-20571.
11. Toropov AA, Leszczynska D, Leszczynski J (2007a) QSPR study on solubility of fullerene C60 in organic solvents using optimal descriptors calculated with SMILES. Chem Phys Lett 441: 119-122.
12. Toropov AA, Leszczynska D, Leszczynski J (2007b) Predicting water solubility and octanol water partition coefficient for carbon nanotubes based on the chiral vector. Comput Biol Chem 31: 127-128.
13. Toropov AA, Rasulev BF, Leszczynska D, Leszczynski J (2007c) Additive SMILES based optimal descriptors: QSPR modeling of fullerene C60 solubility in organic solvents. Chem Phys Lett 444: 209-214.
14. Toropov AA, Rasulev BF, Leszczynska D, Leszczynski J (2008) Multiplicative SMILES-based optimal descriptors: QSPR modeling of fullerene C60 solubility in organic solvents. Chem Phys Lett 457: 332-336.
15. Toropov AA, Toropova AP, Benfenati E, Leszczynska D, Leszczynski J (2009) Additive InChI-based optimal descriptors: QSPR modeling of fullerene C60 solubility in organic solvents. J Math Chem 46: 1232-1251.
16. El-Oker MM, Mekky ABH, Elhaes H, Medhat A (2012) Ibrahim " Electronic Properties of Group V Substituted Fullerene: DFT Approach "International Journal of Scientific & Engineering Research, 3.
17. Mekky ABH, Elhaes H, El-Oker MM, Ibrahim MA (2012) " Electronic Properties of Substituted C59X (X= B, Al,Ga, In) Fullerene " Materials Science : An Indian Journal 8: 10.
18. Tomasi J, Persico M (1994) ' Molecular Interactions in Solution: An Overview of Methods Based on Continuous Distributions of the Solvent '. Chem Rev 94: 2027-2094.
19. Frisch MJ (2001) GAUSSIAN 98 (Revision A 11), Gaussian, Pittsburgh PA.
20. Barone V, Cossi M (1998) 'Quantum Calculation of Molecular Energies and Energy Gradients in Solution by a Conductor Solvent Model', Journal of Physical Chemistry A 102: 1995-2001.

6

Synchronization of Discrete Chaotic Systems via Double Scaling Matrix in Different Dimensions

Adel Ouannas*

LAMIS Laboratory, Department of Mathematics and Computer Science, University of Tebessa, Algeria

Abstract

In this paper, a new type of chaos synchronization in discrete-time is proposed when the scaling functions in Q-S synchronization are replaced by two suitable matrices. This new synchronization type allows us to study synchronization between different dimensional discrete chaotic systems in different dimensions. By using Lyapunov stability theory and nonlinear controllers, new results are derived. Numerical simulations are used to verify the effectiveness of the theoritical results and the proposed schemes derived in this paper.

Keywords: Q-S synchronization; Chaos; Dynamical systems; Discrete-time; Scaling matrix

Introduction

Since the pioneering work of Pecora and Carroll [1], many types of synchronization have been presented [2-5] and synchronization of chaotic dynamical systems has attracted a great deal of interest from many fields such as ecology [6], physics [7], chemistry [8], secure communications [9], and so forth. From then, various methods have been devloped to study chaos synchronization [10-13]. Discrete-time dynamical systems plays an important role in mathematical modelization of many problems in sciences and engineering [14]. Therefore, it plays also an important role to consider chaos (hyperchaos) synchronization in discrete-time dynamical systems. Recently, another interesting type of synchronization has received a great deal of attention, called Q-S synchronization [15-17]. In Q-S synchronization, the drive chaotic system and the response chaotic system synchronize up to two scaling functions Q-S synchronization. a few results about Q-S synchronization of chaotic systems in discrete-time [18,19].

In this paper, we propose a new type of synchonization when the scaling functions in Q-S synchronization are replaced by two scaling matrice to study the problem of synchronization in different dimension for different dimensional discrete-time chaotic system. Based on new control laws and Lyapunov stability theory, a constructive schemes to investigate special and new synchronization type with double scaling matrice between typical chaotic dynamical systems in discrete-time with different dimensions are presented. To verify the validity and feasibility of the new synchronization results, the proposed controllers are applied to the drive 2D Hénon map [20] and the controlled 3D hyperchaotic Baier-Klein map [21].

This paper is organized as follows. In Section 2, the problem of double scaling matrix synchronization is formulated by given the de.nition of the new synchronization type. In Section 3, synchronization in 2D between 2D drive system and 3D response system in discrete-time is studied. In Section 4, synchronization in 3D between 2D drive system and 3D response system in discrete-time is investigated. In Section 5, synchronization in 4D between 2D drive system and 3D response system in discrete-time is proposed. In section 6, the new synchronization schemes derived in this paper are applied to some typical drive response chaotic systems and numerical simulations are given to illustrate the effectiveness of the main results. Finally, conclusions are drawn in Section 7.

Problem Formulation and Definition

Consider the following drive chaotic system described by

$$X(k+1) = f(X(k)) \tag{1}$$

where $X(k) \in R^n$ is the state vector of the drive system and $f : R^n \to R^n$. As the response system we consider the following chaotic system described by

$$Y(k+1) = g(Y(k)) + U \tag{2}$$

Where $Y(k) \in R^m$ the state vector of the response system is, $g : R^m \to R^m$ is the nonlinear part of the response system and $U = (u_i)_{1 \le i \le m} \in R^m$ is the vector controller.

Definition 1: The drive system (1) and the rsponse system (2) are said to be synchronized in dimension d, with respect to scaling matrice Θ and Φ, respectively, if there exists a controller $U = (u_i)_{1 \le i \le m} \in R^m$ and a given matrice $\Theta = (\Theta)_{d\times m}$ and $\Phi = (\Phi)_{d\times n}$ such that the synchronization error

$$e(k) = \Theta Y(k) - \Phi X(k) \tag{3}$$

Satisfies that $\lim_{k\to+\infty} \| e(k) \| = 0$

Because in real world all chaotic maps are described in 2D and 3D, we restrict our study about double scaling matrice synchronization to 2D chaotic maps and 3D discrete hyperchaotic systems this restriction does not lose the generality of our main results.

Synchronization of 2D Drive System and 3D Response System in 2D

In this section, the drive and the response chaotic systems are in

***Corresponding author:** Adel Ouannas, LAMIS Laboratory, Department of Mathematics and Computer Science, University of Tebessa, Algeria
E-mail: ouannas_adel@yahoo.fr

the following forms

$$X(k+1) = AX(k) + f(X(k)) \quad (4)$$

$$Y(k+1) = g(Y(k)) + U \quad (5)$$

Where $X(k) = (x_1(k), x_2(k))^T \in R^2, Y(k) = (y1(k), y2(k), y3(k))^T \in R^3$ are state vectors of the drive system and the response system, respectively, $A \in R^{2\times2}, f : R^2 \to R^2$ is the nonlinear part of the drive system (4), $g : R^3 \to R^3$ and $U \in R^3$ is the vector controller.

The synchronization error between the drive system (4) and the response system (5) can be derived as

$$e(k+1) = \Theta BY(k) + \Theta_g + \Theta U - \Phi AX(k) - \Phi f \quad (6)$$

Where $\Theta = (\Theta_{ij}) \in R^{2\times3}$ and $\Phi = (\Phi_{ij}) \in R^{2\times2}$ are scaling matrices. To achieve synchronization between systems (4) and (5), we assume that

$$\Phi A = A\Phi \quad (7)$$

And

$$\Theta = \begin{pmatrix} \Theta_{11} & \Theta_{12} & 0 \\ \Theta_{21} & \Theta_{22} & 0 \end{pmatrix} \quad (8)$$

The error system between the drive system (4) and the response system (5), can be written as

$$e(k+1) = (A - L_1)e(k) + R + \widehat{\Theta}\widehat{U} \quad (9)$$

$$R = (\Theta B - A\Theta - L_1\Theta)Y(k) + L_1 \Phi X(k) + \Theta g - \Phi f \quad (10)$$

$$\Theta = \begin{pmatrix} \Theta_{11} & \Theta_{12} \\ \Theta_{21} & \Theta_{22} \end{pmatrix} \quad (11)$$

$$\widehat{U} = (u_1, u_2)^T \quad (12)$$

And L is 2×2 control matrix to be determined. The controller is chosen as

$$\widehat{U} = -\widehat{\Theta}^{-1} R \quad (13)$$

Where $\widehat{\Theta}^{-1}$ is the inverse matrix of $\widehat{\Theta}$

By substituting Eq. (13) in Eq. (9), the error system can be written as.

$$e(k+1) = (A - L_1)e(k) \quad (14)$$

Theorem 2: If there exists a positive definite matrix P, such that

$$(A - L_1)^T(A - L_1) - I = -P \quad (15)$$

Then, the drive system (4) and the response system (5) are globally synchronized, with respect to scaling matrice Θ and Φ,. under the controller. (13).

Proof: Construct the candidate Lyapunov function in the form

$$V(e(k)) = e^T(k)e(k) \quad (16)$$

We obtain

$$\Delta V(e(k)) = e^T(k+1)e(k+1) - e^T(k)e(k)$$

$$= e^T(k(A - L_1)^T(A - L_1)e(k) - e^T(k)e(k)$$

$$= e^T(k)[(A - L_1)^T(A - L_1) - I]e(k) = -e^T(k)Pe(k)$$

$$< 0$$

Thus, from the Lyapunov stability theory, it is immediate that $\lim_{k\to\infty ei}(k) = 0, (i = 1,2)$. : That is the zero solution of the error system (14) is globally asymptotically stable and therefore, systems (4) and (5) are globally synchronized.

Synchronization of 2D Drive System and 3D Response System in 3D

In this section, the drive and the response chaotic systems are in the following forms

$$X(k+1) = f(X(k)) \quad (17)$$

$$Y(k+1) = BY(k) + g(Y(k)) + U \quad (18)$$

Where $X(k) = (x_1(k), x_2(k))^T \in R^2, Y(k) = (y_1(k), y_2(k), y_3(k))^T \in R^3$ state vectors of the drive system and the response system are, respectively, $f : R^2 \to R^2, B \in R^{3\times3}, g : R^3 \to R^3$ is nonlinear part response system and $U \in R^3$ is the vector controller. To achieve synchronization between systems (17) and (18), we assume that

$$\Theta B = B\Theta \quad (19)$$

Then, the error system between the drive system (17) and the response system (18), can be derived as

$$e(k+1) = (B - L_2)e(k) + R + \Theta U \quad (20)$$

Where

$$R = L_2\Theta Y(k) + (B\Phi - \Phi A = L_2\Theta)X(k) + \Theta_g - \Phi f \quad (21)$$

And L_2 *is* 3×3 control matrix to be determined. We choose the controller U as

$$U = -\Theta^{-1}R \quad (22)$$

Where Θ^{-1} is the inverse matrix of Θ

Theorem 3 If L_2 is chosen such that all eigenvalues of $B - L_2$ are strictly inside the unit disk, then the drive system (17) and the response system (18) are globally synchronized with respect to Θ and Φ under the control law (22).

Proof: By substituting Eq. (22) in Eq. (20), the error system can be written as

$$e(k+1) = (B - L_2)e(k) \quad (23)$$

Thus, by asymptotic stability of linear discrete-time systems, if all eigenvalues of $B - L_2$ are strictly inside the unit disk, it is immediate that all solution of error system (23) go to zero as $k \to \infty$. Therefore, systems (8) and (9) are globally synchronized.

Synchronization of 2D Drive System and 3D Response System in 4D

In this section, the drive and the response chaotic systems are in the following forms

$$X(k+1) = f(X(k)) \quad (24)$$

$$Y(k+1) = g(Y(k)) + U \quad (25)$$

Where $X(k) = (x_1(k), x_2(k))^T \in R^2, Y(k) = (y_1(k), y_2(k), y_3(k))^T \in R^3$ state vectors of the drive system and the response system, respectively, $f : R2 \to R2, g : R3 \to R3 and U \in R^3$ is the vector controller.

The error system between the drive system (24) and the response system (25) can be derived as

$$e(k+1) = L_{3e}(k) + R + \Theta U \tag{26}$$

Where

$$R = L_3 \Theta Y(k) + \Theta g(Y(k)) - L_3 \Phi X(k) - \Phi f(X(k)), \tag{27}$$

$\Phi \in R^{4\times2}$ is arbitrary scaling matrix, $\Theta = (\Theta_{ij}) \in R^{4\times3}$ is scaling matrix described as

$$\Theta = \begin{pmatrix} \Theta_{11} & 0 & 0 \\ 0 & \Theta_{22} & 0 \\ 0 & 0 & \Theta_{33} \\ -\Theta_{11}\dfrac{R_4}{R_1} & 0 & 0 \end{pmatrix} \tag{28}$$

Where $\Theta_{ii} \neq 0, (i = 1,2,3)$ and $L_3 = diag(l_1, l_2, l_3, l_4)$ is control matrix. Then, the error system between systems (24) and (25), can be written as

$$\begin{cases} e_1(k+1) = l_1 e_1(k) + R_1 + \Theta_{11} u_1 \\ e_2(k+1) = l_2 e_2(k) + R_2 + \Theta_{22} u_2 \\ e_3(k+1) = l_3 e_3(k) + R_3 + \Theta_{33} u_3 \\ e_4(k+1) = l_4 e_4(k) + R_4 - \Theta_{11}\dfrac{R_4}{R_1} u_1 \end{cases} \tag{29}$$

To achieve synchronization, we choose the controller U as

$$u_i = -\frac{R_i}{\Theta_{ii}}, \qquad i = 1,2,3. \tag{30}$$

By substituting Eq. (30) in Eq. (26), the error system can be written as

$$e_i(k+1) = l_i e_i(k), \qquad i = 1,2,3,4. \tag{31}$$

Theorem 4: If L_3 is chosen such that

$$| l_i | < 1, \quad i = 1,2,3,4. \tag{32}$$

Then, the drive system (24) and the response system (25) are globally synchronized, with respect to Θ and Φ, under the control law (30).

Proof: We take as a candidate Lyapunov function:

$$V(e(k)) = \sum_{i=1}^{4} e_i^2(k) \tag{33}$$

We get:

$$\Delta V(e(k)) = V(e(k+1)) - V(e(k))$$

$$= \sum_{i=1}^{4} e_i^2(k=1) - \sum_{i=1}^{4} e_i^2(k)$$

$$= \sum_{i=1}^{4} (l_i^2 - 1) e_i^2(k)$$

By using (32), we obtain: $\Delta V(e(k)) < 0$ Thus, by Lyapunov stability it is immediate that $\lim_{k\to\infty} e_i(k) = 0, (i = 1,2,3,4)$, and from the fact $\lim_{k\to\infty} \| e(k) \| = 0$. We conclude that the systems (24) and (25) are globally synchronized.

Numerical Application

Now, we consider 2D Hénon map as the drive system and the controlled 3D hyperchaotic Baier-Klein map as the response system. The Hénon map is can be described as

$$\begin{cases} x_1(k+1) = x_2(k) + 1 - ax_1^2(k) \\ x_2(k+1) = bx_1(k) \end{cases} \tag{34}$$

Which has a chaotic attractor, for example, when $(\alpha, \beta) = (1.4, 0.2)$ [20]. The Henon map chaotic attractor is shown in Figure 1.

The controlled the Baier-Klein map can be described as [21]:

$$\begin{cases} (k+1) = -0.1_{y3}(k) - y_2^2(k) + 1.76 + u_1 \\ y_2(k+1) = y_1(k) + u_2 \\ y_3(k+1) = y_2(k) + u_3 \end{cases} \tag{35}$$

Where $U = (u_1, u_2, u_3)^T$ is the vector controller. The chaotic attractors of Baier Klein map is shown in Figure 2.

Case 1: Synchronization of Hénon map and Baier-Klein map in 2D

According to our approach presented in section 3, we obtain

$$A = \begin{pmatrix} 0 & 1 \\ b & 0 \end{pmatrix} \tag{36}$$

$$\Phi = \begin{pmatrix} 1 & 1 \\ b & 1 \end{pmatrix} \tag{37}$$

$$\Phi = \begin{pmatrix} 2 & 0 & 0 \\ 0 & 2 & 0 \end{pmatrix} \tag{38}$$

and the control matrix L as

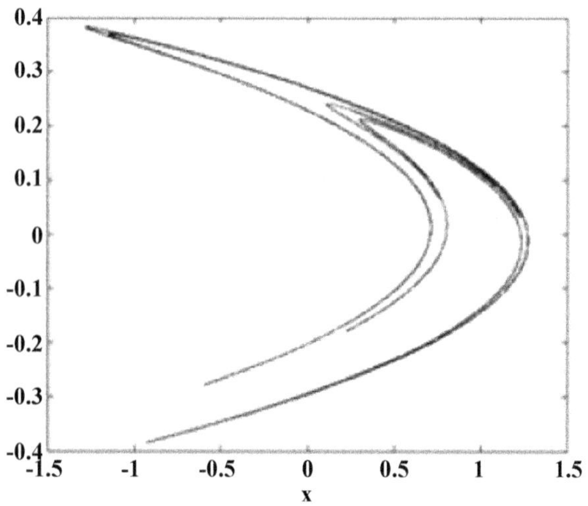

Figure 1: The Henon map chaotic attractor.

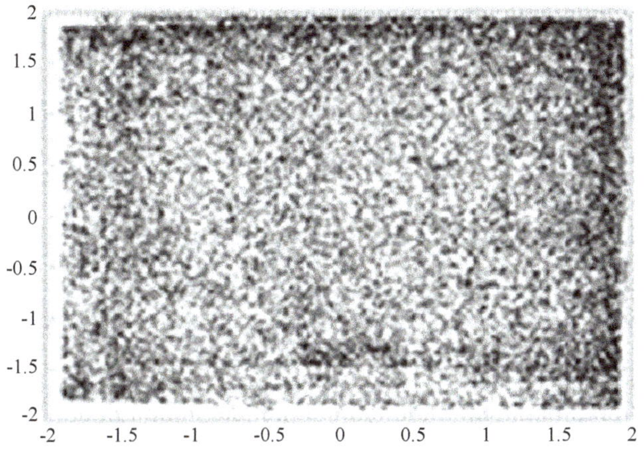

Figure 2: The Baier-Klein map chaotic attractor.

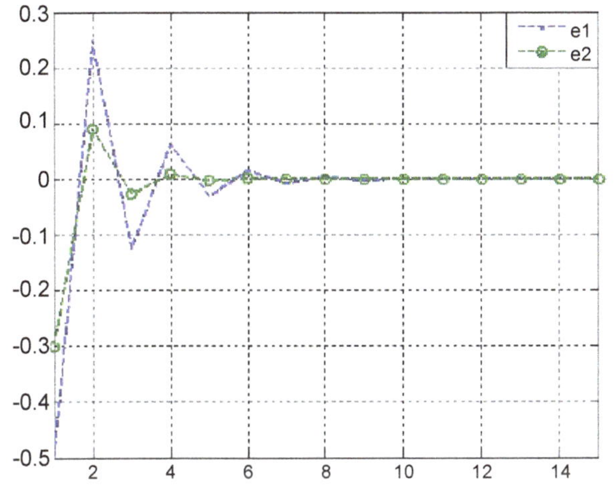

Figure 3: Time evolution of synchronization errors: e1 and e2 between 2D Henon map and 3D hyperchaoticBaier-Klein map in 2D.

$$L_1 = \begin{pmatrix} 2 & 1 \\ b & 2 \end{pmatrix} \tag{39}$$

Using simple calculations, we can show that $-((A-L_1)^T(A-L_1)-I)$ is a positive de...nite matrix. Therefore, in this case, systems (34) and (35) are synchronized in 2D. The error functions evolution is shown in Figure 3.

Case 2: Synchronization of Hénon map and Baier-Klein map in 3D

According to our approach presented in section 4, we obtain

$$B = \begin{pmatrix} 0 & 0 & -0.1 \\ 1 & 0 & 0 \\ 0 & 1 & 0 \end{pmatrix} \tag{40}$$

$$\Theta = \begin{pmatrix} 1 & -0.1 & -0.1 \\ 1 & 1 & -0.1 \\ 1 & 1 & 1 \end{pmatrix} \tag{41}$$

And the control matrix L_2 as

$$L_2 = \begin{pmatrix} 1 & 0 & -0.1 \\ 1 & 1 & 0 \\ 0 & 1 & 1 \end{pmatrix} \tag{42}$$

Simply, we can show that all eigenvalues of $B - L_2$ are strictly inside the unit disk. Therefore, in this case, systems (34) and (35) are synchronized. The error functions evolution is shown in Figure 4.

Case 3: Synchronization of Hénon map and Baier-Klein map in 4D

According to our approach presented in section 5, we obtain

$$L_3 = \begin{pmatrix} 0.5 & 0 & 0 & 0 \\ 0 & 0.5 & 0 & 0 \\ 0 & 0 & 0.5 & 0 \\ 0 & 0 & 0 & 0.5 \end{pmatrix} \tag{43}$$

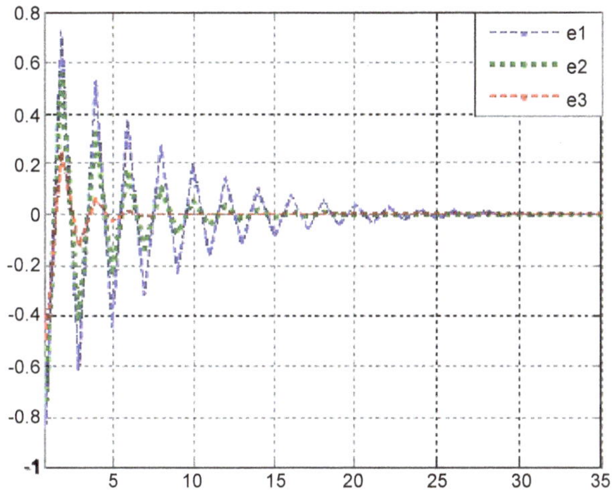

Figure 4: Time evolution of synchronization errors e1, e2 and e3 between 2D Henon map and 3D hyperchaotic Baier-Klein map in 3D.

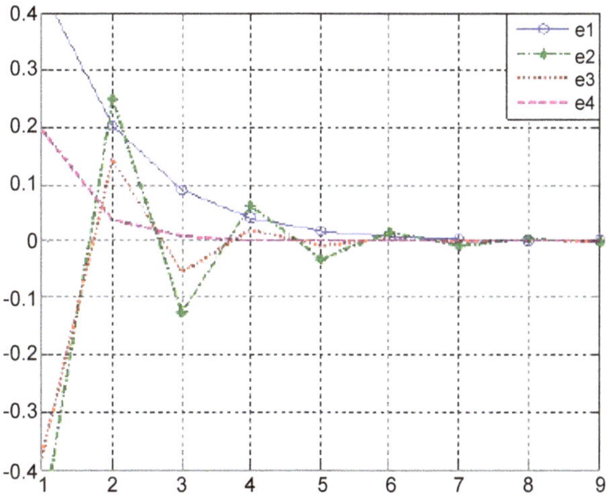

Figure 5: Time evolution of synchronization errors e1, e2, e3 and e4 between 2D Henon map and 3D hyperchaotic Baier-Klein map in 4D.

Finally, it is easy to know that the conditions of Theorem 3 are satisfied. Therefore, in this case, systems (34) and (35) are synchronized. The error functions evolution is shown in Figure 5.

Conclusion

In this paper, a new type of synchronization with double scaling matrice was proposed and new synchronization result are derived using new control schemes and Lyapunov stability theory. Firstly, when the dimension of synchronization is the same of the response system the synchronization control is achieved by controlling the linear part of the response system. Secondly, if synchronization is made in dimension of the drive system, the linear part of the drive system is controlled. Finally, synchronization is guaranteed by new diagonal matrix when the synchronization dimension is greater than the dimensions of drive and response systems. Numerical example and simulations results were used to verify the effectiveness of the proposed schemes.

References

1. Pecora L, Carrol T (1990) Synchronization in chaotic systems. Phys Rev Lett 64: 821-824.
2. Li X, Leung A, Han X, Liu X, Chu Y (2011) Complete (anti-)synchronization of chaotic systems with fully uncertain parameters by adaptive control. Nonlinear Dyn 63: 263-275.
3. Zhang G, Liu Z, Ma Z (2007) Generalized synchronization of different dimensional chaotic dynamical systems. Chaos Soliton Fract 32: 773-779.
4. Qiang J (2007) Projective synchronization of a new hyperchaotic Lorenz system chaotic systems. Phys Lett A 370: 40-45.
5. Li X (2009) Generalized projective synchronization using nonlinear control method. Int J Nonlinear Sci 8: 79-85.
6. Blasius B, Stone L (2000) Chaos and phase synchronization in ecological systems. Int J Bifur Chao 10: 2361-2380.
7. Lakshmanan M, Murali K (1996) Chaos in Nonlinear Oscillators: Controlling and Synchronization. Singapore: World Scientific.
8. Han SK, Kerrer C, Kuramoto Y (1995) Dephasing and bursting in coupled neural oscillators. Phys Rev Lett 75: 3190-3193.
9. Mengue AD, Essimbi BZ (2012) Secure communication using chaotic synchronization in mutually coupled semiconductor lasers. Nonlinear Dyn 70: 1241-1253.
10. Lu J, Wu X, Han X, Lü J (2004) Adaptive feedback synchronization of a uni.ed chaotic system. Phys Lett A 329: 327-333.
11. Wu X, Lu J (2003) Parameter identi.cation and backstepping control of uncertain Lü system. Chaos Soliton Fract 18: 729-721.
12. Zhang X, Zhu H (2008) Anti-synchronization of two di¤erent hyperchaotic systems via active and adaptive control. Int J Nonlinear Sci 6: 216-223.
13. Yang C, Lin C (2012) Robust adaptive sliding mode control for synchronization of space-clamped Fitz Hugh. Nagumo neurons. Nonlinear Dyn 69: 2089-2096.
14. Strogatz SH (2001) Nonlinear Dynamics and Chaos: With Applications to Physics, Biology, Chemistry, And Engineering. Studies in Nonlinearity, Westview Press.
15. Manfeng H, Zhenyuan X (2008) A general scheme for Q-S synchronization of chaotic systems. Nonlinear Analysis: Theory, Methods and Applications 69: 1091-1099.
16. Yang Y, Chen Y (2009) The generalized Q-S synchronization between the generalized Lorenz canonical form and the Rössler system. Chaos Soliton Fract 39: 2378-2385.
17. Zhao J, Zhang K (2010) A general scheme for Q-S synchronization of chaotic systems with unknown parameters and scaling functions. Appl Math Comput 7: 2050-2057.
18. Yan Z (2005) Q-S. Synchronization in 3D Hénon-like map and generalized Hénon map via a scalar controller. Phys Lett A 342: 309-317.
19. Yan Z (2006) Q-S (complete or anticipated) synchronization backstepping scheme in a class of discrete-time chaotic (hyperchaotic) systems: A symbolic-numeric computation approach. Chaos 16: 19-25.
20. Hénon M (1976) A Two-dimensional Mapping with Strange Attractor. Commun Math Phys 50: 69-77.
21. Baier G, Klein M (1990) Maximum hyperchaos in generalized Hénon maps. Physics Letters A 51: 281-284.

Volume Estimation of Various Brain Components Using MR Images - A Technical Report

Fuhua Chen* and Katherine Hastings

West Liberty University, West Virginia, USA

Abstract

In this paper, we present and discuss issues related to volume estimation of various brain components using MR brain images. We discuss pre-processing techniques to remove elements such as skull, blood vessels and fats from the MR images since these are non-essential to the volume calculations. The volume estimation is based on image segmentation. A challenge in MR brain image segmentation is to distinguish central gray matter from surrounding matters. This paper provides a frame work from pre- processing stage through volume estimation for white matter, gray matter and cerebrospinal fluid (CSF). The main contribution of this paper contains two parts. First, it provides a software-based method for interactive image pre-processing; second, it introduces a software-based, supervised interactive image segmentation method to deal with the segmentation of different matters, especially the central gray matter. Experiments with real data demonstrate the efficiency of our frame work.

Keywords: Image segmentation; Magnetic Resonance (MR) images; Central gray matter

Introduction

Traumatic brain injury happens when brain is hurt by an external force. Many imaging techniques, such as Computed Tomography (CT), Magnetic Resonance imaging (MR), and Positron Emission Tomography (PET), have been used in the diagnosis and treatment of such injuries. In the project "Biochemical Markers of Traumatic Brain Injury", we attempted a pathological analysis of diseases related to the brain by using MR images to estimate the change in volumes of different brain tissues. This paper is a technical report of part of the project focusing on volume estimations of white matter, gray matter, and Cerebrospinal Fluid (CSF). Volume estimation of different matters is especially important in this project. The key step in volume estimation based on images for the brain matters is image segmentation [1]. However, there are two main challenges that arise during this process.

The first challenge is how to pre-process the raw data so that it can be easily segmented and the volumes of different matters can be easily estimated. The raw MR data for a person's brain is a series of two-dimensional DICOM data that contains the skull, cerebrum, cerebellum and other organ matters in addition to the white matter, gray matter and CSF of interest. This data needs to be converted to a format that is easier to process using softwares such as C++ or MATLAB. In addition, it is necessary to remove the non-essential part pertaining to tissues and matter other than white matter, gray matter and CSF.

The second main challenge is the differentiation of various matters only based on intensities. The big trouble is of the central gray matter. Central gray matter is the gray matter located around the cerebral aqueduct within the tegmentum of the midbrain [2]. In most settings of MR imaging, white matter has the strongest intensity, gray matter a mild intensity, and cerebrospinal fluid the weakest intensity. However, the intensity of central gray matter in a MR brain image is often very similar to the intensity of white matter in the surface layer of the brain. As a result, it is difficult to approximately distinguish central gray matter from white matter. If the threshold for gray matter is set higher, then the white matter at surface layer may be misclassified as gray matter; if the threshold for gray matter is set lower, then some central gray matter may be misclassified as white matter. Many unsupervised segmentation methods have been developed to address this issue, such as non-local methods [3,4] and bias-correction methods [5-7]. However, researchers have not yet overcome such a dilemma efficiently due to the inherent closeness between the intensity of central gray matter and the intensity of white matter located at the surface layer of the cerebrum.

The intensities of blood and fat can also be very strong in MR brain images, which can lead to their misclassification as white matter. In order to estimate the volume for white matter more precisely, those high-intensity parts that are not white matter must be removed prior to unsupervised image segmentation. Lastly, the cerebellum must also be separated since only the functions of cerebrum is concerned in this project.

In this paper, we deal with the first challenge by integrating some existing soft- wares, and deal with the second challenge by developing a new software so that the resulted data after unsupervised segmentation can be refined using supervised and interactive way.

Materials

Most of the raw MR brain images used in this paper were provided by Shands Hospital in Gainesville, Florida, while some were obtained from various hospitals in Jacksonville, Florida. All the data has a DICOM format and is provided by either patients (as abnormal data) or volunteers (as normal data).

Procedures

The overall procedure can be broken into three stages. In the first stage, we pre-process the data so that the brain images contain only matters of interest which will be applied to an unsupervised

***Corresponding author:** Fuhua Chen, West Liberty University, West Virginia, USA, E-mail: fuhua.chen@westliberty.edu

image segmentation algorithm; In the second stage, an unsupervised image segmentation is applied to the data, and then a supervised segmentation is used to adjust and refine the segmentation. Finally, the segmentations are converted to a volume calculation. The overall procedures can be outlined below:

Overall steps:

1. Pre-processing

2. Image segmentation and adjustment and/or refinement

3. Volume estimation

Pre-processing

The first step of pre-processing is to remove skull from original MR brain images. We used the software FSL. It is a comprehensive library of analysis tools for FMRI, MR and DTI brain imaging data. The software is developed by Oxford University and can be easily download and installed. It is free for non-profitable use. As the input data is required to be either Analyze-format or NIFTI-format, we first use MRIcro to convert DICOM data to either of Analyze-format or of NIFTI-format. The detailed steps are listed below.

Steps for pre-processing

1. Use MRIcro (Version: 1.40 build 1) to transform a series of 2D DICOM data into a 3D Analyze data (.img). The software was developed by Chris Rorden's Neuropsychology Lab and can be downloaded from their official website.

2. Use FSL (Version: 4.1.7) to remove the skull from the given 3D Analyze-format brain file (.img) or NIFTI-format brain file (.nii). The software FSL is developed by Oxford University and can be free download from their official website.

3. Use the developed software "BrainImg Processing" to further process the data. Several slices of the original DICOM data of a set of MR images from one brain are shown in Figure 1.

Figure 2 shows the data after the skull was removed in Step 2. Note there are still some parts that are not main matters of interest, such as blood and fats, which can drastically affect our volume estimations. The removing of these parts is carried out in Step 3 using the developed software. Figure 3 highlights these elements with the red circle identifying fat, the blue circle identifying blood, and the yellow circle identifying the cerebellum.

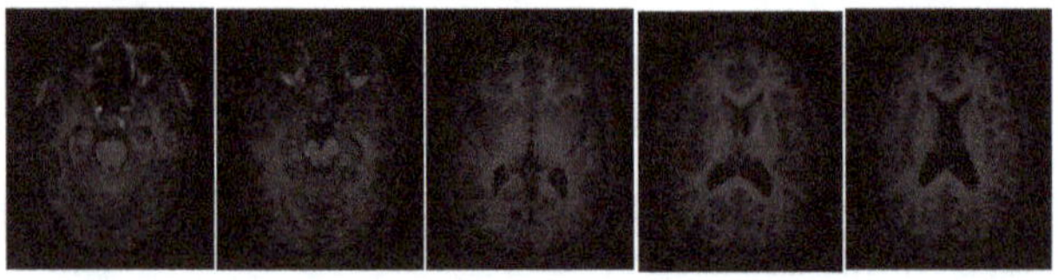
(a) (b) (c) (d) (e)

Figure 1: Original data.

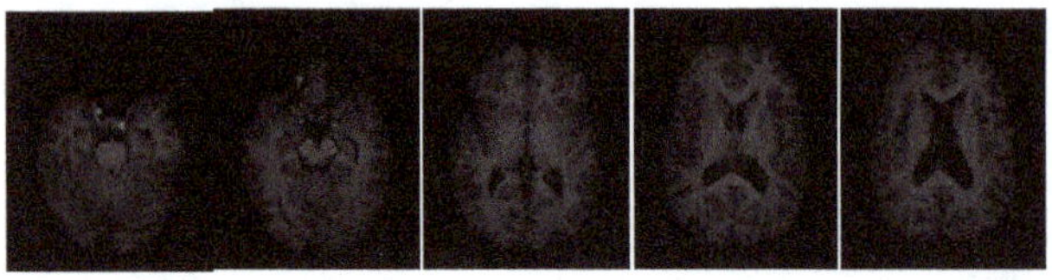
(a) (b) (c) (d) (e)

Figure 2: Brain images without skull.

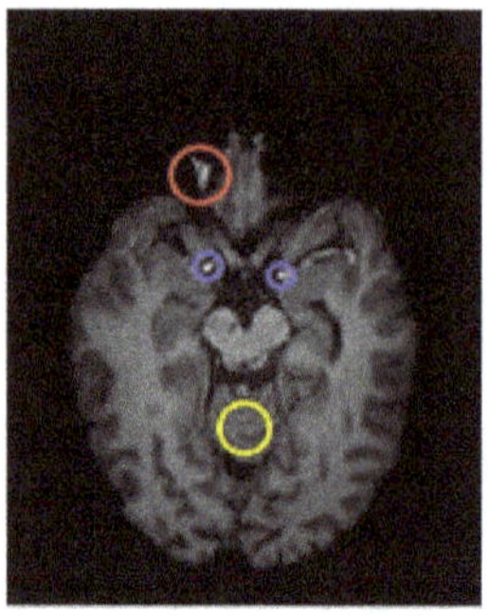

Figure 3: Brain images with non-essential matters identified.

Although there have been papers addressing automatic cerebellum segmentation [8,9], the removal is usually neither clear nor precise for real images. In our software BrainImg Processing, we provide an interactive way to remove blood and fats, as well as the cerebellum. Figure 2 shows the main images following the removal of blood, fat, and the cerebellum.

Image segmentation

After pre-processing the raw MR brain images, we apply image segmentation to the resulting images using a multiphase soft image segmentation method [6,10]. This process is embedded in the BrainImg Processing software. The image segmentation model is given below.

where I (x) is the image to be segmented, ci and σi are the mean and the deviation of the i-th phase, K is the number of phases, pi (x) denotes the percentage of pixel x belonging to the i-th phase, and ∇pi is the gradient of pi. In this application, K=3 because there is one phase for each brain matter of interest, namely, white matter, gray matter, and CSF. The first term in the equation is used to force the segmentation to fit the given image, and the second term (called total variation of pi) is used to force smoothness for each phase (or, in this context, each type of brain matter) (Figure 4).

Adjustment and refinement

As we mentioned at the beginning of the paper, the intensity of central gray matter is usually close to the intensity of white matter in the surface layer of cerebrum. Therefore, the unsupervised segmentation may not be ideal. The central gray matters of images Figure 5a, b and c are highlighted in Figure 4a, b and c, respectively.

After applying the unsupervised segmentation, most of the matters were classified correctly. However, due to the closeness of intensities between central gray matter and white matter, much central gray matter was misclassified to white matter. There have been some papers addressing the problem of inhomogeneity of intensities, such as bias-correction based multiphase image segmentations [5,7,11-13] and non-local information or global information based image segmentations [3,4]. However, all of those methods, including the unsupervised segmentation method employed inBrainImg Processing, do not work very well for real MR brain images due to the existence of central gray matter. Therefore, in BrainImg Processing, we added two functions, namely, adjustment and refinement which will be briefly described. These functions are especially useful for experienced doctors who can distinguish between white matter and central gray matter on MR brain images.

1. The adjustment function provides a supervised image segmentation framework.

It assigns some regions to each matter based on prior knowledge and then uses the assigned matter as a reference to determine to which matter its neighborhood belongs.

2. The refinement function provides a way for local image segmentation. The reason that gray matter could not be segmented correctly using an unsupervised segmentation is that the intensity of central gray matter is much higher than the intensity of gray matter at the surface layer of the brain, and is close to (or even higher than) the intensity of white matter in the outer layer (not in the central region). The refinement function allows users to choose a central region and then apply a local image segmentation only for the chosen region. In this way, a more precise segmentation can be achieved.

The segmentation results of images in Figure 5a, b and c using unsupervised segmentation are shown in the first row in Figure 6. From the images, we can see that much of the central gray matter is misclassified as white matter, while some white matter near the surface layer of the cerebrum are misclassified as gray matter. The results after refinement are shown in the second row in Figure 6.

Introduction to BrainImg Processing

BrainImg Processing is a software that was developed in our project to address two main issues. First, it can be used to manually remove some parts of an image. Second, when the image segmentation is not precise enough, it can be used to refine or adjust the segmentation.

Although there have been papers addressing automatic (or unsupervised) cerebellum segmentation [8,9] the removing is usually neither clear nor precise for real images. Moreover, when using FSL to remove skull from an given image, it is sometime either over-removed or under-removed. In addition, all fats and blood vessels must be removed from the images in our project. All of these tasks can be achieved with the software BrainImg Processing. For instance, if the skull is under-removed, it can be polished manually using BrainImg Processing by selecting unnecessary parts and then choosing "Removing". Fats and blood vessels can be similarly removed. When a proper set of parameters are chosen to apply an image segmentation and those parts near the surface layer of the cerebrum are segmented

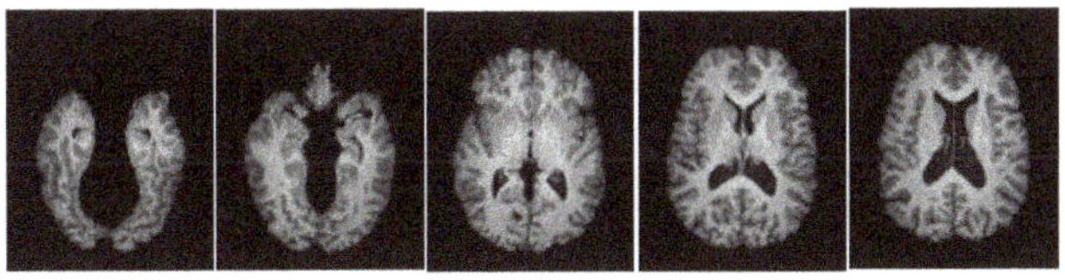

(a) (b) (c) (d) (e)

Figure 4: Matters of interest.

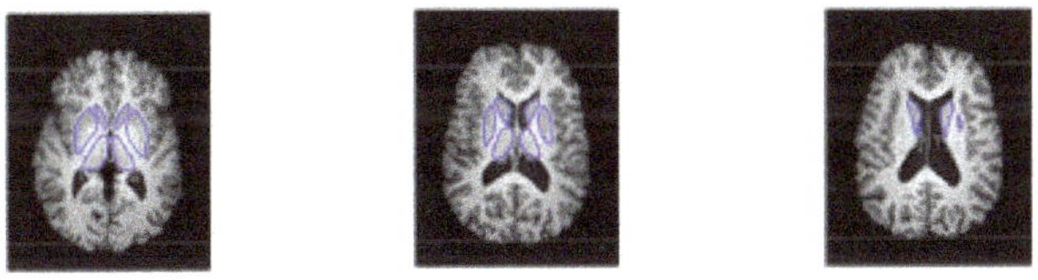

(a) (b) (c)

Figure 5: Central gray matter.

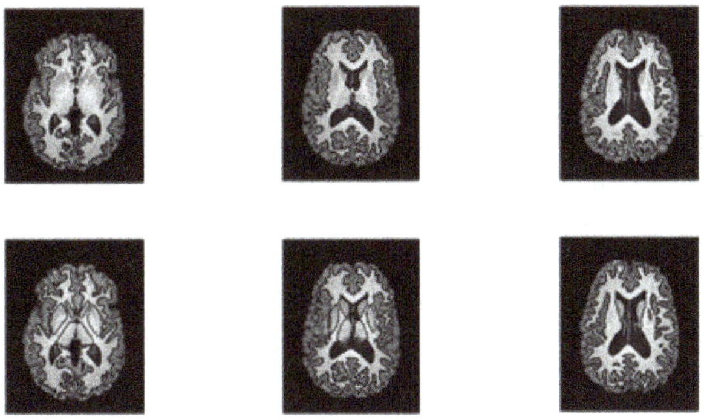

Figure 6: Comparison between refined and non-refined brain images.

ideally, the central gray matter is usually under-segmented. That is, much of the central gray matter is misclassified as white matter. In this case, the software provides two options for the refinement or adjustment of the segmentation. One option is to choose a region that is not segmented correctly, and then apply the segmentation locally using adjusted parameters. The second option is to use the software to assign some regions to each matter, and then run supervised image segmentation. Both options can achieve a better result.

Volume estimations

After the image segmentation, three phases are obtained. Roughly speaking, the highest-intensity phase (denoted by P1) corresponds to white matter, the median-intensity phase (denoted by P2) corresponds to gray matter, and the lowest-intensity phase (denoted by P3) corresponds to cerebrospinal fluid. The P3 phase, however, contains more than just cerebrospinal fluid. In addition to the dark background of the image (whose intensity is similar to that of CSF), when fat and blood are removed from an image, the result is also a dark region. Therefore, the P3 phase is a combination of CSF, background, and "missing" fat and blood.

Suppose that the resolution of the 3D MR brain image is l millimeters×m millimeters×h millimeters (In our project, the resolution is 1 mm×1 mm×1 mm). Then the algorithm for calculating the volumes for different matters is as below.

Steps for volume estimation for different matters:

1. Find the total number of voxels for the sum of all three matters, denoted by N; this can be easily obtained by setting a threshold a little bigger than the intensity of the background.

2. Calculate the total number of voxels for Phase P1, denoted by N1;

3. Calculate the total number of voxels for Phase P2, denoted by N2;

4. The volume of white matter is obtained by $N1 \times l\ mh$ millimeters;

5. The volume of gray matter is obtained by $N2 \times l\ mh$ millimeters;

6. The volume of cerebrospinal fluid is obtained by $(N-N1\ N2) \times l\ mh$ millimeters. The volumes for different matters can be shown automatically by the software after each segmentation or adjustment.

Conclusion

In this paper, we give a technical report on a previous project related to MR brain image processing. The first part of the paper discusses pre-processing of data using existing softwares. We then focus on issues related to volume estimation of various brain matters in MR brain images. Namely, we discuss and analyze the problem of central gray matter segmentation. We point out that, based on experience with real MR brain images, the intensity of white matter at the surface layer of the cerebrum is usually very close to the intensity of the gray matter in the central region of the brain, which leads to imprecise unsupervised image segmentations. Therefore, some refinement must be made using either local segmentation or a manual adjustment. In our project, we developed a specific software for the processing of such interactive image segmentation.

Acknowledgment

The work reported in this paper is a part of the project "Biochemical Markers of Traumatic Brain Injury." We were responsible for the calculation of different matters in a cerebrum. Throughout the project, I received a great deal of help for which I am greatly appreciative. I would like to thank my adviser, Dr. Yunmei Chen at the University of Florida. It is under her advisement that this project was completed. Secondly, I want to express my thanks to Dr. Ilona Schmalfuss, a Neuroradiological doctor at Shands Hospital. She spent a lot of time helping to verify the segmentation results. Without her help, it would have been impossible to finish the project. Finally, thank you to Dr. Xiaojing Ye and Dr. Feng Huang for their consultation and advice during the project. Thanks are given to all of them!

Funding

The project is funded by NIH/R01(#: 7095364).

References

1. http://en.wikipedia.org/wiki/WikiGrey_matter
2. http://en.wikipedia.org/wiki/WikiPeriaqueductal_gray
3. Bresson X, Chan TF (2008) Non-local unsupervised varia-tional image segmentation models.
4. Houhou, etc. Semi-Supervised Segmentation based on Non-local Continuous Min-Cut. Lecture Notes in Computer Science 5567: 112-123
5. Ahmed MN, Yamany SM, Mohamed N, Farag AA, Moriarty T (2002) A modified fuzzy C-means algorithm for bias field estimation and segmentation of MRI data. IEEE Trans Med Imag 21: 193-199.
6. Chen F, Chen Y (2010) A Stochastic Variational Model for Multi-phase Soft Seg- mentation with Bias correction. Advanced Modeling and Optimization 12: 336-345.
7. Li C, Huang R, Ding Z, Gatenby C, Metaxas D (2008) A variational level set ap- proach to segmentation and bias correction of images with intensity inhomogeneity. Miccai Lncs 5242: 1083-1091.
8. Powell S, Magnotta V, Johnson H, Vamsi K, Ronald P, et al. (2008) Registration and Machine Learning Based Automated Segmentation of Subcortical and Cerebellar Brain Structures. Neuroimage 39: 238-247.
9. Saeed N, Puri BK (2002) Cerebellum segmentation employing texture properties and knowledge based image processing: applied to normal adult controls and patients. Magnetic Resonance Imaging 20: 425-429.
10. Wang H, Chen F, Chen Y (2013) A New Multiphase Soft Segmen- tation with Adaptive Variants. Applied Computational Intelligence and Soft Computing 6.
11. Fang Li, Michael KNG, Chunming Li (2010) Variational Fuzzy Mumford-Shah Model for Image Segmentation. SIAM J Appl Math 70: 2750-2770.
12. Pham DL, Prince JL (1998) An adaptive fuzzy C-means algorithm for the image segmentation in the presence of intensity inhomogeneities. Pattern Recognit Lett 20: 57-68.
13. Wells W, Grimson E, Kikinis R, Jolesz F (1996) Adaptive segmentation of MRI data. IEEE Trans Med Imag 15: 429-442.

8

More-For-Less Paradox in a Transportation Problem under Fuzzy Environments

Debiprasad Acharya[1]*, Manjusri Basu[2] and Atanu Das[2]

[1]*Department of Mathematics, N.V.College, Nabadwip, Nadia-741302, W.B, India*

[2]*Department of Mathematics, University of Kalyani, Kalyani-741235, India*

Abstract

In this paper, we discuss more-for-less paradox in a transportation problem under fuzzy environment with linear constraints. In some cases of the transportation problem, an increase in the supplies and demands or in other words, increase in the flow results a decrease in the optimum transportation cost. This type of behavior which means paradoxical, is called transportation paradox. Thereby, we establish a sufficient condition for the existence of paradox in a transportation problem under fuzzy environment. Also we illustrate a numerical example in support of the theory.

Key words: Fuzzy number; Fuzzy transportation problem; Paradox; Paradoxical range of flow

Introduction

The basic transportation problem is one of the special class of linear programming problem, which was first formulated by Hitchcook [1] Charnes et al. [2] Appa [3] Klingman and Russel [4] developed further the basic transportation problem. Basically, the papers of Charnes and Klingman [5] and Szwarc [6] are treated as the sources of transportation paradox for the researchers. In the paper of Charnes and Klingman, they name it "morefor-less" paradox and wrote "The paradox was first observed in the early days of linear programming history (by whom no one knows) and has been a part of the folklore known to some (e.g. A.Charnes and W.W.Cooper), but unknown to the great majority of workers in the field of linear programming". Subsequently, in the paper of Appa, he mentioned that this paradox is known as "Doig Paradox" at the London School of Economics, named after Alison Doig. Gupta et al. [7] established a sufficient condition for a paradox in a linear fractional transportation problem with mixed constraints. Adlakha and Kowalski [8] derived a sufficient condition to identify the cases where the paradoxical situation exists.

Ryan [9] developed a goal programming approach to the representation and resolution of the more for less and more for nothing paradoxes in the distribution problem. Deineko et al. [10] developed a necessary and sufficient condition for a cost matrix which is immuned against the transportation paradox. Dahiya and Verma [11] considered paradox in a nonlinear capacitated transportation problem. Adlakha et al. [12] developed a simple heuristic algorithm to identify the demand destinations and the supply points to ship more for less in fixed-charge transportation problems. Storoy [13] considered the classical transportation problem and studied the occurrence of the so-called transportation paradox (also called the more-for-less paradox). Joshi and Gupta [14] studied an efficient heuristic algorithm for solving more-for-less paradox and algorithm for finding the initial basic feasible solution for linear plus linear fractional transportation problem. Schrenk et al. [15] analyzed degeneracy characterizations for two classical problems (1) the transportation paradox in linear transportation problems and (2) the pure constant fixed charge transportation problem. asu et al. [16] considered the algorithm of finding all paradoxical pairs in a linear transportation problem.

Fuzzy sets and fuzzy logic were introduced by Lotfi A. Zadeh in 1965. Zadeh [17] was almost single-handedly responsible for the early development in this field. A fuzzy transportation problem is an extension of linear transportation problem, where at least one of the transportation costs, supply and demand quantities are fuzzy quantities. The objective function of the fuzzy transportation problem is to determined the total fuzzy minimum transportation cost by shipping the fuzzy supply and fuzzy demand. Bellman and Zadeh [18], Liu [19] developed further. Dinagar and Palanivel [20] investigated fuzzy transportation problem with the aid of trapezoidal fuzzy numbers. Dutta and Murthy [21] investigated the transportation problem with additional impurity restrictions where costs are not deterministic numbers but imprecise ones, also the elements of the cost matrix are subnormal fuzzy intervals with strictly increasing linear membership functions. Ojha et al. [22] considered capacitated-multi-objective, solid transportation problem which formulated in fuzzy environment with non-linear varying transportation charge and an extra cost for transporting the amount to an interior place through small vehicles. In this paper, we present more-foe-less paradox in a transportation problem under fuzzy environment with linear constraints. To solve such type of problem we consider the transporting cost per unit product, supply and demand quantities are described in trapezoidal fuzzy. Thereby, we state the sufficient condition of existence of paradox. We also justify the theory by illustrating a numerical example.

Definition. Fuzzy Set: Let A be a classical set and $\mu A(x)$ be a function defined over $A \rightarrow [0, 1]$. A fuzzy set $A*$ with membership function $\mu A(x)$ is defined by $A* = \{(x, \mu A(x)) : x \in A \text{ and } \mu A(x) \in [0, 1]\}$ Definition 1.2. Fuzzy Number: A real fuzzy number $\tilde{a} \approx (a1, a2, a3, a4)$, where $a1, a2, a3, a4 \in R$ and two functions f(x) and $g(x) : R \rightarrow [0, 1]$, where f(x) is non-decreasing and g(x) is non-increasing, such that

***Corresponding author:** Debiprasad Acharya, Department of Mathematics, N.V. College, India, E-mail: debiipsitacharya@gmail.com

we can define membership function $\mu\tilde{a}$ (x) satisfying the following conditions

$$\mu_{\tilde{a}}(x) = \begin{cases} f(x) \text{ if } a_1 \le x \le a_2 \\ 1 \text{ if } a_2 \le x \le a_3 \\ g(x) \text{ if } a_3 \le x \le a_4 \\ 0 \text{ otherwise} \end{cases}$$

Trapezoidal Membership Function: The trapezoidal membership function of trapezoidal fuzzy number $\tilde{a} \approx$ (a1, a2, a3, a4) is defined by

$$\mu_{\tilde{a}}(x) = \begin{cases} \dfrac{x-a_1}{a_2-a_1} \text{ for } a_1 \le x \le a_2 \\ 1 \text{ for } a_2 \le x \le a_3 \\ \dfrac{a_4-x}{a_4-a_3} \text{ for } a_3 \le x \le a_4 \\ 0 \text{ otherwise} \end{cases}$$

Arithmetic operations: Let $\tilde{a} \approx (a_1, a_2, a_3, a_4)$ and $\tilde{b} \approx (b_1, b_2, b_3, b_4)$ be two trapezoidal fuzzy numbers, where a1, a_2, a_3, a_4, b_1, b_2, b_3, $b_4 \in R$ then the arithmetic operation on $\tilde{a}$ and $\tilde{b}$ are:

Addition: The addition of two fuzzy numbers and $\tilde{b}$ is $\tilde{a} \oplus \tilde{b} \approx (a_1+b_1, a_2+b_2, a_3+b_3, a_4+b_4)$.

Subtraction: The negative fuzzy number of $\tilde{b}$ is $\ominus \tilde{b} \approx (-b_4, -b_3, -b_2, -b_1)$, then the

subtraction of two fuzzy numbers $\tilde{a}$ and $\tilde{b}$ is $\tilde{a} \ominus \tilde{b} \approx (a_1 - b_4, a_2 - b_3, a_3 - b_2, a_4 - b_1)$.

Multiplication:

(i) The multiplication of an arbitrary number _ and a fuzzy number $\tilde{a}$ is

$$\rho \otimes \tilde{a} \begin{cases} (\rho a_1, \rho a_2, \rho a_3, \rho a_4) \text{ for } \rho \ge 0 \\ (\rho a_4, \rho a_3, \rho a_2, \rho a_1) \text{ for } \rho \le 0 \end{cases}$$

(ii) The multiplication of two fuzzy numbers $\tilde{a}$ and $\tilde{b}$ is $\tilde{a} \otimes \tilde{b} \approx$ (t1, t2, t3, t4),

where t1 = min{a1b1, a1b4, a4b1, a4b4}, t2 = min{a_2b_2, a_2b_3, a_3b_2, a_3b_3},

t3 = max{a_2b_2, a_2b_3, a_3b_2, a_3b_3}, t_4 =max{a_1b_1, a_1b_4, a_4b_1, a_4b_4}.

Definition: The magnitude of the trapezoidal fuzzy number $\tilde{a} \approx (a_1, a_2, a_3, a_4)$ is defined by Mag($\tilde{a}$) = $\dfrac{a_1 + 2a_2 + 2a_3 + a_4}{6}$

Problem Formulation

Let $\tilde{x}$ ij$\approx (x_{ij}^1, x_{ij}^2, x_{ij}^3, x_{ij}^4)$ is the uncertain number of units transported from the ith origin to the jth destination, $\tilde{c}$ ij $\approx (c_{ij}^1, c_{ij}^2, c_{ij}^3, c_{ij}^4)$ is the uncertain cost involved in transporting per unit product from the ith origin to the jth destination, $\tilde{a}$i $\approx (a_i^1, a_i^2, a_i^3, a_i^4)$ is the uncertain number of units available at the ith origin, $\tilde{b}_j \approx (b_j^1, b_j^2, b_j^3, b_j^4)$ is the uncertain number of units required at the jth destination. Then the cost minimizing fuzzy transportation problem be

$$P: Min\ \tilde{Z} \approx \sum_{i=1}^{m}\sum_{j=1}^{n} \tilde{c}_{ij}\tilde{x}_{ij}$$

subject to the constraints,

$$\sum_{j=1}^{n} \tilde{x}_{ij} \approx \tilde{a}_i;\ \forall\ i \in I = (1, 2, ...m)$$

$$\sum_{j=1}^{n} \tilde{x}_{ij} \approx \tilde{b}_i;\ \forall\ j \in J = (1, 2, ...n)$$

And

$$\tilde{x}_{ij} \succeq 0\ \forall\ (i, j) \in IxJ$$

Let B be the basis of the problem P and $X^0 \approx \{\tilde{x}_{ij}^0 \mid (i, j) \in IxJ\}$ be its basic feasible solution. The value of the objective function is $\tilde{Z}$ 0 and the flow $\tilde{F}$ 0 corresponding to the basic feasible solution $\tilde{x}$ 0 are $\tilde{Z}0 \approx \sum_{i=1}^{m}\sum_{j=1}^{n} \tilde{c}_{ij}\tilde{x}_{ij}^0$ and $\tilde{F} \approx \sum_{i \in I} \tilde{a}_i \approx \sum_{i \in I} \tilde{b}_j$ We consider the dual variables $\tilde{u}_i$ for $i \in I$ and $\tilde{V}_j$ for $j \in J$ such that $\tilde{u}_i \oplus \tilde{v}_j \approx \tilde{c}_{ij}$ corresponding to the basis B. Also $\forall$ (i, j) $\notin$ **B** let $\tilde{c}_{ij} \approx (\tilde{u}_i \oplus \tilde{v}_j) \odot \tilde{c}_{ij}$ and if $\tilde{c}_{ij} \preceq 0 \forall (i, j) \notin B$ then solution of the fuzzy transportation problem is optimum.

Definitions

Paradox in a fuzzy transportation problem

In a fuzzy transportation problem if we can obtain more flow ($\tilde{F}$ 1) with lesser transportation cost ($\tilde{Z}$ 1) than the optimum flow ($\tilde{F}$ 0) corresponding to the optimum transportation cost ($\tilde{Z}$ 0) i.e. $\tilde{F}$ 1 $\succ$ $\tilde{F}$ 0 and $\tilde{Z}$ 1 $\prec$ $\tilde{Z}$ 0, then we say that a paradox occurs in a fuzzy transportation problem.

Fuzzy cost-flow pair

If the value of the objective function is $\tilde{Z}$ i and the flow is $\tilde{F}$ i corresponding to the feasible solution $\tilde{x}$ i of a fuzzy transportation problem, then the pair ($\tilde{Z}$ i, $\tilde{F}$ i) is called the fuzzy cost-flow pair corresponding to the feasible solution $\tilde{x}$ i.

Fuzzy paradoxical pair

A fuzzy cost-flow pair ($\tilde{Z}$, $\tilde{F}$) of an objective function is called fuzzy paradoxical pair if $\tilde{Z} \prec \tilde{Z}$ 0 and $\tilde{F} \succ \tilde{F}$ 0 where $\tilde{Z}$ 0 is the optimum transportation cost and $\tilde{F}$ 0 is the optimum flow of the fuzzy transportation problem.

Best fuzzy paradoxical pair

The fuzzy paradoxical pair ($\tilde{Z}$ *, $\tilde{F}$ *) is called the best fuzzy paradoxical pair of a fuzzy transportation problem if for all fuzzy paradoxical pair ($\tilde{Z}$, $\tilde{F}$), either $\tilde{Z}* \prec \tilde{Z}$ or $\tilde{Z}* \approx \tilde{Z}$ but $\tilde{F}* \succ \tilde{F}$

Fuzzy paradoxical range of flow

If $\tilde{F}$ 0 be the optimum flow and $\tilde{F}$ * be the flow corresponding to the best fuzzy paradoxical pair of a fuzzy transportation problem then [$\tilde{F}$ 0, $\tilde{F}$ *] is called fuzzy paradoxical range of flow.

Theorem 2.1. The sufficient condition for the existence of paradoxical solution of (P) is that in the optimum table of (P), ∃ at least one cell (r, s) $\notin$ B where we have $(\tilde{u}_r \oplus \tilde{v}_s) \prec 0$ if $\tilde{a}$ r and $\tilde{b}$ s are replaced by $\tilde{a}$ r $\oplus \tilde{l}$ and $\tilde{b}$ s $\oplus \tilde{l}$ ($\tilde{l} \succ \tilde{0}$) respectively.

Proof: Let $\tilde{Z}$ 0 be the value of the objective function and $\tilde{F}$ 0 be the optimum flow corresponding to the optimum solution ($\tilde{x}$ 0) of the problem (P). The dual variables $\tilde{u}$ i and $\tilde{v}$ j are given by $\tilde{u}$ i $\oplus$ $\tilde{v}$ j $\approx$ $\tilde{c}$ ij ∀ (i, j) ∈ B Then the value of the objective function in terms of the dual variables is given by

$$\tilde{Z}^0 \approx \sum_i \sum_j \tilde{c}_{ij} \tilde{x}_{ij}{}^0$$
$$\approx \sum_i \sum_j (\tilde{u}_i \oplus \tilde{v}_j) \tilde{x}_{ij}{}^0$$
$$\approx \sum_i (\sum_j \tilde{x}_{ij}{}^0) \tilde{u}_i \oplus \sum_j (\sum_i \tilde{x}_{ij}{}^0) \tilde{v}_j$$
$$\approx \sum_i \tilde{a}_i \tilde{u}_i \oplus \sum_j \tilde{b}_j \tilde{v}_j$$

And $\tilde{F}^0 \approx \sum_i \tilde{a}_i \approx \sum_j \tilde{b}_j$

Now, let ∃ at least one cell (r, s) $\notin$ B, where if we replace $\tilde{a}$ r and $\tilde{b}$ s by $\tilde{a}$ r $\oplus \tilde{l}$ and $\tilde{b}$ s $\oplus \tilde{l}$ respectively ($\tilde{l} \succ \tilde{0}$), in such a way that the optimum basis remains same, then the value of the objective function $\hat{\tilde{Z}}$ is given by

$$\hat{\tilde{Z}} \approx [\sum_{i,i \neq r} \tilde{a}_i \tilde{u}_i \oplus \sum_{j,j \neq s} \tilde{b}_j \tilde{v}_j \oplus \tilde{u}_r (\tilde{a}_r \oplus \tilde{1}) \oplus \tilde{v}_s (\tilde{b}_s \oplus \tilde{1})]$$
$$\approx [\tilde{Z}^0 \oplus \tilde{1}(\tilde{u}_r \oplus \tilde{v}_s)]$$

The new flow $\hat{\tilde{F}}$ is given by

$$\hat{\tilde{F}} \approx \sum_i \tilde{a}_i \oplus \tilde{l} \approx \sum_j \tilde{b}_j \oplus \tilde{l} \approx \tilde{F}^0 \oplus \tilde{l}$$
$$\hat{\tilde{F}} \odot \tilde{F}^0 \approx \tilde{l} \succ \tilde{0}$$

Therefore, for the existence of paradox we must have $\hat{\tilde{Z}} \Theta Z^0 \prec \tilde{0}$ since $\hat{\tilde{F}} \Theta \tilde{F}^0 \succ \tilde{0}$ i.e. $\tilde{u}_r \oplus \tilde{v}_s \prec \tilde{0}$ because $\tilde{l} \succ \tilde{0}$

Hence, the theorem.

Now we state the following algorithm to find all the paradoxical pairs of the problem

(P).

3 Algorithm : To obtain all the paradoxical pairs

Step 1: i = 0.

Step 2: Find the cost-flow pair ($\tilde{Z}$ 0, $\tilde{F}$ 0) for the optimum solution $\tilde{x}$ 0.

Step 3: Find all cells (r, s) $\notin$ B such that ($\tilde{u}$ r $\oplus$ $\tilde{v}$ s) $\prec$ $\tilde{0}$ if it exists, otherwise go to step 8.

Step 4: Find min flow for $\tilde{l} \approx (1, 0, 0, 0)$, $\tilde{l} \approx (0, 1, 0, 0)$, $\tilde{l} \approx (0, 0, 1, 0)$, $\tilde{l} \approx (0, 0, 0, 1)$ or $\tilde{l} \approx (1, 1, 1, 1)$ which enters into the existing basis whose corresponding cost is minimum. Let ($\tilde{Z}$ i, $\tilde{F}$ i) be the new cost flow pair corresponding to the optimum solution Xi.

Step 5: i = i + 1.

Step 6: Write ($\tilde{Z}$ i, $\tilde{F}$ i).

Step 7: Find all cells (r, s) $\notin$ B such that ($\tilde{u}$ r $\oplus$ $\oplus$ s) $\prec$ $\tilde{0}$ if it exists go to step 4, otherwise go to step 9.

Step 8: Write paradox does not exist and go to step 10.

Step 9: Write paradox exists and the best paradoxical pair ($\tilde{Z}$ *, $\tilde{F}$ *) ≈ ($\tilde{Z}$ i, $\tilde{F}$ i) for the optimum solution $\tilde{x}$ * ≈ $\tilde{x}$ i.

Step 10: End.

Numerical Example

We consider a numerical example which consists of three origins and four destinations, the uncertain numbers of supply, demand and cost per unit are tabulated in Table 1.

Solving the fuzzy transportation problem given in Table 1, the optimum solution is given in Table 2.

Table 2 gives the optimum solution $\tilde{x}$ 0 = {$\tilde{x}$ 11 = (18, 19, 21, 22), $\tilde{x}$ 13 = (6, 8, 12, 14), $\tilde{x}$ 14 = (4, 7, 13, 16), $\tilde{x}$ 22 = (23, 24, 26, 27), $\tilde{x}$ 24 = (21, 23, 27, 29), $\tilde{x}$ 33 = (38, 39, 41, 42)} and the cost-flow pair is ($\tilde{Z}$ 0, $\tilde{F}$ 0) = ((72, 216, 512, 714), (124, 127, 133, 136)).

Now we have $\tilde{u}$ 2 + $\tilde{v}$ 1 < 0 and $\tilde{u}$ 2 + $\tilde{v}$ 3 < 0 where the cells (2, 1) and (2, 3) are not in the basis, so paradox exists. We take $\tilde{l}$ = (1, 1, 1, 1) and we have; for the cell (2, 1) the optimum cost-flow pair is ($\tilde{Z}$, $\tilde{F}$) = ((68, 213, 511, 714), (123, 125, 135, 139)) given in Table 3, for the cell (2, 3) the optimum cost-flow pair is ($\tilde{Z}$, $\tilde{F}$) = ((69, 214, 512, 715), (123, 125, 135, 139)) given in Table 4, The paradoxical cost-flow pair is ($\tilde{Z}$ 1, $\tilde{F}$ 1) = ((68, 213, 511, 714), (125, 128, 134, 137)) given in Table 4.

Some of the fuzzy paradoxical pairs and best fuzzy paradoxical pair obtained by algorithm in section 3, are given in Table 5.

Dest → ↓Orifagin	D_1	D_2	D_3	D_4	$\tilde{a}_i$
O_1	(0, 1, 3, 4)	(5, 6, 8, 9)	(1, 2, 4, 5)	(6, 7, 9, 10)	(38, 39, 41, 42)
O_2	(4, 5, 7, 8)	(0, 1, 1, 2)	(7, 8, 10, 11)	(2, 3, 5, 6)	(48, 49, 51, 52)
O_3	(1, 2, 4, 5)	(6, 7, 9, 10)	(0, 1, 3, 4)	(8, 9, 11, 12)	(38, 39, 41, 42)
	(18, 19, 21, 22)	(23, 24, 26, 27)	(48, 49, 51, 52)	(33, 34, 36, 37)	

Table 1: Demand and cost per unit.

Dest → ↓Origin	D_1	D_2	D_3	D_4	$\tilde{a}_i$ $\tilde{u}_i$
O_1	(0, 1, 3, 4) [18,19,21,22]	(5, 6, 8, 9)	(1, 2, 4, 5) [6,8,12,14]	(6, 7, 9, 10) [4, 7, 13, 16]	(38, 39, 41, 42) (0, 0, 0, 0)
O_2	(4, 5, 7, 8)	(0, 1, 1, 2) [23,24,26,27]	(7, 8, 10, 11)	(2, 3, 5, 6) [21,23,27,29]	(48, 49, 51, 52) (-8,-6,-2, 0)
O_3	(1, 2, 4, 5)	(6, 7, 9, 10)	(0, 1, 3, 4) [38,39,41,42]	(8, 9, 11, 12)	(38, 39, 41, 42) (-5,-3, 1, 3)
	(18, 19, 21, 22) (0, 1, 3, 4)	(23, 24, 26, 27) (0, 3, 7, 10)	(48, 49, 51, 52) (1, 2, 4, 5)	(33, 34, 36, 37) (6, 7, 9, 10)	

Table 2: The optimum solution.

Dest → ↓Origin	D_1	D_2	D_3	D_4	$\tilde{a}_i$
O_1	(0, 1, 3, 4) (19, 20, 22, 23)	(5, 6, 8, 9)	(1, 2, 4, 5) (6, 8, 12, 14)	(6, 7, 9, 10) (3, 6, 12, 15)	(38, 39, 41, 42)
O_2	(4, 5, 7, 8)	(0, 1, 1, 2) (23, 24, 26, 27)	(7, 8, 10, 11)	(2, 3, 5, 6) (22, 24, 28, 30)	(48, 49, 51, 52)
O_3	(1, 2, 4, 5)	(6, 7, 9, 10)	(0, 1, 3, 4) (38, 39, 41, 42)	(8, 9, 11, 12)	(38, 39, 41, 42)
	(18, 19, 21, 22)	(23, 24, 26, 27)	(48, 49, 51, 52)	(33, 34, 36, 37)	

Table 3: For the cell (2, 3) the optimum cost-flow pair.

Dest → ↓Origin	D_1	D_2	D_3	D_4	$\tilde{a}_i$
O_1	(0, 1, 3, 4) (18, 19, 21, 22)	(5, 6, 8, 9)	(1, 2, 4, 5) (7, 9, 13, 15)	(6, 7, 9, 10) (3, 6, 12, 15)	(38, 39, 41, 42)
O_2	(4, 5, 7, 8)	(0, 1, 1, 2) (23, 24, 26, 27)	(7, 8, 10, 11)	(2, 3, 5, 6) (22, 24, 28, 30)	(48, 49, 51, 52)
O_3	(1, 2, 4, 5)	(6, 7, 9, 10)	(0, 1, 3, 4) (38, 39, 41, 42)	(8, 9, 11, 12)	(38, 39, 41, 42)
	(18, 19, 21, 22)	(23, 24, 26, 27)	(48, 49, 51, 52)	(33, 34, 36, 37)	

Table 4: The paradoxical cost-flow pair.

Ĩ	Cost $\tilde{Z}$ i	Mag ($\tilde{Z}$ i)	Flow $\tilde{F}$ i	Mag ($\tilde{Z}$ i)
(0, 0, 0, 0)	(72, 216, 512, 714)	373.67	(124, 127, 133, 136)	130
(1, 0, 0, 0)	(74, 216, 512, 704)	372.33	(125, 127, 133, 136)	130.17
(0, 1, 0, 0)	(72, 220, 503, 714)	372	(124, 128, 133, 136)	130.33
(1, 1, 1, 1)	(68, 213, 511, 714)	371.67	(125, 128, 134, 137)	131
(16, 0, , 0)	(64, 216, 512, 690)	368.3	(140, 127, 133, 136)	132.667
(0, 13, 0, 0)	(72, 181, 536, 714)	370	(124, 140, 133, 136)	134.33
(16, 0, 7, 0)	(64, 167, 568, 690)	370.667	(140, 127, 140, 136)	135
(0, 13, 7, 0)	(72, 209, 473, 714)	358.33	(124, 140, 140, 136)	136.667
(16, 0, 0, 4)	(72, 216, 512, 650)	363	(140, 127, 133, 140)	133.33
(0, 13, 0, 4)	(48, 181, 536, 754)	372.667	(124, 140, 133, 140)	135
……	…….	………	……..	……………
(16, 13, 7, 4)	(72, 209, 473, 650)	347.67	(140, 140, 140, 140)	140

Table 5: Algorithm in section 3.

Conclusion

We have developed an efficient algorithm for finding paradoxical solution, if paradox exists, in a transportation problem under fuzzy environments. Adlakha and Kowalski demonstrated the practicality of identifying cases where the paradoxical situation exists in crisp environment. Klingman and Russel's approach, Adlakha and Kowalski absolute point procedure provide only best paradoxical pair whereas this method gives step by step development of this solution procedure for finding all paradoxical pairs. 6 Acknowledgements The authors would like to thank the referees for their useful comments which have improved the paper significantly.

References

1. Hitchcock FL (1941) The distribution of a product from several resources to numerous localities, Journal of Mathematical Physics 20: 224-230.
2. Charnes A, Cooper WW, Henderson A (1953) An introduction to linear programming, Wiley New Work.
3. Appa GM (1973) The Transportation Problem and Its Variants, Operational Research Quarterly 24: 79-99.
4. Klingman D, Russel R (1947) The transportation problem with mixed constraints, Operational Research Quaterly 25: 447-455.
5. Charnes A, Klingman D (1971) The more-for-less paradox in the distribution model, Cachiers du Centre dEtudes de Recherche Operationelle 13: 11-22.
6. Szwarc W (1973) The Transportation Paradox, Naval Research Logistics Quarterly 18: 185-202.
7. Gupta A, Khanna S, Puri MC (1993) A paradox in linear fractional transportation problems with maxed constraints, Optimazation 27: 375-387.
8. Adlakha V, Kowalski K (1998) A quick sufficient solution to the more-for-less paradox in a transportation problem. Omega 26: 541-547.
9. Ryan MJ (2000) The distribution problem, the more for less (nothing) paradox and economies of scale and scope, European Journal of Operational Research 121: 92-104.
10. Dinagar DS, Palanivel K (2009) On trapezoidal membership function in solving transportation problem under fuzzy environment, International Journal of Computational Physical Sciences 1: 1-12.
11. Dahiya K, Verma V (2006) Paradox in a non-linear capacitated transportation problem, Yugoslav Journal of Operations Research 16: 189-210.
12. Adlakha V, Kowalski K, Vemugantia RR, Levc B (2007) More-for-less algorithm for fixed-charge transportation problems. Omega 35: 116-127.
13. Storoy S (2007) The Transportation Paradox Revisited, N-5020 Bergen, NORWAY.
14. Joshi VD, Gupta N (2010) On a paradox in linear plus fractional transportation problem, Mathematika 26: 167-178.
15. Schrenk S, Finke G, Cung VD (2011) Two classical transportation problems revisited: Pure constant fixed charges and the paradox, Mathematical and Computer Modelling 54: 2306-2315.
16. Basu M, Acharya D, Das A (2012) The algorithm of finding all paradoxical pairs in a linear transportation problem, Discrete Mathematics, Algorithm and Application 4.
17. Zadeh LA (1965) Fuzzy sets, Information and Control 8: 338-353.
18. Bellman RE, Zadeh LA (1970) Decision making in a fuzzy environments, Management Science 17: 141-164.
19. Liu B (2002) Theory and practices of uncertain programming, physica-verlag, New York (2002).
20. Dinagar DS, Palanivel K (2009) On I membership function in solving transportation problem under fuzzy environment 1: 1-12.
21. Dutta D, Murthy AS (2010) Fuzzy transportation problem with additional restrictions, ARPN Journal of Engineering and Applied Sciences 2: 36-40.
22. Ojha A, Mondal SK, Maiti M (2014) A solid transportation problem with partial nonlinear transportation cost, Journal of Applied and Computational Mathematics.

9

High Order Numerical Solutions to Convection Diffusion Equations with Different Approaches

Don Liu* **and Yifan Wang**

Mathematics and Statistics, Louisiana Tech University Ruston, LA 71272, USA

Abstract

Convection diffusion equations, one of the fundamental kinds of nonlinear partial differential equations (PDE), describe a variety of phenomena such as fluid convection and diffusion and traffic flow etc. This paper uses four different approaches to handle nonlinearity in convection diffusion equations. High order accuracy numerical solutions to nonlinear PDE were acquired with several discretization methods: spectral modal and nodal element, finite element, and compact difference methods. Numerical results were compared with exact solutions to demonstrate the error convergence rate. For irregular domains where exact solutions are unavailable, very high accuracy numerical solutions were used for error comparisons. Pros and cons of these four approaches for handling nonlinearity were reviewed and discussed. This study provides a reference in solving nonlinear PDE with different options.

Keywords: Spectral element method; Finite element method; Compact difference method; Convection diffusion equation; Lagrange interpolation

Introduction

Numerically solving nonlinear PDE has been on the center stage for computational sciences, especially in computational fluid dynamics (CFD), a mature interdisciplinary area with many applications. The key component of CFD is obtaining reliable and accurate numerical solutions to governing equations such as Burgers' equations, convection diffusion equations, and Navier-Stokes equations under a variety of boundary and initial conditions in different dimensions. Convection diffusion equation and Burgers' equation have been widely used to test numerical algorithms as they are simple nonlinear PDE describing convection and diffusion of a fluid, gas dynamics, shock waves, acoustic waves, and traffic flows etc. Handling nonlinear terms is central to solving convection diffusion equations and Burgers' equations.

Previous studies on convection diffusion and Burgers' equations are mainly analytical and numerical. Courant et al. [1] proposed the method of characteristics to solve Burgers' equation. But it has a limitation and would not work for the inviscid Burgers' equation because the characteristics may intersect at certain time and may result in discontinuities in solution. Hopf et al. [2] introduced a nonlinear transformation to convert an original PDE into a linear heat equation which can be solved with less effort. Zhu et al. [3] reported a discrete Adomain decomposition method, which derived a fully implicit finite difference scheme for Burgers' equation, and nonlinear terms was defined by the infinite series of Adomian's polynomials. Srivastava et al. [4,5] used finite difference method to discretize spatial derivatives and adopted Newton's root-finding method [6,7] to solve a system of nonlinear equations without linearizing Burgers' equation. Although this method has the advantage of solving Burgers' equation implicitly, generating Jacobian matrix of the system is computational expensive.

Nonlinear terms could be numerically solved either with a linearization or without a linearization such as using conservative forms. Linearization could relieve the constraint on the size of time steps due to stability condition because a linearized PDE could be solve implicitly with much larger time step than the time step of an explicit method. For very large systems of nonlinear PDE, linearization is especially useful in decoupling systems of PDE and making them possible to be solved individually and sequentially. For nonlinear PDE in multiple dimensions, linearization also facilitates decoupling and makes it possible to solve one component ahead of time. On the other hand, a relatively recent and successful approach is to cast a nonlinear term into a conservative form and use numerical fluxes to solve conservation laws [8-13]. This approach is very effective for hyperbolic equation where convection is dominant and information propagates along certain direction.

There are many methods to solve convection diffusion equations. Compact difference method (CDM) [14-18], due to its simple nature and convenience in implementation, offers high order algebraic convergence and uses smaller stencils than finite difference method. But the linear system is usually twice or even three times bigger than the one from finite difference. Finite element method (FEM) is known for handling geometrical complexity. Fletcher et al. obtained finite element solution to convection diffusion equations [19,20]. However, FEM is limited in acquiring higher order accuracy as basis functions become less orthogonal mutually once the order of basis functions increases beyond fifth and mass matrices become ill-conditioned. For high accuracy resolution, although one could use h-type discretization refinement, i.e., smaller and more elements, rounding errors would accumulate and even defeat accuracy at certain point. Spectral element method (SEM), first appeared in [21], could be a good alternative. It achieves very high accuracy (e.g. 15th) by using orthogonal polynomial bases and zeros of orthogonal polynomials as quadrature points. SEM is capable of hp-refinement, and especially the p-type refinement which enhances resolution without extra numbers of elements [22-24]. Due to the capability of handling complexity geometry and both hp-type refinement for high accuracy, SEM has successfully appeared

***Corresponding author:** Don Liu, Mathematics and Statistics, Louisiana Tech University Ruston, LA 71272 USA, E-mail: DonLiu@ LATech.edu

in computational fluid dynamics [21,25-34], especially in simulating microfluidic devices [35-41]. SEM could also be used to model advanced microfluidic biosensors [42] with applications in magnetic labeling, sorting, medical diagnostics, and weak magnetic field detection [43].

Another relatively new approach, discontinuous Galerkin finite element method (DG- FEM), has been successful in handling nonlinear PDE in which nonlinear terms are written in conservative forms [22,44-54]. Using discontinuous basis functions and appropriate numerical fluxes at boundaries, DG-FEM is free from the C0 constraint on basis functions and is capable of capturing discontinuity in the solution. How to choose numerical fluxes is crucial to some strong hyperbolic equations.

In this paper, we discuss four different approaches, namely Method I: Taylor expansion, Method II: Cole-Hopf transformation, Method III: Compact Difference Preprocessing, and Method IV: Conservative Form, to handle nonlinearity. With these approaches, we produce numerical solutions to one and two dimensional convection diffusion equations using a blend of CDM, FEM, and SEM. Convergence rates are demonstrated in h-type and/or p-type refinement tests. Pros and cons of different approaches are discussed to provide a reference for solving nonlinear PDE in multiple dimensions and systems of nonlinear PDE.

Methodology and Numerical Examples

We consider convection diffusion equations in one or two dimensional domains:

$$\frac{\partial \mathbf{u}}{\partial t}+(\mathbf{u}\cdot\nabla)\mathbf{u}=\mu\Delta\mathbf{u}, \qquad in\ \Omega\ and\ dt\geq 0, \tag{1}$$

where μ is the dynamic viscosity. In some occasions, for the convenience of discussion, we treat the viscosity as a small value and drop the diffusion term; therefore, the convection diffusion becomes the inviscid Burgers' equation. In order to verify the accuracy of numerical solutions, we need an exact solution to compute the norm of errors at all quadrature points. Therefore, the convection diffusion equation was modified with the diffusion term replaced by a known function of space and time. In some examples, we solve the following modified convection diffusion equations instead:

$$\frac{\partial \mathbf{u}}{\partial t}+(\mathbf{u}\cdot\nabla)\mathbf{u}=\mathbf{f}(\mathbf{x},t). \tag{2}$$

In the following subsections, four different approaches are discussed to handle nonlinearity in one dimensional (1D) or two dimensional (2D) situations. Numerical solutions are provided with SEM, FEM, CDM or a blend of the above.

Linearization with Taylor expansion (Method I)

We start with the following simplest nonlinear PDE:

$$\frac{\partial u}{\partial t}+uu_x=f(x,t), \quad 1\leq x\leq 2, \quad t\geq 0, \tag{3}$$

$$u(x, 0) = u_0(x), \tag{3a}$$

$$u(1, t) = f_1(t), \tag{3b}$$

and use two different discretization methods in time and space in subsequent discussions.

Backward Euler and finite element method

Discretizing Equ. (3) implicitly in time, we obtain:

$$\frac{u^{n+1}-u^n}{\Delta t}+u^{n+1}u_x^{n+1}=f^{n+1}. \tag{4}$$

Due to the nonlinear term, we could not directly treat the entire convective term implicitly, because after a Galerkin projection, there will be three different indices for the nonlinear term. Therefore, we attempted to linearize Equ. (4). The first approach to handle nonlinearity, denoted as Method I, applies the Taylor expansion to the coefficient of u^{n+1} in time and repetitively replace time derivatives with space derivatives using the exact information in the governing PDE. This linearization is similar to the general idea of Lax-Wendroff scheme [55,56,57], except we use first or higher order Taylor expansion [58] to achieve higher accu- racy and implement a finite element or high order difference method to compute the spatial derivatives. The CFL number here is no more than 1. The first order approximation for the nonlinear term has the second order truncation error in time:

$$u^{n+1}u_x^{n+1}=(u^n+\Delta t u_t^n+O(\Delta t^2))u_x^{n+1}\approx(u^n+\Delta t u_t^n)u_x^{n+1}. \tag{5}$$

Substitute Equ. (5) into Equ. (4), we have:

$$\frac{u^{n+1}-u^n}{\Delta t}+(u^n+\Delta t u_t^n)u_x^{n+1}=f^{n+1}. \tag{6}$$

Using the exact information from the PDE: $u_t^n=f^n-u^n u_x^n$, we substitute it into Equ. (6):

$$\frac{u^{n+1}-u^n}{\Delta t}+(u^n+\Delta t f^n-\Delta t u^n u_x^n)u_x^{n+1}=f^{n+1}. \tag{7}$$

Let $(u^n)*=u^n+\Delta t f^n-\Delta t u^n u^n$, which is known at the time step n, then Equ. (7) becomes a linearized ordinary differential equation:

$$\frac{u^{n+1}-u^n}{\Delta t}+(u^n)^* u_x^{n+1}=f^{n+1}. \tag{8}$$

Next we apply the nodal Galerkin projection to form a linear system $A\ \hat{U}=B$. We divide the domain into N elements. Within each element Ω_e, we expand u and u_x in k^{th} order Lagrangian basis functions over k+1 uniform points in order to simplify the evaluation of $(un)^*$, although an alternative way could use Lagrangian basis function over Gauss-Lobattolegendre points. In a typical element Ω_e:

$$u^n=\sum_{i=0}^{k}\hat{U}_i^n\phi_i(x), u^{n+1}=\sum_{i=0}^{k}\hat{U}_i^{n+1}\phi_i(x), \tag{9}$$

$$u_x^{n+1}=\sum_{i=0}^{k}\hat{U}_i^{n+1}\phi_{i'}(x). \tag{10}$$

Substituting Equs. (9) and (10) into Equ. (8) and applying Galerkin projection, we obtain the weak form of Equ. (8) within one element Ω_e:

$$\sum_{i=0}^{k}\int_{\Omega_e}\phi_i(x)\phi_j(x)dx\frac{\hat{U}_i^{n+1}-\hat{U}_i^n}{\Delta t}+\sum_{i=0}^{k}\int_{\Omega_e}\phi_i'(x)\phi_j(x)dx(U^n)_i^*\hat{U}_i^{n+1}=\int_{\Omega_e}f^{n+1}\phi_j(x)dx, j=0,1...k-1,k, \tag{11}$$

where $(U^n)*$ denotes the expansion term for $(u^n)*$ at the point of i. After global assembling, the matrix-vector form of Equ. (8) is:

$$\underline{\underline{M}}\frac{\hat{\underline{U}}^{n+1}-\hat{\underline{U}}^n}{\Delta t}+\underline{K}\hat{\underline{U}}^{n+1}=\underline{F}^{n+1}, \tag{12}$$

$$\underline{F}^{n+1}=\int_{\Omega_e}f(x,n+1)\phi_j(x)dx, \tag{13}$$

where, M is the global mass matrix; F^{n+1} is obtained by global assembly; the matrix K is obtained by the scalar multiplies global advection matrix. If we choose a linear expansion basis, the matrix K has a tridiagonal form as below, while if quadratic and cubic basis function and pentadiagonal and seven-diagonal matrix for using respectively:

$$\begin{bmatrix} K_{22}U_2 & K_{22}U_2 & 0 & \cdots & \cdots & \cdots & 0 \\ K_{32}U_3 & K_{33}U_3 & K_{34}U_3 & 0 & \ddots & \ddots & \vdots \\ 0 & K_{43}U_4 & K_{44}U_4 & K_{45}U_4 & \ddots & \ddots & \vdots \\ \vdots & \cdots & \cdots & \cdots & \cdots & \cdots & \vdots \\ \vdots & \cdots & \cdots & \cdots & \cdots & \cdots & 0 \\ \vdots & \ddots & \ddots & 0 & K_{N-1N-2}U_{N-1} & K_{N-1N-1}U_{N-1} & K_{N-1N}U_{N-1} \\ 0 & \cdots & \cdots & \cdots & 0 & K_{NN-1}U_N & K_{NN}U_N \end{bmatrix}$$

In which, $U_l = \hat{U}_l^n + \Delta t F_l^n - \Delta t(\hat{U}^n \hat{U}_x^n)_l$, l stands for the numbering of global points, and each Klm corresponds to the entry of global advection matrix at l, m. Here, l and m is from 2 to N, since boundary points at 1 and N+1 are known. The key idea here is to treat $(u^n)^*$ explicitly and take it as a scalar multiplier for u^{n+1}, which is treated implicitly, then include it into the global advection matrix to form matrix K. Therefore, we obtain a set of discretized equations with only two indices which are represent Table 1 in a linear system.

Finally, Equ. (12) becomes:

$$(\underline{M} + \Delta t \underline{K})\underline{\hat{U}}^{n+1} = \underline{M}\underline{\hat{U}}^n + \Delta t \underline{F}^{n+1}. \quad (14)$$

To examine the accuracy of this approach, we set up the problem as below:

$$\frac{\partial u}{\partial t} + uu_x = f(x,t) = -e^{(x-t)} + e^{(2x-2t)}, 1 \le x \le 2, t \ge 0, \quad (15)$$

subject to the boundary and initial conditions:

$$u(1, t) = e^{1-t}, u(x, 0) = e^x. \quad (16)$$

The analytic solution to this well-posed problem is:

$$u(x, t) = e^{x-t}. \quad (17)$$

Since this problem is time-dependent, time integrations are crucial in numerical solutions. We attempted both implicit and explicit time integration. The CFL condition (for the latter) and the diffusion condition were satisfied by restricting the time step to be small enough:

$$\Delta t < \frac{\Delta x}{U_{max}}, \quad (18)$$

$$\Delta t < \frac{\Delta x^2}{\mu}, \quad (19)$$

where U_{max} is the maximum of absolute phase velocity, Δt and Δx are time step and distance between quadrature points, respectively, and μ is the diffusion coefficient Table 2.

Order	α	β	A	B	C
4th	1/4	0	3/2	0	0
6th	1/3	0	14/2	1/9	0
8th	4/9	1/36	40/27	25/54	0

Table 1: Where the coefficients are.

order	α	β	γ	A	B	C	D	E	G	H
4th	1	3	-17/6	-17/6	3/2	3/2	-1/6	0	0	0
5th	1	4	-37/12	-37/12	2/3	3	-2/3	1/12	0	0
6th	1	5	-197/60	-197/60	-5/12	5	-5/3	5/12	-1/20	0
7th	1/10	1	-227/600	-227/600	-13/12	7/6	1/3	-1/24	1/300	0
8th	1/12	1	-79/240	-79/240	-77/60	55/48	5/9	-5/48	1/60	-1/720

Table 2: Where the coefficients are.

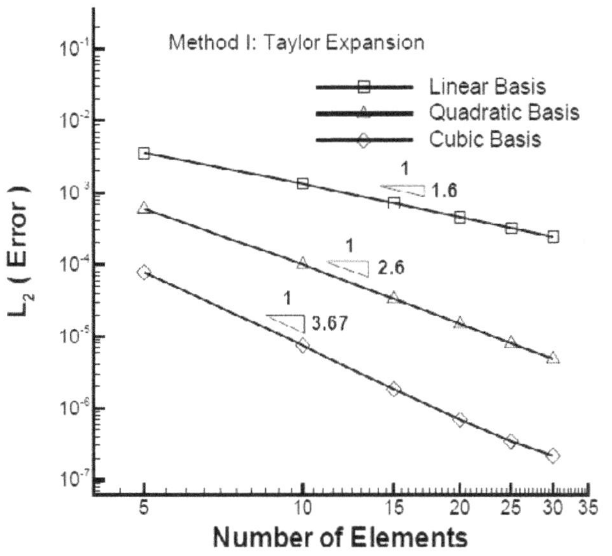

Figure 1: Convective Equ. (15) linearized with Method I: Algebraic convergence in space at a fixed time step $\Delta t=10^{-4}$.

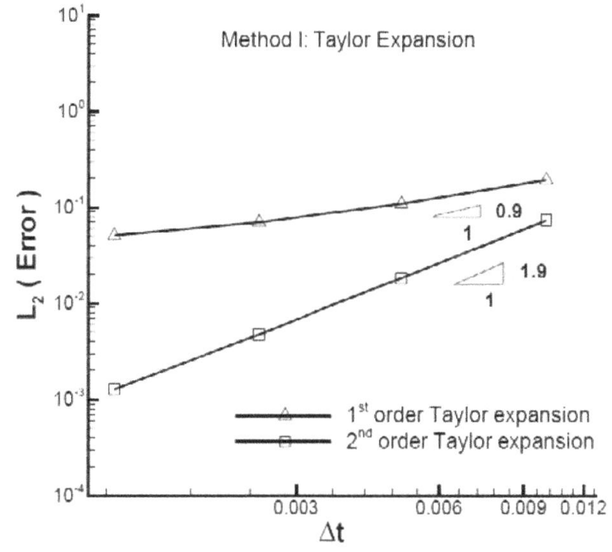

Figure 2: Convective Equ. (15) linearized with Method I: Temporal convergence at a fixedspatial resolution of using 10 elements for a total time of 0.2 dimensionless unit.

When we choose linear, quadratic, and cubic bases in space and backward Euler in time, the error estimations are order of $O(\Delta t^1+\Delta x^2)$, $O(\Delta t^1+\Delta x^3)$, $O(\Delta t^1+\Delta x^4)$, respectively. We compute the Euclidean norm of errors (L2 Error) at all nodes. Figure 1 presents the L_2 Error versus the number of elements in the log-log scale at a fixed time step $\Delta t=10^{-4}$. The convergence rates are basically one order higher than the order of expansion functions. To investigate the order of accuracy in time, we fix element number N=10, and vary Δt=0.01, 0.005, 0.0025, 0.00125. Using a first order Taylor expansion in the log-log scale, the first order convergence in time is observed in Figure 2. When we switch

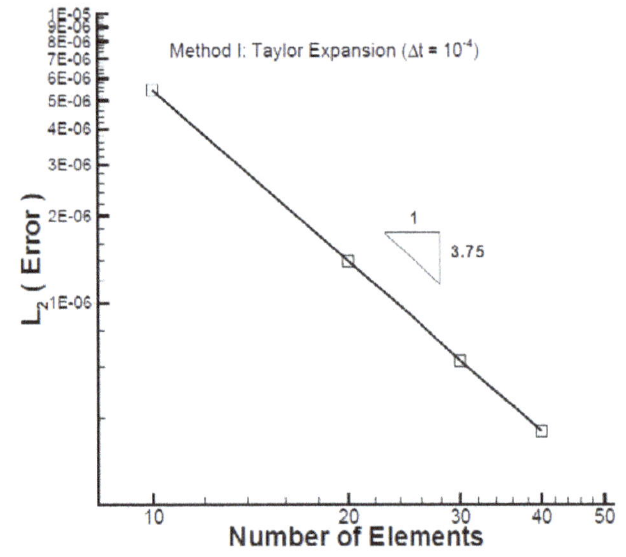

Figure 3: The convective equation (15) linearized with Method I: Convergence in space at afixed time step.

it to the second order Taylor expansion, a second order accuracy in time is obtained as shown in Figure 3.

Crank-Nicholson and compact difference method

We test the same approach for nonlinearity but using the Crank-Nicholson scheme in time and compact method in space for Equ. (3):

$$\frac{u_i^{n+1}-u_i^n}{\Delta t}+\frac{u_i^{n+1}u_{xi}^{n+1}+u_i^n u_{xi}^n}{2}=\frac{f_i^{n+1}+f_i^n}{2}. \quad (20)$$

To linearize the term $u_i^{n=1}u_{xi}^{n=1}$, we use the Taylor expansion in time again to express un+1 with some terms in the time level n, such that $u_i^{n=1}$ is replaced with u_i^*. Therefore, Equ. (20) becomes:

$$\frac{u_i^{n+1}-u_i^n}{\Delta t}+\frac{\overset{*}{u_i}u_{xi}^{n+1}+u_i^n u_{xi}^n}{2}=\frac{f_i^{n+1}+f_i^n}{2}. \quad (21)$$

In order to keep the 2nd order temporal accuracy, we use a few more terms in the expansion such that

$$u_i^{n+1}u_{xi}^{n+1}\approx[u_i^n+\Delta t u_{ti}^n+\frac{\Delta t^2}{2}u_{tti}^n+O(\Delta t^3)]u_{xi}^{n+1}. \quad (22)$$

After truncating the 2nd order term, we have:

$$u_i^{n+1}u_{xi}^{n+1}\approx[u_i^n+\Delta t u_{ti}^n]u_{xi}^{n+1}, \quad (23)$$

in which, u_t^n can be obtained from the PDE exactly:

$$u_i^n u_{xi}^n=f_i^n-u_{ti}^n. \quad (24)$$

Hence, Equ. (20) becomes linearized as below:

$$\frac{u_i^{n+1}-u_i^n}{\Delta t}+\frac{[u_i^n+\Delta t(f_i^n-u_i^n u_{xi}^n)]u_{xi}^{n+1}+u_i^n u_{xi}^n}{2}=\frac{f_i^{n+1}+f_i^n}{2}. \quad (25)$$

Rearrange the above equation, we have:

$$u_i^{n+1}+[\frac{\Delta t}{2}(u_i^n+\Delta t(f_i^n-u_i^n u_{xi}^n))]u_{xi}^{n+1}=u_i^n-[\frac{\Delta t}{2}u_i^n]u_{xi}^n+\frac{\Delta t}{2}(f_i^{n+1}+f_i^n). \quad (26)$$

We use the following 4th order Pa´de scheme to compute u^{n+1} and u_x^{n+1} simultaneously:

$$u_{xi-1}+4u_{xi}+u_{xi+1}=\frac{3u_{i+1}-3u_{i-1}}{h}+O(h^4). \quad (27)$$

To examine the spatial and temporal accuracy of using Crank-Nicholson and Compact Difference, we use the same problem as before, Equs. (15) and (16), with the same exact solution Equ. (17). For the spatial error only, we fix Δt=0.0001, compute u(x, 0.0001), and compare it with the exact solution in L_2 error. We notice that the number of elements N=10, 20, 30, 40 increases, the error goes down consistently. We observed that the order of convergence rate is close to 4th in space, as shown in Figure 3.

Figure 4 illustrates a temporal convergence rate close to the second order at the total time of 1. In this Figure 5, we set the time step to be Δt=0.001, 0.0005, 0.00025, 0.000125, use only 21 elements in space.

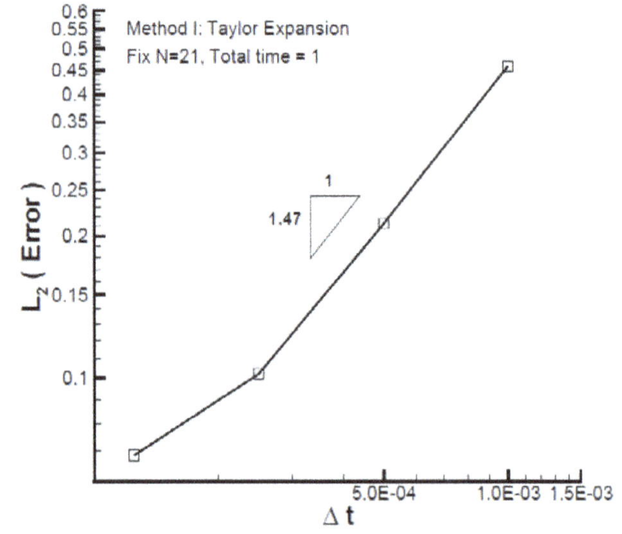

Figure 4: The convective equation (15) linearized with Method I: Convergence in time at afixed spatial resolution.

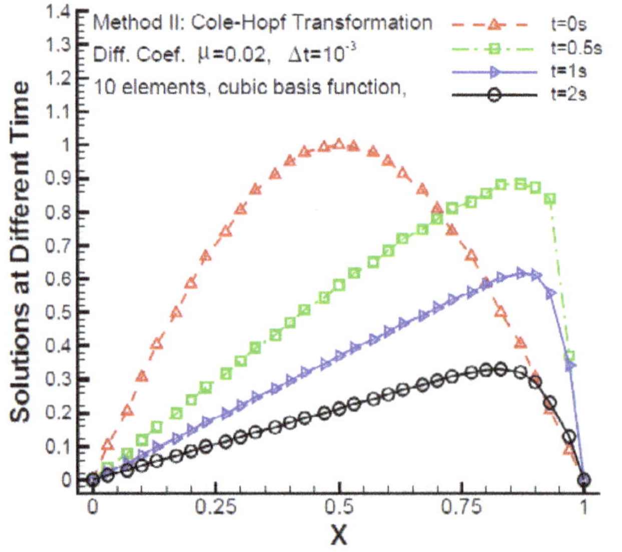

Figure 5: Evolution of numerical solution of 1D convection diffusion equation (28) linearizedwith Method II: Cole-Hopf transformation.

Since the Crank-Nicholson was used in time and 4th order Pa´de schemes in space, the second order convergence in time is anticipated.

In summary, the idea of using Taylor expansion to express an unknown of time level n+1 explicitly to approximate nonlinearity is convenient and stable, provided that CFL condition is obeyed. This idea is rather simple and could be used as long as an explicit expression for time derivative is possible or other approaches are difficult to use.

Linearization with Cole-Hopf transformation (Method II)

One-dimensional convection diffusion equation: Now we include the diffusion and consider the one dimensional convection diffusion equation:

$$\frac{\partial u}{\partial t}+uu_x=\mu u_{xx}, \tag{28}$$

with the initial condition and two Dirichlet boundary conditions below:

$$u(x, 0) = f(x), 0 \le x \le 1, \tag{28a}$$

$$u(0, t) = \alpha(t), u(1, t) = \beta(t), t \ge 0. \tag{28b}$$

With the Cole-Hopf transformation [2]:

$$u(x,t)=-2\mu\frac{w_x(x,t)}{w(x,t)}, \tag{29}$$

the original equation becomes a linear heat equation of in w(x, t),

$$\frac{\partial w}{\partial t}=\mu w_{xx}, 0\le x\le 1, t\ge 0 \tag{30}$$

with new initial and boundary conditions as below:

$$w(x,0)=e^{-\int_0^x \frac{f(s)}{2\mu}ds}, \quad 0\le x\le 1, \tag{30a}$$

$$2\mu w_x(0,t)+\alpha(t)w(0,t)=0, \quad t>0, \tag{30b}$$

$$2\mu w_x(1,t)+\beta(t)w(1,t)=0, \quad t>0. \tag{30c}$$

We solve the transformed linear equation, then use the inverse transformation, i.e., Equ. (29), to acquire the solution to the original problem.

Two-dimensional convection diffusion equation: For two-dimensional domains, the Cole-Hopf transformation [59] is given below:

$$u(x,y,t)=-2\mu\frac{w_x(x,y,t)}{w(x,y,t)}, \tag{31}$$

$$v(x,y,t)=-2\mu\frac{w_y(x,y,t)}{w(x,y,t)} \tag{32}$$

We could transform the vector form of convection diffusion equation, Equ. (1), into a scalar equation,

$$\frac{\partial w}{\partial t}=\mu\Delta w, \quad in\,\Omega\, and\, t\ge 0 \tag{33}$$

To obtain numerical solutions, we compute w in Equ. (33) with the nodal spectral element method using high order Lagrangian interpolants. Then we use high order finite difference schemes to compute wx and wy from w. Finally, u and v could be obtained from w using Equs. (31) and (32), respectively. When we use the kth order Lagrangian interpolants as the basis functions, in order to guarantee the exponential convergence, we used k + 1st order difference schemes to compute wx and wy .

Numerical results: One-dimensional Example:

To validate Method II in one dimension (1D), we solve Equ. (28) with μ=0.02, u(x,0)=sin(πx), $0 \le x \le 1$, and $\alpha(t)=\beta(t)=0$. After the Cole-Hopf transformation, we have the heat equation in the new variable w(x) with new initial and boundary conditions:

$$w(x,0)=e^{\frac{\cos(\pi x)-1}{2\pi\mu}}, \quad 0\le x\le 1, \tag{34}$$

$$w_x(0,t)=w_x(1,t)=0, \quad t\ge 0. \tag{35}$$

Since the original nonlinear PDE becomes linear, we use implicit treatment in time and finite element method in space. After we obtain the solution in w(x), we use Cole-Hopf inverse transformation of to obtain the numerical solution to the original problem. Figure 5 shows the numerical solution to the 1D problem at different time. The number of elements is fixed to be 10. As in the figure, the wave propagates towards right and forms a wave front, and the magnitude of the wave decreases with the time due to the viscous dissipation. We used cubic basis functions, as the number of elements varying from 10, 20, 30, 40, to 50, the 4th order convergence rate in space was obtained as shown in Figure 6, where the time step was Δt=0.0001. For the following two dimensional (2D) situations, we test two examples using different initial and boundary conditions.

Two-dimensional Example I:

We set the viscosity μ=0.02 and the initial and boundary conditions for Equ. (1) as below:

$$u(x, y, 0)=\sin(\pi x)\sin(\pi y), v(x,y,0)=y, \Omega: 0 \le x, y \le 1, t \ge 0. \tag{36}$$

$$u(x,y,t)=0, v(x, y, t) = y, x, y \in \partial\Omega, \tag{37}$$

We perform the Cole-Hopf transformation to reduce a vector PDE

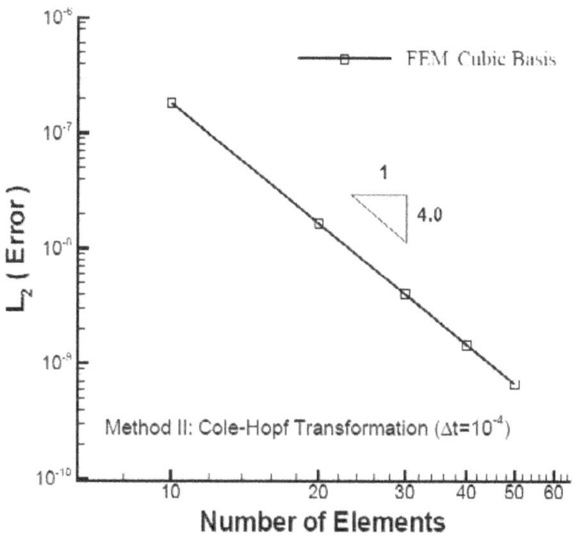

Figure 6: Spatial convergence for the convection diffusion equation (28) linearized withMethod II: Cole-Hopf transformation.

in u(x,y,t) into a scalar one in w(x,y,t). The domain was divided into 4 elements. Nodal SEM with 5th order basis function and Lagrangian interpolants over Gauss-Lobatto-Legendre points were used to solve the 2D scalar heat equation. We still use Lagrangian interpolation and difference method to compute wx and wy. By Equs. (31) and (32), we found the solutions to u(x,y) for the original problem. Figure 7a and b show the contour lines for the velocity component u and v at t=0.1, respectively. Figure 7c illustrates the elements and quadrature points. Figure 7d shows the vector field of u=(u,v). Because velocity components u and v were asymmetric and coupled together, the vector field convects in both x and y directions.

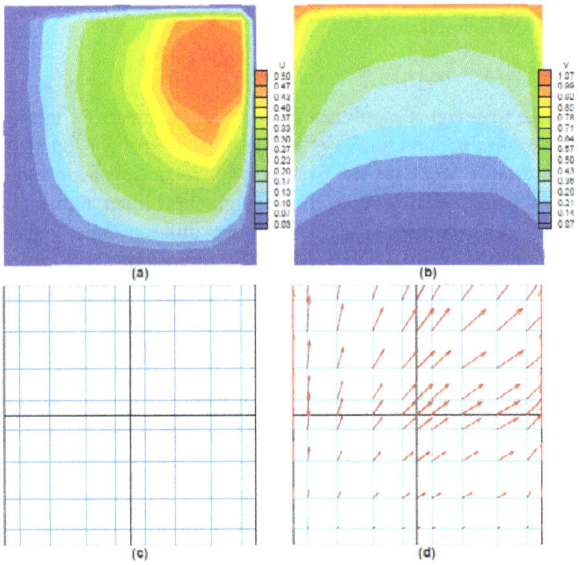

Figure 7: The two-dimensional convection diffusion equation (28) (Example I) linearized withMethod II: Velocity components u, v, the computational elements and quadrature points.

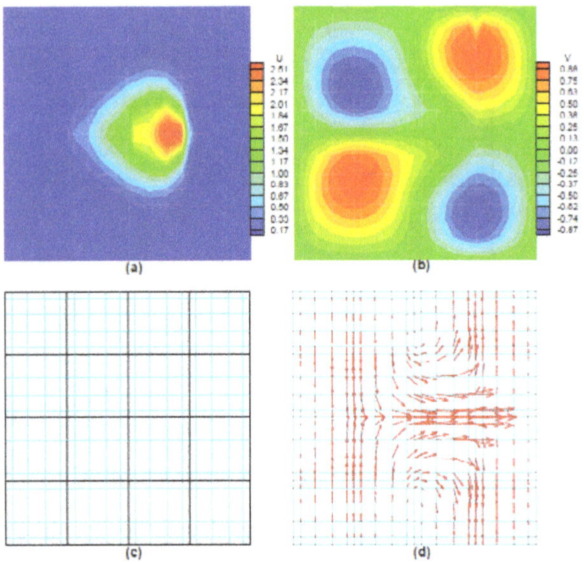

Figure 8: The two-dimensional convection diffusion equation (28) (Example II) linearizedwith Method II: Velocity components u, v, the mesh with 16 elements and quadrature points.

Two-dimensional Example II:

In the second test, we change the domain into Ω={(x, y): 0 ≤ x, y ≤ 2π}, set μ=0.005, and use new initial conditions

$$u(x,y,0)=e^{(1-(x-\pi)2-(y-\pi)2)}, \tag{38}$$

$$v(x,y,0)=\sin(x)\sin(y), \tag{39}$$

and new boundary conditions on the boundary ∂Ω

$$u(x,y,t)=0, v(x,y,0)=0, x,y \in \partial\Omega. \tag{40}$$

The domain was divided into 16 elements. We used the nodal SEM with 5th order basis function in each direction in each element to solve the 2D heat equation. Then, we used Lagrange interpolation and compact schemes to compute the first derivatives. Finally, solu- tions for u and v were obtained from Equs. (31) and (32). Numerical results were shown at time t=0.5. Figure 8a and 8b show contour lines for u and v, respectively. Figure 8c illustrates 16 elements and quadrature points. Figure 8d is the vector field for u=(u, v).

Generally speaking, Cole-Hopf transformation offers two advantages: reducing a vector equation into a linear scalar one and allowing an implicit temporal treatment in which the time step is unrestricted by the stability condition and time integration is unconditionally stable. However, Cole-Hopf transformation could only be applied to certain nonlinear PDE [58], such as hyperbolic PDE which contains second or higher spatial derivatives [60] and the Korteweg-de Vries equation (KdV equation) [61]. Chu et al. [62] listed a system of PDE with second order spatial derivatives that Cole-Hopf transformation is applicable.

Compact Difference Preprocessing (Method III)

Consider the 1D nonlinear problem in Equs. (3), (3a), (3b), and (3c), we linearize it by treating ux explicitly as the coefficient of u. This is called Method III in this paper, which is an explicit and non-conservative method [58]. The linearized equation below could be solved with an explicit scheme:

$$\frac{u^{n+1}-u^n}{\Delta t}+((u_x)^n)u^n=f^n, \tag{41}$$

or implicit scheme:

$$\frac{u^{n+1}-u^n}{\Delta t}+((u_x)^n)u^{n+1}=f^n. \tag{42}$$

The success of Method III relies on the accuracy of evaluating ux. We compute the nodal values of ux with high order compact difference [15-17,63] schemes before multiplying u. To demonstrate the idea of Method III, we compute ux at interior points with sixth order compact difference scheme as shown below:

$$12(u_x)_{i-1}+36(u_x)_i+12(u_x)_{i+1}=h^{-1}(-u_{i-2}-28u_{i-1}+28u_{i+1}+u_{i+2})+o(h^6), \tag{43}$$

where h=1 . For boundary points, we use Dirichlet boundary conditions (for simplicity) and the following sixth order scheme:

$$-60(u_x)_{i-1}-300(u_x)_i=h^{-1}(197u_{i-1}+25u_i-300u_{i+1}+100u_{i+2}-25u_{i+3}+3u_{i+4})+o(h^6). \tag{44}$$

Once we computed the values of $(u_x)_i^n$ at all points, Equs. (41) and (42) are linearized.We divide the whole domain into a total of N elements (Ωe) and express un and un+1 with expansions of basis functions (for example, linear) on each element:

$$u^n = \sum_{i=0}^{1} \hat{U}_i^n \phi_i(x), \tag{45}$$

$$u^{n+1} = \sum_{i=0}^{1} \hat{U}_i^{n+1} \phi_i(x). \tag{46}$$

Then we use the Galerkin projection to acquire the weak form of Equ. (42):

$$\sum_{i=0}^{1}\int_{\Omega_e}\phi_i(x)\phi_j(x)dx\frac{\hat{U}_i^{n+1}-\hat{U}_i^n}{\Delta t}+\sum_{i=0}^{1}\int_{\Omega_e}\phi_i(x)\phi_j(x)dx(u_x)_i^n\hat{U}_i^{n+1}=\int_{\Omega_e}f^n\phi_j(x)dx, j=0,1. \tag{47}$$

After global assembling procedure, the linear system of Equ. (42) in vector form is:

$$\underline{\underline{M}}\frac{\underline{\hat{U}}^{n+1}-\underline{\hat{U}}^n}{\Delta t}+\underline{\underline{M}}^*\underline{\hat{U}}^{n+1}=\underline{F}^n, \tag{48}$$

in which, $\underline{\underline{M}}$ is the global mass matrix; $\underline{F}^n$ is the assembled RHS of the Equ. (47) over all element. The matrix $\underline{\underline{\ }}$ consists of the products of the global mass matrix and the explicit values of $(u_x)_i^n$ given by Equs. (43) and (44). The matrix M∗ has the following form:

$$\begin{bmatrix} M_{22}(\mathrm{u_x})_2^n & M_{23}(\mathrm{u_x})_2^n & 0 & \cdots & \cdots & 0 \\ M_{32}(\mathrm{u_x})_2^n & M_{33}(\mathrm{u_x})_2^n & M_{34}(\mathrm{u_x})_2^n & 0 & \cdots & \cdots \\ 0 & M_{43}(\mathrm{u_x})_2^n & M_{44}(\mathrm{u_x})_2^n & M_{45}(\mathrm{u_x})_2^n & \cdots & \cdots \\ \cdots & \cdots & \cdots & \cdots & \cdots & \cdots \\ \cdots & \cdots & \cdots & \cdots & \cdots & 0 \\ \cdots & \cdots & 0 & M_{N-1N-2}(\mathrm{u_x})_{N-1}^n & M_{N-1N-1}(\mathrm{u_x})_{N-1}^n & M_{N-1N}(\mathrm{u_x})_{N-1}^n \\ 0 & \cdots & \cdots & 0 & M_{NN-1}(\mathrm{u_x})_N^n & M_{NN}(\mathrm{u_x})_N^n \end{bmatrix}$$

For the convenience of time integration, Equ. (48) can be written as:

$$(\underline{\underline{M}}+\Delta t\underline{\underline{M}}^*)\underline{\hat{U}}^{n+1}=\underline{\underline{M}}\underline{\hat{U}}^n+\Delta t\underline{F}^n. \tag{49}$$

Method III can readily be a high order scheme by adopting higher order compact schemes to evaluate ux on equispaced grids and use higher order polynomials as basis functions in a Galerkin finite element method. Using cubic or quartic basis functions could provide solu- tions with decent algebraic convergence. Beyond fifth order, the convergence rate starts to deteriorate due to the Runge's phenomenon

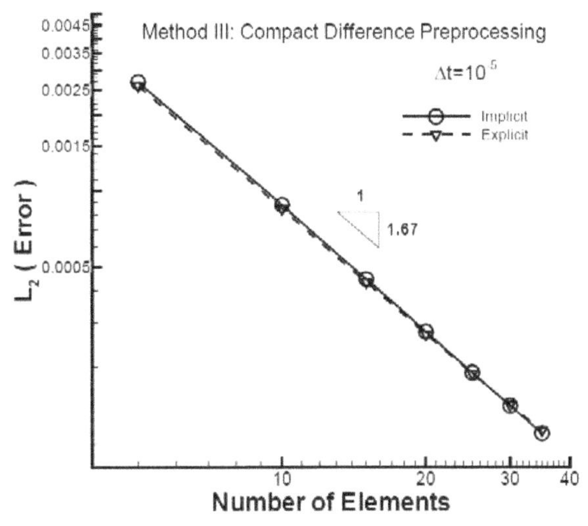

Figure 9: The 1D convective equation linearized with Method III: Convergence in space ata fixed time step.

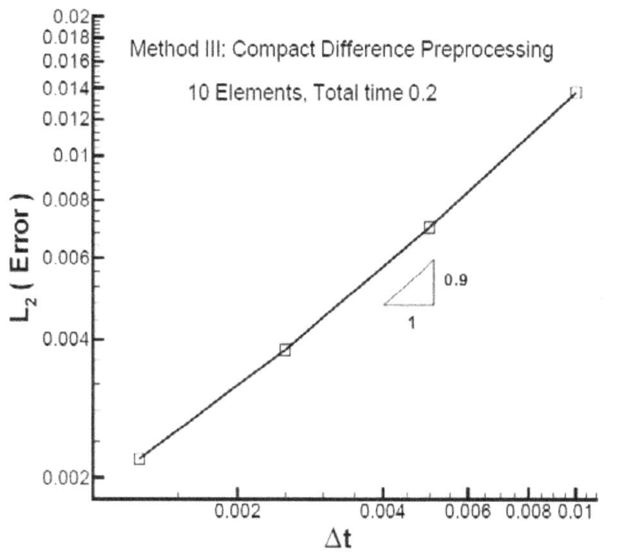

Figure 10: The 1D convective equation linearized with Method III: Convergence in time ata fixed spatial resolution.

and a spectral element method using orthogo- nal bases is a better alternative. We demonstrate this by using Lagrangian interpolation for ux from a uniform finite difference grid to Gauss-Lobatto-Legendre quadrature points. For two dimensional situation, the same idea works since Lagrangian interpolation in 2D can be utilized to calculate first derivatives of u at Gauss-Lobatto-Legendre points.

To test the accuracy of Method III in 1D, we consider the previous problem with the initial and boundary conditions in Equs. (15) and (16). We choose linear basis functions in space and forward Euler method in time, and use the 2nd order central difference scheme to compute value of ux. Therefore, the estimated error is of order $O(\Delta t^1 + \Delta x^2)$. The CFL condition is obeyed with $\Delta t = 10^{-4}$. By varying the spatial resolution, the number of elements from N=5 to 35, we present the spatial convergence in the log-log scale in Figure 9. The order of convergence is almost 2.

To examine the order of accuracy in time, we fix N=10 and decrease Δt=0.01, 0.005, 0.0025, 0.00125. The time convergence rate is almost one, as shown in the log-log scale in Figure 10. This convergence rate is expected because the forward Euler method was used.

For two-dimensional convection diffusion equations, we denote velocity components with u, v along x and y and let f, g be corresponding forcing terms. We perform linearization and set up the following implicit schemes:

$$\frac{u^{n+1}-u^n}{\Delta t}+(u_x)^n u^{n+1}+(u_y)^n v^{n+1}=f^{n+1}, \tag{50}$$

$$\frac{u^{n+1}-u^n}{\Delta t}+(u_x)^n u^{n+1}+(u_y)^n v^{n+1}=f^{n+1},\frac{v^{n+1}-v^n}{\Delta t}+(v_x)^n u^{n+1}+(v_y)^n v^{n+1}=g^{n+1}. \tag{51}$$

The derivative terms, ux, uy , vx and vy , at interior points, were computed from u and v at time level n using compact difference method given below [64] and were treated as known coefficients in Equs. (50) and (51):

$$\beta f'_{i-2}+\alpha f'_{i-1}+f'_i+\alpha f'_{i+1}+\beta f'_{i+2}=C\frac{f_{i+3}-f_{i-3}}{6h}+B\frac{f_{i+2}-f_{i-2}}{4h}+A\frac{f_{i+1}-f_{i-1}}{2h}, \tag{52}$$

$$\alpha f_1' + \beta f_2' + \gamma f_3' = \frac{1}{h}(Af_1 + Bf_2 + Cf_3 + Df_4 + Ef_5 + Gf_6 + Hf_7), \quad (53)$$

After a Galerkin projection, Equs. (50) and (51) could form a 2N by 2N system:

$$\begin{bmatrix} \underline{\underline{M}} + \Delta t * \underline{\underline{M_1}} & \Delta t * \underline{\underline{M_2}} \\ \Delta t * \underline{\underline{M_3}} & \underline{\underline{M}} + \Delta t * \underline{\underline{M_4}} \end{bmatrix} \begin{bmatrix} \hat{U}^{n+1} \\ \hat{V}^{n+1} \end{bmatrix} = \begin{bmatrix} \underline{\underline{M}} * \hat{U}^n + \Delta t * \underline{F}^{n+1} \\ \underline{\underline{M}} * \hat{V}^n + \Delta t * \underline{G}^{n+1} \end{bmatrix} \quad (54)$$

where M is global mass matrix. M1, M2, M3 and M4 are constructed by following the same pattern as the products of global mass matrix and corresponding spatial differentiation (ux)n, (uy)n, (vx)n and (vy)n respectively.

To facilitate validation, we choose exact solutions of u, v as following:

$$u = sin(x-t)sin(y-t), \quad (55)$$

$$v = sin(\frac{x}{2}-t), \quad (56)$$

$$f(x,y,t) = -cos(x-t)sin(y-t) - sin(x-t)cos(y-t) + sin(x-t)cos(x-t)sin^2(y-t) + sin(\frac{x}{2}-t)sin(x-t)cos(y-t), \quad (57)$$

$$g(x,y,t) = -cos\Big(\frac{x}{2}-t\Big) + \frac{1}{2}sin(x-t)sin(y-t)cos\Big(\frac{x}{2}-t\Big), \quad (58)$$

and the initial and boundary conditions are acquired from Equs (55) and (56). After we computed the first derivatives of velocity components with respect to x and y, we use two dimensional Lagrangian interpolants to compute their values at Gauss-Lobatto-Legendre quadrature points. Figure 11a and 11b show the contour lines of u and v at time t=0.5. Figure 11c shows four elements and the interior quadrature points. Figure 11d shows the velocity vector field. For all elements, the highest order of polynomial is 8. We pre-compute ux, uy, vx and vy with compact difference schemes of 8th order accuracy. Then we interpolate values of u and v on uniform grids to Gauss-Lobatto-Legendre points and use the nodal SEM with 8th order basis expansions, which are Lagrangian interpolants on Gauss-Lobatto-Legendre points. The values of u and v could be computed with high accuracy.

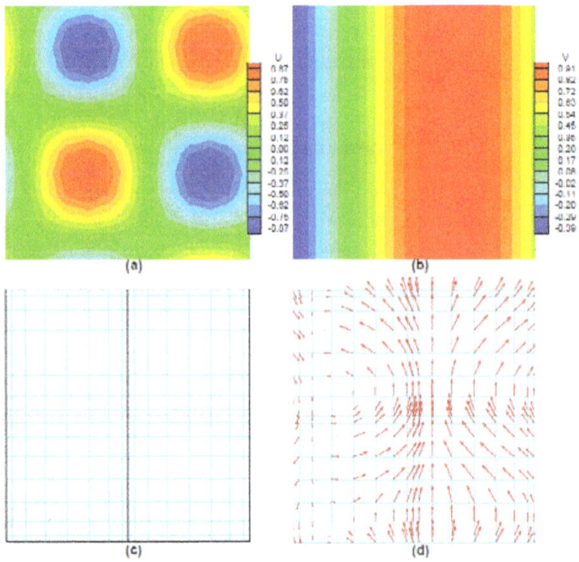

Figure 11: Two dimensional convection diffusion equation linearized with Method III: Velocity components u, v and computational mesh.

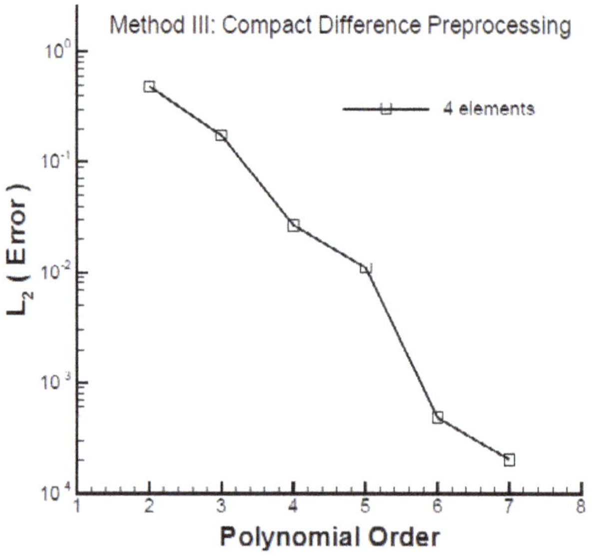

Figure 12: Two dimensional convection diffusion equations linearized with Method III: p-Refinement.

Figure 12 presents exponential convergence of L^2 norm of errors of u as the polynomial order varying from 2 to 8. We only derived and tested 8th compact difference schemes for pre-processing first derivatives. It has been observed that Method III works well for convection diffusion equations. For irregular domains, we have to use a mapping and pre-compute the first derivatives on a mapped grid or using full inclusion of metrics [17] which could add difficulty to Method III.

In summary, the approach of Compact Difference Preprocessing uses previously obtained values in prior time steps to compute unknowns explicitly. This way of handling nonlinearity provides convenience in decoupling PDE and especially for systems of PDE or multiple dimensions. It opens up the possibility of using an implicit method in time in which a large time step is feasible and is especially good for long time integration.

Using conservative form (Method IV): The one dimensional convection diffusion equation, Equ. (28), could be written in conservative form [65], since it is always conservative in one dimensional situation (See the Appendix in Section 4). The approaching of casting it into conservative form is called Method IV.

$$\frac{\partial u}{\partial t} + \Big(\frac{u^2}{2}\Big)_x = u_{xx}. \quad (59)$$

We use the forward Euler scheme in time for the above equation:

$$\frac{u^{n+1}-u^n}{\Delta t} + \Big(\frac{u^2}{2}\Big)_x^n = u_{xx}^n, \quad (60)$$

and discretize the whole domain into N elements. Within a typical element Ω_e, we expand terms such as un, (u^2)n and un with the kth order Lagrangian interpolants φ on the k+1 2 xxx Gauss-Lobatto-Legendre points:

$$u^n = \sum_{i=0}^{k} \hat{U}_i^n \phi_i(x), \tag{61}$$

$$(\frac{u^2}{2})_x^n = \sum_{i=0}^{k} (\frac{\hat{U}_i^2}{2})^n \phi_{i'}(x), \tag{62}$$

$$u_{xx}^n = \sum_{i=0}^{k} \hat{U}_i^n \phi_{i''}(x). \tag{63}$$

We substitute the above expansions into Equ. (60) and apply a Galerkin projection to obtain the weak form of Equ. (59) within Ω_e:

$$\sum_{i=0}^{k} \int_{\Omega_e} \phi_i(x)\phi_j(x)dx \frac{\hat{U}_i^{n+1} - \hat{U}_i^n}{\Delta t} + \sum_{i=0}^{k} \int_{\Omega_e} \phi_{i'}(x)\phi_j(x)dx)(\frac{\hat{U}_i^2}{2})^n = -\sum_{i=0}^{k} \int_{\Omega_e} \phi_{i'}(x)\phi_{j'}(x)dx)\hat{U}_i^n, \quad j = 0,1...k-1,k, \tag{64}$$

Note that within each time step, (u2)n is obtained explicitly from un. Assembling Equ. (64) over all elements, we obtain the global form of Equ. (60):

$$\underline{\underline{M}}\hat{\underline{U}}^{n+1} = \underline{\underline{M}}\hat{\underline{U}}^n - \Delta t(\underline{\underline{K}}(\frac{\hat{\underline{U}}^2}{2})^n + \underline{\underline{L}}\hat{\underline{U}}^n). \tag{65}$$

In which, M , K and L are global mass, convection and stiffness matrices, respectively.

Two dimensional convection diffusion equation: For two dimensional convection diffusion equations, Equ. (1), subject to some initial and boundary conditions, we assume u and v are velocity components along x and y directions, respectively, and write Equ. (1) in the component form:

$$\frac{\partial u}{\partial t} + uu_x + vu_y = \mu(u_{xx} + u_{yy}), \tag{66}$$

$$\frac{\partial v}{\partial t} + uv_x + vv_y = \mu(v_{xx} + v_{yy}). \tag{67}$$

For Equs. (66) and (67), Method IV is only conditionally conservative [58]. The proof of this property is given in the Appendix in Section 4. Now we change Equs. (66) and (67) into the partial conservation form:

$$\frac{\partial u}{\partial t} + (\frac{u^2}{2})_x + vu_y = \mu(u_{xx} + u_{yy}), \tag{68}$$

$$\frac{\partial v}{\partial t} + uv_x + (\frac{v^2}{2})_y = \mu(v_{xx} + v_{yy}). \tag{69}$$

Unless we know that u=(u, v) is a conservative vector field, i.e., $u_y=v_x$, or incompressible flow field, i.e., $u_x+v_y=0$, we cannot take it for granted to change coupled terms vu_y and uv_x into conservative forms. In Equ. (68), the coupled term vu_y remains nonlinear. We could linearize it by treating v explicitly as known coefficient for u_y in Method IV. Similarly, the coupled term uv_x in Equ. (69) is linearized by treating u explicitly as the known coefficient for v_x.

In terms of spatial discretization, we first use rectangular elements (irregular elements are discussed later) and choose the kth order Lagrangian interpolants, Φ, on Gauss-Lobatto-Legendre points as the basis functions in both x and y directions, although different orders of bases could be used in x and y. In a typical element Ω_e, we expand u(x, y) as below:

$$u(x,y) = \sum_{i=0}^{k}\sum_{j=0}^{k} \hat{U}_{ij}\Phi_i(x)\Phi_j(y). \tag{70}$$

Similarly for v. The derivatives of u with respect to x and y, respectively, are:

$$\frac{\partial u(x,y)}{\partial x} = \sum_{i=0}^{k}\sum_{j=0}^{k} \hat{U}_{ij} \frac{\partial \Phi_i(x)}{\partial x} \Phi_j(y), \tag{71}$$

$$\frac{\partial u(x,y)}{\partial y} = \sum_{i=0}^{k}\sum_{j=0}^{k} \hat{U}_{ij}\Phi_i(x) \frac{\partial \Phi_j(y)}{\partial y}. \tag{72}$$

We perform a Galerkin projection with the test function w:

$$(x,y) = \Phi_{pq}(x,y) = \sum_{p=0}^{k}\sum_{q=0}^{k} \Phi_p(x)\Phi_q(y) \tag{73}$$

to acquire a linear algebraic system about global variables.

Numerical examples one dimensional example: Consider the same 1D problem, Equs. (15) and (16) as in Method I, we use Method IV to solve it. We discretize the whole domain into 2 elements and use the forward Euler method in time. To achieve exponential convergence in space, we use SEM and perform a p-refinement by varying the order of polynomial basis functions. Figure 13 shows the exponential convergence with the order of polynomial basis functions versus the L2 norm of point-wise errors. Notice that, when polynomial order is less than 4, the exponential convergence of L2 norm of errors is not reached, because the approximation space have not been well set up yet. When the polynomial order is greater than 16, the L2 norm of error will not decrease any more and oscillate around 10^{-13} due to double decision rounding error.

Two Dimensional Rectangular Domain

The two dimensional computational domain is set to be Ω={(x, y) : 0 ≤ x, y ≤ 1, t ≥ 0}. Equ. (1) is the governing equation and the initial and boundary conditions are generated from the exact solution of u and v which are given below:

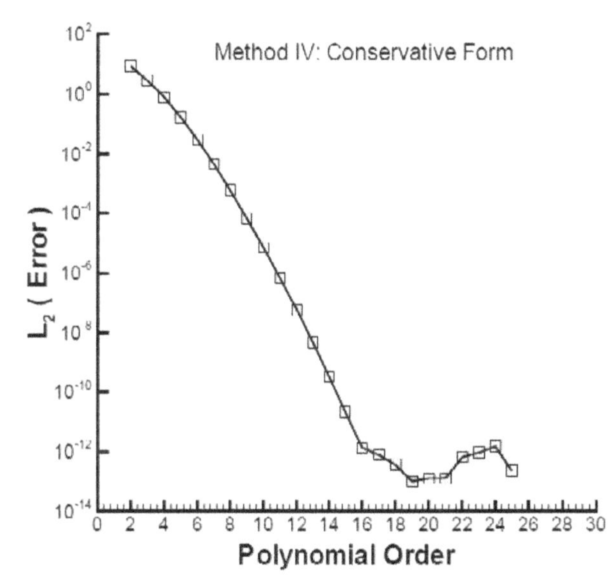

Figure 13: One dimensional convection diffusion equation (28) with Method IV: p-Refinement.

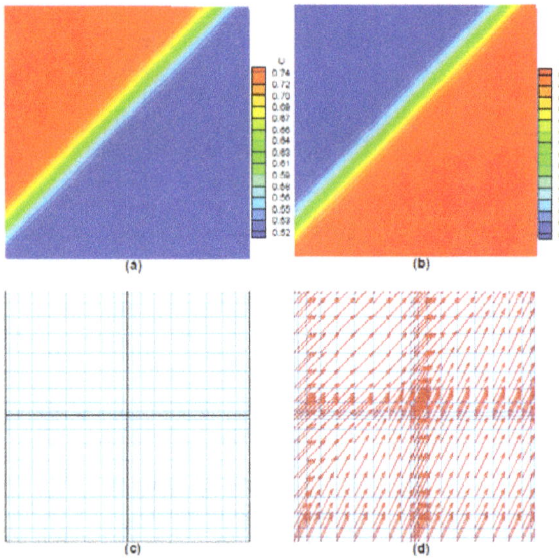

Figure 14: Two dimensional convection diffusion equation, Equ. (1) with Method IV: Velocity components u, v with 4 elements N=4 and the polynomial order Po=10.

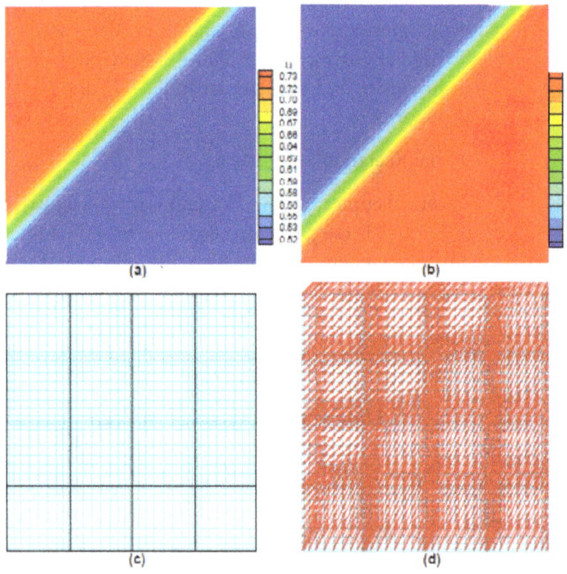

Figure 15: Two dimensional convection diffusion equation, Equ. (1) with Method IV: Velocity components u, v with 16 elements N=16 and the polynomial order Po=10.

$$u(x,y,t)=\frac{3}{4}-\frac{1}{4(1+e^{200(-t-4x+4y)/32})} \quad (74)$$

$$v(x,y,t)=\frac{3}{4}+\frac{1}{4(1+e^{200(-t-4x+4y)/32})}. \quad (75)$$

We use Method IV to solve the 2D problem on a rectangular domain, which was divided into 4 and 16 elements, respectively. Numerical results on 4 elements are shown in Figure 14 and on 16 elements are shown in 15. In both figures, the top plots ((a) and (b)) show contour lines of the velocity component u and v, separately at t = 0.5. The plot (c) illustrates the elements and the quadrature points. The plot (d) shows the velocity vector field. The Reynolds number is about 200; therefore, the magnitude of solution waves decrease with time due to viscous dissipation. Figure 15 gives more details about the convective flow since it has higher resolution than Figure 14 using 4 elements.

The exponential convergence in space is shown in Figure 16. Polynomial order were up to 40. The time step Δt was chosen to satisfy the CFL and diffusion conditions. Specifically, we chose Δt=0.001 and compared numerical result of u at time t=0.001 with the exact solution in the infinity norm. Using 16 elements gives faster convergence rate than using 4 elements and the error reaches machine zero at the polynomial order of 24.

Two-dimensional Irregular Domain

For the same problem as given in Equs. (66) and (67) but defined on an irregular domain with similar initial and boundary conditions that we designed here, we map the physical coordinates of (x, y) of a quadrilateral element Ωe to the coordinates ($\xi 1$, $\xi 2$) of a standard element (square) Ωst with the Jacobian:

$$J_{2D}=\begin{vmatrix}\frac{\partial x}{\partial \xi_1} & \frac{\partial x}{\partial \xi_2}\\ \frac{\partial y}{\partial \xi_1} & \frac{\partial y}{\partial \xi_2}\end{vmatrix}=\frac{\partial x}{\partial \xi_1}\cdot\frac{\partial y}{\partial \xi_2}-\frac{\partial x}{\partial \xi_2}\cdot\frac{\partial y}{\partial \xi_1}. \quad (76)$$

For any convex quadrilateral element, we denote its vertices as A, B, C and D, and their coordinates as A_x, A_y, B_x, B_y, C_x, C_y, D_x and D_y. The components of the Jacobian are:

$$\frac{\partial x}{\partial \xi_1}=\frac{\xi_2(A_x-B_x+C_x-D_x)+(-A_x+B_x+C_x-D_x)}{4}, \quad (77)$$

$$\frac{\partial y}{\partial \xi_1}=\frac{\xi_2(A_y-B_y+C_y-D_y)+(-A_y+B_y+C_y-D_y)}{4}, \quad (78)$$

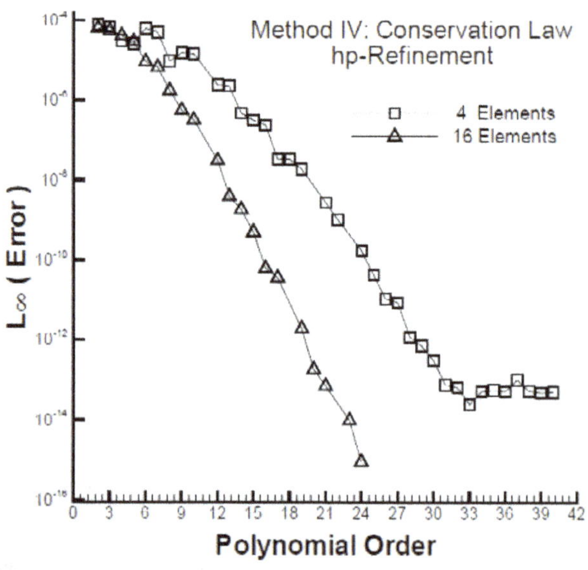

Figure 16: Two dimensional convection diffusion equation, Equ. (1) with Method IV: hp-refinement in rectangular domain

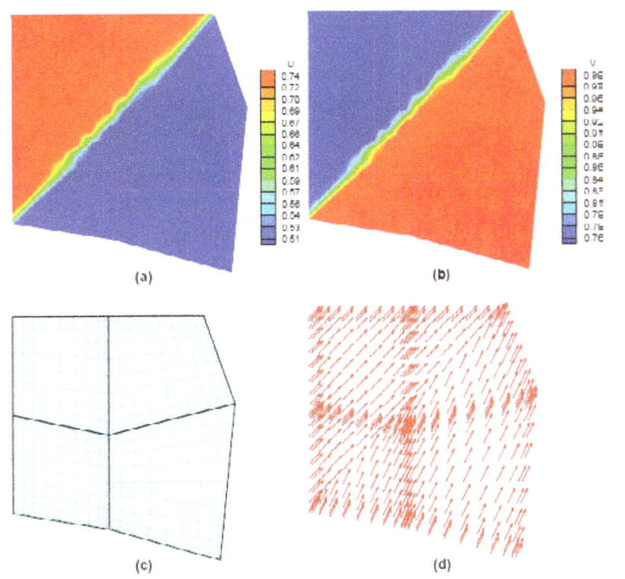

Figure 17: Two dimensional convection diffusion equation, Equ. (1) with Method IV: Velocity components u, v and quadrature points in irregular domain

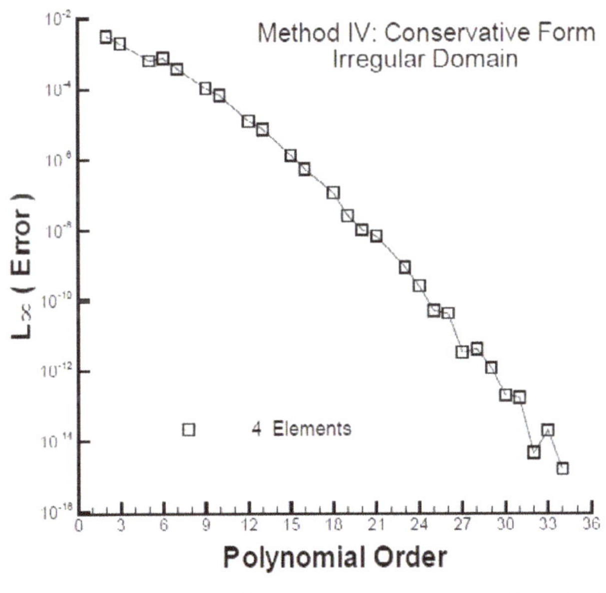

Figure 18: Two dimensional convection diffusion equation, Equ. (1) with Method IV: p-refinement in irregular domain

$$\frac{\partial x}{\partial \xi_2} = \frac{\xi_1(A_x - B_x + C_x - D_x) + (-A_x - B_x + C_x + D_x)}{4}, \quad (79)$$

$$\frac{\partial y}{\partial \xi_2} = \frac{\xi_1(A_y - B_y + C_y - D_y) + (-A_y - B_y + C_y + D_y)}{4}. \quad (80)$$

With a known mapping and Jacobian, we could construct the mass, convection, and stiff- ness matrices for any convex quadrilateral element. Similar procedures were performed and Method IV was used to obtain numerical solutions.

Figure 17 shows the velocity component u, v at time t=0.5, quadrature points, and the velocity vector field. In this figure, the domain was divided into 4 elements and the spectral nodal element method with polynomials of order Po=10 in each direction and each element was used as in the simulation. We completed a very high order solution with nodal basis polynomial functions of order Po=35 using 36 nodes, i.e., Gauss-Legendre-Lobatto points in each direction. This high accuracy solution was used as our "exact" solution to compute nodal errors for lower order runs. Figure 18 shows the L∞ norm of errors at all nodes versus the order of expansion polynomial order. A spectral convergence rate was shown in this figure. The solutions are consistent as the order increases.

In summary, Method IV is capable of capturing discontinuity developed in the solution and preserves good convergence rates which is unaffected by irregularity in geometry. This approach is very effective for strong hyperbolic equations provided that the conservative form exists.

Conclusion

In this paper, four different approaches for solving convection diffusion equations in 1D and 2D are discussed. Method I, II and III use linearization procedures which give the advantage of using an implicit time discretization and a large time step without the stability constraint. For the linearized problem, an implicit method in time such as Crank-Nicholson or backward Euler is preferred especially for long time integration. Spatial discretization methods used in this paper include spectral modal and nodal element, finite element, and compact difference. Of course other methods such as discontinuous Galerkin finite element or finite volume etc. are possible.

Since Method I uses a Taylor expansion in time to linearize a convective term, the trun- cation rror in time is consistent with the temporal discretization method. This method is simple and effective for strongly nonlinear terms such as (un+1)2un+1 as long as the first temporal derivative could be expressed in terms of the rest terms in the original PDE. The drawback of Method I is that it requires an explicit expression for the time derivative which could be difficult to derive for PDE with second or higher order temporal derivative and may introduce complexity in linearized equation.

Method II is an excellent approach for convection diffusion equations and possibly for other nonlinear equations if a Cole-Hopf transformation is applicable. The success of this method relies on performing a nonlinear transformation, which could reduce the original nonlinear vector form of PDE into a linear one. This is the major advantage of Method.

II. Apparently, both explicit and implicit time schemes can be used afterwards, although implicit is better in time integration. Nevertheless, the Cole-Hopf transformation is difficult for higher order spatial derivatives. Depending on the specific equation, it could be difficult to find analytical forms of transformed initial and boundary conditions. In general, it is not a trivial work to find a suitable Cole-Hopf transformation for a nonlinear equation.

Method III is a good idea in this paper which uses compact difference method to pre-compute first order spatial derivatives so that a convective term could be approximated by the pre-computed value multiplying the unknown which could be treated implicitly. Basically, Method III uses previously obtained values to compute a part of the unknown explicitly. Handling nonlinearity in this approach provides convenience in decoupling PDE and especially for systems of PDE or multiple dimensions. It gives the convenience to use an implicit method in time in which a large time step is feasible and is especially good for long time integration. However, for irregular domains, Method III becomes complicated due to the spatial discretization and a mapping may be necessary. If SEM is used for numerical solutions, Method III requires deriving difference schemes on uniform grids to compute the first derivatives and then interpolate them to Gauss-Lobatto-Legendre nodes. Method III could be improved by using schemes of full inclusion of metrics [16,17] on non-uniform grids or an efficient mapping between grids to compute the first derivatives.

Compared with the above approaches, in terms of convenience and efficiency, Method IV could be superior in dealing with strong nonlinearity and complex geometry. It is straight forward for general weighted residual methods and is simple for multiple dimensions, since generating conservative form does not rely on mesh grids. However, the disadvantage is that using conservative forms to treat nonlinear terms could make the state vector and the matrix system much larger than without using it. This could rely on more memory in computation.

In general, for nonlinear PDE, if possible, one could use Method II and design a Cole-Hopf Transformation, to reduce a multi-dimensional nonlinear PDE into a scalar linear PDE so as to bypass nonlinearity. Depending on the dimension and problem, alternatively, one could select an appropriate linearization with controlled errors to open an option for implicit time treatment, which relax the constraint on time step size and could be beneficial to long time integration. For solutions develops discontinuity or there is discontinuous initial or boundary conditions, Method IV, using conservative form is very effective and capable of capturing discontinuity. The accuracy is usually unaffected by complexity of geometry or discontinuity in the domain.

Appendix

For Method IV, the one dimensional convection diffusion equation, Equ. (28), is always conservative. To prove this claim, we use the chain rule backward for the convective term:

$$\frac{\partial u}{\partial t}+(\frac{u^2}{2})_x = u_{xx}, \tag{81}$$

then we introduce new variables F and G, respectively:

$$F=\frac{u^2}{2}, \tag{82}$$

$$G=u_x, \tag{83}$$

then we change Equ. (81) into the following:

$$\frac{\partial u}{\partial t}+F_x-G_x=0, \tag{84}$$

which is conservative. Therefore, the one dimensional convection diffusion equation is always conservative.

However, the two dimensional convection diffusion equations are not always conservative [58] because of the existence of coupled terms. Again, we use the chain rule backward in the two dimensional component-wise convection diffusion equation for the convective terms to obtain the partial conservative forms:

$$\frac{\partial u}{\partial t}+(\frac{u^2}{2})_x+vu_y=\mu(u_{xx}+u_{yy}), \tag{85}$$

$$\frac{\partial u}{\partial t}+(\frac{u^2}{2})_x+vu_y=\mu(u_{xx}+u_{yy}), \tag{86}$$

Focusing on two coupled terms above, we use the chain rule to form below equations:

$$(vu)_y=v_y u + vu_y, \tag{87}$$

$$(uv)_x=u_x v + uv_x. \tag{88}$$

If the irrotaional condition is valid for the vector field which is defined as u=(u, v), i.e., conservation of the total pressure in an irrotational flow [66], then we have the following:

$$\frac{\partial u}{\partial y}=\frac{\partial v}{\partial x}\Leftrightarrow u_y=v_x, \tag{89}$$

which is equivalent to the vorticity-free condition in fluid mechanics:

$$\nabla\times u=0, \tag{90}$$

By Equ. (89), we replace vuy with vvx in Equ. (85) and put vvx in conservative form:

$$\frac{\partial u}{\partial t}+(\frac{u^2}{2})_x+(\frac{v^2}{2})_x=\mu(u_{xx}+u_{yy}). \tag{91}$$

By Equ. (89), we replace uvx with uuy in Equ. (86) to and put uuy in conservative form:

$$\frac{\partial v}{\partial t}+(\frac{u^2}{2})_y+(\frac{v^2}{2})_y=\mu(v_{xx}+v_{yy}). \tag{92}$$

Now Equs. (91) and (92) are both conservative.

Alternatively, if the incompressible flow condition is satisfied by the vector field which is defined as u = (u, v), i.e., conservation of mass, then we have the following equations:

$$\frac{\partial u}{\partial x}+\frac{\partial v}{\partial y}=0\Leftrightarrow\frac{\partial u}{\partial x}=-\frac{\partial v}{\partial y}\Leftrightarrow u_x=-v_y \tag{93}$$

which is equivalent to the divergence-free condition in fluid mechanics:

$$\nabla\cdot u=0, \tag{94}$$

then we replace –vy with ux in the variation of the convective term vuy in Equ. (85) so that vuy becomes:

$$vu_y=(vu)_y-v_yu=(vu)_y+u_xu=(vu)_y+(\frac{u^2}{2})_x, \tag{95}$$

then Equ. (85) becomes:

$$\frac{\partial u}{\partial t}+(\frac{u^2}{2})_x+[(vu)_y+(\frac{u^2}{2})_x]=\frac{\partial u}{\partial t}+(u^2)_x+(uv)_y=\mu(u_{xx}+u_{yy}). \tag{96}$$

Now the new form of Equ. (85) is already in conservative form. Similarly, we replace ux with –vy in the variation of the convective term uvx in Equ. (86) so that uvx becomes:

$$uv_x=(uv)_x-u_xv=(uv)_x+vv_y=(uv)_x+(\frac{v^2}{2})_y, \tag{97}$$

then Equ. (86) becomes:

$$\frac{\partial v}{\partial t}+[(uv)_x+(\frac{v^2}{2})_y]+(\frac{v^2}{2})_y=\frac{\partial v}{\partial t}+(uv)_x+(v^2)_y=\mu(v_{xx}+v_{yy}). \tag{98}$$

Now the new form of Equ. (86) is also in conservative form. To summarize these proofs:

1. The one dimensional convection diffusion equation is conservative.

2. If ∇×u=0 is valid, then the dimensional convection diffusion equation is in conservative form:

$$\frac{\partial u}{\partial t}+\left(\frac{u^2}{2}\right)_x+\left(\frac{v^2}{2}\right)_x=\mu(u_{xx}+u_{yy}), \quad (99)$$

$$\frac{\partial v}{\partial t}+\left(\frac{u^2}{2}\right)_y+\left(\frac{v^2}{2}\right)_y=\mu(v_{xx}+v_{yy}). \quad (100)$$

If $\nabla \cdot u = 0$ is valid, then the two dimensional Burger's equation is in conservative form:

$$\frac{\partial u}{\partial t}+(u^2)_x+(uv)_y=\mu(u_{xx}+u_{yy}), \quad (101)$$

$$\frac{\partial v}{\partial t}+(uv)_x+(v^2)_y=\mu(v_{xx}+v_{yy}). \quad (102)$$

Acknowledgement

This study was supported by National Science Foundation under grants DMS-1115546, DMS-1115527, and Louisiana Board of Regents grant LEQSF (2007-10)-RDA-22. We thank the support from the Louisiana Optical Network Initiative (LONI) and Center for Computation and Technology, Louisiana State University. We are especially grateful to the encouragement from Professor Chi-Wang Shu (Brown University) and his continued constructive comments and insightful advice to improve the quality of this research.

Reference

1. Richard C, David H (1962) Methods of mathematical physics.
2. Hopf E (1950) The partial differential equation ut + uux=μuxx, Comm. Pure Appl Math 3: 201-230.
3. Zhu H (2010) Numerical solutions of two-dimensional Burgers' equation by discrete Adomian decomposition method. Com Math App 60: 840-848.
4. Bahadir (2003) A fully implicit finite-difference scheme for two-dimensional Burgers' equations. Appl Math Comput 131-137.
5. Srivastava V (2011) A comparison of finite element and finite difference solutions of the one-and two-dimensional Burgers' equations, Int J Scient Eng Res 2
6. Burden R, Faires J (2005) Numerical Analysis, 8th Edition, Thomson Brooks/ Cole, Belmont, CA, USA
7. Pepper D, Heinrich J (2006) The Finite Elements Method: Basic Concepts and Applications. Taylor and Francis, USA
8. Jiang G, Shu C (2000) Efficient implementation of weighted ENO schemes.
9. Shu C (1987) TVB boundary treatment for numerical solutions of conservation laws. Mathematics of Computation 49: 123-134.
10. Shu C (2001) An overview on high order numerical methods for convection dominated pdes, Hyperbolic Problems: Theory, Numerics, Applications.
11. Shu C (1987) TVB uniformly high order schemes for conservation laws. Mathematics of Computation 49: 105-121.
12. Xing Y, Shu C (2005) High order finite difference WENO schemes with the exact conservation property for the shallow water equations, J Com Phy 208: 206-227.
13. LeVeque R (1992) Numerical Methods for Conservation Laws. Birkhauser Verlag, Zurich.
14. Dai W, Nassar R (2001) A compact finite difference scheme for solving a one-dimensional heat transport equation at the microscale. J Com App Mathe 132: 431-441.
15. Gamet L, Ducros F, Nicoud F, Poinsot T (1999) Compact finite difference schemes on non-uniform meshes. Application to direct numerical simulations of compressible flows. Int J Numerical Methods in Fluids 29: 159-191.
16. Lele S (1992) Compact finite difference schemes with spectral-like resolution. J com phy.
17. Liu D, Kuang W, Tangborn A (2009) High-order compact implicit difference methods for parabolic equations in Geodynamo simulation.
18. Mahesh K (1989) A family of high order finite difference schemes with good spectral resolution, J Com Phy 145: 332-358.
19. Aksan E (2005) A numerical solution of Burgers' equation by finite element method constructed on the method of discretization in time. App Mathe Com 170: 895-904.
20. Fletcher C (1983) Crank-Nicholson scheme for numerical solution of two-dimensional coupled Burgers' equations. J Com Phy 51: 159-188.
21. Patera A (1984) A spectral element method for fluid dynamics-Laminar flow in a channel expansion. J Com Phy 54: 468-488.
22. Karniadakis G, Sherwin S (2005) Spectral/hp Element Methods for Computational Fluid Dynamics. Oxford University Press,USA
23. Mathews J, Fink K (2004) Numerical Method Using Matlab. Prentice Hall, USA
24. Pozrikidis C (2005) Introduction to Finite and Spectral Element Method using MATLAB. Taylor and Francis Group, USA.
25. Beskok A, Warburton T (2002) An unstructured hp fnite-element scheme for fluid flow and heat transfer in moving domains. J Comp Phys 174: 492-509.
26. Dong S, Liu D, Maxey M, Karniadakis G (2004) Spectral distributed Lagrange multiplier method: Algorithm and Benchmark test. J Com Phy 195: 695-717.
27. Giraldo F, Warburton T (2005) A nodal triangle-based spectral element method for the shallow water equations on the sphere. J Comp Phys 207: 129-150.
28. Hesthaven J, Gottlieb D (1999) Stable spectral methods for conservation laws on triangles with unstructured grids. J Comput Methods Appl Mech Engg 175: 361-381.
29. Huang L, Chen Q (2009) A spectral collocation model for solitary wave attenuation and mass transport over viscous fluid mud. J Eng Mechanics 135: 881-891.
30. Karniadakis G, Israeli M, Orszag S (1991) High-order splitting methods for the incompressible Navier-Stokes equations. J Comp Phys 97: 414-443.
31. Liu D, Chen Q, Wang Y (2011) Spectral element modeling of sediment transport in shear flows. J Comput Methods Appl Mech Engrg 200: 1691-1707.
32. Liu D, Keaveny E, Maxey M, Karniadakis G (2009) Force-coupling method for flows with ellipsoidal particles. J Computational Physics 228: 3559-3581.
33. Xia M, Symeonidis V, Karniadakis G (2003) A spectral vanishing viscosity method for stabilizing viscoelastic flows, J Non-Newtonian Fluid Mechanics 115: 125-155.
34. Xia M, Karniadakis G (2002) A low-dimensional model for simulating three-dimensional cylinder flow, Journal of Fluid Mechanics 458: 181-190.
35. Beskok A, Karniadakis G (1999) A model for flows in channels, pipes and ducts at micro and nano scales, J Microscale Thermophysical Engineering 3: 43-77.
36. Karniadakis G, Beskok A (2002) Microflows: Fundamentals and Simulation. Springer-Verlag, New York.
37. Karniadakis G, Beskok A, Aluru N (2005) Microflows and Nanoflows: Fundamentals and Simulation. Springer, New York.
38. Liu D, Lvov Y, Dai W (2011) Joint simulations of confined diffusion inside nanotubules. J Computational and Theoretical Nanoscience 8: 1-11.
39. Liu D, Maxey M, Karniadakis G (2002) A fast method for particulate microflows. J Micro-electromechanical System 11: 691-702.
40. Liu D, Maxey M, Karniadakis G (2004) Modeling and optimization of colloidal micro-pumps. J Micromechanics and Microengineering 14: 567-575.
41. Liu D, Maxey M, Karniadakis G (2005) Simulations of dynamic self-assembly of paramagnetic microspheres in confined microgeometries. J Micromechanics and Microengineering 15: 2298-2308.
42. Bellamkonda R, John T, Mathew B, DeCoster M, Hegab H, Davis D (2010)

Fabrication and testing of a CoNiCu/Cu CPP-GMR nanowire-based microfluidic biosensor. J Micromechanical and Microengineering 20: 025012.

43. Cox B, Liu D, Davis D (2012) GMR sensors: Technologies and medical applications, Journal of Recent Patents on Nanomedicine 1: 130-137.

44. Cockburn B, Karniadakis G, Shu C (2000) Discontinuous Galerkin Methods. Theory, Computation and Applications, Springer-Verlag, Berlin.

45. Hesthaven J, Warburton T (2008) Nodal Discontinuous Galerkin Methods: Algorithms,Analysis, and Applications. Springer-Verlag, New York.

46. Pietro DD, Ern A (2011) Mathematical Aspects of Discontinuous Galerkin Methods. Springer-Verlag, Berlin.

47. Cockburn B, Lin S, Shu C (1989) TVB Runge-Kutta local projection discontinuous Galerkin finite element method for conservation laws III: one dimensional systems. J Computational Physics 84: 90-113.

48. Cockburn B, Shu C (1989) TVB Runge-Kutta local projection discontinuous Galerkin finite element method for conservation laws II: general framework, Mathematics of Computation 52: 411-435.

49. Engsig-Karup A, Hesthaven J, Bingham H, Madsen P (2006) Nodal DG-FEM solutions of high-order Boussinesq-type equations. J Eng Math 56: 351-370.

50. Engsig-Karup A, Hesthaven J, Bingham H, Warburton T (2008) Nodal DG-FEM solutions of high-order Boussinesq-type equations. Coast. Eng. 55: 197-208.

51. Eskilsson C, Sherwin S (2002) A discontinuous spectral element model for Boussinesq-type equations. J Sci Comp 17: 143-153.

52. Eskilsson C, Sherwin S (2006) Spectral/hp discontinuous Galerkin methods for modelling 2D Boussinesq equations, J Comp Phys 212: 566-589.

53. Jiang B, Carey G (1988) A stable least-squares finite element method for non-linear hyper-bolic problems, International journal for numerical methods in fluids 8: 933-942.

54. Li B (2006) Discontinuous Finite Elements in Fluid Dynamics and Heat Transfer.

55. Lax P, Wendroff B (1960) Systems of conservation laws, Comm. Pure Appl Mathematics 13: 217-237.

56. Lax P, Wendroff B (1964) Difference schemes for hyperbolic equations with high-order of accuracy. Comm Pure Appl Mathematics 17: 381-398.

57. Richtmyer R, Morton K (1967) Difference Methods for Initial-Value Problems. John Wiley and Sons, New York.

58. Sheu T, Chen C, Hsieh L (2001) Development of a sixth-order two-dimensional convection-diffusion scheme via Cole-Hopf transformation, Comput Methods Appl Mech Engrg 191: 2979-2995.

59. Sachdev P (1978) A generalized Cole-Hopf transformation for nonlinear parabolic and hyper-bolic equations, J Applied Mathematics and Physics 29: 963-970.

60. Whitham G (1974) Linear and nonlinear wave. John Wiley and Sons, New York.

61. Chu C (1965) A class of reducible system of quasi-linear partial differential equations, Quart Appl Math 23: 275-278.

62. Roache P (1976) Computational Fluid Dynamics. Hermosa Publishers, Albuquerque.

63. Gaitonde D, Visbal M (1998) High-order schemes for Navier-Stokes equations: Algorithm and implementation into FDL3DI.

64. Shu C (2012) Efficient algorithm for solving partial differential equations with discontinuous solutions.

65. Mohanty A (1994) Fluid Mechanics. Prentice-Hall of India Private Limited, New Delhi.

10

Classifying Users of a Topic Recommendation System with a Restricted Boltzmann Machine with Nearest Neighbor Interactions

James Tesiero*

University of Maine, Orono, USA,

Abstract

In this work, we discover groups of similar users of a topic recommendation system. The data sets used are from Tipster Newsgroups, which are used in the annual TREC (Text Retrieval Conference) competition. The users are simulated, represented by tag/rating pairs. The documents in the Tipster data sets are clustered with a topic clustering algorithm that is a subset of the topic recommendation system being proposed in this paper. The users query the clusters derived from the topic clustering algorithm with tags, then rate the degree of relevance the content returned by the system has to their tag. It is shown in this work that, starting from a random sampling of the clusters by the users, and a random initial distribution of ratings per user, that a topic recommendation engine powered by a Boltzmann machine with nearest neighbor interactions results in two distinct clusters of users: those that converge quickly to a particular single topic, and those who explore a few different topics in a way that is periodic in time. This allows new users entering the system to be clustered and hence given a more relevant experience earlier in the process.

Keywords: Topic clustering; Cluster entropy; Information retrieval; Boltzmann machines

Introduction

The problem of topic recommendation applied to an evolving set of users is a subset of the general problem of user classification. Here, we focus on user acquisition of knowledge due to an evolving set of questions based on the interaction of users of various topic interests and levels of expertise. The structure of this paper is as follows. First, we describe the business problem which gives context to the problem we are solving. Next, the data sets and how they are used in the model is described. We then explain the components of the model, first the topic clustering followed by the user simulation/topic recommendation process. The findings of this research and their significance are then discussed, and the article finishes with a description of future work stemming from this research.

Brief review of the model

Recommendation systems fall into three main categories:

1) Content based recommendation systems

2) Collaborative Filtering Techniques

3) Hybrid Recommendation Systems

This model falls into the hybrid category since it contains aspects of both content (via the document clustering model) and collaborative filtering (using ratings of other users). However, it is unique in the hybrid category as it is the first system to our knowledge that combines a Restricted Boltzmann Machine with a fuzzy clustering technique to define a single new metric (the CR divergence) to classify users based on their content selection, rating of that content, and topic exploration behavior.

Business problem

The business problem that is addressed in this paper is that of increasing the quantity and quality of shared knowledge in the topic areas, and identifying sources of authority as well as building shared knowledge between related topic areas. The implementation of this concept is an intelligent database of knowledge and users that evolves over time and continuing user interaction.

In this paper, we discuss the simulation of this intelligent database, through the use of simulated users. A simulated user in our system is represented by a tag-rating pair. The tag represents a query to the database of knowledge stored in the clustered content which results from applying our supervised topic clustering algorithm to the data. We use inverse cluster entropy as a metric of similarity; from this the probability distribution of cluster membership for the tag is found.

In real-time, new users will enter the system at different rates, and existing users are likely to return. Also, different documents in the content clusters will be visited multiple times. Since each user is represented by a tag-rating pair, and a tag is a query of the clustered database, then each user effectively is represented by the probability distribution of cluster membership given by its tag and the ratings that it gives the items returned as a result of the query.

The users then, form a graph over the underlying content clusters. From the ratings of content items by the users and their nearest neighbors, and the distance on the graph, a cost function is derived which determines the next tag used to query the clusters.

The next tag can either be from the same user at a different time or another user in the same user cluster. This tag then results in more content being displayed to the user (or user in the same cluster), which is in turn rated, and results in a new set of nearest neighbors.

The users enter and exit the system in batches. Each batch represents a new time step in the simulation, with a noise parameter

***Corresponding author:** James Tesiero, Principal Data Scientist Consultant, University of Maine, Orono, USA, E-mail: jimtes_jim@aol.com

introduced analogous to the inverse temperature in simulated annealing applications. It is shown that at a critical value of the noise parameter, the system becomes ordered into at least two distinct user modes.

Data sets

The data sets used for this work were from the Tipster Newsgroup Collection. These data sets are used in the annual TREC competition (Text Retrieval Competition) centered on the TREC conference. There were 10 K documents selected randomly from 20 different topic areas used in this research. The topic areas were the following: atheism, computer graphics, computer operating systems, computer PC hardware, computer MacIntosh hardware, computer windows OS, miscellaneous, automobiles, motorcycles, baseball, hockey, encryption, electronics, medicine, space, Christian religion, guns, Middle East politics, general politics, and general religion. They covered the broad topic areas of religion, science, politics, computers, and sports. The documents in the Tipster dataset were classified manually by a team of human editors.

Related work

Other research related to this work are: [1] "A Unified Approach to Building Hybrid Recommendation Systems", A. Guadawardana and C. Meek, which describes a method of classifying users based on a Restricted Boltzmann Machine [2], "Recommendation as Classification: Using Social and Content Based Information in Recommendation", C. Basu, H. Hirsh, and W. Cohen, which uses hybrid features that contain elements of content and rating behavior to classify users.

Components of the Model

We now describe the components of the model. First, the topic clustering model is discussed. This model clusters the documents comprising the Tipster data collection. Then, the Boltzmann Machine is described, which simulates the users with tag-rating pairs that query the clusters for content, then rate that content, and share the content with their nearest neighbors as defined by the Boltzmann Machine [3].

Topic clustering

The first step in classifying users that query a clustered corpus to obtain, rate, and share content is to cluster the documents in the corpus. In order to do this, there must be a model of the documents that represents them in an N dimensional space. The model that we use to do this is used often in information retrieval, the tf-idf model (term frequency-inverse document frequency). However, we define the term frequency and the inverse document frequency in a different manner than is often encountered in the information retrieval literature. We also define the cluster term frequency and use it to filter terms that have little or no discriminative power.

Term frequency: The standard definition of the term frequency is the relative number of times a term in the corpus of documents appears in a particular document.

$$tf(i,j) = \frac{n(i,j)}{N(j)} \tag{1}$$

(Where $tf(i,j)$ is the term frequency of the ith term in the jth document, $n(i,j)$ is the number of times that the ith term appears in the jth document, and $N(j)$ is the total number of terms in the jth document)

There are two significant problems with this definition, in particular with short documents. First, for terms that occur at least once in short documents, the formula overestimates the probability because there are fewer discrete choices.

For example, in a twelve term document, if a term appears once, according to the above formula its term frequency is 0.083, which is in the 90th percentile at least in most corpuses. The second problem associated with the simple definition above is that for terms those do not appear at all in short documents, they potentially may have appeared if the document was longer.

We can exploit the fact that the documents have been pre-classified by treating documents in the clusters as if they were random selections of terms in their cluster, where the document length is also a random variable. The term frequency can be treated as a state whose value depends upon the selection (document) from the reservoir (cluster).

Our definition of the term frequency, which normalizes the frequency to document length, is as follows:

$$tf(i,j,k) = tf_{raw}(i,j) * \exp\left(-\left(\frac{\left(tf_{raw}(i,j) - \langle tf(k,j)\rangle\right)}{\sigma_{tf(k,j)}}\right)^{1}\right) \tag{2}$$

where $tf_{raw}(i,j)$ is the term frequency as it is usually defined in (1) above, k denotes the cluster that the Ith document that contains the jth term is in, $\langle tf(k,j)\rangle$ is the mean term frequency for the jth term in the kth cluster over the documents in that cluster other than the ith document. $\sigma_{tf}(k,j)$ denotes the standard deviation of the term frequency for the jth term in the kth cluster over the documents in that cluster other than the ith document.

Weighted inverse document frequency: The inverse document frequency as typically defined in the information retrieval literature is:

$$idf(j) = \frac{1}{N(d, tf(j) > 0)} \tag{3}$$

(where $idf(j)$ is the inverse document frequency of the jth term in the corpus, and $N(d,tf(j)>0)$ is the number of documents in the corpus where the raw term frequency is greater than zero (or, all of the documents where that term occurs at least once))

However, the occurrence of a term just once, especially in a longer document, can occur by chance. This artificially lowers the idf and makes a term appear less relevant than it actually is. We define an inverse document frequency metric that is robust to noise fluctuations in term occurrence.

We do this by again exploiting the fact that we have a supervised classification of the documents. This allows us to define a cluster term frequency. The cluster term frequency is similar in form to the raw term frequency, except that we count the relative frequency of terms in a particular cluster.

$$cf(j,k) = n(j,k) / N(k) \tag{4}$$

(where $cf(j,k)$ is the cluster frequency of the jth term in the kth cluster, $n(j,k)$ *is* the number of occurrences of the jth term in the kth cluster, and $N(k)$ *is* the number of terms in the kth cluster)

With the above definition of the cluster term frequency, we can build a definition of the idf that conceptually treats documents in the same cluster as equivalent and those outside as different. The cluster term entropy can be defined as the following, when the cluster frequency is interpreted as a probability:

$$\begin{aligned} &cs(j,k) = -cf(j,k) * \log(cf(j,k)) \\ &for\ \forall\, cf(j,k)\ s.t. 0 < cf(j,k) \le 1 \end{aligned} \tag{5}$$

The value at $cf(j,k) = 0$ is excluded since the logarithm diverges to negative infinity there. We define a smoothed inverse cluster entropy to avoid divergence when $cf(j,k) = 1$ as the following:

$$ics(j,k) = exp(-cs(j,k)) \tag{6}$$

This equation yields a value for $ics(j,k)$ near 1 when the entropy is low (when the term is highly discriminating between the clusters), and near 0 where the entropy is high (a uniform distribution over the clusters), and without diverging anywhere. This metric is used to weight the traditional idf (equation 3) to produce a weighted idf:

$$widf(j,k) = ics(j,k) * idf(j) \tag{7}$$

which is then multiplied by the robust term frequency (equation 2) to obtain the relevance score:

$$rel(i,j,k) = tf(i,j,k) * widf(j,k) \tag{8}$$

This equation yields the relevance of the jth term to the ith document in the kth cluster. The relevance is then summed over the terms in a document and used in a logistic equation to calculate the probability of the ith document in the kth cluster.

$$p(i,k) = 2.0 * \left(\frac{1}{1 + \exp(-\sum_j rel(i,j,k)} \right) - 1.0 \tag{9}$$

Since $rel(i,j,k)$ is constrained to lie between zero and one, we use the constants in the equation (9) above to allow $p(i,k)$, which is the probability of the ith document in the kth cluster, to lie between 0 and 1 inclusive. The probability distribution of each document over the clusters is then put into a graph, which serves as the content database which simulated users query with tags.

Simulating users

Now that we have demonstrated how documents get classified, we turn our focus to the user querying process, and the rating and sharing of content amongst users.

Ratings map and distance function: Each user is simulated as a tag-rating pair. The user queries the clusters with a tag and is returned a set of documents to rate (in the most general case, the documents can be other tags). The ratings are done with the original Netflix system (scale from 1-5), which is then mapped to a binary variable (like/dislike). Netflix ratings from 1-3 map to -1 (dislike) and ratings 4-5 map to 1 (like). The documents that are returned for rating are based on the highest document probability conditioned on the cluster frequency of the user tag.

$$p(i,j)_u = p(i,k) * cf(j,k) \tag{10}$$

(where $p(i,k)$ is given by equation (9) and $cf(j,k)$ is given by equation (4), and $p(i,j)u$ is the probability of the ith document given the jth term associated with the tag of the uth user)

For each user making a query to the clusters, we return the top three documents. These correspond to the documents having the top three values of $p(i,j)u$ in equation (10) above. The users are related to each other by the probability distribution across the clusters created by their tag. For example, if there are C clusters, then users A and B lie on the following points of the C dimensional graph at time t:

User A: {cf(tag(A(t)),1),cf(tag(A(t)),2),…,cf(tag(A(t)),C)}

User B: {cf(tag(B(t)),1),cf(tag(B(t)),2),…,cf(tag(B(t)),C)}

Using the city block distance metric, we can construct a C dimensional graph such that the nearest neighbors of any user on the graph at any time can be calculated.

$$d(A(t),B(t)) = \sum_{c=1}^{C} abs\left(cf\left(tag\left(A(t)\right),c\right) - cf\left(tag\left(B(t)\right),c\right)\right) \tag{11}$$

Using the above distance function, we can construct a graph for the users of the system at any point in time. The nearest neighbors of a user have an influence on the user that is inversely related to the above distance in the following way:

$$w(A(t),B(t)) = \exp(-d(A,B)) \tag{12}$$

Where $w(A(t),B(t))$ is the mutual influence that users A and B have on each other. The symmetry of the weight function is directly related to the symmetry of the distance function (11). This weight function has two nice properties: when A and B are the same user at the same time, the value is 1 (which is the maximum value it can have), and since the weight function goes to zero asymptotically, large numbers of nearest neighbors can still have some small influence.

Cost and partition functions: In this section, we develop the concepts of the cost and partition functions, and use them to define the metric which will be used to classify the users, the CR divergence (content-rating divergence).

The pairwise cost between users A and B above (or any two users of the system at any time) is defined as:

$$\varphi(A(t),B(t)) = -w(A(t),B(t)) * r(A(t)) * r(B(t)) \tag{13}$$

where $\varphi(A(t),B(t))$ is the cost of putting users A and B in the same neighborhood, $w(A(t),B(t))$ is the weight, or influence of user B on A as defined in (12), and $r(A(t)),r(B(t))$ are the ratings given to the content seen by users A and B at time t. Note that the larger the weight, the greater the effect on the cost function for any two users at any time. Since the goal is to minimize the cost, with the presence of the minus sign in (13) the larger the weight, the lower the pairwise cost becomes. Also, users that give the same rating lower the cost function, while those giving opposite ratings raise it.

The cost function as defined above is between a single nearest neighbor B with A; if we sum this over all nearest neighbors, we obtain the single user cost φ $(A(t))$ defined as the following:

$$\varphi(A(t)) = \sum_{b=1}^{B} \varphi(A(t),b(t)) \tag{14}$$

Where b=1,…,B are the nearest neighbors of A

The single user cost function (14) is then used to derive a single user partition function, with a Boltzmann probability distribution.

$$Z(A(t)) = \exp(-\beta(t) * \varphi(A(t))) \tag{15}$$

Where $\beta(t)$ is an inverse noise parameter and $\varphi(A(t))$ is as defined in

Substituting equations (13) and (14) in (15) expresses the single user partition function in terms of the influence and ratings of its nearest neighbors at that time:

$$Z(A(t)) = \exp(\beta(t) * \sum_{b=1}^{B} w(A(t),b(t)) * r(b(t)) * r(A(t))) \tag{16}$$

The sum over the product of the weights and the ratings of the nearest neighbor's yields a metric we call the content-rating divergence (or, CR divergence).

$$CR(A(t)) = \sum_{b=1}^{B} w(A(t),b(t)) * r(b(t)) \tag{17}$$

The CR divergence can be conceptualized as a "mean field" on the user A at time t. If the neighboring users that like content similar to that viewed by user A at time t have more influence as measured by the weight function at that time than neighboring users that dislike content similar to that viewed by user A at time t, then the overall CR divergence is positive, for example.

Substituting the CR divergence (17) into the single user partition function (16), we have:

$$Z\left(A(t)\right) = \exp\left(\beta(t) * CR\left(A(t)\right) * r\left(A(t)\right)\right) \quad (18)$$

Since the users are acting independently of each other, the partition function *Z(G(t))* over the graph at time t can be written as a product of single user partition functions.

$$Z\left(G(t)\right) = \prod_{g \varepsilon G} Z\left(g(t)\right) \quad (19)$$

User classification: In this section, we use the cost and partition function derived above to show how users can be classified with the correlation of their CR divergences.

The free energy function is related to the partition function by the equation

$$F\left(G(t)\right) = -\left(\frac{1}{\beta(t)}\right) * \ln\left(Z\left(G(t)\right)\right) \quad (20)$$

The partition function *Z(g(t))* for a single user at a particular time has two possible states, corresponding to the two possible ways the user can rate the content item being viewed at that time. Therefore, the explicit form of *Z(G(t))* is:

$$Z\left(g(t)\right) = 2\cosh(\beta(t) * CR\left(g(t)\right) * r\left(g(t)\right)) \quad (21)$$

Substituting this explicit form of the single user partition function into equation (20) and differentiating with respect to r yields:

$$\left\langle CR\left(g(t)\right)\right\rangle = \tanh\left(\beta(t) * CR\left(g(t)\right) * r\left(g(t)\right)\right) \quad (22)$$

Since this quantity can be calculated anywhere on the graph, the correlation between any two points on the graph can be measured. The points on the graph represent a single user at a particular time, so the two user partition function can be written as:

$$Z\left(g(t), h(t)\right) = Z\left(g(t)\right) * Z\left(h(t)\right) \quad (23)$$

The mixed partial derivative of this partial derivative with respect to g(t),h(t) then yields the correlation between the users corresponding to those points:

$$Corr\left(g(t), h(t)\right) = \partial^2 Z\left(g(t), h(t)\right) / \partial g(t) \partial h(t) \quad (24)$$

Finding the zeros of this equation between all pairwise comparisons on the graph yields boundaries separating different user classes. In the beginning of the simulated process, the "inverse temperature" parameter β is close to 0 (analogous to high temperature in physical systems). This yields a two point correlation function that is zero almost everywhere. As the simulated annealing process evolves and β increases, two classes of users form from the boundary that forms on the graph, creating two distinct regions. The boundary separates users that focus in a single topic area from those that periodically visit multiple, but related topic areas. Data to support these conclusions are provided in the Findings section below.

Findings and Conclusions

Using the Tipster data set alluded to previously, and the model developed by the author, two distinct user classes have been discovered in a simulated set of 1000 users. These classes emerge at a critical beta value that occurs during the simulated annealing process. One of the classes consists of users who focus on a particular topic, while the other consists of users that periodically hop between different topics. The users in both classes started from an initial state in which they received random content, and were producing random ratings (Table 1).

The data below are from the final 100 users in the data set, after the critical value of beta was reached and the graph was close to equilibrium. Of these users, 60 were single topic focused users, split evenly between the computer operating systems content cluster and computer windows operating systems content cluster. The remaining 40 users consisted of two subgroups. There were 30 users who hopped periodically between hockey, encryption, and medicine, and 10 users who hopped periodically between guns and religion. But both subgroups demonstrated a periodic time dependence, hence they were classified as the same user group although the content that they sampled was different. Because of the simple types of time dependence (static and periodic), it is practical and relatively simple to classify new users in the system, based on their CR divergence, which is shown in the graph below (Figure 1).

The cells in the chart above refer to the cluster visited by a user at a particular time. The numbers are in the order referenced in the Data Sets section above. Note the periodic pattern in the last 4 rows.

The user groups referred to in the CR divergence graph correspond to the rows from top to bottom in the equilibrium visitor trajectories chart. Note that there is a clear distinction between user groups 1-6 that have a CR divergence different than zero, and positive, while groups 7-10 have a CR divergence statistically equivalent to zero. This demonstrates the two phases of user classification alluded to previously.

A comparison of the performance of this model to related models mentioned in the Related Work section is difficult due to the use of simulated users in this work, even if the same content collection (the Tipster data set collection) was used.

3	3	3	3	3	3	3	3	3	3
3	3	3	3	3	3	3	3	3	3
3	3	3	3	3	3	3	3	3	3
6	6	6	6	6	6	6	6	6	6
6	6	6	6	6	6	6	6	6	6
6	6	6	6	6	6	6	6	6	6
12	11	14	12	11	14	12	11	14	12
14	12	11	14	12	11	14	12	11	14
20	17	20	17	20	17	20	17	20	17
11	14	12	11	14	12	11	14	12	11

Table 1: Clusters visited by members of the user groups (rows are User Groups).

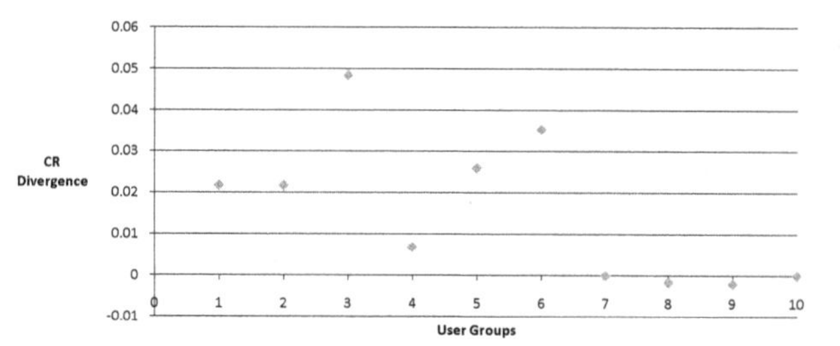

Figure 1: CR Divergence vs. User Groups.

Future Works

The scope of future work is twofold at this point. First, we will further study how the algorithm and grouping of users evolves with greater scale of users and content, and with increased content and user diversity. The second focus will be to study what happens more closely at the critical value of beta, where the transition into clusters actually takes place, to better understand that process.

References

1. Guadawardana A, Meek C (2009) A Unified Approach to Building Hybrid Recommendation Systems. Association for Computing Machinery, Inc.
2. Basu C, Hirsh H, Cohen W (1998) Recommendation as Classification: Using social and content -based information in recommendation. AAAI.
3. Goldenfeld N (1992) Lectures on Phase Transitions and the Renormalization Group. Addison-Wesley Publishing Company.

The Relations between Characterized Fuzzy Proximity, Fuzzy Compact, Fuzzy Uniform Spaces and Characterized Fuzzy T_s – Spaces and Fuzzy R_k – Spaces

Abd-Allah AS[1] and Al-Khedhairi A[2*]

[1]Department of Mathematics, College of Science, El-Mansoura University, El-Mansoura, Egypt
[2]Department of Statistics and Operations Research, College of Science, King Saud University, PO Box 2455, Riyadh 11451, Saudi Arabia

Abstract

In this research work, we study the relations between the characterized fuzzy T_s–spaces and characterized fuzzy R_k–spaces presented in old papers, for $s \in \{0,1,2,3,3\frac{1}{2},4\}$ and $k \in \{1,2,2\frac{1}{2},3\}$ and the characterized fuzzy proximity spaces presented. We also study the relations between the characterized fuzzy T_s–spaces, the characterized fuzzy R_k–spaces and the characterized fuzzy compact spaces which is presented in old paper, as a generalization of the weaker and stronger forms of the G–compactness defined by Gähler. Moreover, we show here the relations between these characterized fuzzy T_s–spaces, characterized fuzzy R_k–spaces and the characterized fuzzy uniform spaces introduced and studied by Abd-Allah in 2013 as a generalization of the weaker and stronger forms of the fuzzy uniform spaces introduced by Gähler.

Keywords: Fuzzy filter; Fuzzy topological space; Operationsl; Isotone and idempotent; Characterized fuzzy space; $\varphi_{1,2}$–fuzzy neighborhood filters; Fuzzy uniform structure; Characterized fuzzy proximity space; Characterized fuzzy compact space; Characterized fuzzy uniform space; Characterized FT_s–space; $F\varphi_{1,2}$–T_s space; Characterized FR_k–space and $F\varphi_{1,2}$–R_k space for $s \in \{0,1,2,3,3\frac{1}{2},4\}$; $k \in \{1,2,2\frac{1}{2},3\}$

Introduction

The notion of fuzzy filter has been introduced by Eklund et al. By means of this notion the point-based approach to fuzzy topology related to usual points has been developed. The more general concept for fuzzy filter introduced by Gähler [1] and fuzzy filters are classified by types. Because of the specific type of fuzzy filter however the approach of Eklund is related only to the fuzzy topologies which are stratified, that is, all constant fuzzy sets are open. The more specific fuzzy filters considered in the former papers are now called homogeneous. The operation on the ordinary topological space (X,T) has been defined by Kasahara [2] as the mapping φ from T into 2^X such that $A \subseteq A^\varphi$, for all $A \in T$. In 1983, Abd El-Monsef et al. [3] extend Kasahara operation to the power set $P(X)$ of a set X. In 1999, Kandil [4] and the author extended Kasahars's and Abd El-Monsef's operations by introducing an operation on the class of all fuzzy subsets endowed with an fuzzy topology τ as the mapping $\varphi : L^X \to L^X$ such that $int\, \mu \le \mu^\varphi$ for all $\mu \in L^X$, where μ^φ denotes the value of φ at μ.

The notions of the fuzzy filters and the operations on the class of all fuzzy subsets on X endowed with a fuzzy topology τ are applied by Abd-Allah in [5-7] to introduce a more general theory including all the weaker and stronger forms of the fuzzy topology. By means of these notions the notion of $\varphi_{1,2}$-fuzzy interior of a fuzzy subset, $\varphi_{1,2}$-fuzzy convergence and $\varphi_{1,2}$-fuzzy neighborhood filters are defined and applied to introduced many special classes of separation axioms. The notion of $\varphi_{1,2}$-interior operator for a fuzzy subset is defined as a mapping $\varphi_{1,2}.int: L^X \to L^X$ which fulfill (I1) to (I5) in Abd-Allah [5]. There is a one-to-one correspondence between the class of all $\varphi_{1,2}$-open fuzzy subsets of X and these operators, that is, the class $\varphi_{1,2}OF(X)$ of all $\varphi_{1,2}$-open fuzzy subsets of X can be characterized by these operators. Then the triple $(X,\varphi_{1,2}.int)$ as will as the triple $(X,\varphi_{1,2}OF(X))$ will be called the characterized fuzzy space [5] of $\varphi_{1,2}$-open fuzzy subsets. The characterized fuzzy spaces are identified by many of characterizing notions in Abd-Allah [5-7], for example by the $\varphi_{1,2}$-fuzzy neighborhood filters, $\varphi_{1,2}$-fuzzy interior of the fuzzy filters and by the set of $\varphi_{1,2}$-inner points of the fuzzy filters. Moreover, the notions of closeness and compactness in the characterized fuzzy spaces are introduced and studied by Abd-Allah in [7]. The notions of characterized FT_s-spaces, $F\varphi_{1,2}$-T_s spaces, characterized FR_k-spaces and $F\varphi_{1,2}$-R_k spaces are introduced and studied in Abd-Allah [9-11] for all $s \in \{0,1,2,2\frac{1}{2},3,3\frac{1}{2},4\}$ and $k \in \{0,1,2,2\frac{1}{2},3\}$. The notions of characterized fuzzy compact spaces, characterized fuzzy proximity spaces and characterized fuzzy uniform spaces are introduced and studied by the author in 2004 and 2013 in [7,12]. This paper is devoted to introduce and study the relations between the characterized FT_s and FR_k-spaces, for $s \in \{0,1,2,3,3\frac{1}{2},4\}$ and $k \in \{1,2,2\frac{1}{2},3\}$, the characterized fuzzy proximity spaces and the characterized fuzzy compact spaces. Moreover, we show here the relations between these characterized FT_s and FR_k-spaces and the characterized fuzzy uniform spaces. In section 2, some definitions and notions related to the fuzzy subsets, fuzzy topologies, fuzzy filters, fuzzy proximity, operations on fuzzy subsets, $\varphi_{1,2}$-fuzzy neighborhood filters, characterized fuzzy space, characterized FT_s-spaces, $F\varphi_{1,2}$-T_s spaces, characterized FR_k-spaces and $F\varphi_{1,2}$-R_k spaces are given for $s \in \{0,1,2,3,3\frac{1}{2},4\}$ and $k \in \{2,2\frac{1}{2},3\}$. Section 3, is devoted to

***Corresponding author:** Al-Khedhairi A, Mathematics Department, College of Science and Humanity Studies, Prince Sattam Bin Abdul-Aziz University, PO Box 132012, Hotat Bani Tamim 11941, Saudi Arabia, E-mail: akhediri@ksu.edu.sa

introduce and study the relation between the characterized fuzzy proximity spaces and our classes of the characterized FT_s-spaces and of the characterized FR_k-spaces. It will be shown that in the characterized fuzzy space $(X,\varphi_{1,2}.\text{int})$, the fuzzy proximity δ will be identified with the finer relation on the $\varphi_{1,2}$-fuzzy neighborhood filters. Also, we will show that any fuzzy proximity is separated if and only if the associated characterized fuzzy proximity space is characterized FT_0 and to each fuzzy proximity is associated a characterized FR_2-space in our sense. Generally, it will be shown that the associated characterized fuzzy proximity space $(X,\varphi_{1,2}.\text{int}_\delta)$ is characterized FR_2-space if the related fuzzy topological space (X,τ) is $F\varphi_{1,2}$-R_2 space. Moreover, for each characterized FR_3-space the binary relation on L^X defined by means the $\varphi_{1,2}$-fuzzy closure operator $\varphi_{1,2}.cl$ of τ in Equation (3.6), is fuzzy proximity on X and conversely to each fuzzy proximity δ which has a $\varphi_{1,2}$-fuzzy closure operator fulfills the binary relation given in (3.6) is associated a characterized FR_3-space $(X,\varphi_{1,2}.\text{int}_\delta)$. Moreover, when L is a complete chain, $\varphi_2 \geq 1_{L^X}$ is isotone and φ_1 is wfip with respect to φ_1 OF (X), then we show that the associated characterized fuzzy space $(X,\varphi_{1,2}.\text{int}_\tau)$ from the fuzzy normal topological space (X,τ) is finer than the associated characterized fuzzy proximity space $(X,\varphi_{1,2}.\text{int}_\delta)$ by the fuzzy proximity δ defined by (3.6) and they identical if and only if $(X,\varphi_{1,2}.\text{int}_\tau)$ is characterized FT_4-space. At the end of this section we prove that the associated characterized fuzzy proximity space $(X,\varphi_{1,2}.\text{int}_\delta)$ is characterized $FR_{2\frac{1}{2}}$- space and therefore it is characterized $FT_{3\frac{1}{2}}$- space. There is a good notion of $\varphi_{1,2}$-fuzzy compactness of the fuzzy filters and of the fuzzy topological spaces introduced and studied by Abd-Allah et al. [7]. This notion fulfills many properties, for example, it fulfills the Tychonoff Theorem. In section 4, we used this notion to study the relations between the characterized fuzzy compact spaces and our classes of the characterized FT_s-spaces and of the characterized FR_k-spaces. It will be shown that every $\varphi_{1,2}$-closed subset of a characterized fuzzy compact space is $\varphi_{1,2}$-fuzzy compact and each $\varphi_{1,2}$-fuzzy compact subset of the characterized FT_2-space is $\varphi_{1,2}$-closed. Also, it will be shown that each characterized fuzzy compact FT_2-space is characterized FT_4-space. Specially, we prove that the characterized fuzzy unit interval space $(I_L,\psi_{1,2}.\text{int}_I)$ is characterized fuzzy compact FT_2-space and characterized $FT_{3\frac{1}{2}}$- space. Generally, we show that every characterized fuzzy compact space is characterized FT_2-space if and only if it is characterized $FT_{3\frac{1}{2}}$- space. We show that, if $(X,\psi_{1,2}.\text{int}_\sigma)$ is characterized fuzzy compact space finer than the characterized FT_2-space $(X,\varphi_{1,2}.\text{int}_\tau)$, then $(X,\varphi_{1,2}.\text{int}_\tau)$ is $\varphi_{1,2}\psi_{1,2}$-fuzzy isomorphic to $(X,\psi_{1,2}.\text{int}_\sigma)$. Moreover, if τ is finer than σ, $(X,\varphi_{1,2}.\text{int}_\tau)$ is characterized fuzzy compact space and $(X,\psi_{1,2}.\text{int}_\sigma)$ is characterized $FT_{3\frac{1}{2}}$- space, then $(X,\psi_{1,2}.\text{int}_\tau)$ and $(X,\psi_{1,2}.\text{int}_\sigma)$ are $\varphi_{1,2}\psi_{1,2}$-fuzzy isomorphic. The notion of fuzzy uniform structure had been introduced and studied by Gähler et al. [13]. This notion with the notion of the operations on the class of all fuzzy subsets are applied to introduce and study the notion of characterized fuzzy uniform spaces. In section 5, we introduce and study the relations between the characterized fuzzy uniform spaces and our classes of the characterized FT_s-spaces and of the characterized FR_k-spaces. We show that the fuzzy uniform space $(X,\mathcal{U})$ is separated if and only if the associated characterized fuzzy uniform space $(X,\varphi_{1,2}.\text{int}_{\mathcal{U}})$ is characterized FT_i-space but the fuzzy uniform space $(X,\mathcal{U})$ is separated if and only if the associated stratified fuzzy topological space $(X,\tau_{\mathcal{U}})$ is $F_{1,2}$-T_i space for all $i\in\{0,1\}$. For each fuzzy uniform structure on a set X, we prove that there is an induced stratified fuzzy proximity on L^X. Moreover, both the fuzzy uniform structure and this induced stratified fuzzy proximity are associated with the same stratified characterized fuzzy uniform space. Finally, for each fuzzy uniform space $(X,\mathcal{U})$ we prove that the associated stratified characterized fuzzy uniform space $(X,\varphi_{1,2}.\text{int}_{\mathcal{U}})$ with the fuzzy uniform structure $\mathcal{U}$ is characterized $FR_{2\frac{1}{2}}$- space and it is characterized $FT_{3\frac{1}{2}}$- space if $(X,\mathcal{U})$ is separated.

Preliminaries

We begin by recalling some facts on fuzzy subsets and on fuzzy filters. Let L be a completely distributive complete lattice with different least and last elements 0 and 1, respectively. Let $L_0 = L \setminus \{0\}$. Sometimes we will assume more specially that L is a complete chain, that is, L is a complete lattice whose partial ordering is a linear one. For a set X, let L^X be the set of all fuzzy subsets of X, that is, of all mappings $f: X \to L$. Assume that an order-reversing involution $\alpha \mapsto \alpha'$ of L is fixed. For each fuzzy subset $\mu \in L^X$, let μ' denote the complement of μ and it is given by the relation $\mu'(x) = \mu(x)'$ for all $x \in X$. Denote by $\bar{\alpha}$, the constant fuzzy subset of X with value is $\alpha \in L$. For all $x \in X$ and for all $\alpha \in L_0$, the fuzzy subset x_α of X whose value α at x and 0 otherwise is called a fuzzy point in X. The set of all fuzzy points of a set X will be denoted by $S(X)$.

Fuzzy filters

The fuzzy filter on X [1] is the mapping $M: L^X \to L$ such that the following conditions are fulfilled:

(F1) $\mathcal{M}(\bar{\alpha}) \leq \alpha$ for all $\alpha \in L$ and $\mathcal{M}(\bar{1}) = 1$.

(F2) $\mathcal{M}(\mu \wedge \rho) = \mathcal{M}() \wedge \mathcal{M}(\rho)$ for all $\mu,\rho \in L^X$.

The fuzzy filter $\mathcal{M}$ is called homogeneous [14] if $\mathcal{M}(\bar{\alpha}) = \alpha$ for all $\alpha \in L$. For each $x \in X$, the mapping $\dot{x}(\mu) = \mu(x)$ defined by $\dot{x}(\mu) = \mu(x)$ for all $\mu \in L^X$ is a homogeneous fuzzy filter on X. For each $\mu \in L^X$, the mapping $\dot{\mu}: L^X \to L$ defined by $\dot{\mu}(\eta) = \bigwedge_{0<\eta(x)} \eta(x)$ for all $\eta \in L^X$ is a homogeneous fuzzy filter on X, called homogenous fuzzy filter at the fuzzy subset $\mu \in L^X$. Let $\mathcal{F}_L X$ and $\mathcal{F}_L X$ be the sets of all fuzzy filters and of all homogeneous fuzzy filters on X, respectively. If $\mathcal{M}$ and $\mathcal{N}$ are fuzzy filters on a set X, $\mathcal{M}$ is said to be finer than $\mathcal{N}$, denoted by $\mathcal{M} \leq \mathcal{N}$, provided $\mathcal{M}(\mu) \geq \mathcal{N}(\mu)$ holds for all $\mu \in L^X$. Noting that if L is a complete chain then M is not finer than N, denoted by $\mathcal{M}$ $\mathcal{N}$, provided there exists $\mu \in L^X$ such that $\mathcal{M}(\mu) < \mathcal{N}(\mu)$ holds.

Lemma 2.1

If $\mathcal{M}$, $\mathcal{N}$ and $\mathcal{L}$ are fuzzy filters on a set X. Then the following sentences are fulfilled [1].

$$\mathcal{M} \neq \mathcal{L} \geq \mathcal{N} \text{ implies } \mathcal{M} \neq \mathcal{N} \text{ and } \mathcal{M} \geq \mathcal{L} \neq \mathcal{N} \text{ implies } \mathcal{M} \neq \mathcal{N}$$

Proposition 2.1

For all $\mu,\rho \in L^X$, we have $\mu \leq \rho$ if and only if $\dot{\mu} \leq \dot{\rho}$ [15].

For each non-empty set $\mathcal{A}$ of the fuzzy filters on X the supremum $\bigvee_{\mathcal{M}\in\mathcal{A}} \mathcal{M}$ exists [1] and given by

$$\Big(\bigvee_{\mathcal{M}\in\mathcal{A}} \mathcal{M}\Big)(\mu) = \bigwedge_{\mathcal{M}\in\mathcal{A}} \mathcal{M}(\mu),$$

for all $\mu \in L^X$. Whereas the infimum $\bigwedge_{\mathcal{M}\in\mathcal{A}} \mathcal{M}$ of A does not exists in general as an fuzzy filter. If the infimum $\bigwedge_{\mathcal{M}\in\mathcal{A}} \mathcal{M}$ exists, then we have

$$\Big(\bigwedge_{\mathcal{M}\in\mathcal{A}} \mathcal{M}\Big)(\mu) = \bigvee_{\substack{\mu_1\wedge\cdots\wedge\mu_n \leq \mu,\\ \mathcal{M}_1,\ldots,\mathcal{M}_n \in \mathcal{A}}} (\mathcal{M}_1(\mu_1) \wedge \cdots \wedge \mathcal{M}_n(\mu_n)),$$

for all $\mu \in L^X$, where n is an positive integer, $\mu_1,\ldots,\mu_n$ is a collection such that $\mu_1 \wedge \ldots \wedge \mu_n \leq \mu$ and $\mathcal{M}_1,\ldots,\mathcal{M}_n$ are fuzzy filters from $\mathcal{A}$. Let X be a set and $\mu \in L^X$, then the homogeneous fuzzy filter $\dot{\mu}$ at μ is the fuzzy filter on X given by:

$$\dot{\mu} = \bigvee_{0<\mu(x)} \dot{x}, \tag{2.1}$$

Fuzzy filter bases

The family $(\beta_\alpha)\alpha\in(\mathscr{B}_\alpha)_{\alpha\in L_0}$ of a non-empty subsets of L^X is called a valued fuzzy filter base [1] if the following conditions are fulfilled:

(V1) $\in\beta_\alpha$ implies $\alpha\leq\sup\mu$.

(V2) For all α,β L_0 with $\alpha\wedge\beta\in L_0$ and all $\in\beta_\alpha$ and $\rho\in\beta_\beta$ there are $\gamma\geq\alpha\wedge\beta$ and $\eta\leq\mu$ ρ such that $\eta\in\beta_\gamma$.

As shown in Gähler [1], each valued fuzzy filter base $(\mathscr{B}_\alpha)_{\alpha\in L_0}$ defines a fuzzy filter $\mathcal{M}$ on X by $\mathcal{M}(\mu)=\bigvee_{\rho\in\mathscr{B}_\alpha,\rho\leq\mu}\alpha$ for all $\mu\in L^X$. Conversely, each fuzzy filter M can be generated by a valued fuzzy filter base, e.g. by $(\alpha-\text{pr}\mathcal{M})_{\alpha\in L_0}$ with α-pr $\mathcal{M}=\{\mu\in L^X \mid \alpha\leq\mathcal{M}(\mu)\}$. The $(\alpha\text{-pr}\,\mathcal{M})_{\alpha\in L_0}$ is the family of pre filters on X and it is called the large valued fuzzy filter base of $\mathcal{M}$. Recall that the pre filter on X [16] is a non-empty proper subset $\mathcal{F}$ of L^X such that (1) $\mu,\rho\in\mathcal{F}$ implies $\mu\wedge\rho\in\mathcal{F}$ and (2) from $\mu\in\mathcal{F}$ and $\mu\leq$ it follows $\rho\in\mathcal{F}$.

Valued and superior principal fuzzy filters

Let a non-empty set X be fixed, $\mu\in L^X$ and $\alpha\in L$ such that $\alpha\leq\sup\mu$, the valued principal fuzzy filter [20] generated by μ and α, will be denoted by $[\mu,\alpha]$, is the fuzzy filter on X which has $(\mathscr{B}_\beta)_{\beta\in L_0}$ with $\beta_\beta=\{\mu\}$ if $0\leq\beta\leq\alpha$ and $\mathscr{B}_\beta=\{\bar{1}\}$ otherwise as a valued fuzzy filter base. For all $\eta\in L^X$, we have $[\mu,\alpha](\eta)=0$ if $\mu\nleq\eta$, $[\mu,\alpha](\eta)=\alpha$ if $\mu\leq\eta\neq\bar{1}$ and $[\mu,\alpha](\eta)=1$ if $\eta=\bar{1}$. Moreover, for each $\beta\in L_0$ we have β-pr$[\mu,\alpha]=\{\eta \mid \mu\leq\eta\}$ if $\beta\leq\alpha$ and β-pr$[\mu,\alpha]=\{\bar{1}\}$ otherwise. The superior principal fuzzy filter [1] generated by μ, written $[\mu]$, is the homogeneous fuzzy filter on X which has $\beta=\{\mu\wedge\bar{\alpha} \mid \alpha\in L\}\cup\{\bar{\alpha} \mid \alpha\in L\}$ as a superior fuzzy filter base. As shown in Katsaras [18], the superior principal fuzzy filter $[\mu]$ is representable by a fuzzy pre filter if and only if $\sup\mu=1$.

Fuzzy filter functors and fuzzy filter monads

The fuzzy filter functor $\mathcal{F}_L$: $SET\to SET$ is the covariant functor from the category SET of all sets to this category which assigns to each set X the set $\mathcal{F}_L X$ and to each mapping $f: X^{\circ}Y$ the mapping $\mathcal{F}_L f:\mathcal{F}_L X\to\mathcal{F}_L Y$. The homogeneous fuzzy filter functor F_L: $SET\to SET$ is the sub fuzzy filter functor of F_L which assigns to each set X the set $F_L X$ and to each mapping f $:X^{\circ}Y$ the domain-range restriction $F_L f$:$F_L X\to F_L Y$ of the mapping $\mathcal{F}_L f$:$\mathcal{F}_L X\to\mathcal{F}_L Y$. For each set X, let η_X:$X\to\mathcal{F}_L X$ be the mapping defined by $\eta_X(x)=\dot{x}$ for all $x\in X$, and let $e_X: L^X\to L^{\mathcal{F}_L X}$ be the mapping for which $e_X(\mu)(M)=M(\mu)$ for all μ L^X and $M\in F_L X$. Moreover, let $\mu_X:\mathcal{F}_L(\mathcal{F}_L X)\to\mathcal{F}_L X$ be the mapping which assigns to each fuzzy filter $\mathcal{L}$ on $\mathcal{F}_L X$ the fuzzy filter $\mu_X(\mathcal{L})=\mathcal{L}\circ e_X$ on X. $\eta=(\eta_X)_{X\in\text{Ob}(SET)}:\text{id}\to\mathcal{F}_L$ with id the identity set functor and $\mu=(\mu_X)_{X\in\text{Ob}(SET)}:\mathcal{F}_L\circ\mathcal{F}_L\to\mathcal{F}_L$ are natural transformations. $(\mathcal{F}_L,\eta,\mu)$ is a monad in the categorical sense, called the fuzzy filter monad [1], that is, $\mu_X\circ\mathcal{F}_L(\eta_X)=\mu_X\circ\eta_{\mathcal{F}_L X}=1_{\mathcal{F}_L X}$ and $\mu_X\circ\mathcal{F}_L(\mu_X)=\mu_X\circ\mu_{\mathcal{F}_L X}$ for each set X. Related to the sub functor F_L of $\mathcal{F}_L$, there are analogous natural transformations as η and μ, denoted η' and μ', respectively. η' consists of the range-restrictions $\eta'_X:X\to F_L X$ of the mappings η_X. μ' is the family of all mappings $\mu'_X:F_L(F_L X)\to F_L X$ defined by $\mu'_X(\mathcal{L})=\mathcal{L}\circ e'_X$ for all homogeneous fuzzy filters L on $F_L X$, where $e'_X:L^X\to L^{F_L X}$ is the mapping given by $e'_X(\mu)(\mathcal{M})=\mathcal{M}(\mu)$ for all $\mu\in L^X$ and $\mathcal{M}\in F_L X$. As has been shown in Gähler et al. [13], (F_L,η',μ') is a sub monad of (F_L,η,μ) $(\mathcal{F}_L,\eta,\mu)$, that is, for the inclusion mappings $i_X:F_L X\to\mathcal{F}_L X$ we have $\eta_X=i_X\circ\eta'_X$ and $\mu_X\circ\mathcal{F}_L i_X\circ i_{F_L X}=i_X\circ\mu'_X$ for all sets X.

Fuzzy topologies

By a fuzzy topology on a set X [20,21], we mean a subset of L^X which is closed with respect to all suprema and all finite infima and contains the constant fuzzy sets $\bar{0}$ and $\bar{1}$. A set X equipped with an fuzzy topology τ on X is called fuzzy topological space. For each fuzzy topological space (X,τ), the elements of τ are called open fuzzy subsets of this space. If τ_1 and τ_2 are two fuzzy topologies on a set X, then τ_2 is said to be finer than τ_1 and τ_1 is said to be coarser than τ_2, provided $\tau_1\subseteq\tau_2$ holds. The fuzzy topological space (X,τ) and also τ are said to be stratified provided $\bar{\alpha}\in$ holds for all $\alpha\in L$, that is, all constant fuzzy subsets are open [17].

Fuzzy proximity spaces

The binary relation δ on L^X is called fuzzy proximity on X [18], provided it fulfill the following conditions:

(P1) $\mu\,\bar{\delta}\,\rho$ implies $\rho\,\bar{\delta}\,\mu$ for all $\mu,\rho\in L^X$, where $\bar{\delta}$ is the negation of δ.

(P2) $(\mu\vee\rho)\,\bar{\delta}\,\eta$ if and only if $\mu\,\bar{\delta}\,\eta$ and $\rho\,\bar{\delta}\,\eta$ for all $\mu,\rho,\eta\in L^X$.

(P3) $\mu=\bar{0}$ or $\rho=\bar{0}$ implies $\mu\,\bar{\delta}\,\rho$ for all $\mu,\rho\in L^X$.

(P4) $\mu\,\bar{\delta}\,\rho$ implies $\mu\leq\rho'$ for all $\mu,\rho\in L^X$.

(P5) If $\mu\,\bar{\delta}\,\rho$, then there is an $\eta\in L^X$ such that $\mu\,\bar{\delta}\,\eta$ and $\eta'\,\bar{\delta}\,\rho$.

The set X equipped with an fuzzy proximity δ on X is said to be fuzzy proximity space and will be denoted by (X,δ). Every fuzzy proximity δ on a set X is associated an fuzzy topology on X denoted by τ_δ. The fuzzy proximity δ on a set X is said to be separated if and only if for all $x,y\in X$ such that $x\neq y$ we have $x_\alpha\,\bar{\delta}\,y_\beta$ for all $\alpha,\beta\in L_0$.

Operation on fuzzy sets

In the sequel, let a fuzzy topological space (X,τ) be fixed. By the operation [4] on a set X, we mean the mapping $\varphi: L^X\to L^X$ such that $int\,\mu\leq\mu^\varphi$ holds, for all $\mu\in L^X$, where μ^φ denotes the value of φ at μ. The class of all operations on X will be denoted by $O_{(L^X,\tau)}$. By the identity operation on $O_{(L^X,\tau)}$, we mean the operation $1_{L^X}:L^X\to L^X$ such that $1_{L^X}(\mu)=\mu$ for all $\mu\in L^X$. Also by the constant operation on $O_{(L^X,\tau)}$, we mean the operation $c_{L^X}:L^X\to L^X$ such that $c_{L^X}(\mu)=\bar{1}$ for all $\mu\in L^X$. If $\leq$ is a partially ordered relation on $O_{(L^X,\tau)}$ defined as by $\varphi_1\leq\varphi_2\Leftrightarrow\mu^{\varphi_1}\leq\mu^{\varphi_2}$ for all $\mu\in L^X$, then obviously, $(O_{(L^X,\tau)},\leq)$ is a completely distributive lattice. As an application on this partially ordered relation, the operation φ:$L^X\to L^X$ will be called:

(i) Isotone if $\mu\leq\rho$ implies $\mu^\varphi\leq\rho^\varphi$ holds, for all $\mu,\rho\in L^X$.

(ii) Weakly finite intersection preserving (wfip, for short) with respect to $\mathcal{A}\subseteq L^X$ if $\rho\wedge\mu^\varphi\leq(\rho\wedge\mu)^\varphi$ holds, for all $\rho\in A$ and $\mu\in L^X$.

(iii) Idempotent if $\mu^\varphi=(\mu^\varphi)$, for all $\mu\in L^X$.

The operations $\varphi,\psi\in O_{(L^X,\tau)}$ are said to be dual if $\mu=co\,((co\mu))$ or equivalently $\varphi\mu=co\,(\psi\,(co\mu))$ for all $\mu\in L^X$, where $co\mu$ denotes the complement of μ. The dual operation of φ is denoted by $\tilde{\varphi}$. In the classical case of $L=\{0,1\}$, by the operation on the set X [3], we mean the mapping $\varphi: P(X)\to P(X)$ such that $int\,A\leq A^\varphi$ for all A in the power set $P(X)$ and the identity operation on the class of all ordinary operations $O_{(P(X),T)}$ on X will be denoted by $i_{P(X)}$, where $i_{P(X)}(A)=A$ for all $A\in P(X)$.

φ-open fuzzy subsets

Let a fuzzy topological space (X,τ) be fixed and $\varphi \in O_{(L^X,\tau)}$. The fuzzy subset $\mu : X \to L$ is said to be φ-open fuzzy subset if $\mu \le \mu^{\varphi}$ holds. We will denote the class of all φ-open fuzzy subsets on X by OF (X). The fuzzy subset μ is called φ-closed if its complement $co\mu$ is φ-open. The two operations $\varphi, \psi \in O_{(L^X,\tau)}$ are equivalent and written $\varphi \sim \psi$ if and only if φ OF $(X) = \psi$ OF (X).

$\varphi_{1,2}$–interior of fuzzy subsets

Let a fuzzy topological space (X,τ) be fixed and $\varphi_1, \varphi_2 \in O_{(L^X,\tau)}$. Then the $\varphi_{1,2}$-interior of the fuzzy subset μ: $X \to L$ is the mapping $\varphi_{1,2}$.int $\mu : X \to L$ defined by:

$$\varphi_{1,2}.\text{int}\,\mu = \bigvee_{\rho \in \varphi_1 OF(X),\, \rho^{\varphi_2} \le \mu} \rho, \tag{2.2}$$

As easily seen that $\varphi_{1,2}$.int μ is the greatest φ_1-open fuzzy subset ρ such that ρ^{φ_2} less than or equal to μ [5]. The fuzzy subset μ is said to be $\varphi_{1,2}$-open if $\mu \le \varphi_{1,2}$*.int* μ. The class of all $\varphi_{1,2}$-open fuzzy subsets of X will be denoted by $\varphi_{1,2}$ $OF(X)$. The complement $co\mu$ of a $\varphi_{1,2}$-open fuzzy subset μ will be called $\varphi_{1,2}$-closed and the class of all $\varphi_{1,2}$-closed fuzzy subsets of X will be denoted by $\varphi_{1,2}CF(X)$. In the classical case of $L = \{0,1\}$, we note that the fuzzy topological space (X,τ) is up to an identification by the ordinary topological space (X,T) and $\varphi_{1,2}$.int μ is the classical one. Hence, in this case the ordinary subset A of X is $\varphi_{1,2}$-open if A $\subseteq \varphi_{1,2}$.int A. The complement of the $\varphi_{1,2}$-open subset A of X will be called $\varphi_{1,2}$-closed. The class of all $\varphi_{1,2}$-open and the class of all $\varphi_{1,2}$-closed subsets of X will be denoted by $\varphi_{1,2}O$ (X) and $\varphi_{1,2}C$ (X), respectively. Clearly, F is $\varphi_{1,2}$-closed if and only if $\varphi_{1,2}.\text{cl}_T\, F = F$.

Proposition 2.2

If (X,τ) is an fuzzy topological space and $\varphi_1, \varphi_2 \in O_{(L^X,\tau)}$. Then, the mapping $\varphi_{1,2}$*.int* $\mu : X \to L$ fulfills the following axioms [5]:

(i) If $\varphi_2 \ge 1_{L^X}$, then $\varphi_{1,2}$.int $\mu \le \mu$ holds.

(ii) $\varphi_{1,2}$.int is isotone operator, that is, if $\mu \le \rho$ then, $_{1,2}$.int $\mu \le \varphi_{1,2}$.int ρ holds for all $\mu, \rho \in L^X$.

(iii) $\varphi_{1,2}.\text{int}\bar{1} = \bar{1}$.

(iv) If $\varphi_2 \ge 1_{L^X}$ is isotone and φ_1 is wfip with respect to φ_1 OF (X), then $_{1,2}$.int $(\mu \wedge \rho) = \varphi_{1,2}$.int $\mu \wedge \varphi_{1,2}$.int ρ for all $\mu, \rho \in L^X$.

(v) If $_2$ is isotone and idempotent operation, then $\varphi_{1,2}$.int $\mu \le \varphi_{1,2}$.int $(\varphi_{1,2}$.int $\mu)$ holds.

(vi) $\varphi_{1,2}.\text{int}(\bigvee_{i \in I} \mu_i) = \bigvee_{i \in I} \varphi_{1,2}.\text{int}\,\mu_i$ for all $\mu_i \in \varphi_{1,2}$ OF (X).

Proposition 2.3

Let (X,τ) be an fuzzy topological space and $\varphi_1, \varphi_2 \in O_{(L^X,\tau)}$. Then the following are fulfilled [5]:

(i) If $\varphi_2 \ge 1_{L^X}$, then the class $\varphi_{1,2}$ $OF(X)$ forms extended fuzzy topology on X [19].

(ii) If $\varphi_2 \ge 1_{L^X}$ and $\varphi_{1,2}.\text{int}\bar{1} = \bar{1}$, then the class $\varphi_{1,2}$ $OF(X)$ forms a supra fuzzy topology on X [19].

(iii) If $\varphi_2 \ge 1_{L^X}$ is isotone and φ_1 is wfip with respect to $\varphi_1 OF(X)$, then $\varphi_{1,2}$ $OF(X)$ is fuzzy pre topology on X [19].

(iv) If $\varphi_2 \ge 1_{L^X}$ is isotone and idempotent operation and φ_1 is wfip with respect to $\varphi_1 OF(X)$, then $\varphi_{1,2}$ $OF(X)$ is a fuzzy topology on X [20,21].

From Propositions 2.2 and 2.3, if the fuzzy topological space (X,τ) be fixed and $\varphi_1, \varphi_2 \in O_{(L^X,\tau)}$. Then

$$\varphi_{1,2}OF(X) = \{\mu \in L^X \mid \mu \le \varphi_{1,2}.\text{int}\mu\}, \tag{2.3}$$

and the following conditions are fulfilled:

(I1) If $\varphi_2 \ge 1_{L^X}$, then $\varphi_{1,2}$.int $\mu \le \mu$ holds, for all $\mu \in L^X$.

(I2) If $\mu \le \rho$, then $\varphi_{1,2}$.int $\mu \le \varphi_{1,2}$.int ρ for all $\mu, \rho \in L^X$.

(I3) $\varphi_{1,2}.\text{int}\bar{1} = \bar{1}$.

(I4) If $\varphi_2 \ge 1_{L^X}$ is isotone and φ_1 is wfip with respect to $\varphi_1 OF(X)$, then $\varphi_{1,2}$.int $\mu \wedge \varphi_{1,2}$.int $\rho = \varphi_{1,2}$.int $(\mu \wedge \rho)$ for all $\mu, \rho \in L^X$.

(I5) If $_1$ is isotone and idempotent, then $\varphi_{1,2}$.int $(\varphi_{1,2}$.int $\mu) = \varphi_{1,2}$.int μ for all $\mu \in L^X$.

Characterized fuzzy spaces

Independently on the fuzzy topologies, the notion of $\varphi_{1,2}$-interior operator for fuzzy subsets can be defined as a mapping $\varphi_{1,2}$.int : $L^X \to L^X$ which fulfill (I1) to (I5). It is well-known that (2.2) and (2.3) give a one-to-one correspondence between the class of all $\varphi_{1,2}$-open fuzzy subsets and these operators, that is, $\varphi_{1,2}OF(X)$ can be characterized by the $\varphi_{1,2}$-interior operators. In this case the triple $(X,\varphi_{1,2}.\text{int})$ as will as the triple $(X,\varphi_{1,2}OF(X))$ will be called *characterized fuzzy space* [5] of the $\varphi_{1,2}$-open fuzzy subsets of X. For each characterized fuzzy space $(X,\varphi_{1,2}.int)$, the elements of $\varphi_{1,2}$ of (X) are called $\varphi_{1,2}$–open fuzzy subsets of this space. If $(X,\varphi_{1,2}.\text{int})$ and $(X,\psi_{1,2}.\text{int})$ are two characterized fuzzy spaces, then $(X,\varphi_{1,2}.\text{int})$ is said to be *finer than* $(X,\psi_{1,2}.\text{int})$ and denoted by $\varphi_{1,2}.\text{int} \le \psi_{1,2}.\text{int}$ provided $\varphi_{1,2}.\text{int}\,\mu \ge \psi_{1,2}.\text{int}\,\mu$ holds for all $\mu \in L^X$. The characterized fuzzy space $(X,\varphi_{1,2}.\text{int})$ is said to be *stratified* if and only if $\varphi_{1,2}.\text{int}\,\bar{\alpha} = \bar{\alpha}$ for all $\alpha \in L$. As shown in Abd-Allah [5], the characterized fuzzy space $(X,\varphi_{1,2}.\text{int})$ is stratified if the related fuzzy topology is stratified. Moreover, the characterized fuzzy space $(X,\varphi_{1,2}.\text{int})$ is said to have the *weak infimum property* [19] provided $\varphi_{1,2}.\text{int}\,(\mu \wedge \bar{\alpha}) = \varphi_{1,2}.\text{int}\,\mu \wedge \varphi_{1,2}.\text{int}\,\bar{\alpha}$ for all $\mu \in L^X$ and $\alpha \in L$. The characterized fuzzy space $(X,\varphi_{1,2}.\text{int})$ is said to be *strongly stratified* provided $\varphi_{1,2}$.int is stratified and have the weak infimum property.

Fuzzy unit interval

The fuzzy unit interval will be denoted by I_L and it is defined in Gähler [24] as the fuzzy subset $I_L = \{x \in \mathbb{R}_L^* \mid x \le 1^\sim\}$, where $I = [0,1]$ is the real unit interval and $\mathbb{R}_L^* = \{x \in \mathbb{R}_L \mid x(0) = 1 \text{ and } 0^\sim \le x\}$ is the set of all positive fuzzy real numbers. Note that, the binary relation $\le$ is defined on $\mathbb{R}_L$ as follows: $x \le y \Leftrightarrow x_{\alpha_1} \le y_{\alpha_1}$ and $x_{\alpha_2} \le y_{\alpha_2}$, for all x, $y \in \mathbb{R}_L$, where $x_{\alpha_1} = \inf\{z \in \mathbb{R} \mid x(z) \ge \alpha\}$ and $x_{\alpha_2} = \sup\{z \in \mathbb{R} \mid x(z) \ge \alpha\}$ for all $\alpha \in L_0$. Note that the family Ω which is defined by: $\Omega = \{R_\delta \mid I_L \mid \delta \in I\} \cup \{R^\delta \mid I_L \mid \delta \in I\} \cup \{0^\sim \mid I_L\}$ is a base for a fuzzy topology I on I_L and the order pair (I_L,I) is said to be fuzzy unit interval topological space, where R_δ and R^δ are the fuzzy subsets of $\mathbb{R}_L$ defined by $R_\delta(x) = \bigvee_{\alpha > \delta} x(\alpha)$ and $R^\delta(x) = (\bigvee_{\alpha \ge \delta} x(\alpha))'$ for all $x \in \mathbb{R}_L$ and $\delta \in \mathbb{R}$. \The restrictions of R_δ and R^δ on I_L are the fuzzy subsets $R^\delta| I_L$ and $R^\delta|I_L$, respectively. Recall that the inequality $R^\delta(x) \wedge R^\gamma(y) \le R^{\delta+\gamma}(x+y)$ holds, where $x + y$ is the fuzzy real number defined by: $(x+y)(\xi) = \bigvee_{\gamma,\zeta \in \mathbb{R},\, \gamma+\zeta=\xi} (x(\gamma) \wedge y(\zeta))$ for all $\xi \in \mathbb{R}$. Consider a fuzzy unit interval topological space (I_L,I) be given and $\psi_1, \psi_2 \in O_{(I_L,\Im)}$, then in this work the characterized fuzzy space $(I\ ,\psi\ .\text{int})$ will be called *characterized fuzzy unit interval space* and we define the cartesian product of a number of copies of the fuzzy unit interval I_L equipped with the product of the characterized fuzzy unit interval spaces generated by $\psi_{1,2}$.int$_1$ on it as a *characterized fuzzy cube*.

$\varphi_{1,2}$-fuzzy neighborhood filters

An important notion in the characterized fuzzy space $(X,\varphi_{1,2}.\text{int})$ is that of the $\varphi_{1,2}$-fuzzy neighborhood filter at the point and at the ordinary subset in this space. Let (X,τ) be a fuzzy topological space and $\varphi_1,\varphi_2 \in O_{(L^X,\tau)}$. As follows by (I1) to (I5) for each $x\in X$, the mapping $\mathcal{N}_{\varphi_{1,2}}(x) : L^X \to L$ which is defined by:

$$\mathcal{N}_{\varphi_{1,2}}(x)(\mu) = (\varphi_{1,2}.\text{int}\,\mu)(x)\,, \tag{2.4}$$

for all $\mu \in L^X$ is a fuzzy filter, called *$\varphi_{1,2}$-fuzzy neighborhood filter* at x [5]. If $\emptyset\neq F\in P(X)$, then the *$\varphi_{1,2}$-fuzzy neighborhood filter* at F will be denoted by $\mathcal{N}_{\varphi_{1,2}}(F)$ and it will be defined by:

$$\mathcal{N}_{\varphi_{1,2}}(F) = \bigvee_{x\in F} \mathcal{N}_{\varphi_{1,2}}(x).$$

Since $\mathcal{N}_{\varphi_{1,2}}(x)$ is fuzzy filter for all $x\in X$, then $\mathcal{N}_{\varphi_{1,2}}(F)$ is also fuzzy filter on X. Because of $[\chi_F] = \bigvee_{x\in F} \dot{x}$, then we have $\mathcal{N}_{\varphi_{1,2}}(F) \geq [\chi_F]$ holds. If the related $\varphi_{1,2}$-interior operator fulfill the axioms (I1) and (I2) only, then the mapping $\mathcal{N}_{\varphi_{1,2}}(x) : L^X \to L$, defined by (2.4) is an fuzzy stack, called *$\varphi_{1,2}$-fuzzy neighborhood stack* at x. Moreover, if the $\varphi_{1,2}$-interior operator fulfill the axioms (I1), (I2) and (I4) such that in (I4) instead of $\rho \in L^X$ we take $\bar{\alpha}$, then the mapping $\mathcal{N}_{\varphi_{1,2}}(x) : L^X \to L$, is an fuzzy stack with the cutting property, called *$\varphi_{1,2}$-fuzzy neighborhood stack with the cutting property* at x. Obviously, the $\varphi_{1,2}$-fuzzy neighborhood filters fulfill the following conditions:

(N1) $\dot{x} \leq \mathcal{N}_{\varphi_{1,2}}(x)$ holds for all $x \in X$.

(N2) $\mathcal{N}_{\varphi_{1,2}}(x)(\mu) \leq \mathcal{N}_{\varphi_{1,2}}(x)(\rho)$ holds for all $\mu, \rho \in L^X$ and $\mu \leq \rho$.

(N3) $\mathcal{N}_{\varphi_{1,2}}(x)(y \mapsto \mathcal{N}_{\varphi_{1,2}}(y)(\mu)) = \mathcal{N}_{\varphi_{1,2}}(x)(\mu)$, for all $x\in X$ and $\mu \in L^X$.

Clearly, $y \mapsto \mathcal{N}_{\varphi_{1,2}}(y)(\mu)$ is the fuzzy subset $\varphi_{1,2}$.int μ.

The characterized fuzzy space $(X,\varphi_{1,2}.\text{int})$ is characterized as the *fuzzy filter pre topology* [5], that is, as a mapping $\mathcal{N}_{\varphi_{1,2}} : X \to \mathcal{F}_L X$ such that the conditions (N1) to (N3) are fulfilled.

$\varphi_{1,2}\psi_{1,2}$-fuzzy continuity

Let now the fuzzy topological spaces (X,τ_1) and (Y,τ_2) are fixed, $\varphi_1,\varphi_2 \in O_{(L^X,\tau_1)}$ and $\psi_1,\psi_2 \in O_{(L^Y,\tau_2)}$. The mapping $f : (X,\varphi_{1,2}.\text{int}) \to (Y,\psi_{1,2}.\text{int})$ is said to be *$\varphi_{1,2}\psi_{1,2}$-fuzzy continuous* [5] if the inequality

$$(\psi_{1,2}.\text{int}\eta)\circ f \leq \varphi_{1,2}.\text{int}(\eta \circ f)\,, \tag{2.5}$$

holds for all $\eta\in L^Y$. If an order reversing involution $'$ of L is given, then we have that f is a fuzzy continuous if and only if $\varphi_{1,2}.\text{cl}(\eta\circ f) \leq (\psi_{1,2}.\text{cl}\eta)\circ f$ holds for all $\eta\in L^Y$. Note that $\varphi_{1,2}$.cl and $\psi_{1,2}$.cl, means that the closure operators related to $\varphi_{1,2}$.int and $\psi_{1,2}$.int, respectively which are defined by $\varphi_{1,2}.\text{cl}\ \mu = co\ (\varphi_{1,2}.\text{int}\ co\mu)$ for all $\mu \in\in L^X$. Obviously if f is $\varphi_{1,2}\psi_{1,2}$-fuzzy continuous and the inverse $f^{-1}$of f exists, then f^{-1}: $(Y,\psi_{1,2}.\text{int}) \to (X,\varphi_{1,2}.\text{int})$ is $\psi_{1,2}\varphi_{1,2}$-fuzzy continuous, that is, $(\varphi_{1,2}.\text{int}\mu)\circ f^{-1} \leq \psi_{1,2}.\text{int}(\mu\circ f^{-1})$ holds for all $\mu\in L^X$. By means of characterizing the $\varphi_{1,2}$-fuzzy neighborhoods $\mathcal{N}_{\varphi_{1,2}}(x)$ of $\varphi_{1,2}$.int and $\mathcal{N}_{\psi_{1,2}}(x)$ of $\psi_{1,2}$.int which are defined by (2.4), the fuzzy continuity of f can also be characterized as follows:

The mapping f: $(X,\varphi_{1,2}.\text{int}) \to (Y,\psi_{1,2}.\text{int})$ is $\varphi_{1,2}\psi_{1,2}$-fuzzy continuous if the inequality $\mathcal{N}_{\psi_{1,2}}(f(x)) \geq \mathcal{F}f(\mathcal{N}_{\varphi_{1,2}}(x))$ holds for each $x\in X$. Obviously, in case of $L=\{0,1\}$, $\varphi_1=\psi_2=\text{int}$, $\varphi_2 = 1_{L^X}$ and $\psi_2 = 1_{L^Y}$, the $\varphi_{1,2}\psi_{1,2}$-fuzzy continuity coincides with the usual fuzzy continuity.

$\varphi_{1,2}$-fuzzy convergence

Let an fuzzy topological space (X,τ) be fixed and $\varphi_1,\varphi_2 \in O_{(L^X,\tau)}$. If x is a point in the characterized fuzzy space $(X,\varphi_{1,2}.\text{int})$, F $\subseteq$ X and $\mathcal{M}$ is a fuzzy filter on X. Then $\mathcal{M}$ is said to be *$\varphi_{1,2}$-fuzzy convergence* [5] to x and written $\mathcal{M} \xrightarrow[\varphi_{1,2}.\text{int}]{} x$, provided $\mathcal{M}$ is finer than the $\varphi_{1,2}$-fuzzy neighborhood filter $\mathcal{N}_{\varphi_{1,2}}(x)$. Moreover, $\mathcal{M}$ is said to be $\varphi_{1,2}$-fuzzy convergence to F and written $\mathcal{M} \xrightarrow[\varphi_{1,2}.\text{int}]{} F$, provided $\mathcal{M}$ is finer than the $\varphi_{1,2}$-fuzzy neighborhood filter $\mathcal{N}_{\varphi_{1,2}}(x)$ for all $x\in F$, that is, $\mathcal{M}$ is finer than the $\varphi_{1,2}$-fuzzy neighborhood filter $\mathcal{N}_{\varphi_{1,2}}(F)$.

Internal $\varphi_{1,2}$-closure of fuzzy sets and $\varphi_{1,2}$-closure operators

Let a fuzzy topological space (X,τ) be fixed and $\varphi_1,\varphi_2 \in O_{(L^X,\tau)}$. The internal $\varphi_{1,2}$-closure of the fuzzy set μ: $X \to L$ is the mapping $\varphi_{1,2}.\text{cl}\ \mu : X \to L$ defined by:

$$(\varphi_{1,2}.\text{cl}\mu)(x) = \bigvee_{\mathcal{M}\leq\mathcal{N}_{\varphi_{1,2}}(x)} \mathcal{M}(\mu)\,, \tag{2.6}$$

$$(\varphi_{1,2}.\text{cl}\mu)(x) = \bigvee_{\mathcal{M}\leq\mathcal{N}_{\varphi_{1,2}}(x)} \mathcal{M}(\mu)\,, \tag{2.6}$$

for all $x\in X$. In (2.6), the fuzzy filter M my have additional properties, e.g, we my assume that is homogeneous or even that is ultra. Obviously, $\varphi_{1,2}.\text{cl}\mu \leq \mu$ holds for all $\mu\in L^X$. The mapping $\varphi_{1,2}.\text{cl}\ F_L X \to F_L X$ which assigns $\varphi_{1,2}.\text{cl}\ M$ to each fuzzy filter M on X, that is,

$$\varphi_{1,2}.\text{cl}\mathcal{M}(\mu) = \bigvee_{\varphi_{1,2}.\text{cl}\rho\leq\mu} \mathcal{M}(\rho)\,, \tag{2.7}$$

is called $\varphi_{1,2}$-closure operator [7] of the characterized fuzzy space $(X,\varphi_{1,2}.\text{int})$ with respect to the related fuzzy topology τ. Obviously, the $\varphi_{1,2}$-closure operator $\varphi_{1,2}$.cl is isotone hull operator, that is, for all $\mathcal{M},\mathcal{N} \in\mathcal{F}_L X$ we have $\mathcal{M}\leq\mathcal{N}$ implies $\varphi_{1,2}.\text{cl}\mathcal{M} \leq \varphi_{1,2}.\text{cl}N$ and that $\mathcal{M} \leq\varphi_{1,2}.\text{cl}\mathcal{M}$.

Lemma 2.2

Let (X,τ) be a fuzzy topological space and $\varphi_1,\varphi_2 \in O_{(L^X,\tau)}$. Then for each $x\in X$, we have $\varphi_{1,2}.\text{cl}\,\dot{x} = \dot{x}$ implies that $\varphi_{1,2}.\text{cl}\{x\} = \{x\}$ [10].

Characterized fuzzy R_k and fuzzy $\varphi_{1,2}R_k$-spaces

The notions of characterized fuzzy R_k and fuzzy $\varphi_{1,2}R_k$-spaces are introduced and studied in Abd-Allah [9,11] for all $k \in \{0,1,2\frac{1}{2}\}$. Moreover, the notion of $\varphi_{1,2}$-fuzzy neighborhood filter at the point x and at the ordinary subset of the characterized fuzzy space $(X,\varphi_{1,2}.\text{int})$ is applied by Abd-Allah [10], to introduced and studied the notions of characterized fuzzy R_k-spaces for $k\in\{2,3\}$. However, the notions of fuzzy $_{1,2}R_k$-spaces are also given by means of the $\varphi_{1,2}$-fuzzy convergence at the point x and at the ordinary subset in the space. We will denote by characterized FR_k and $F\varphi_{1,2}R_k$-spaces to the characterized fuzzy R_k and fuzzy $\varphi_{1,2}R_k$-spaces for shorts, respectively.

Let a fuzzy topological space (X,τ) be fixed and $\varphi_1,\varphi_2 \in O_{(L^X,\tau)}$. Then the characterized fuzzy space $(X,\varphi_{1,2}.\text{int})$ is said to be:

(1) Characterized FR_2-space (resp. Characterized FR_3-space), if for

all $x \in X$, F$\varphi_{1,2}C$ (X) such that x Ï F (resp. F_1, F_2 $\in \varphi_{1,2}C$ (X) such that $F_1 \cap F_2 = \emptyset$), the infimum $\mathcal{N}_{\varphi_{1,2}}(x) \wedge \mathcal{N}_{\varphi_{1,2}}(F)$ (resp. $\mathcal{N}_{\varphi_{1,2}}(F_1) \wedge \mathcal{N}_{\varphi_{1,2}}(F_2)$) does not exists. The related fuzzy topological space (X,τ) is said to be F$\varphi_{1,2}R_2$-space (resp. F$\varphi_{1,2}R_3$-space) if for all $x \in X$ (resp. F $\in \varphi_{1,2}.C$ (X)) and $\mathcal{M} \in \mathcal{F}_L X$ such that $\mathcal{M} \xrightarrow[\varphi_{1,2}.\text{int}]{} x$ (resp. $\mathcal{M} \xrightarrow[\varphi_{1,2}.\text{int}]{} F$) we have $\varphi_{1,2}.\text{cl}\mathcal{M} \xrightarrow[\varphi_{1,2}.\text{int}]{} x$ (resp. $\varphi_{1,2}.\text{cl}\mathcal{M} \xrightarrow[\varphi_{1,2}.\text{int}]{} F$).

(2) Characterized $FR_{2\frac{1}{2}}$- space if for all $x \in X$, F$\in\varphi_{1,2}C$ (X) such that x Ï F, there exists an $\varphi_{1,2}\psi_{1,2}$-fuzzy continuous mapping f: $(X,\varphi_{1,2}.\text{int}) \to (I_L,\psi_{1,2}.\text{int}_I)$ such that $f(x) = \overline{1}$ and $f(y) = \overline{0}$ for all $y \in F$. The related fuzzy topological space (X,τ) is said to be $F\varphi_{1,2}R_{2\frac{1}{2}}$- space if and only if $(X,\varphi_{1,2}.\text{int})$ is characterized $FR_{2\frac{1}{2}}$- space.

Characterized fuzzy T_s and fuzzy $\varphi_{1,2}$-T_S spaces

The notions of characterized fuzzy T_s and fuzzy $\varphi_{1,2}$-T_S spaces are investigated and studied by Abd-Allah and by Abd-Allah and Al-Khedhairi in [8,9,11] for all $s \in \{0,1,2,2\frac{1}{2},3,3\frac{1}{2},4\}$. These characterized fuzzy spaces depend only on the usual points and the operation defined on the class of all fuzzy subsets of X endowed with a fuzzy topological space (X,τ). We will denote by characterized FT_s and F$\varphi_{1,2}$-T_s spaces to the characterized fuzzy T_s and fuzzy $\varphi_{1,2}$-T_S spaces for shorts, respectively.

Let a fuzzy topological space (X,τ) be fixed and $\varphi_1,\varphi_2 \in O_{(L^X,\tau)}$. Then the characterized fuzzy space $(X,\varphi_{1,2}.\text{int})$ is said to be:

(1) Characterized FT_0-space (resp. Characterized FT_1-space) if for all $x,y \in X$ such that $x \neq y$ there exist $\mu \in L^X$ and $\alpha \in L_0$ such that $\mu(x) < \alpha \leq (X,\varphi_{1,2}.\text{int}\mu)(y)$ holds or (resp. and) there exist $\rho \in L^X$ and $\beta \in L_0$ such that $\rho(y) < \beta \leq (\varphi_{1,2}.\text{int}\rho)(x)$ holds. The related fuzzy topological space (X,τ) is said to be F$\varphi_{1,2}$-T_0 space (resp. F$\varphi_{1,2}$-T_1 space) if for all $x,y \in X$ such that $x \neq y$ we have $\dot{x} \nleq \mathcal{N}_{\varphi_{1,2}}(y)$ or (resp. and) $\dot{y} \nleq \mathcal{N}_{\varphi_{1,2}}(x)$.

(2) Characterized FT_2-space if for all $x,y \in X$ such that $x \neq y$, the infimum $\mathcal{N}_{\varphi_{1,2}}(x) \wedge \mathcal{N}_{\varphi_{1,2}}(y)$ does not exists. The related fuzzy topological space (X,τ) is said to be F$\varphi_{1,2}$-T_2 space if $\mathcal{M} \xrightarrow[\varphi_{1,2}.\text{int}]{} x, y$ implies $x = y$ for all $\mathcal{M} \in \mathcal{F}_L X$ and for all $x,y \in X$.

(3) Characterized FT_s space if and only if it is characterized FR_k-space and characterized FT_1-space for $k \in \{2,2\frac{1}{2},3\}$ and $s \in \{3,3\frac{1}{2},4\}$. The related fuzzy topological space (X,τ) is said to be F$\varphi_{1,2}$-T_s if and only if it is F$\varphi_{1,2}$-R_k and F$\varphi_{1,2}$-T_1.

Proposition 2.4

Let (X,τ) be an fuzzy topological space and $\varphi_1,\varphi_2 \in O_{(L^X,\tau)}$. Then the characterized fuzzy space $(X,\varphi_{1,2}.\text{int})$ is characterized FT_1-space if and only if $\varphi_{1,2}.\text{cl}\,\dot{x} = \dot{x}$ for all $x \in X$ [8].

Proposition 2.5

If $(X,\varphi_{1,2}.\text{int})$ is characterized FT_2-space and $\varphi_{1,2}.\text{int}$ is finer than $\psi_{1,2}.\text{int}$, then $(X,\psi_{1,2}.\text{int})$ is also characterized FT_2-space [8].

Proposition 2.6

Let a fuzzy topological space (X,τ) be fixed and $\varphi_1,\varphi_2 \in O_{(L^X,\tau)}$. Then the following are fulfilled [8,22]:

(1) Every characterized FT_i-space $(X,\varphi_{1,2}.\text{int})$ is characterized FT_{i-1}-space for each i$\in\{2,3,4\}$.

(2) The characterized fuzzy subspace and the characterized fuzzy product space of a family of characterized FT_2-spaces are also characterized FT_2-spaces .

New Relations between Characterized FT_s, Characterized FR_k and Characterized Fuzzy Proximity Spaces

In this section we are going to introduce and study the relations between the characterized FT_s-spaces, the characterized FR_k-spaces and the characterized fuzzy proximity spaces presented by Abd-Allah in [12]. We make at first the relation between the farness on fuzzy sets and the finer relation on fuzzy filters. So, we show some results for the notion of the $\varphi_{1,2}$-fuzzy neighborhood filter $\mathcal{N}_{\varphi_{1,2}}(\mu)$ at the fuzzy subset $\mu \in L^X$. The notion of homogeneous fuzzy filter $\dot{\mu}$ which is defined in (2.1) and the notion of $\varphi_{1,2}$-fuzzy neighborhood filter $\mathcal{N}_{\varphi_{1,2}}(\mu)$ at the fuzzy subset $\mu \in L^X$ are applied at first to study the relation between the fuzzy proximity δ defined by Katsaras in [18] and our fuzzy separation axioms [8-10]. Moreover, the relations between characterized fuzzy proximity spaces and the characterized FT_s-spaces and characterized FR_k-spaces are introduced for $s \in \{0,1,2,3,3\frac{1}{2},4\}$ and $k \in \{1,2,2\frac{1}{2},3\}$.

Proposition 3.1

Let a fuzzy topological space (X,τ) be fixed and $\varphi_1,\varphi_2 \in O_{(L^X,\tau)}$ such that $\varphi_2 \geq 1_{L^X}$ is isotone and idempotent and φ_1 is wfip with respect to φ_1OF (X). Then the supremum of the $\varphi_{1,2}$-fuzzy neighborhood filters $\mathcal{N}_{\varphi_{1,2}}(x)$ at $x \in X$ which is given by:

$$\mathcal{N}_{\varphi_{1,2}}(\mu) = \bigvee_{0<\mu(x)} \mathcal{N}_{\varphi_{1,2}}(x), \tag{3.1}$$

for all $\mu \in L^X$ is a fuzzy filter on X called a $\varphi_{1,2}$–fuzzy neighborhood filter at μ.

Proof: Fix an $\alpha \in L_0$, then because of (2.4) and the condition $\varphi_2 \geq 1_{L^X}$, we have

$$\mathcal{N}_{\varphi_{1,2}}(\overline{\alpha}) = \bigwedge_{0<\mu(y)} \mathcal{N}_{\varphi_{1,2}}(y)(\overline{\alpha}) = \bigwedge_{0<\mu(y)} (\varphi_{1,2}.\text{int}\overline{\alpha})(y) \leq \bigwedge_{0<\mu(y)} \overline{\alpha}(y) = \alpha$$

and

$$\mathcal{N}_{\varphi_{1,2}}(\overline{1}) = \bigwedge_{0<\mu(y)} (\varphi_{1,2}.\text{int}\,\overline{1})(y) = \bigwedge_{0<\mu(y)} \overline{1}(y) = 1.$$

Thus, condition (F_1) is fulfilled. To prove condition (F_2), let $\rho,\eta \in L^X$, then because of Proposition 2.4 and (2.4) we have

$$\begin{aligned}\mathcal{N}_{\varphi_{1,2}}(\mu)(\rho \wedge \eta) &= \bigwedge_{0<\mu(y)} \varphi_{1,2}.\text{int}(\rho \wedge \eta)(y) \\ &= \bigwedge_{0<\mu(y)} (\varphi_{1,2}.\text{int}\rho)(y) \wedge \bigwedge_{0<\mu(y)} (\varphi_{1,2}.\text{int}\eta)(y) \\ &= \mathcal{N}_{\varphi_{1,2}}(\mu)(\rho) \wedge \mathcal{N}_{\varphi_{1,2}}(\mu)(\eta).\end{aligned}$$

Hence, $\mathcal{N}_{\varphi_{1,2}}(\mu)$ is a fuzzy filter on X. Since $(\varphi_{1,2}.\text{int})(x) \leq \rho(x)$ holds for all $x \in X$ and $\rho \in L^X$, then $\mathcal{N}_{\varphi_{1,2}}(\mu)(\rho) \leq \dot{\mu}(\rho)$ holds for all $\rho \in \in L^X$. Thus, $\dot{\mu} \leq \mathcal{N}_{\varphi_{1,2}}(\mu)$ and therefore $\mathcal{N}_{\varphi_{1,2}}(\mu)$ fulfills condition (N_1). For condition (N_2), let $\rho,\eta \in L^X$ such that $\rho \leq \eta$. Because of Proposition 2.4, we have $\varphi_{1,2}.\text{int}\rho \leq \varphi_{1,2}.\text{int}\eta$ holds and which implies that $\bigwedge_{0<\mu(y)} (\varphi_{1,2}.\text{int}\rho)(y) \leq \bigwedge_{0<\mu(y)} (\varphi_{1,2}.\text{int}\eta)(y)$ holds for all $y \in X$. Hence $\mathcal{N}_{\varphi_{1,2}}(\mu)(\rho) \leq \mathcal{N}_{\varphi_{1,2}}(\mu)(\eta)$ and therefore condition (N_2) is fulfilled. Since for any $y \in X$ we have $\bigwedge_{0<\mu(y)} y \mapsto \bigwedge_{0<\mu(y)} (\varphi_{1,2}.\text{int}\rho)(y)$ represents the mapping $\varphi_{1,2}$int ρ. Then from Proposition 2.4 we have

$$\mathcal{N}_{\varphi_{1,2}}(\mu)(\varphi_{1,2}.\text{int}\,\rho) = \bigwedge_{0<\mu(x)} \varphi_{1,2}.\text{int}(\varphi_{1,2}.\text{int}\,\rho)(x) = \bigwedge_{0<\mu(x)} (\varphi_{1,2}.\text{int}\,\rho)(x),$$

and then $\mathcal{N}_{\varphi_{1,2}}(\mu)(\bigwedge_{0<\mu(y)} y \mapsto \bigwedge_{0<\mu(y)} \varphi_{1,2}.\text{int}\,\rho(y)) = \mathcal{N}_{\varphi_{1,2}}(\mu)(\rho)$ for all $y \in X$ and $\rho \in L^X$. Thus, condition (N3) is also fulfilled and therefore $\mathcal{N}_{\varphi_{1,2}}(\mu)$ fulfilled the conditions (N_1) to (N_3) of the $\varphi_{1,2}$–fuzzy neighborhood filters.

Not that in Bayoumi et al. [15], the supremum of the empty set of the fuzzy filters is the finest fuzzy filter. This means $\mathcal{N}_{\varphi_{1,2}}(\overline{0}) \leq \dot{\mu}$ for all $\mu \in L^X$. Because of (2.4) the equations (2.1) and (2.2) can be written as in the following:

$$\dot{\mu}(\rho) = \bigwedge_{0<\mu(x)} \rho(x), \tag{3.2}$$

$$\mathcal{N}_{\varphi_{1,2}}(\mu)(\rho) = \bigwedge_{0<\mu(x)} \mathcal{N}_{\varphi_{1,2}}(x)(\rho) = \bigwedge_{0<\mu(x)} (\varphi_{1,2}.\text{int}\,\rho)(x), \tag{3.3}$$

for all $\rho \in L^X$. Here a useful remark is given

Remark 3.1: The homogeneous fuzzy filter $\dot{x}$ at the ordinary point x is nothing that a homogeneous fuzzy filter $\dot{x}$ at the fuzzy point x_α, that is, $\dot{x}_\alpha = \dot{x}$ for all $x \in X$ and $\alpha \in L_0$. Moreover, the $\varphi_{1,2}$–fuzzy neighborhood filter $\mathcal{N}_{\varphi_{1,2}}(x)$ at x is itself the $\varphi_{1,2}$–fuzzy neighborhood filter $\mathcal{N}_{\varphi_{1,2}}(x_\alpha)$ at x_α.

The $\varphi_{1,2}$–fuzzy neighborhood filter $\mathcal{N}_{\varphi_{1,2}}(\mu)$ at the fuzzy subset $\mu \in L^X$ and the homogeneous fuzzy filter $\dot{\mu}$ fulfill the following properties.

Lemma 3.1

Let (X,τ) be a fuzzy topological space and $\varphi_1, \varphi_2 \in O_{(L^X,\tau)}$. Then for all $\mu,\rho \in L^X$ the following properties are fulfilled:

(1) $\dot{\mu} \leq \dot{\rho}$ implies $\mathcal{N}_{\varphi_{1,2}}(\rho') \leq \dot{\mu}'$ and $\mathcal{N}_{\varphi_{1,2}}(\mu) \leq \dot{\rho}$ implies $\mathcal{N}_{\varphi_{1,2}}(\rho') \leq \dot{\mu}'$.

(2) $\mu \leq$ implies $\mathcal{N}_{\varphi_{1,2}}(\mu) \leq \mathcal{N}_{\varphi_{1,2}}(\rho)$.

(3) $\mathcal{N}_{\varphi_{1,2}}(\mu \vee \rho) = \mathcal{N}_{\varphi_{1,2}}(\mu) \wedge \mathcal{N}_{\varphi_{1,2}}(\rho)$.

(4) $\mathcal{N}_{\varphi_{1,2}}(\mu) \leq \dot{\rho}$ implies $\mu \leq \rho$.

(5) $\mathcal{N}_{\varphi_{1,2}}(\mu) \leq \dot{\rho}$ implies there is an $\eta \in L^X$ such that $\mathcal{N}_{\varphi_{1,2}}(\mu) \leq \dot{\eta}$ and $\mathcal{N}_{\varphi_{1,2}}(\eta) \leq \dot{\rho}$.

Proof: Let $\dot{\mu} \leq \dot{\rho}$. From condition (N1) we have $\dot{\mu} \leq \mathcal{N}_{\varphi_{1,2}}(\rho)$ holds and therefore for all $\eta \in L^X$ we have $\bigwedge_{0<\mu(x)} \eta(x) \geq \bigwedge_{0<\rho(y)} (\varphi_{1,2}.\text{int}\,\eta)(y)$ holds. Hence, $\bigwedge_{0<\mu'(x)} \eta(x) \leq \bigwedge_{0<\rho'(y)} (\varphi_{1,2}.\text{int}\,\eta)(y)$ holds also. Thus, $\dot{\mu}'(\eta) \leq \mathcal{N}_{\varphi_{1,2}}(\rho')(\eta)$ and $\mathcal{N}_{\varphi_{1,2}}(\rho') \leq \dot{\mu}'$ are hold. Similarly, if $\mathcal{N}_{\varphi_{1,2}}(\mu) \leq \dot{\rho}$, then from (N1) we have $\dot{\mu} \leq \dot{\rho}$ which implies $\mathcal{N}_{\varphi_{1,2}}(\rho') \leq \dot{\mu}'$. Thus, (1) is fulfilled. Since $\mu \leq \rho$ implies $\mu(x) \leq \rho(x)$ for all xX, then

$$\bigwedge_{0<\mu(x)} (\varphi_{1,2}.\text{int}\,\eta)(x) \geq \bigwedge_{0<\rho(x)} (\varphi_{1,2}.\text{int}\,\eta)(x).$$

Hence, $\mathcal{N}_{\varphi_{1,2}}(\mu)(\eta) \geq \mathcal{N}_{\varphi_{1,2}}(\rho)(\eta)$ holds for all $\eta \in L^X$ and therefore $\mathcal{N}_{\varphi_{1,2}}(\mu) \leq \mathcal{N}_{\varphi_{1,2}}(\rho)$ holds. Hence, (2) is fulfilled.

Since μ, $\rho \leq \mu$ Ú ρ, then from (2) we have $\mathcal{N}_{\varphi_{1,2}}(\mu) \wedge \mathcal{N}_{\varphi_{1,2}}(\rho) \leq \mathcal{N}_{\varphi_{1,2}}(\mu \vee \rho)$. Now, let $\eta \in L^X$ then

$$\begin{aligned}(\mathcal{N}_{\varphi_{1,2}}(\mu) \wedge \mathcal{N}_{\varphi_{1,2}}(\rho))(\eta) &= \bigvee_{k_1 \wedge k_2 \leq \eta} (\mathcal{N}_{\varphi_{1,2}}(\mu)(k_1) \wedge \mathcal{N}_{\varphi_{1,2}}(\rho)(k_2)) \\ &= \bigvee_{k_1 \wedge k_2 \leq \eta} \Big(\bigwedge_{0<\mu(x)} \varphi_{1,2}.\text{int}\,k_1(x) \wedge \bigwedge_{0<\rho(y)} \varphi_{1,2}.\text{int}\,k_2(y)\Big) \\ &\leq \bigvee_{k_1 \wedge k_2 \leq \eta} \bigwedge_{0<(\mu\vee\rho)(z)} \varphi_{1,2}.\text{int}(k_1 \wedge k_2)(z) \\ &\leq \bigwedge_{0<(\mu\vee\rho)(z)} \varphi_{1,2}.\text{int}\,\eta(z) = \mathcal{N}_{\varphi_{1,2}}(\mu \vee \rho)(\eta).\end{aligned}$$

Hence, $\mathcal{N}_{\varphi_{1,2}}(\mu) \wedge \mathcal{N}_{\varphi_{1,2}}(\rho) \geq \mathcal{N}_{\varphi_{1,2}}(\mu \vee \rho)$ holds and therefore (3) is fulfilled. To prove (4), let $\mathcal{N}_{\varphi_{1,2}}(\mu) \leq \dot{\rho}$ holds. Because of (2.1), (3.1) and (N1), we have $\dot{\mu} \leq \mathcal{N}_{\varphi_{1,2}}(\mu)$ and then $\dot{\mu} \leq \dot{\rho}$ holds. Hence, Proposition 2.1 implies $\mu \leq \rho$. Thus, (4) is fulfilled. Finally, let $\mathcal{N}_{\varphi_{1,2}}(\mu) \leq \dot{\rho}$. Then $\bigwedge_{0<\mu(x)} (\varphi_{1,2}.\text{int}\lambda)(x) \geq \bigwedge_{0<\rho(y)} \lambda(y)$ holds for all $\lambda \in L^X$. Hence, there is $\eta \in L^X$ such that

$$\bigwedge_{0<\mu(x)} (\varphi_{1,2}.\text{int}\lambda)(x) \geq \bigwedge_{0<\eta(z)} \lambda(z) \geq \bigwedge_{0<\eta(z)} (\varphi_{1,2}.\text{int}\lambda)(z) \geq \bigwedge_{0<\rho(y)} \lambda(y).$$

This means there is $\eta \in L^X$ such that $\mathcal{N}_{\varphi_{1,2}}(\mu)(\lambda) \geq \dot{\eta}(\lambda)$ and $\mathcal{N}_{\varphi_{1,2}}(\eta)(\lambda) \geq \dot{\rho}(\lambda)$ are hold for all $\lambda \in L^X$. Thus, $\mathcal{N}_{\varphi_{1,2}}(\mu) \leq \dot{\eta}$ and $\mathcal{N}_{\varphi_{1,2}}(\eta) \leq \dot{\rho}$ are also hold. Consequently, (5) is fulfilled.

In the characterized fuzzy space $(X,\varphi_{1,2}.\text{int})$, the fuzzy proximity will be identified with the finer relation on the fuzzy filters, specially with the finer relation on the $\varphi_{1,2}$-fuzzy neighborhood filters. This shown in the following proposition.

Proposition 3.2

Let (X,τ) be a fuzzy topological space and $\varphi_1, \varphi_2 \in O_{(L^X,\tau)}$. Then the binary relation δ on L^X which is defined by:

$\mu\, \overline{\delta}\, \rho$ if and only if $\mathcal{N}_{\varphi_{1,2}}(\rho) \leq \dot{\mu}'$,

for all $\mu,\rho \in L^X$ is fuzzy proximity on X.

Proof: Let $\mu,\rho \in L^X$ such that $\mu \overline{\delta} \rho$, then $\mathcal{N}_{\varphi_{1,2}}(\rho) \leq \dot{\mu}'$. Because of (1) in Lemma 3.1, we have $\mathcal{N}_{\varphi_{1,2}}(\mu) \leq \dot{\rho}'$ and therefore $\rho\, \overline{\delta}\, \mu$. Hence, condition (P1) is fulfilled.

Since $\mathcal{N}_{\varphi_{1,2}}(\mu) \leq \mathcal{N}_{\varphi_{1,2}}(\mu \vee \rho)$ and $\mathcal{N}_{\varphi_{1,2}}(\rho) \leq \mathcal{N}_{\varphi_{1,2}}(\mu \vee \rho)$ are hold for all $\mu,\rho \in L^X$, then $\mathcal{N}_{\varphi_{1,2}}(\mu \vee \rho) \leq \dot{\eta}'$ implies $\mathcal{N}_{\varphi_{1,2}}(\mu) \leq \dot{\eta}'$ and $\mathcal{N}_{\varphi_{1,2}}(\rho) \leq \dot{\eta}'$ are hold for all $\eta \in L^X$. This means $\eta\, \overline{\delta}(\mu \vee \rho)$ implies $\eta\, \overline{\delta}\, \mu$ and $\eta\, \overline{\delta}\, \rho$. Conversely, let $\eta\, \overline{\delta}\, \mu$ and $\eta\, \overline{\delta}\, \rho$ for all $\mu,\rho,\eta \in L^X$, then $\mathcal{N}_{\varphi_{1,2}}(\mu) \leq \dot{\eta}'$ and $\mathcal{N}_{\varphi_{1,2}}(\rho) \leq \dot{\eta}'$ are hold. Hence, (3) in Lemma 3.1 implies $\mathcal{N}_{\varphi_{1,2}}(\mu \vee \rho) = \mathcal{N}_{\varphi_{1,2}}(\mu) \wedge \mathcal{N}_{\varphi_{1,2}}(\rho) \leq \dot{\eta}'$ holds and therefore $\eta\, \overline{\delta}(\mu \vee \rho)$. Consequently, (P2) is fulfilled. To prove (P3), since $\mathcal{N}_{\varphi_{1,2}}(\overline{0}) \leq \dot{\mu}'$ holds for all $\mu \in \in L^X$. Then, $\mu\, \overline{\delta}\, \overline{0}$ for all $\mu \in L^X$. Hence, $\mu = \overline{0}$ or $\rho = \overline{0}$ implies $\mu\, \overline{\delta}\, \rho$ for all $\rho,\eta \in L^X$. Thus, (P3) is fulfilled.

Let $\mu,\rho \in L^X$ such that $\mu\, \overline{\delta}\, \rho$, then $\mathcal{N}_{\varphi_{1,2}}(\rho) \leq \dot{\mu}'$. Because of (1) and (4) in Lemma 3.1, we have $\mathcal{N}_{\varphi_{1,2}}(\mu) \leq \dot{\rho}'$ and therefore $\mu \leq \rho'$, that is, (P4) is fulfilled. Finally, let $\mu,\rho \in L^X$ such that $\mu\, \overline{\delta}\, \rho$, then

$\mathcal{N}_{\varphi_{1,2}}(\rho) \le \dot{\mu}'$ which implies $\mathcal{N}_{\varphi_{1,2}}(\mu) \le \dot{\rho}'$. Because of (5) in Lemma 3.1, there is an $\eta \in L^X$ such that $\mathcal{N}_{\varphi_{1,2}}(\mu) \le \dot{\eta}$ and $\mathcal{N}_{\varphi_{1,2}}(\eta) \le \dot{\rho}'$ are hold. Hence, $\mathcal{N}_{\varphi_{1,2}}(\eta') \le \dot{\mu}'$ and $\mathcal{N}_{\varphi_{1,2}}(\rho) \le \dot{\eta}'$ are also hold, that is, $\mu \,\bar{\delta}\, \eta'$ and $\eta \,\bar{\delta}\, \rho$. Thus, (P5) holds and consequently, δ is fuzzy proximity on X.

If a fuzzy topological space (X,τ) be fixed and $\varphi_1, \varphi_2 \in O_{(L^X,\tau)}$. Then each fuzzy proximity δ on X is associated a set of all $\varphi_{1,2}$-open fuzzy subsets of X with respect to δ denoted by $\varphi_{1,2}OF(X)_\delta$. In this case the triple $(X,\varphi_{1,2}OF(X)_\delta)$ as will as $(X,\varphi_{1,2}.\text{int}_\delta)$ is said to be *characterized fuzzy proximity space.* The related $\varphi_{1,2}$-interior and $\varphi_{1,2}$-closure operators $\varphi_{1,2}.\text{int}_\delta$ and $\varphi_{1,2}.\text{cl}_\delta$ are given by:

$$\varphi_{1,2}.\text{int}_\delta \mu = \bigvee_{\mu' \bar{\delta} \rho} \rho \text{ and } \varphi_{1,2}.\text{cl}_\delta \mu = \bigwedge_{\mu' \bar{\delta} \rho} \rho, \tag{3.5}$$

respectively, for all $\mu \in L^X$. Consider the characterized fuzzy proximity space $(X,\varphi_{1,2}.\text{int}_\delta)$ be fixed and $\mu \in L^X$, then μ is said to be $\varphi_{1,2}\delta$-fuzzy neighborhood for the point $x \in X$ if and only if $x_1 \,\bar{\delta}\, \mu'$. Moreover, the mapping $f:(X,\varphi_{1,2}.\text{int}_\delta) \to (Y,\psi_{1,2}.\text{int}_{\delta^*})$ is said to be $\varphi_{1,2}\psi_{1,2}\delta$-*fuzzy continuous*, provided $\eta \,\bar{\delta}^*\, \rho$ implies $(\eta \circ f)\,\bar{\delta}\,(\rho \circ f)$ for all $\eta,\rho \in L^Y$.

In the following we will show that the characterized fuzzy proximity space $(X,\varphi_{1,2}.\text{int}_\delta)$ is characterized FT_0-space as in sense of [8] if and only if δ is separated.

Proposition 3.3

Let (X,τ) be a fuzzy topological space, $\varphi_1\ \varphi_2 \in$ (,) and is a fuzzy proximity on X. Then the characterized fuzzy proximity space $(X,\varphi_{1,2}.\text{int}_\delta)$ is characterized FT_0-space if and only if δ is separated.

Proof: Let $(X,\varphi_{1,2}.\text{int}_\delta)$ is characterized FT_0-space and let $x,y \in X$ such that $x \neq y$. Then $\dot{x} \not\le \mathcal{N}^\delta_{\varphi_{1,2}}(y)$ and therefore there is $\mu \in L^X$ such that $\varphi_{1,2}.\text{int}_\delta\, \mu\,(y) > \mu\,(x)$. Because of (3.4), we have $\bigvee_{\mu' \bar{\delta} \rho} \rho(y) > \mu(x)$ and hence $\mu\,(x) > \rho\,(y)$ holds for all ρL^X with $\mu' \,\bar{\delta}\, \rho$, that is, $\mu\,(x) > \rho\,(y)$ holds for all $\rho \in L^X$ with $\mathcal{N}^\delta_{\varphi_{1,2}}(\rho) \le \dot{\mu}$. Choose $\mu = x_1'$ and $\mathcal{N}^\delta_{\varphi_{1,2}}(y_1) \le x_1'$, then because of Remark 3.1, we get $\mathcal{N}^\delta_{\varphi_{1,2}}(y_1) \le x_1'$. Using Proposition 3.2 we get $x_1 \bar{\delta} y_1$ and therefore $x_\alpha \bar{\delta} y_\beta$ holds for all $\alpha,\beta \in L_0$. Thus, δ is separated.

Conversely, let δ is separated fuzzy proximity and let $x,y \in X$ such that $x \neq y$. Then, $x_1 \bar{\delta} y_1$ and because of Proposition 3.2 and Remark 3.1, we have $\mathcal{N}^\delta_{\varphi_{1,2}}(y) \le \dot{x}'$. Therefore, $\varphi_{1,2}.\text{int}_\delta \mu(y) \ge \bigwedge_{z \neq x} \mu(z)$ holds for all $\mu \in L^X$. Consider, $\mu = x_1'$ we get $\varphi_{1,2}.\text{int}_\delta x_1'(y) = 1$ and $x_1'(x) = 0$. Hence, there exists $\mu = x_1' \in L^X$ such that $\varphi_{1,2}.\text{int}_\delta\, \mu\,(y) = 1 > \mu\,(x)$, that is, $\dot{x} \not\le \mathcal{N}^\delta_{\varphi_{1,2}}(y)$ and therefore $(X,\varphi_{1,2}.\text{int}_\delta)$ is characterized FT_0-space.

In the following proposition, the $_{1,2}$-closure of the fuzzy subsets in the characterized fuzzy space $(X,\varphi_{1,2}.\text{int}_\delta)$ are equivalent with the fuzzy subsets by the fuzzy proximity δ on X.

Proposition 3.4

Let (X,τ) be a fuzzy topological space, $\varphi_1, \varphi_2 \in O_{(L^X,\tau)}$ such that $\varphi_2 \ge 1_{L^X}$ and δ is a fuzzy proximity on X. Then, $\mu \bar{\delta} \rho$ if and only if $\varphi_{1,2}.\text{cl}_\delta\ \mu \,\bar{\delta}\, \varphi_{1,2}.\text{cl}_\delta\ \rho$ for all $\mu, \rho \in L^X$.

Proof: Let $\mu, \rho \in L^X$ such that $\varphi_{1,2}.\text{cl}_\delta\ \mu \,\bar{\delta}\, \varphi_{1,2}.\text{cl}_\delta\ \rho$, then Proposition 3.2 implies $\mathcal{N}_{\varphi_{1,2}}(\varphi_{1,2}.\text{cl}_\delta \rho) \le (\varphi_{1,2}.\text{cl}_\delta \mu)'$. Since $\varphi_2 \ge 1_{L^X}$ and $\mathcal{N}_{\varphi_{1,2}}(\eta)$ is isotone operator, then $\mu \le \varphi_{1,2}.\text{cl}_\delta\, \mu$ and $\mathcal{N}_{\varphi_{1,2}}(\rho) \le \mathcal{N}_{\varphi_{1,2}}(\varphi_{1,2}.\text{cl}_\delta \rho)$ are hold for all $\mu,\rho \in L^X$. Hence, $\mathcal{N}_{\varphi_{1,2}}(\rho) \le \dot{\mu}'$ and therefore $\mu \,\bar{\delta}\, \rho$.

Conversely, Let μ, $\rho \in L^X$ such that $\mu \,\bar{\delta}\, \rho$. Because of Proposition 3.2 we have $_{1,2}(\rho) \le \mu'$. Since $\varphi_2 \ge 1_{L^X}$ and $\mathcal{N}^\delta_{\varphi_{1,2}}(\eta)$ is isotone operator, then $\mu' \le \varphi_{1,2}.\text{cl}_\delta\, \mu'$ holds for all $\mu' \in L^X$ and therefore $\dot{\rho} \le \mathcal{N}^\delta_{\varphi_{1,2}}(\mu') \le \mathcal{N}^\delta_{\varphi_{1,2}}(\varphi_{1,2}.\text{cl}_\delta \mu')$. From Lemma 3.1, we have $\mathcal{N}^\delta_{\varphi_{1,2}}(\varphi_{1,2}.\text{cl}_\delta \mu) \le \dot{\rho}'$ and then $\rho \,\bar{\delta}\, \varphi_{1,2}.\text{cl}_\delta\, \mu$. Therefore, $\varphi_{1,2}.\text{cl}_\delta\, \mu \le \mathcal{N}^\delta_{\varphi_{1,2}}(\rho')$ holds. Using Lemma 3.1 we get $\varphi_{1,2}.\text{cl}_\delta\, \mu \le \mathcal{N}^\delta_{\varphi_{1,2}}(\rho') \le \mathcal{N}^\delta_{\varphi_{1,2}}(\varphi_{1,2}.\text{cl}_\delta \rho')$. Thus, $\mathcal{N}^\delta_{\varphi_{1,2}}(\varphi_{1,2}.\text{cl}_\delta \rho) \le (\varphi_{1,2}.\text{ci}_\delta \mu)'$ and therefore $\varphi_{1,2}.\text{cl}_\delta \mu \,\bar{\delta}\, \varphi_{1,2}.\text{cl}_\delta \rho$ for all $\mu,\rho \in L^X$.

In the following proposition, we show that the associated characterized fuzzy proximity space $(X,\varphi_{1,2}.\text{int}_\delta)$ is characterized FR_2-space if the related fuzzy topological space (X,τ) is $F\varphi_{1,2}$-R_2 space.

Proposition 3.5

Let (X,τ) be a fuzzy topological space, $\varphi_1, \varphi_2 \in O_{(L^X,\tau)}$ and is an fuzzy proximity on X. Then the associated characterized fuzzy proximity space $(X,\varphi_{1,2}.\text{int}_\delta)$ is characterized FR_2-space if (X,τ) is $F\varphi_{1,2}$-R_2 space.

Proof: Let xX and $\mu \in L^X$ with $\mathcal{N}_{\varphi_{1,2}}(x) \le \dot{\mu}$. Because of Proposition 3.2, we have $\mu' \,\bar{\delta}\, \rho$ and from (P5), there is $\rho \in L^X$ such that $\mu' \,\bar{\delta}\, \rho$ and $\rho' \,\bar{\delta}\, x_1$. Therefore Proposition 3.4 implies $\varphi_{1,2}.\text{cl}_\delta\, \mu' \,\bar{\delta}\, \varphi_{1,2}.\text{cl}_\delta\, \rho$ and hence $\mathcal{N}^\delta_{\varphi_{1,2}}(\varphi_{1,2}.\text{cl}_\delta \rho) \le (\varphi_{1,2}.\text{cl}_\delta \mu')'$ and $\mathcal{N}^\delta_{\varphi_{1,2}}(x) \le \dot{\rho}$ are hold. Hence, $\mathcal{N}_{\varphi_{1,2}}(x) \le \dot{\mu}$ implies there is $\rho \in L^X$ such that $\mathcal{N}^\delta_{\varphi_{1,2}}(x) \le \dot{\rho}$ and $\mathcal{N}^\delta_{\varphi_{1,2}}(\varphi_{1,2}.\text{cl}\rho) \le \dot{\mu}$ are hold. Since $(X,\varphi_{1,2}.\text{int}_\delta)$ is $F\varphi_{1,2}$-R_2 space, then from Theorem 3.1 in Abd-Allah [12], we have $(X,\varphi_{1,2}.\text{int}_\delta)$ is characterized FR_2-space.

The binary relation << on L^X is said to be *fuzzy topogeneous order* on X [23], if the following conditions are fulfilled:

(1) $\bar{\alpha} \ll \bar{\alpha}$ for all $\alpha \in \{0,1\}$.

(2) If $\mu \ll \eta$, then $\mu \le \eta$ holds for all μ, $\in L^X$.

(3) If $_1 \le \mu \ll \eta \le \eta_1$, then $\mu_1 \ll \eta_1$ holds.

(4) If $_1 \ll \eta_1$ and $\mu_2 \ll \eta_2$, then $\mu_1 \wedge \mu_2 \ll \eta_1 \wedge \eta_2$ and $\mu_1 \vee \mu_2 \ll \eta_1 \vee \eta_2$ are hold for $\mu_i, \eta_j \in L^X$, where $i,j \in \{1,2\}$.

The fuzzy topogeneous order $\ll$ is said to be *fuzzy topogeneous structure* if it fulfilled the condition:

(5) If $\ll \eta$, then there is $\sigma \in L^X$ such that $<<\sigma$ and $\sigma << \eta$ are hold for all $\mu,\eta \in L^X$.

The fuzzy topogeneous structure $\ll$ is said to be *fuzzy topogenous complementarily symmetric* if it fulfilled the condition:

(6) If $<<\eta$, then $\eta' <<\mu$ holds for all $\mu,\eta \in L^X$.

As shown in Katsaras [23], every fuzzy topogeneous structure $\ll$ is identify with the mapping $N: L^X \to P(L^X)$ such that $\eta \in N(\mu)$ if and only if $\mu << \eta$ holds for all $\mu,\eta \in L^X$. The fuzzy topogeneous structures are classified by these mappings. As is easily seen, each fuzzy topogeneous order N can be associated a fuzzy pre topology int_N on a set X by defining $\mathrm{int}_{\mathcal{N}}\,\mu = \bigvee_{\mu \in \mathcal{N}(\eta)} \eta$ for all $\mu \in L^X$. In case of N is fuzzy topogeneous structure, int_N is interior operator for fuzzy topology τ_N on X associated to$N$. Obviously, there is an identification between the fuzzy proximity δ and the complementarily symmetric fuzzy topogenous structure $\ll$ on the same set X given by:

$$\mu \ll \eta' \Leftrightarrow \mu \bar{\delta} \eta \qquad (3.5)$$

for all $\mu,\eta \in L^X$. If $\{\ll_n\}_{n=1}^{\infty}$ is a sequence of fuzzy topogenous structure on the set X and $\{\prec_n\}_{n=1}^{\infty}$ is a sequence of fuzzy topogenous structure on I_L, then the fuzzy real function $f: X \to I_L$ is said to be *associated* with the sequence $\{\ll_n\}_{n=1}^{\infty}$ if and only if $\eta \prec_n \rho$ implies that $(\eta \circ f) \ll_{n+1} (\rho \circ f)$ holds for all $\eta, \rho \in L^{I_L}$ and $n \in \mathbb{Z}^+$, where $\mathbb{Z}^+$ is the set of all positive integer numbers.

Remark 3.2

Given that $\{\ll_n\}_{n=1}^{\infty}$ and $\{\prec_n\}_{n=1}^{\infty}$ are two sequence of complementarily symmetric fuzzy topogenous structures $\ll$ and $\prec$ on X and I_L, respectively. If δ and δ^* are two fuzzy proximities on X and I_L identified with δ and δ^* by the equation (3.5), then the associated fuzzy real function $f:(X, \varphi_{1,2}.\mathrm{int}_{\delta}) \to (Y, \varphi_{1,2}.\mathrm{int}_{\delta^*})$ with the complementarily symmetric fuzzy topogenous structures $\ll$ is $\varphi_{1,2}\psi_{1,2}\delta$-fuzzy continuous, because from (3.5) we get that $\eta \bar{\delta}^* \rho$ implies $(\eta \circ f)\,\bar{\delta}(\rho \circ f)$ for all $\eta, \rho \in L^{I_L}$.

Lemma 3.2

Consider $\ll_n$ for n $\{0,1,\ldots\}$ are complementarily symmetric fuzzy topogenous structures on a set X. Then, for each $F,G \in P(X)$ such that $\chi_F \ll_0 \chi_G$ there exists a fuzzy real function $f: X \to I_L$ associated with the sequence $\{\ll_n\}_{n=0}^{\infty}$ for which $f(x) = \bar{0}$ for all xF and $f(y) = \bar{1}$ for all yG' [23].

Because of equation (3.5), Remark 3.2 and Lemma 3.2, we can easily deduce the following proposition.

Proposition 3.6

Let $(X,\varphi_{1,2}.\mathrm{int}_{\delta})$ is a characterized fuzzy proximity space and $F,G \in P(X)$ such that $\chi_F \bar{\delta} \chi_G$. *If* Φ *is the family of all*$_{1,2}\psi_{1,2}\delta$-fuzzy continuous mappings $f:(X, \varphi_{1,2}.\mathrm{int}_{\delta}) \to (I_L, \psi_{1,2}.\mathrm{int}_{\delta^*})$ for which $x \in X$ implies $\bar{0} \le f(x) \le \bar{1}$, then χ_F and χ_G are Φ-separable.

Proof: Let $\ll$ be a complementarily symmetric fuzzy topogenous structure identified with δ. Because of (3.5), $\chi_F \bar{\delta} \chi_G$ implies that. Since $f \in \Phi$ is $\varphi_{1,2}\psi_{1,2}\delta$-fuzzy continuous, then because of Remark 3.2, we have that f is associated with $\ll$. Hence, Lemma 3.2 implies that χ_F and χ_G are separated by f and therefore χ_F and χ_G are Φ-separable.

Proposition 3.7

Let $(X,\psi_{1,2},\mathrm{int}_{\delta})$ and $(Y, \psi_{1,2}.\mathrm{int}_{\delta^*})$ are two characterized fuzzy proximity spaces. If the mapping $f:(X, \varphi_{1,2}.\mathrm{int}_{\delta}) \to (Y, \psi_{1,2}.\mathrm{int}_{\delta^*})$ is $_{1,2}\psi_{1,2}\delta$-fuzzy continuous, then the mapping f: $(X,\varphi_{1,2}.\mathrm{int}) \to (X,\psi_{1,2}.\mathrm{int})$ is $\varphi_{1,2}\psi_{1,2}$-fuzzy continuous.

Proof: Similar to the proof of Proposition 11.2 in Gähler [13].

In the following we are going to show an important relation between the associated characterized fuzzy proximity space and the characterized FR_3-space.

Proposition 3.8

Let (X,τ) be a fuzzy topological space and $\varphi_1, \varphi_2 \in O_{(L^X,\tau)}$ such that $\varphi_2 \ge 1_{L^X}$ is isotone and φ_1 is wfip with respect to $\varphi_1\mathrm{OF}(X)$, where L is complete chain. If (X,τ) is a fuzzy normal topological space, then the binary relation δ on X which is defined by:

$$\mu \bar{\delta} \rho \Leftrightarrow \mathcal{N}_{\varphi_{1,2}}(\varphi_{1,2}.\mathrm{cl}\,\mu) \le (\varphi_{1,2}.\mathrm{c\dot{i}}\,\rho)', \qquad (2.6)$$

for all $\mu,\rho \in L^X$ is a fuzzy proximity on X and (X,δ) is a fuzzy proximity space. On other hand if (X,δ) is a fuzzy proximity space with δ fulfills (3.6), then the associated characterized fuzzy proximity space $(X,\psi_{1,2},\mathrm{int}_{\delta})$ is characterized FR_3-space.

Proof: Let (X,τ) is fuzzy normal topological space and δ a binary relation on X defined by (3.6). Then, $\mu \bar{\delta} \rho$ implies $\mathcal{N}_{\varphi_{1,2}}(\varphi_{1,2}.\mathrm{cl}\,\mu) \le (\varphi_{1,2}.\mathrm{c\dot{i}}\,\rho)'$ and from Lemma 3.1 part (1) we get $\mathcal{N}_{\varphi_{1,2}}(\varphi_{1,2}.\mathrm{cl}\,\rho) \le (\varphi_{1,2}.\mathrm{c\dot{i}}\,\mu)'$ and then $\rho \bar{\delta} \mu$. Hence, condition (P1) is fulfilled. For showing condition (P2), let $(\mu \vee \rho)\bar{\delta}\eta$ for a fixed fuzzy subsets $\mu,\rho,\eta \in L^X$. Then, $\mathcal{N}_{\varphi_{1,2}}(\varphi_{1,2}.\mathrm{cl}(\mu \vee \rho)) \le (\varphi_{1,2}.\mathrm{c\dot{i}}\,\eta)'$. Since L is complete chain, $\varphi_2 \ge 1_{L^X}$ is isotone and φ_1 is wfip with respect to $\varphi_1 OF(X)$, then $\varphi_{1,2}.\mathrm{cl}(\mu \vee \rho) = \varphi_{1,2}.\mathrm{cl}\,\mu \vee \varphi_{1,2}.\mathrm{cl}\,\rho$ and therefore $\mathcal{N}_{\varphi_{1,2}}(\varphi_{1,2}.\mathrm{cl}\,\mu \vee \varphi_{1,2}.\mathrm{cl}\,\rho) \le (\varphi_{1,2}.\mathrm{c\dot{i}}\,\eta)'$. Because of Lemma 3.1 part (3), we have $\mathcal{N}_{\varphi_{1,2}}(\varphi_{1,2}.\mathrm{cl}\,\mu) \le (\varphi_{1,2}.\mathrm{c\dot{i}}\,\eta)'$ and $\mathcal{N}_{\varphi_{1,2}}(\varphi_{1,2}.\mathrm{cl}\,\rho) \le (\varphi_{1,2}.\mathrm{c\dot{i}}\,\eta)'$ are hold and therefore $\mu \bar{\delta} \eta$ and $\rho \bar{\delta} \eta$. Thus, $(\mu \vee \rho)\bar{\delta}\eta$ implies $\mu \bar{\delta} \eta$ and $\rho \bar{\delta} \eta$. On the other hand let $\mu \bar{\delta} \eta$ and $\rho \bar{\delta} \eta$. Then from Lemma 3.1 we have tha the inequalities $\mathcal{N}_{\varphi_{1,2}}(\varphi_{1,2}.\mathrm{cl}\,\mu) \le (\varphi_{1,2}.\mathrm{c\dot{i}}\,\eta)'$ and $\mathcal{N}_{\varphi_{1,2}}(\varphi_{1,2}.\mathrm{cl}\,\rho) \le (\varphi_{1,2}.\mathrm{c\dot{i}}\,\eta)'$ are hold and therefore $\mathcal{N}_{\varphi_{1,2}}(\varphi_{1,2}.\mathrm{cl}(\mu \vee \rho)) = \mathcal{N}_{\varphi_{1,2}}(\varphi_{1,2}.\mathrm{cl}\,\mu) \wedge \mathcal{N}_{\varphi_{1,2}}(\varphi_{1,2}.\mathrm{cl}\,\rho) \le (\varphi_{1,2}.\mathrm{c\dot{i}}\,\eta)'$, that is, $\mu \bar{\delta} \eta$ and $\rho \bar{\delta} \eta$ imply $(\mu \vee \rho)\bar{\delta}\eta$. Hence, (P2) is fulfilled. Now, let μ, ρ $\in L^X$ such that $\mu = \bar{0}$ or $\rho = \bar{0}$. Since $\mathcal{N}_{\varphi_{1,2}}(\bar{0})$ is the finest fuzzy filter on X and from the fact $\varphi_{1,2}.\mathrm{cl}\,\bar{0} = \bar{0}$, we get $\mathcal{N}_{\varphi_{1,2}}(\bar{0}) = \mathcal{N}_{\varphi_{1,2}}(\varphi_{1,2}.\mathrm{cl}\,\bar{0}) \le (\varphi_{1,2}.\mathrm{cl}\,\rho)'$ holds for all $\rho \in L^X$. Thus, $\bar{0}\,\bar{\delta}\,\rho$ for all $\rho \in L^X$. Since $\mu = \bar{0}$ or $\rho = \bar{0}$, then we have $\mu \bar{\delta} \rho$, that is, (P3) is also fulfilled. Since $\mu \bar{\delta} \rho$ implies $\mathcal{N}_{\varphi_{1,2}}(\varphi_{1,2}.\mathrm{cl}\,\mu) \le (\varphi_{1,2}.\mathrm{c\dot{i}}\,\rho)'$ which means by the inequality $(\varphi_{1,2}.\mathrm{c\dot{i}}\,\mu) \le \mathcal{N}_{\varphi_{1,2}}(\varphi_{1,2}.\mathrm{cl}\,\mu)$ that $(\varphi_{1,2}.\mathrm{c\dot{i}}\,\mu) \le (\varphi_{1,2}.\mathrm{c\dot{i}}\,\rho)'$. Because of Proposition 2.1 and the fact that $\varphi_{1,2}$.cl is hull operator we get $\mu \le \varphi_{1,2}.\mathrm{cl}\,\mu \le (\varphi_{1,2}.\mathrm{cl}\,\rho)' \le \rho'$. Thus, (P4) is fulfilled. Let $\mu, \rho \in L^X$ such that $\mu \bar{\delta} \rho$, then $\mathcal{N}_{\varphi_{1,2}}(\varphi_{1,2}.\mathrm{cl}\,\mu) \le (\varphi_{1,2}.\mathrm{c\dot{i}}\,\rho)'$. Consider, $F = S_0(\varphi_{1,2}.\mathrm{cl}\,\mu)$, hence $F \in \varphi_{1,2}C(X)$ and therefore $\mathcal{N}_{\varphi_{1,2}}(F) \le (\varphi_{1,2}.\mathrm{c\dot{i}}\,\rho)'$ holds. Since (X, τ) is characterized fuzz normal space, then from Theorem 3.2 in Abd-Allah [12], there exists $\eta' \in L^X$ with arbitrary choice such that $\mathcal{N}_{\varphi_{1,2}}(F) \le \eta'$ and $\mathcal{N}_{\varphi_{1,2}}(\varphi_{1,2}.\mathrm{cl}\,\eta') \le (\varphi_{1,2}.\mathrm{c\dot{i}}\,\rho)'$ are hold. Therefore, there exists $\in L^X$ such that $\mathcal{N}_{\varphi_{1,2}}(\varphi_{1,2}.\mathrm{cl}\,\mu) \le (\varphi_{1,2}.\mathrm{c\dot{i}}\,\eta)'$ and $\mathcal{N}_{\varphi_{1,2}}(\varphi_{1,2}.\mathrm{cl}\,\mu) \le (\varphi_{1,2}.\mathrm{c\dot{i}}\,\eta)'$, which means that $\mu \bar{\delta} \eta$ and $\eta' \bar{\delta} \rho$. Hence, (P5) is also fulfilled. Consequently, δ is a fuzzy proximity on X.

Conversely, let $F_1, F_2 \in \varphi_{1,2}C(X)$ such that $F_1 \cap F_2 = \emptyset$. Then, $F_1 \subseteq F_2'$ and therefore $\dot{\chi}_{F_1} \le \dot{\chi}_{F_2'} = \dot{\chi}_{F_2}'$. Hence because of Lemma 3.1 part (1) we have $\mathcal{N}^{\delta}_{\varphi_{1,2}}(\chi_{F_2}) \leq \dot{\chi}'_{F_1}$. Since $F_1, F_2 \in \varphi_{1,2}C(X)$, then $\mathcal{N}^{\delta}_{\varphi_{1,2}}(\varphi_{1,2}.\mathrm{cl}_{\delta}\,\chi_{F_2}) = \mathcal{N}^{\delta}_{\varphi_{1,2}}(\chi_{F_2}) \leq \dot{\chi}'_{F_1} = (\varphi_{1,2}.\mathrm{cl}_{\delta}\,\chi_{F_1})'$ and therefore $\chi_{F_1}\,\bar{\delta}\,\chi_{F_2}$. From (P5), there exists $\rho \in L^X$ such that $\mathcal{N}^{\delta}_{\varphi_{1,2}}(\chi_{F_2}) = \mathcal{N}^{\delta}_{\varphi_{1,2}}(F_2) \leq \dot{\rho}$ and $\mathcal{N}^{\delta}_{\varphi_{1,2}}(\rho) \leq \dot{\chi}'_{F_1} = F'_1$ are hold. Because of Lemma 3.1 part (1), we have $\mathcal{N}^{\delta}_{\varphi_{1,2}}(F_1) \leq \dot{\rho}'$. Hence, $\mathcal{N}^{\delta}_{\varphi_{1,2}}(F_1)(\mu) \wedge \mathcal{N}^{\delta}_{\varphi_{1,2}}(F_2)(\eta) \geq \dot{\rho}'(\mu) \wedge \dot{\rho}'(\eta)$ holds for all. Consider $\eta = \chi_{F_2} \vee x_1 = \rho$ and $\mu = (\chi_{F_2} \vee x_1)' = \rho'$ for all $x \in F'_1 \setminus F_2$, then we get $\sup(\mu\wedge) = 0$ and $\mathcal{N}^{\delta}_{\varphi_{1,2}}(F_1)(\mu) \wedge \mathcal{N}^{\delta}_{\varphi_{1,2}}(F_2)(\eta) \geq 0$. Hence, there exist $\mu,\eta \in L^X$ such that $\mathcal{N}^{\delta}_{\varphi_{1,2}}(F_1)(\mu) \wedge \mathcal{N}^{\delta}_{\varphi_{1,2}}(F_2)(\eta) \geq \sup(\mu \wedge \eta)$, that is, the infimum $\mathcal{N}^{\delta}_{\varphi_{1,2}}(F_1) \wedge \mathcal{N}^{\delta}_{\varphi_{1,2}}(F_2)$ does not exists. Consequently, $(X,\psi_{1,2}.\mathrm{int}_{\delta})$ is characterized FR_3-space.

In the following we are going to show an important relation between the associated characterized fuzzy proximity space $(X,\psi_{1,2}.\mathrm{int}_{\delta})$ by the fuzzy proximity δ defined by (3.6) and the associated characterized fuzzy space $(X,\psi_{1,2}.\mathrm{int})$ that introduced form the fuzzy normal topological space (X,τ).

Proposition 3.9

Let (X,τ) is a fuzzy normal topological space and $\varphi_1,\varphi_2 \in O_{(L^X,\tau)}$ such that $\varphi_2 \geq 1_{L^X}$ is isotone and φ_1 is wfip with respect to $\varphi_1 OF(X)$. If δ is the fuzzy proximity on X defined by (3.6) and L is a complete chain, then $(X,\varphi_{1,2}.\mathrm{int}_{\tau})$ is finer than $(X,\varphi_{1,2}.\mathrm{int}_{\delta})$. Moreover, $(X,\varphi_{1,2}.\mathrm{int}_{\tau}) = (X,\varphi_{1,2}.\mathrm{int}_{\delta})$ if and only if $(X,\varphi_{1,2}.\mathrm{int}_{\tau})$ is characterized FR_4-space.

Proof: Let (X,τ) is fuzzy normal topological space and μ is $\varphi_{1,2}\delta$-fuzzy neighborhood for the point xX, then $x_1\,\bar{\delta}\,\mu'$ and because of (3.6), we have $\mathcal{N}_{\varphi_{1,2}}(\varphi_{1,2}.\mathrm{cl}_{\tau}x_1) \leq (\varphi_{1,2}.\mathrm{c\dot{l}}_{\tau}\,\mu')'$. Therefore, $\dot{x} \leq \mathcal{N}_{\varphi_{1,2}}(x) \leq \mathcal{N}_{\varphi_{1,2}}(\varphi_{1,2}.\mathrm{cl}_{\tau}\{x\}) = \mathcal{N}_{\varphi_{1,2}}(\varphi_{1,2}.\mathrm{cl}_{\tau}(x_1)) \leq (\varphi_{1,2}.\mathrm{c\dot{l}}_{\tau}\,\mu')' \leq \dot{\mu}$. Because of Proposition 2.1, we get $x_1 \leq (\varphi_{1,2}.\mathrm{cl}_{\tau}\,\mu')' \leq \dot{\mu}$ and $(\varphi_{1,2}.\mathrm{cl}_{\tau}\,\mu')' \in \varphi_{1,2}OF(X)$. Then, μ is $_{1,2}$-fuzzy neighborhood of x and therefore the family $(\varphi_1 OF(X))_{\delta}$ is coarser than the family $(\varphi_1 OF(X))$, that is, $(X,\varphi_{1,2}.\mathrm{int}_{\tau})$ is finer than $(X,\varphi_{1,2}.\mathrm{int}_{\delta})$.

Now, let $(X,\varphi_{1,2}.\mathrm{int}_{\tau})$ is characterized FR_4-space, $\mathcal{N}_{\varphi_{1,2}}$ and $\mathcal{N}^{\delta}_{\varphi_{1,2}}(x)$ denote for the $\varphi_{1,2}$-fuzzy neighborhood filters at x in the characterized fuzzy space $(X,\varphi_{1,2}.\mathrm{int}_{\tau})$ and in the associated characterized fuzzy proximity space $(X,\varphi_{1,2}.\mathrm{int}_{\delta})$, respectively. Then, $(X,\varphi_{1,2}.\mathrm{int}_{\tau})$ is characterized FR_3 and FR_1-space. Therefore, $(\varphi_{1,2}OF(X))_{\delta} \subseteq (\varphi_{1,2}OF(X))$ and $\mathcal{N}_{\varphi_{1,2}}(x) \leq \mathcal{N}^{\delta}_{\varphi_{1,2}}(x)$ holds for all $y \neq x$ in X. Hence, $\mathcal{N}_{\varphi_{1,2}}(x) \leq \mathcal{N}^{\delta}_{\varphi_{1,2}}(x)$ holds for all $x \in X$ and then $\mathcal{N}^{\delta}_{\varphi_{1,2}}(x) \geq \mathcal{N}_{\varphi_{1,2}}(x) \not\leq \dot{y}$ holds for all $y \neq x$ in X. Because of Lemma 2.1, we have that $\mathcal{N}^{\delta}_{\varphi_{1,2}}(x) \not\leq \dot{y}$ holds for all $y \neq x$ in X and therefore $(X,\varphi_{1,2}.\mathrm{int}_{\delta})$ is characterized FT_1-space. Because of Proposition 2.4 and Lemma 2.2, we get $\varphi_{1,2}.\mathrm{cl}_{\tau}(x_1) = x_1$ for all $x \in X$ and therefore $x_1 \in (\varphi_{1,2}CF(X))_{\delta}$ for all xX. Consider μ is the $\varphi_{1,2}$-fuzzy neighborhood of x in $(X,\varphi_{1,2}.\mathrm{int}_{\tau})$, then $\mu' \leq x'_1$ and since $x'_1 \in (\varphi_{1,2}OF(X))_{\delta}$, then x'_1 is a $_{1,2}$-fuzzy neighborhood for every $y \in X$ such that $y_1 \leq \mu'$. Thus, $\mu'\,\bar{\delta}\,x_1$ and hence μ is $\varphi_{1,2}\delta$-fuzzy neighborhood of x in $(X,\varphi_{1,2}.\mathrm{int}_{\delta})$. Thus, $(\varphi_{1,2}OF(X)) \subseteq (\varphi_{1,2}OF(X))_{\delta}$, that is, $\mathcal{N}^{\delta}_{\varphi_{1,2}}(x) \leq \mathcal{N}_{\varphi_{1,2}}(x)$ holds for all $x \in X$ and therefore $(X,\varphi_{1,2}.\mathrm{int}_{\delta})$ is finer than $(X,\varphi_{1,2}.\mathrm{int}_{\tau})$. Consequently, $(X,\varphi_{1,2}.\mathrm{int}_{\tau})$ is characterized FR_4-space implies that $(X,\varphi_{1,2}.\mathrm{int}_{\tau}) = (X,\varphi_{1,2}.\mathrm{int}_{\delta})$.

Conversely, let $(X,\varphi_{1,2}.\mathrm{int}_{\tau}) = (X,\varphi_{1,2}.\mathrm{int}_{\delta})$, $x \in X$ and μ is $\varphi_{1,2}$-fuzzy neighborhood of x in $(X,\varphi_{1,2}.\mathrm{int}_{\tau})$. Then, $\mu \in (\varphi_{1,2}OF(X))_{\delta}$ and $x_1 \leq \mu$, this means that $(\varphi_{1,2}.\mathrm{c\dot{l}}_{\tau}x_1) \leq \mathcal{N}_{\varphi_{1,2}}(\varphi_{1,2}.\mathrm{cl}_{\tau}x_1) \leq (\varphi_{1,2}.\mathrm{c\dot{l}}_{\tau}\,\mu')' \leq \dot{\mu}$. Because of Proposition 2.1, we get $\varphi_{1,2}.\mathrm{cl}_{\tau}x_1 \leq \mu$ and therefore $\varphi_{1,2}.\mathrm{cl}_{\tau}x_1 \leq x_1$ holds for all $x \in X$. Thus, $\varphi_{1,2}.\mathrm{cl}_{\tau}x_1 = x_1$ for all $x \in X$. Hence, Proposition 2.4 implies that, $(X,\varphi_{1,2}.\mathrm{int}_{\tau})$ is characterized FR_1-space. Because of Proposition 3.7, $(X,\varphi_{1,2}.\mathrm{int}_{\delta})$ is characterized FR_3-space and the hypothesis that $(X,\varphi_{1,2}.\mathrm{int}_{\tau}) = (X,\varphi_{1,2}.\mathrm{int}_{\delta})$, implies that $(X,\varphi_{1,2}.\mathrm{int}_{\tau})$ is characterized FR_3-space. Consequently, $(X,\varphi_{1,2}.\mathrm{int}_{\tau})$ is characterized FR_4-space.

In the following we are going to introduce some important relations joining our characterized $FR_{2\frac{1}{2}}$- spaces, characterized $FT_{2\frac{1}{2}}$-spaces and the associated characterized fuzzy proximity spaces.

Proposition 3.10

Let (X,τ) be an fuzzy topological space and $\varphi_1,\varphi_2 \in O_{(L^X,\tau)}$. If δ is an fuzzy proximity on X, then the associated characterized fuzzy proximity space $(X,\varphi_{1,2}.\mathrm{int}_{\delta})$ is characterized $FR_{2\frac{1}{2}}$- space.

Proof: Let xX and $F\varphi_{1,2}C(X)$ such that $x \notin F$. Since $\chi_{F'}$ is $_{1,2}\delta$-fuzzy neighborhood of x, then $x_1\,\bar{\delta}\,\chi_F$. Because of Proposition 3.2, we get that x_1 and χ_F are Φ-separated by the $\varphi_{1,2}\psi_{1,2}\delta$-fuzzy continuous mapping $f:(X,\varphi_{1,2}.int_{\delta}) \to (I_L,\psi_{1,2}.int_{\delta^*})$ for which $\bar{0} \leq f(x) \leq \bar{1}$, that is, $f(x) = \bar{1}$ and $f(y) = \bar{1}$ for all $y \in F$. Consequently, $(X,\varphi_{1,2}.\mathrm{int}_{\delta})$ is characterized $FR_{2\frac{1}{2}}$- space.

Corollary 3.1

Let (X,τ) be a fuzzy topological space, $\varphi_1,\varphi_2 \in O_{(L^X,\tau)}$ and δ is a fuzzy proximity on X. Then the associated characterized fuzzy proximity space $(X,\varphi_{1,2}.\mathrm{int}_{\delta})$ is characterized $FT_{2\frac{1}{2}}$- space.

Proof: Immediately from Propositions 2.4 and 3.10.

Now, we introduce an example of an fuzzy proximity δ on a set X and show that it is induces an associated characterized $FT_{2\frac{1}{2}}$- space compatible with the related characterized fuzzy space.

Example 3.1

Let $L = \{0,\frac{1}{2},1\}$, $X = \{x,y\}$ and $= \{\bar{1},\bar{0},x_1,y_1\}$ is a fuzzy topology on X. Choose $\varphi_1 = \mathrm{int}_{\tau}$, $\varphi_2 = \mathrm{cl}_{\tau}$, $\psi_1 = \mathrm{int}_I$, and $\psi_2 = \mathrm{cl}$. Hence, $x \neq y$ and there is only two cases, the first is $x \notin F = \{y\} \in \varphi_{1,2}C(X)$ and the second is $y \notin F = \{x\} \in \varphi_{1,2}C(X)$. We shall consider the first case and the second case is similar. Consider the mapping $f:(X,\varphi_{1,2}.\mathrm{int}_{\tau}) \to (I_L,\psi_{1,2}.\mathrm{int}_I)$ defined by $f(x) = \bar{1}$ and $f(y) = \bar{0}$, then f is $_{1,2}\psi_{1,2}$-fuzzy continuous and therefore $(X,\varphi_{1,2}.\mathrm{int}_{\tau})$ is characterized $FR_{2\frac{1}{2}}$- space and obviously $(X,\varphi_{1,2}.\mathrm{int}_{\tau})$ is also characterized FR_1-space, that is, $(X,\varphi_{1,2}.\mathrm{int}_{\tau})$ is characterized $FT_{3\frac{1}{2}}$- space. Now, consider δ is a binary relation on L^X defied as follows:

$$\mu\,\bar{\delta}\,\eta \Leftrightarrow \exists\, \varphi_{1,2}\psi_{1,2}\text{-fuzzy continuous mapping } f:(X,\varphi_{1,2}.\mathrm{int}_{\tau}) \to (I_L,\psi_{1,2}.\mathrm{int}_I) \ni f(x) = \bar{1} \text{ for all } x \in X$$

$$\text{with } x_1 \leq \mu \text{ and } f(y) = \bar{0} \text{ for all } y_1 \leq \eta,$$

for all $\mu,\eta \in L^X$. Hence obviously, δ is a fuzzy proximity on X and $(X,\varphi_{1,2}.\mathrm{int}_{\tau}) = (X,\varphi_{1,2}.\mathrm{int}_{\delta})$, that is, the associated characterized fuzzy proximity space $(X,\varphi_{1,2}.\mathrm{int}_{\delta})$ with δ is characterized $FT_{3\frac{1}{2}}$- space and compatible with $(X,\varphi_{1,2}.\mathrm{int}_{\tau})$.

Some Relations between Characterized FT_s and Characterized Fuzzy Compact Spaces

The notion of $\varphi_{1,2}$-fuzzy compactness of the fuzzy filters and of the fuzzy topological spaces are introduced by Abd-Allah in [7] by means of the $\varphi_{1,2}$-fuzzy convergence in the characterized fuzzy spaces. Moreover, the fuzzy compactness in the characterized fuzzy spaces is also introduced by means of the $\varphi_{1,2}$-fuzzy compactness of the fuzzy filters and therefore it will be suitable to study here the relations between the characterized fuzzy compact spaces and some of our classes of separation axioms in the characterized fuzzy spaces.

Let (X,τ) be an fuzzy topological space, $F \subseteq X$ and $\varphi_1, \varphi_2 \in O_{(L^X,\tau)}$. Then $x \in X$ is said to be $\varphi_{1,2}$-adherence point for the fuzzy filter $\mathcal{M}$ on X [7], if the infimum $\mathcal{M} \wedge \mathcal{N}_{\varphi_{1,2}}(x)$ exists for all $\varphi_{1,2}$-fuzzy neighborhood filters $\mathcal{N}_{\varphi_{1,2}}(x)$ at $x \in X$. As shown in Abd-Allah [7], the point $x \in X$ is said to be $\varphi_{1,2}$-adherence point for the fuzzy filter $\mathcal{M}$ on X if and only if there exists an fuzzy filter $\mathcal{K} \in \mathcal{F}_L X$ finer than M and $\mathcal{K} \xrightarrow[\varphi_{1,2}.\text{int}]{} x$, that is, $\mathcal{K} \le \mathcal{M}$ and $\mathcal{K} \le \mathcal{N}_{\varphi_{1,2}}(x)$ are hold for some $\mathcal{K} \in \mathcal{F}_L X$. The subset F of X is said to be $\varphi_{1,2}$-fuzzy closed with respect to $\varphi_{1,2}$.int if $\mathcal{M} \le \mathcal{N}_{\varphi_{1,2}}(x)$ implies $x \in F$ for some $\mathcal{M} \in \mathcal{F}_L F$. The subset F is said to be $\varphi_{1,2}$-fuzzy compact [7], if every fuzzy filter on F has a finer $\varphi_{1,2}$–fuzzy converging filter, that is, every fuzzy filter on F has $\varphi_{1,2}$-adherence point in F. Moreover, the fuzzy topological space (X,τ) is said to be $\varphi_{1,2}$-fuzzy compact if X is $\varphi_{1,2}$-fuzzy compact. More generally, the characterized fuzzy space $(X,\varphi_{1,2}.\text{int})$ is said to be fuzzy compact space if the related fuzzy topological space (X,τ) is $\varphi_{1,2}$-fuzzy compact.

At first, in the following we shall benefit from these facts. Consider the fuzzy unit interval topological space $(I_L,\Im)$ be given and $\psi_1, \psi_2 \in O_{(I_L,\Im)}$. Then:

(1) The usual topological space (I,T_I) and the ordinary characterized usual space $(I,\psi_{1,2}.\text{int}_{T_I})$ on the closed unite interval $I = [0,1]$ are $\psi_{1,2}$-compact T_2 space and characterized compact T_2-space, respectively in the classical sense.

(2) The closed unite interval I is identified with the fuzzy number $[0,1]^\sim$ in Gähler [24] defined by $[0,1]^\sim(\alpha) = 0$ for all $\alpha \in I$ and $[0,1]^\sim(\alpha) = 0$ for all α Ï I.

(3) The characterized fuzzy unite space $(I_L,\psi_{1,2}.\text{int}_\Im)$ is up to an identification the characterized usual space $\varphi_1, \varphi_2 \in O_{(L^X,\tau)}$ in the classical sense.

In the following proposition, we show that every $\varphi_{1,2}$-fuzzy compact subset in the characterized FT_2–space $(X,\varphi_{1,2}.\text{int}_\tau)$ is $\varphi_{1,2}$-fuzzy closed with respect to the $\varphi_{1,2}$-interior operator $\varphi_{1,2}.\text{int}_\tau$.

Proposition 4.1

Let a fuzzy topological space (X,τ) be fixed and $\varphi_1, \varphi_2 \in O_{(L^X,\tau)}$. Then every $\varphi_{1,2}$-fuzzy compact subset of the characterized FT_2-space is $_{1,2}$-closed.

Proof: Let $(X,\varphi_{1,2}.\text{int}_\tau)$ is characterized FT_2-space and F is $\varphi_{1,2}$-fuzzy compact subset of X. Then, for all $\mathcal{M} \in \mathcal{F}_L F$, there exists $\mathcal{K} \in \mathcal{F}_L F$ such that $\mathcal{K} \le \mathcal{M}$ and $\mathcal{K} \le \mathcal{N}_{\varphi_{1,2}}(x)$ for some $x \in F$. Since $\mathcal{K} \in \mathcal{F}_L F \le F_L X$ and $(X,\varphi_{1,2}.\text{int}_\tau)$ is characterized FT_2-space, then $\mathcal{K} \le \mathcal{N}_{\varphi_{1,2}}(x)$ and $\mathcal{K} \le \mathcal{N}_{\varphi_{1,2}}(y)$ imply that $x = y$. Therefore, $y \in F$ for some $\mathcal{K} \in \mathcal{F}_L F$. Hence, F is $\varphi_{1,2}$-fuzzy closed with respect to $_{1,2}.\text{int}_\tau$.

Proposition 4.2

Let $(I_L,\Im)$ be a fuzzy unit interval topological space and $\psi_1, \psi_2 \in O_{(L^{I_L},\Im)}$. Then the characterized fuzzy unit interval space $(I_L,\psi_{1,2}.\text{int}_I)$ is characterized fuzzy compact FT_2-space.

Proof: Let $(I,\psi_{1,2}.\text{int}_{T_I})$ be an ordinary characterized usual space. Then, $(I,\psi_{1,2}.\text{int}_{T_I})$ is characterized compact space in the classical sense, that is, every filter on I has $\psi_{1,2}$-adherence point. Consider the mapping $f:(I,\psi_{1,2}.\text{int}_{T_I}) \to (I_L,\psi_{1,2}.\text{int}_\Im)$ defined by: $f(\alpha) = \tilde{\alpha}$ for all $\alpha \in I$, then it is easily to seen that f is $\psi_{1,2}\psi_{1,2}$-fuzzy homeomorphism between $(I,\psi_{1,2}.\text{int}_{T_I})$ and $(I_L,\psi_{1,2}.\text{int}_I)$. Therefore, $(I_L,\psi_{1,2}.\text{int}_I)$ is characterized fuzzy compact space. Since (I,T_I) is $\psi_{1,2}T_2$-space, then $(I,\psi_{1,2}.int_{T_I})$ is characterized FT_2-space and therefore by using the same $_{1,2}\psi_{1,2}$-fuzzy homeomorphism, we have for all $\tilde{\alpha}, \tilde{\beta} \in I_L$ such that $\tilde{\alpha} \ne \tilde{\beta}$, the infimum $\mathcal{N}_{\psi_{1,2}}(\tilde{\alpha}) \wedge \mathcal{N}_{\psi_{1,2}}(\tilde{\beta})$ does not exists. Consequently, $(I_L,\psi_{1,2}.\text{int}_I)$ is characterized FT_2-space and therefore $(I_L,\psi_{1,2}.\text{int}_I)$is characterized fuzzy compact FT_2-space.

Now, we are going to prove an important relation between the characterized compact FT_2-spaces and the characterized FT_4-spaces. For this reason at first, we give a new property for the characterized FT_2-spaces by using the $\varphi_{1,2}$-fuzzy neighborhood filters for the fuzzy subsets.

Proposition 4.3

Let (X,τ) be n fuzzy topological space and $\varphi_1, \varphi_2 \in O_{(L^X,\tau)}$. Then every disjoint $\varphi_{1,2}$-fuzzy compact subsets F_1 and F_2 of in the characterized FT_2-space $(X,\varphi_{1,2}.\text{int}_\tau)$ have two disjoint $\varphi_{1,2}$-fuzzy neighborhood filters $\mathcal{N}_{\varphi_{1,2}}(F_1)$ and $\mathcal{N}_{\varphi_{1,2}}(F_2)$ for which F_1 and F_2 are separated by them.

Proof: Let F_1 and F_2 are two $\varphi_{1,2}$-fuzzy compact subsets of the characterized FT_2-space $(X,\varphi_{1,2}.\text{int}_\tau)$ such that $F_1 \cap F_2 = \emptyset$. Then, for all $\mathcal{M}_i \in \mathcal{F}_L F_i$ there exists $\mathcal{K}_i \in \mathcal{F}_L F_i$ such that $\mathcal{K}_i \le \mathcal{M}_i$ and $\mathcal{K}_i \le \mathcal{N}_{\varphi_{1,2}}(x_i)$ for some $x_i F_i$, where $i \in \{1,2\}$. Since $\mathcal{F}_L F_i \le \mathcal{F}_L X$ for all $i \in \{1,2\}$, then we can say that $\mathcal{K}_i \le \mathcal{N}_{\varphi_{1,2}}(x_i) \le \mathcal{N}_{\varphi_{1,2}}(F_i)$ and therefore there is $\mathcal{K} = \mathcal{K}_1 \wedge \mathcal{K}_2 \in \mathcal{F}_L X$ such that $\mathcal{K} \le \mathcal{N}_{\varphi_{1,2}}(x_i)$ for some $x_i F_i$. Since $(X,\varphi_{1,2}.\text{int}_\tau)$ is characterized FT_2-space, then $x_1 = x_2$ which contradicts $F_1 \cap F_2 = \emptyset$. Hence, for every $\mathcal{L} \in \mathcal{F}_L X$ we get $\mathcal{L} \nleq \mathcal{N}_{\varphi_{1,2}}(F_1)$ or $\mathcal{L} \nleq \mathcal{N}_{\varphi_{1,2}}(F_2)$ which means that the infimum $\mathcal{N}_{\varphi_{1,2}}(F_1) \wedge \mathcal{N}_{\varphi_{1,2}}(F_2)$ does not exists and therefore F_1 and F_2 can be separated by two disjoint $\varphi_{1,2}$-fuzzy neighborhood filters.

Secondly, the notion of the fuzzy compactness for the characterized fuzzy spaces fulfills the following property which will be also used in the prove of this important result which given in Proposition 4.4.

Lemma 4.1

Let (X,τ) be a fuzzy topological space and $\varphi_1, \varphi_2 \in O_{(L^X,\tau)}$. Then every $\varphi_{1,2}$-fuzzy closed subset of the characterized fuzzy compact space $(X,\varphi_{1,2}.\text{int}_\tau)$ is $\varphi_{1,2}$-fuzzy compact.

Proof: Let F is $\varphi_{1,2}$-fuzzy closed subset of the characterized fuzzy compact space $(X,\varphi_{1,2}.\text{int}_\tau)$ and let $\mathcal{M} \in \mathcal{F}_L F$. Then, $\mathcal{M} \le \mathcal{N}_{\varphi_{1,2}}(x)$ implies that $x \in F$. Since $\mathcal{F}_L F \le \mathcal{F}_L X$, then $\mathcal{M} \in \mathcal{F}_L X$ and hence there exists $\mathcal{K} \in \mathcal{F}_L X$ such that $\mathcal{K} \le \mathcal{M}$ and $\mathcal{K} \le \mathcal{N}_{\varphi_{1,2}}(x)$. Since $\mathcal{M} \in \mathcal{F}_L F$ and $\mathcal{K} \le \mathcal{M}$, then $\mathcal{K} \in \mathcal{F}_L F$. Thus, for all $M \in F_L F$ we get $\mathcal{K} \le \mathcal{M}$ such that $\mathcal{K} \le \mathcal{N}_{\varphi_{1,2}}(x)$. Therefore, $x \in F$ is $\varphi_{1,2}$-adherence point of $\mathcal{M}$, that is, F is $\varphi_{1,2}$-fuzzy compact.

Proposition 4.4

Let (X,τ)be an fuzzy topological space and $\varphi_1,\varphi_2 \in O_{(L^X,\tau)}$. Then every characterized fuzzy compact FT_2-space $(X,\varphi_{1,2}.\text{int}_\tau)$ is characterized FT_4-space.

Proof: Follows directly from Lemma 4.1 and Proposition 4.3.

One of the application of Proposition 4.4, we have more generally the following result to the characterized fuzzy unit interval space.

Proposition 4.5

Let $(I_L,\Im)$ be an fuzzy unit interval topological space and $\psi_1,\psi_2 \in O_{(L^{I_L},\Im)}$. Then the characterized fuzzy unit interval space $(I_L,\psi_{1,2}.\text{int}_I)$ is characterized $FT_{3\frac{1}{2}}$- space.

Proof: Because of Proposition 4.2, the characterized fuzzy unit interval space $(I_L,\psi_{1,2}.\text{int}_I)$ is characterized fuzzy compact FR_2-space. Therefore from Proposition 4.4, we get $(I_L,\psi_{1,2}.\text{int}_I)$ is characterized FR_4-space. Hence, Proposition 4.6 in Abd-Allah [11] gives us that, $(I_L,\psi_{1,2}.\text{int}_I)$ is characterized $FT_{3\frac{1}{2}}$- space.

The $\varphi_{1,2}$-fuzzy compactness in the characterized fuzzy spaces is applied to fulfilled the Generalized Tychonoff Theorem [11] and from (2) in Proposition 2.6, the characterized fuzzy product space of the characterized FR_2-spaces is also characterized FR_2-space. Hence, by means of Propositions 4.2 and 4.4, the following result goes clear.

Proposition 4.6

Let $(I_L,\Im)$ be a fuzzy unit interval topological space and $\psi_1,\psi_2 \in O_{(L^{I_L},\Im)}$. Then the characterized fuzzy cube is characterized FT_2-space and it is characterized FT_4-space.

Proof: Since the characterized fuzzy cube is product of copies of the characterized fuzzy unit interval space $(I_L,\psi_{1,2}.\text{int}_\Im)$ and by means of Proposition 4.2, $(I_L,\psi_{1,2}.\text{int}_\Im)$ is characterized fuzzy compact FT_2-space. Then because of Proposition 2.6, part (3) and Generalized Tychonoff Theorem in Abd-Allah [11], it follows that, the characterized fuzzy cube is characterized FT_2-space. Moreover, Proposition 5.1, it follows that the characterized fuzzy cube is characterized FT_4-space.

Lemma 4.2

Let (X,τ) and (X,σ) are two fuzzy topological spaces such that τ is finer than σ If $\varphi_1,\varphi_2 \in O_{(L^X,\tau)}$, $\psi_1,\psi_2 \in O_{(L^X,\sigma)}$ and $(X,\psi_{1,2}.\ \text{int}_\sigma)$ is characterized fuzzy compact space, then $(X,\varphi_{1,2}.\text{int}_\tau)$ s also characterized fuzzy compact space.

Proof: Let $\mathcal{N}_{\varphi_{1,2}}(x)$ and $\mathcal{N}_{\psi_{1,2}}(x)$ are the $\varphi_{1,2}$-fuzzy neighborhood and $\psi_{1,2}$-fuzzy neighborhood at x with respect to $\psi_{1,2}.\text{int}_\tau$ and $\psi_{1,2}.\text{int}_\sigma$, respectively. Since τ is finer than σ, then $\mathcal{N}_{\psi_{1,2}}(x) \leq \mathcal{N}_{\varphi_{1,2}}(x)$ for all $x \in X$. Because of $\psi_{1,2}.int_\sigma$ is characterized fuzzy compact space, then for all $\mathcal{M} \in \mathcal{F}_L X$, there exists $\mathcal{K} \in \mathcal{F}_L X$ such that $\mathcal{K} \leq \mathcal{M}$ and $\mathcal{K} \leq \mathcal{N}_{\psi_{1,2}}(x)$ for all x X. Therefore $\mathcal{K} \leq \mathcal{N}_{\varphi_{1,2}}(x)$ for all x X. Consequently, $(X,\varphi_{1,2}.\text{int}_\tau)$ is characterized fuzzy compact space.

Proposition 4.7

Let (X,τ) and (X,σ) are two fuzzy topological spaces such that is finer than $\varphi_1,\varphi_2 \in O_{(L^X,\tau)}$ and $\psi_1,\psi_2 \in O_{(L^X,\sigma)}$. If $(X,\psi_{1,2}.\text{int}_\sigma)$ is characterized fuzzy compact space and $(X,\varphi_{1,2}.\text{int}_\tau)$ is characterized FT_2-space, then $(X,\varphi_{1,2}.\text{int}_\tau)$ and $(X,\psi_{1,2}.\text{int}_\sigma)$ are $\varphi_{1,2}\psi_{1,2}$-fuzzy isomorphic.

Proof: Since τ is finer than σ, then $\psi_{1,2}.\text{int}_\sigma \leq \varphi_{1,2}.\text{int}_\tau$. Hence, because of Proposition 2.5, $(X,\psi_{1,2}.\text{int}_\sigma)$ is characterized FT_2-space. From Lemma 4.2, we have $(X,\varphi_{1,2}.\text{int}_\tau)$ is characterized fuzzy compact space. Hence, we can find the identity mapping $id_X : (X,\varphi_{1,2}.\text{int}_\tau) \to (X,\psi_{1,2}.\text{int}_\sigma)$ which is bijective $\varphi_{1,2}\psi_{1,2}$-fuzzy continuous and its inverse is $\varphi_{1,2}\psi_{1,2}$-fuzzy continuous, that is, id_X is $\varphi_{1,2}\psi_{1,2}$-fuzzy isomorphism. Consequently, $(X,\varphi_{1,2}.\text{int}_\tau)$ and $(X,\psi_{1,2}.\text{int}_\sigma)$ are $\varphi_{1,2}\psi_{1,2}$-fuzzy isomorphic.

Proposition 4.8

Let (X,τ) be a fuzzy topological spaces and $\varphi_1,\varphi_2 \in O_{(L^X,\tau)}$. Then every characterized fuzzy compact space $(X,\varphi_{1,2}.\text{int}_\tau)$ is characterized FT_2-space if and only if it is characterized $FT_{3\frac{1}{2}}$- space.

Proof: Let $(X,\varphi_{1,2}.\text{int}_\tau)$ is characterized fuzzy compact FT_2-space. Because of Proposition 4.4 we have $(X,\varphi_{1,2}.\text{int}_\tau)$ is characterized FT_4-space and therefor Proposition 4.6 in Abd-Allah S [11], implies that $(X,\varphi_{1,2}.\text{int}_\tau)$ is characterized $FT_{3\frac{1}{2}}$- space. Conversely, let $(X,\varphi_{1,2}.\text{int}_\tau)$ is characterized $FT_{3\frac{1}{2}}$- space, then because of Proposition 3.2 in Abd-Allah [11] and part (1) of Proposition 2.6, it follows that $(X,\varphi_{1,2}.\text{int}_\tau)$ is characterized fuzzy compact FT_2-space.

From Lemma 4.2 and Corollary 3.3 in [22], we can prove the following result.

Proposition 4.9

Let (X,τ) and (X,σ) are two fuzzy topological spaces such that τ is finer than σ, $\varphi_1,\varphi_2 \in O_{(L^X,\tau)}$ and $\psi_1,\psi_2 \in O_{(L^X,\sigma)}$. If $(X,\varphi_{1,2}.\text{int}_\tau)$ is characterized fuzzy compact space and $(X,\psi_{1,2}.int_\sigma)$ is characterized $FT_{3\frac{1}{2}}$- space, then $(X,\varphi_{1,2}.\text{int}_\tau)$ and $(X,\psi_{1,2}.\text{int}_\sigma)$ are $\varphi_{1,2}\psi_{1,2}$-fuzzy isomorphic.

Proof: Follows directly from Corollary 3.3 in [22] and Lemma 4.2 similar to the proof of Proposition 4.7.

Some Relations Between Characterized FT_s, Characterized FR_k and Characterized Fuzzy Uniform Spaces

In this section, we are going to investigate and study the relations between the characterized FT_s-spaces, the characterized FT_k-spaces and the characterized fuzzy uniform spaces presented in Abd-Allah [12]. For this, we applied the notion of homogeneous fuzzy filter at the point and at the fuzzy set which is defined by (2.1), the superior principal fuzzy filter $[\mu]$ generated by $\mu \in L^X$ and the $\varphi_{1,2}$-fuzzy neighborhoods at the fuzzy set μ which is defined by (3.1) in the characterized fuzzy space $(X,\varphi_{1,2}.\text{int}_\tau)$. Specially, the relation between the separated fuzzy uniform spaces, the associated characterized fuzzy uniform FT_s-spaces, the associated characterized uniform $FR_{2\frac{1}{2}}$- space and the $F\varphi_{1,2}T_s$-space which introduced by Abd-Allah and Abd-Allah et al. in [8,11] are investigated for all $s \in \{0,1,3\frac{1}{2}\}$.

By the fuzzy relation on the set X, we mean the mapping $R : X \times X \to L$, that is, any fuzzy subset of $X \times X$. For each fuzzy relation R on X, the inverse R^{-1} of R is the fuzzy relation on X defined by $R^{-1}(x,y) = R(y,x)$ for all $x,y \in X$. Let $\mathcal{U}$ be a fuzzy filer on $X \times X$. The inverse $\mathcal{U}^{-1}$ of $\mathcal{U}$ is a fuzzy filter on $X \times X$ defined by $\mathcal{U}^{-1}(R) = \mathcal{U}(R^{-1})$ for all $RL^{X\times X}$. The composition $R_1 \circ R_2$ of two fuzzy relations R_1 and R_2 on the set X is a fuzzy relation on X defined by:

$$(R_1 \circ R_2)(x,y) = \bigvee_{z \in X} (R_2(x,z) \wedge R_1(z,y))$$

for all $x,y \in X$. For each pair (x,y) of elements x and y of $X \times X$, the mapping $(x,y)^{\cdot} : L^{X\times X} \to X$ defined by: $(x,y)^{\cdot}(R) = R(x,y)$ for all $R \in X \times X$ is a homogeneous fuzzy filter on $X \times X$. Let $\mathcal{U}$ and $\mathcal{V}$ are fuzzy filers on

$X \times X$ such that $(x,y) \le \mathcal{U}$ and $(y,z) \le V$ hold for some $x,y,z \in X$. Then the composition $\mathcal{V}_\circ \mathcal{U}$ of $\mathcal{V}$ and $\mathcal{U}$ is a fuzzy filter [13] on $X \times X$ defined by:

$$(\mathcal{V} \circ \mathcal{U})(R) = \bigvee_{R_2 \circ R_1 \le R} (\mathcal{U}(R_1) \wedge \mathcal{V}(R_2))$$

for all $R \in L^{X\times X}$.

By the *fuzzy uniform structure* $\mathcal{U}$ on a set X [13], we mean a fuzzy filter on $X \times X$ such that the following axioms are fulfilled:

(U1) $(x,x) \le \mathcal{U}$ for all $x \in X$.

(U2) $\mathcal{U} = \mathcal{U}^{-1}$.

(U3) $\mathcal{U}_\circ \mathcal{U} \le \mathcal{U}$.

The pair $(X,\mathcal{U})$ is called *fuzzy uniform space*. The fuzzy uniform structure $\mathcal{U}$ [13] on a set X is said to be *separated* if for all $x,y \in X$ with $x\ddot{I}y$ there is $R \in L^{X\times X}$ such that $\mathcal{U}(R) = 1$ and $R(x,y) = 0$. In this case the fuzzy uniform space $(X,\mathcal{U})$ is called *separated* fuzzy uniform space. Let $\mathcal{U}$ is a fuzzy uniform structure on a set X such that $(x,x) \le \mathcal{U}$ holds for all $x \in X$ and let $\mathcal{M} \in \mathcal{F}_L X$ then the mapping $\mathcal{U}[\mathcal{M}]: L^X \to L$ which is defined by:

$$\mathcal{U}[\mathcal{M}](\mu) = \bigvee_{R(\eta) \le \mu} (\mathcal{U}(R) \wedge \mathcal{M}(\eta))$$

for all $\mu \in L^X$ is a fuzzy filter on X, called the image of $\mathcal{M}$ with respect to the fuzzy uniform structure $\mathcal{U}$[13], where $\eta, R[\eta] \in L^X$ such that

$$R[\eta](x) = \bigvee_{y \in X} (\eta(y), R(y,x)).$$

Each fuzzy uniform structure $\mathcal{U}$ on the set X is associated a stratified fuzzy topology $\tau_{\mathcal{U}}$ on X. Consider $\varphi_1,\varphi_2 \in O_{(L^X,\tau_{\mathcal{U}})}$, then the set of all $\varphi_{1,2}$-open fuzzy subsets of X related to τ_U forms a base for an characterized stratified fuzzy space on X generated by the $\varphi_{1,2}$-interior operator with respect to $\tau_{\mathcal{U}}$ denoted by $\varphi_{1,2}.\text{int}_{\mathcal{U}}$ and $(X,\varphi_{1,2}.\text{int}_{\mathcal{U}})$ is the related stratified characterized fuzzy space. In this case, $(X,\varphi_{1,2}.\text{int}_{\mathcal{U}})$ will be called the *associated characterized fuzzy uniform space* [12] which is stratified. The related $\varphi_{1,2}$-interior operator $\varphi_{1,2}.\text{int}_U$ is given by:

$$(\varphi_{1,2}.\text{int}_U\mu)(x) = \mathcal{U}[\dot{x}](\mu) \tag{5.1}$$

for all $x \in X$ and $\mu \in L^X$. The fuzzy set μ is said to be $\varphi_{1,2}\mathcal{U}$-fuzzy neighborhood of $x \in X$ in the associated characterized fuzzy uniform space $(X,\varphi_{1,2}.\text{int}_{\mathcal{U}})$, provided $\mathcal{U}[\dot{x}] \le \dot{\mu}$. Because of (2.1), (3.1) and (5.1) we have that

$$\mathcal{U}[\dot{x}] = \mathcal{N}_{\varphi_{1,2}}(x) \text{ and } \mathcal{U}[\dot{\mu}] = \mathcal{N}_{\varphi_{1,2}}(\mu) \tag{5.2}$$

for all $x \in X$ and $\mu \in L^X$. In this case $\mathcal{N}_{\varphi_{1,2}}(x)$ and $\mathcal{N}_{\varphi_{1,2}}(\mu)$ are the $\varphi_{1,2}$-fuzzy neighborhood filters of the associated characterized fuzzy uniform space $(X,\varphi_{1,2}.\text{int}_{\mathcal{U}})$ at x and μ, respectively.

Proposition 5.1

Let X be a non-empty set, U is a fuzzy uniform structure on X and $\varphi_1,\varphi_2 \in O_{(L^X,\tau_{\mathcal{U}})}$. Then the fuzzy uniform space (X,U) is separated if and only if the associated characterized fuzzy uniform space $(X,\varphi_{1,2}.\text{int}_{\mathcal{U}})$ is characterized FT_0-space.

Proof: Let $(X,\mathcal{U})$ is separated and let $x,y \in X$ such that $x \neq y$. Then, there exists $R \in L^{X\times X}$ such that $\mathcal{U}(R) = 1$ and $R(x,y) = 0$. Consider $\mu = R[y_1]$ for which

$$\mu(x) = R[y_1](x) = \bigvee_{z \in X} R(z,x) \wedge y_1(z) = 0 \text{ and}$$

$$(\varphi_{1,2}.\text{int}_{\mathcal{U}}\,\mu)(y) = \mathcal{U}[\dot{y}](\mu) = \bigvee_{R(\eta) \le \mu} \mathcal{U}(R) \wedge \eta(y) = 1$$

for all $\eta \in L^X$. Hence, there exists $\mu \in L^X$ and $\alpha \in L_0$ such that $\mu(x) < \alpha \le (\varphi_{1,2}.\text{int}_{\mathcal{U}}\mu)(y)$, that is, $(X,\varphi_{1,2}.\text{int}_U)$ is characterized FT_0-space.

Conversely, let $(X,\varphi_{1,2}.\text{int}_U)$ is characterized FT_0-space and let $x \neq y$ in X. Then, there exists $\mu \in L^X$ and $\alpha \in L_0$ such that $\mu(x) < \alpha \le (\varphi_{1,2}.\text{int}_{\mathcal{U}}\mu)(y)$. This means that $\bigvee_{R(\eta) \le \mu} \mathcal{U}(R) \wedge \eta(y) > \mu(x)$ holds for all $\eta \in L^X$. Hence, there is $R \in L^{X\times X}$ for which $R(x,y) = (\varphi_{1,2}.\text{int}_U\mu)(x)$, if $x = y$ and $R(x,y) = \mu(x)$, if $x \neq y$ such that $R(x,y) = 0$ and $\mathcal{U}(R) = 1$. Thus, $(X,\mathcal{U})$ is separated.

Corollary 5.1

Let X be a non-empty set, $\mathcal{U}$ is a fuzzy uniform structure on X and $\varphi_1,\varphi_2 \in O_{(L^X,\tau_{\mathcal{U}})}$. Then the fuzzy uniform space $(X,\mathcal{U})$ is separated if and only if the associated stratified fuzzy topological space $(X,\tau_{\mathcal{U}})$ is $F\varphi_{1,2}$-T_0 space.

Proof: Immediate from Proposition 5.1 and Theorem 2.1 in Abd-Allah [8].

Proposition 5.2

Let X be a non-empty set, $\mathcal{U}$ is a fuzzy uniform structure on X and $\varphi_1,\varphi_2 \in O_{(L^X,\tau_{\mathcal{U}})}$. Then the fuzzy uniform space $(X,\mathcal{U})$ is separated if and only if the associated characterized fuzzy uniform space $(X,\varphi_{1,2}.\text{int}_{\mathcal{U}})$ is characterized FT_1-space.

Proof: Let $(X,\mathcal{U})$ is separated and let $x,y \in X$ such that $x \neq y$. Then, there exists $R_1, R_2 \in L^{X\times X}$ such that $\mathcal{U}(R_i) = 1$ and $R_i(x,y) = 0$ for all $i \in \{1.2\}$. Consider $\mu = R[y_1]$ and $\eta = R[x_1]$, then we have $\mu(x) = R_1[y_1](x) = \bigvee_{z \in X} (R_1(z,x) \wedge y_1(z)) = 0$ and $\eta(y) = R_2[x_1](y) = \bigvee_{z \in X} (R_2(z,y) \wedge x_1(z)) = 0.$ Moreover, $(\varphi_{1,2}.\text{int}_{\mathcal{U}}\,\mu)(y) = \mathcal{U}[\dot{y}](\mu) = \bigvee_{R_1(\rho) \le \mu} (\mathcal{U}(R_1) \wedge \rho(y)) = 1$ and $(\varphi_{1,2}.\text{int}_{\mathcal{U}}\,\eta)(x) = \mathcal{U}[\dot{x}](\eta) = \bigvee_{R_1(\rho) \le \eta} (\mathcal{U}(R_2) \wedge \rho(x)) = 1$ for all $\rho \in L^X$. Hence, there exists $\mu,\eta L^X$ and $\alpha, \beta\, L_0$ such that $\mu(x) < \alpha \le (X,\varphi_{1,2}.\text{int}_{\mathcal{U}}\mu)(y)$ and $\eta(y) < \beta \le (\varphi_{1,2}.\text{int}_\eta)(x)$ are hold. Consequently, $(X,\varphi_{1,2}.\text{int}_{\mathcal{U}})$ is characterized FT_1-space.

Conversely, let $(X,\varphi_{1,2}.\text{int}_{\mathcal{U}})$ is characterized FT_1-space and let $x \neq y$ in X. Then, there exists $\mu,\eta \in L^X$ and $\alpha,\beta \in L_0$ such that $\mu(x) < \alpha \le (X,\varphi_{1,2}.\text{int}_\mu)(y)$ and $(y) < \beta \le (\varphi_{1,2}.\text{int}_\eta)(x)$ are hold. This means that $\bigvee_{R_1(\rho) \le \mu} (\mathcal{U}(R_1) \wedge \rho(y)) > \mu(x)$ and $\bigvee_{R_2(\rho) \le \eta} (\mathcal{U}(R_2) \wedge \rho(x)) > \eta(y)$ are also hold for all $\rho \in L^X$. Hence, there is $R_1, R_2 L^{X\times X}$ such that $R_1(x,y) = (X,\varphi_{1,2}.\text{int}_\mu)(x)$ if $x = y$ and $R_1(x,y) = \mu(x)$ if $x \neq y$ such that $R_1(x,y) = 0$ and $\mathcal{U}(R_1) = 1$ and $R_2(x,y) = (X,\varphi_{1,2}.\text{int}_\eta)(x)$ if $x = y$ and $R_2(x,y) = \eta(y)$ if $x \neq y$ such that $R_2(x,y) = 0$ and $\mathcal{U}(R_2) = 1$. Thus, in every case $(X,\mathcal{U})$ is separated.

Corollary 5.2

Let X be a non-empty set, $\mathcal{U}$ is a fuzzy uniform structure on X and $\varphi_1,\varphi_2 \in O_{(L^X,\tau_{\mathcal{U}})}$. Then the fuzzy uniform space $(X,\mathcal{U})$ is separated if and only if the associated stratified fuzzy topological space $(X,\tau_{\mathcal{U}})$ is $F\,\varphi_{1,2}$-T_1 space.

Proof: Immediate from Proposition 5.2 and Theorem 2.2 in Abd-Allah [8].

For each fuzzy uniform structure $\mathcal{U}$ on the set X, the mapping $h: \mathcal{F}_L X \to \mathcal{F}_L X$ which is defined by $h(\mathcal{M}) = [\mathcal{M}]\,\mathcal{U}$ for all $\mathcal{M} \in \mathcal{F}_L X$ is global homogeneous fuzzy neighborhood structure on X [13]. The mapping h will be called global homogeneous fuzzy neighborhood structure *associated to* the fuzzy uniform structure $\mathcal{U}$ and will be denoted by $h_{\mathcal{U}}$.

The global fuzzy neighborhood structure h on the set X is said to be *symmetric* [13], provided that h ($\mathcal{L}$) $\wedge$ $\mathcal{M}$ exists if and only if $L \wedge h$ ($\mathcal{M}$) exists for all $\mathcal{M}$, $\mathcal{L} \in \mathcal{F}_L X$. As shown in Gähler [13], for each fuzzy uniform structure $\mathcal{U}$, the associated homogenous fuzzy neighborhood structure $h_{\mathcal{U}}$ is symmetric and both the global homogenous fuzzy neighborhood structures associated to the fuzzy uniform structures $\mathcal{U}$ and its homogenization $\mathcal{U}^*$ are coincide.

Proposition 5.3

Let f : $(X,\mathcal{U}) \to (Y,\mathcal{V})$ be an fuzzy uniformly continuous mapping between fuzzy uniform spaces. Then the mapping f : $(X,h_{\mathcal{U}}) \to (Y,h_{\mathcal{V}})$ between the associated global homogeneous fuzzy neighborhood spaces is $(h_{\mathcal{U}},h_{\mathcal{V}})$-fuzzy continuous [13].

Proposition 5.4

Let f : $(X,\mathcal{U}) \to (Y, \mathcal{V})$ be an fuzzy uniformly continuous mapping between fuzzy uniform spaces, $\varphi_1,\varphi_2 \in O_{(L^X,\tau_U)}$ and $\psi_1,\psi_2 \in O_{(L^Y,\tau_V)}$. Then the mapping f: $(X,\varphi_{1,2}.\text{int}_{\mathcal{U}}) \to (X,\varphi_{1,2}.\text{int}_{\mathcal{U}})$ between the associated characterized fuzzy uniform spaces is $\varphi_{1,2}\psi_{1,2}$-fuzzy continuous.

Proof: Immediate from Proposition 3.3 in Abd-Allah [11] and Proposition 5.3.

In the following, we prove that for each fuzzy uniform structure on a set X, there is an induced stratified fuzzy proximity on L^X. Moreover, both the fuzzy uniform structure and this induced stratified fuzzy proximity are associated with the same stratified characterized fuzzy uniform space.

Proposition 5.5

Let X be a non-empty set, $\mathcal{U}$ is a fuzzy uniform structure on X and $\varphi_1,\varphi_2 \in O_{(X,\tau_U)}$. Then the binary relation $\delta_{\mathcal{U}}$ on L^X which is defined by:

$$(X,\varphi_{1,2}.\text{int}_{\mathcal{U}}) = (X,\varphi_{1,2}.\text{int}_{\delta_{\mathcal{U}}}) \quad (5.3)$$

for all $\mu,\rho \in L^X$ is a stratified fuzzy proximity on X. Moreover, both the fuzzy uniform structure $\mathcal{U}$ and the induced stratified fuzzy proximity $\delta_{\mathcal{U}}$ are associated with the same stratified characterized fuzzy uniform space, that is, $(X,\varphi_{1,2}.\text{int}_{\mathcal{U}}) = (X,\varphi_{1,2}.\text{int}_{\delta_{\mathcal{U}}})$.

Proof: Immediate from (5.2), (5.3) and Proposition 3.2.

Corollary 5.3

Let $(X,\mathcal{U})$, $(Y,\mathcal{V})$ are two fuzzy uniform spaces, $\varphi_1,\varphi_2 \in O_{(X,\tau_{\delta_{\mathcal{U}}})}$ and $\psi_1,\psi_2 \in O_{(Y,\tau_{\delta_{\mathcal{V}}})}$. Then the mapping f: $(X,\mathcal{U}) \to (Y,\mathcal{V})$ is fuzzy uniformly continuous between fuzzy uniform spaces if and only if the mapping $f:(X,\varphi_{1,2}.\text{int}_{\delta_{\mathcal{U}}}) \to (Y,\psi_{1,2}.\text{int}_{\delta_{\mathcal{V}}})$ is $\varphi_{1,2}\psi_{1,2}$-fuzzy continuous between the associated stratified fuzzy proximity spaces.

Proof: Immediate from Propositions 5.4 and 5.5. □

Because of Propositions 3.7 and 5.5 and Corollary 5.3, we can deduce the result.

Proposition 5.6

Let $(X,\mathcal{U})$ be an fuzzy uniform space, $F,G \in P$ (X) such that $\mathcal{U}[\dot{F}] = \mathcal{U}[\dot{\chi}_F] \le \dot{\chi}_{G'} = \dot{G}'$ and $\varphi_1,\varphi_2 \in O_{(X,\tau_{\delta_U})}$. If Φ is the family of all fuzzy uniformly continuous functions f: $(X,\mathcal{U}) \to (I_L,\mathcal{U})$ for which $x \in X$ implies $\overline{0} \le f(x) \le \overline{1}$, then χ_F and χ_G are Φ-separable.

Proof: Immediate from Propositions 3.7 and 5.5 and Corollary 5.3.

Now, we shall prove that the stratified characterized fuzzy uniform space which associated with an fuzzy uniform structure is characterized $FR_{2\frac{1}{2}}$- space in sense of Abd-Allah et al. [11].

Proposition 5.7

Let X be a non-empty set, $\mathcal{U}$ is a fuzzy uniform structure on X and $\varphi_1,\varphi_2 \in O_{(X,\tau_U)}$. Then the associated stratified characterized fuzzy uniform space $(X,\varphi_{1,2}.\text{int}_{\mathcal{U}})$ with the fuzzy uniform structure $\mathcal{U}$ is characterized $FR_{2\frac{1}{2}}$- space.

Proof: Let xX, $F \in \varphi_{1,2}C(X)$ such that x Ï F. Since $\chi_{F'}$ is $\varphi_{1,2}$ $\mathcal{U}$ –fuzzy neighborhood of x, that is, $\mathcal{U}[\dot{x}] = \mathcal{N}_{\varphi_{1,2}}(x) \le \dot{F}'$. On account of Proposition 5.6, we get that x_1 and χ_F are Φ–separated by the fuzzy uniformly continuous function f: $(X,\mathcal{U}) \to (I_L, \mathcal{U})$. Because of Proposition 5.4, the function $f:(X,\varphi_{1,2}.\text{int}_{\mathcal{U}}) \to (I_L,\psi_{1,2}.\text{int}_{\mathcal{U}^*})$ is $\varphi_{1,2}\psi_{1,2}$–fuzzy continuous. Consequently, $(X,\varphi_{1,2}.\text{int}_{\mathcal{U}})$ is characterized $FR_{2\frac{1}{2}}$- space.

Corollary 5.4

Let $(X,\mathcal{U})$ be a separated fuzzy uniform space and $\varphi_1,\varphi_2 \in O_{(X,\tau_{\mathcal{U}})}$. Then the associated stratified characterized fuzzy uniform space $(X,\varphi_{1,2}.\text{int}_{\mathcal{U}})$ with the fuzzy uniform structure $\mathcal{U}$ is characterized $FT_{3\frac{1}{2}}$- space.

Proof: Immediate from Propositions 5.2 and 5.7.

In the following we give an example of a homogeneous fuzzy uniform structure and we show that the associated stratified characterized fuzzy uniform space is characterized fuzzy uniform $FT_{3\frac{1}{2}}$- space.

Example 5.1

The fuzzy metric in sense of S. Gähler and W. Gähler [24] is canonically generate a homogeneous fuzzy structure as follows: Consider X is non-empty set and d is a fuzzy metric on X, then the mapping $\mathcal{U}_d$:$L^{X\times X} \to L$ which is defined by:

$$\mathcal{U}_d(R) = \bigvee_{0<\delta,\varepsilon_{\alpha,\delta}\circ d \le R} \alpha$$

for all $R \in L^{X \times X}$ is a homogeneous fuzzy uniform structures on X. Moreover, the associated stratified characterized fuzzy uniform space $(X,\varphi_{1,2}.\text{int}_{\mathcal{U}_d})$ is identical with the associated characterized fuzzy metrizable space $(X,\varphi_{1,2}.\text{int}_{\tau_d})$, that is, $(X,\varphi_{1,2}.\text{int}_{\mathcal{U}_d}) = (X,\varphi_{1,2}.\text{int}_{\tau_d})$. Because of Proposition 3.1 in Abd-Allah et al. [22], we have, $(X,\varphi_{1,2}.\text{int}_{\tau_d})$ is characterized FT_4-space and therefore $(X,\varphi_{1,2}.\text{int}_{\mathcal{U}_d})$ is also characterized FT_4-space. Hence from Proposition 4.6 in Abd-Allah et al. [11], we get $(X,\varphi_{1,2}.\text{int}_{\mathcal{U}_d})$ is characterized $FT_{3\frac{1}{2}}$- space.

Conclusion

In this paper, we studied the relations between the characterized fuzzy T_s–spaces, the characterized fuzzy R_k–spaces presented in Abd-Allah and Abd-Allah and Al-Khedhairi [8-10,11] and the characterized fuzzy proximity spaces presented by Abd-Allah [12], for $s \in \{0,1,2,3,3\frac{1}{2},4\}$ and $k \in \{1,2,2\frac{1}{2},3\}$. We also introduced and studied the relations between our characterized fuzzy T_s–spaces, the characterized fuzzy R_k–spaces and the characterized fuzzy compact spaces presented by Abd-Allah [12] as a generalization of the weaker and stronger forms to the G-compactness defined by Gähler in 1995. Moreover, we shows here the relations between these characterized fuzzy T_s–spaces, the characterized fuzzy R_k-spaces and the characterized fuzzy uniform spaces introduced and studied by Abd-Allah in 2013 as a generalization of the weaker and stronger forms of the fuzzy uniform spaces introduced by Gähler et al. in 1998.

References

1. Gähler W (1995) The general fuzzy filter approach to fuzzy topology, I. Fuzzy Sets and Systems, 76(2: 205-224.
2. Kasahara S (1979) Operation-compact spaces. Math Japonica 24: 97-105.
3. Abd El-Monsef ME, Zeyada FM, Mashour AS, El-Deeb SN (1983) Operations on the power set P (X) of a topological space (X, T) In Colloquium on topology, Janos Bolyai Math. Soc. Eger, Hungry.
4. Kandil A, Abd-Allah AS, Nouh AA (1999) Operations and its applications on L-fuzzy bitopological spaces: Part I. Fuzzy sets and systems 106: 255-274.
5. Abd-Allah AS (2002) General notions related to fuzzy filters. Journal of Fuzzy Mathematics 10: 321-358.
6. Abd-Allah AS, El-Essawy M (2003) On characterizing notions of characterized spaces. Journal of Fuzzy Mathematics 11: 835-876.
7. Abd-Allah AS, El-Essawy M (2004) Closedness and compactness in characterized spaces. Journal of Fuzzy Mathematics 12: 591-632.
8. Abd-Allah AS (2007) Separation axioms in characterized fuzzy spaces. Journal of Fuzzy Mathematics 15: 291.
9. Abd-Allah AS (2008) Fuzzy regularity axioms and $T_{2\frac{1}{2}}$ axioms in characterized spaces, J Egypt Math Soci 16: 225-253.
10. Abd-Allah AS (2010) On characterized fuzzy regular and fuzzy normal spaces. J Egypt Math Soci 18: 9-38.
11. Abd-Allah AS, Al-Khedhairi A (2016) Characterized fuzzy $R_{2\frac{1}{2}}$ and characterized fuzzy $T_{3\frac{1}{2}}$-spaces. Indiana Univ Math J pp: 1-31.
12. Abd-Allah AS (2013) Characterized Proximity *L*-spaces, Characterized Compact *L*-spaces and Characterized Uniform *L*-spaces. J Egypt Math Soci 3: 98-119.
13. Gähler W, Bayoumi F, Kandil A, Nouh A (1998) The theory of global fuzzy neighborhood structures (III) Fuzzy uniform structures. Fuzzy Sets and Systems 98: 175-199.
14. Rodabaugh SE, Klement EP, Höhle U (1991) Applications of category theory to fuzzy subsets. Springer Science Business Media 14: 109-136.
15. Bayoumi F, Ibedou I (2004) The relation between GT i-spaces and fuzzy proximity spaces, G-compact spaces, fuzzy uniform spaces. Chaos Solitons Fractals 20: 955-966.
16. Lowen R (1979) Convergence in fuzzy topological spaces. General topology and its applications 10: 147-160.
17. Lowen R (1976) Fuzzy topological spaces and fuzzy compactness. Journal of Mathematical analysis and applications 56: 621-633.
18. Katsaras AK (1980) Fuzzy proximities and fuzzy completely regular spaces. J Anal St Univ Jasi 26: 980.
19. Gähler W, Abd-Allah AS, Kandil A (2000) On extended fuzzy topologies. Fuzzy Sets and Systems 109: 149-172.
20. Chang L (1968) Fuzzy topological spaces. J Math Anal Appl 24: 182-190.
21. Goguen JA (1967) L-fuzzy sets. J Math Anal Appl 18: 145-174.
22. Abd-Allah SA, Al-Khedhairi A (2014) Initial and final characterized $FT_{3\frac{1}{2}}$ and finer characterized $FR_{2\frac{1}{2}}$ spaces, Fuzzy Sets and Systems.
23. Katsaras K, Petalas CG (1984) On Fuzzy syntopogenous structures. J Math Anal Appl 99: 219-236.
24. Gähler S, Gähler W (1994) Fuzzy real numbers. Fuzzy Sets and Systems 66: 137-158.

Refined Estimates on Conjectures of Woods and Minkowski-I

Kathuria L* and Raka M

Centre for Advanced Study in Mathematics, Panjab University, Chandigarh-160014, India

Abstract

Let ^ be a lattice in R^n reduced in the sense of Korkine and Zolotare having a basis of the form (A_1, 0, 0, . . . , 0), ($a_{2,1}$, A_2, 0, . . . , 0), . . . , ($a_{n,1}$, $a_{n,2}$, . . . , $a_{n,n-1}$, A_n) where A_1, A_2, . . . ,An are all positive. A well known onjecture of Woods in Geometry of Numbers asserts that if $A_1 A_2 \ldots A_{n=1}$ and $A_i \leq A_1$ for each i then any closed sphere in R^n of radius $\sqrt{n/2}$ contains a point of ^. Woods' Conjecture is known to be true for $n \leq 9$. In this paper we obtain estimates on the Conjecture of Woods for n=10; 11 and 12 improving the earlier best known results of Hans-Gill et al. These lead to an improvement, for these values of n, to the estimates on the long standing classical conjecture of Minkowski on the product of n non-homogeneous linear forms.

MSC: 11H46; 11 - 04; 11J20; 11J37; 52C15.

Keywords: Lattice; Covering; Non-homogeneous; Product of linear forms; Critical determinant; Korkine and Zolotare reduction; Hermite's constant; Centre density

Introduction

Let $L_i = a_{i1}x_1 + \ldots + a_{in}x_n$; $1 \leq i \leq n$ be n real linear forms in n variables x1; : : : ; xn and having determinant $\Delta = \det(a_{ij}) \neq 0$ The following conjecture is attributed to H. Minkowski:

Conjecture I: For any given real numbers c1; : : : ; cn, there exists integers x1; : : : ; xn such that

$$|(L_1 + c_1)\ldots(L_n + c_n)| \leq \frac{1}{2n}|\Delta| \tag{1.1}$$

Equality is necessary if and only if after a suitable unimodular transformation the linear forms L_i have the form $2c_i x_l$ for $1 \leq i \leq n$

This result is known to be true for $n \leq 9$ For a detailed history and the related results,

Minkowski's Conjecture is equivalent to saying that [1]

$$M_n \leq \frac{1}{2n}|\Delta|$$

where $M_n = M_n(\Delta)$ is given by

$$M_n = \underset{L_1,\ldots,L_n}{Sup} \; \underset{(c_1,\ldots,c_n)\in R^n}{Sup} \; \underset{(u_1,\ldots,u_n)\in z^n}{Inf} \prod_{i=1}^{n} |L_i(u_1,\ldots,u_n) + c_i|$$

Chebotarev proved the weaker inequality

$$M_n \leq \frac{1}{2^{n/2}}|\Delta| \tag{1.2}$$

Since then several authors have tried to improve upon this estimate. The bounds have been obtained in the form

$$M_n \leq \frac{1}{v_n 2^{n/2}}|\Delta| \tag{1.3}$$

where Vn>1. Clearly $v_n \leq 2^{n/2}$ by considering the linear forms Li=xi and $c_i = \frac{1}{2}$ for $1 \leq i \leq n$ During 1949-1986, many authors such as Davenport, Woods, Bombieri, Gruber, Skubenko, Andrijasjan, Il'in and Malyshev obtained V_n for large n. obtained $v_n = 4 - 2(2 - 3\sqrt{2/4})^n - 2^{-n/2}$ for all n [2-4] improved Mordell's estimates for $6 \leq n \leq 31$ Hans-Gill et al. [12,14] got improvements on the results of [5-8] for $9 \leq n \leq 31$ Since recently V_n $9 = 2^{9/2}$ has been established by the authors [9], we study Vn for $10 \leq n \leq 33$ in a series of three papers.

In this paper we obtain improved estimates on Minkowski's Conjecture for n=10; 11 and 12. In next papers [10-12], we shall derive improved estimates on Minkowski's Conjecture for n=13; 14; 15 and for $16 \leq n \leq 33$ respectively [13-16]. For sake of comparison, we give results by our improved Vn in Table 1.

We shall follow the Remak-Davenport approach. For the sake of convenience of the reader we give some basic results of this approach. Minkowski's Conjecture can be restated in the terminology of lattices as : Any lattice ^ of determinant d(^) in Rn is a covering lattice for the set

$$S : |x_1 x_2 \ldots x_n| \leq \frac{d(\wedge)}{2^n}$$

The weaker result (1.3) is equivalent to saying that any lattice ^ of determinant d(^) in Rn is a covering lattice for the set

$$S : |x_1 x_2 \ldots x_n| \leq \frac{d(\wedge)}{v_n 2^{n/2}}$$

Define the homogeneous minimum of ^ as

$$m_H(\wedge) = \inf\{|x_1 x_2 \ldots x_n| : X = (x_1, x_2, \ldots x_n) \in \wedge, X \neq o\}$$

Proposition 1. Suppose that Minkowski Conjecture has been proved for dimensions 1, 2,…., n - 1: Then it holds for all lattices ^ in Rn for which MH(^)=0.

Proposition 2. If ^ is a lattice in Rn for $n \geq 3$ with MH(^) ≠ 0 then there exists an ellipsoid having n linearly independent points of ^ on its boundary and no point of ^ other than O in its interior.

It is well known that using these results, Minkowski's Conjecture would follow from

***Corresponding author:** Kathuria L, Centre for Advanced Study in Mathematics, Panjab University, Chandigarh-160014, India
E-mail: kathurialeetika@gmail.com

	Estimates by Mordell	Estimates by Il'in	Estimates by Hans-Gill et al	Our improved Estimates
n	V_n	V_n	V_n	V_n
10	2.899061	3.47989	24.3627506	27.60348
11	2.973102	3.52291	29.2801145	33.47272
12	3.040525	3.55024	32.2801213	39.59199
13	3.102356	3.57856	34.8475153	45.40041
14	3.159373	3.60209	37.8038391	51.26239
15	3.21218	3.61116	40.905198	57.00375
16	3.261252	3.61908	44.3414913	57.4702
17	3.306972	3.63924	47.2339309	57.67598
18	3.349652	3.66176	46.7645724	57.38876
19	3.389556	3.66734	47.2575897	60.09339
20	3.426907	3.67236	46.8640155	58.48592
21	3.461897	3.67692	46.0522028	56.42571
22	3.494699	3.68408	43.6612034	53.94142
23	3.525464	3.68633	37.8802374	50.98842
24	3.55433	3.68978	32.5852958	47.74632
25	3.581421	3.69295	27.8149432	42.39088
26	3.606852	3.69589	23.0801951	38.8657
27	3.630729	3.70012	17.3895105	31.93316
28	3.653149	3.70263	12.9938763	26.10663
29	3.674203	3.70497	9.5796191	19.96254
30	3.693976	3.70867	6.7664335	16.06884
31	3.712547	3.72558	4.745972	11.23872
32	3.729989			8.325879
33	3.746371			5.411488

Table 1: The weaker result.

Conjecture II. If ^ is a lattice in Rn of determinant 1 and there is a sphere |X| <R which contains no point of ^ other than O in its interior and has n linearly independent points of ^ on its boundary then ^ is a covering lattice for the closed sphere of radius $\sqrt{n/4}$ Equivalently, every closed sphere of radius $\sqrt{n/4}$ lying in Rn contains a point of ^.

They formulated a conjecture from which Conjecture-II follows immediately. To state Woods' conjecture, we need to introduce some terminology [17,18].

Let L be a lattice in Rn. By the reduction theory of quadratic forms introduced by a cartesian co-ordinate system may be chosen in Rn in such a way that L has a basis of the form [19-22],

$$(A_1; 0; 0; : : : ; 0); (a_{2;1};A_2; 0; : : : ; 0); : : : ; (a_{n;1}; a_{n;2}; : : : ; a_{n;n-1},A_n);$$

where A1;A2; : : : ;An are all positive and further for each i=1; 2; : : : ; n any two points of the lattice in R^{n-i+1} with basis

$$(A_i; 0; 0; : : : ; 0); (a_{i+1;i};A_{i+1}; 0; : : : ; 0); : : : ; (a_{n;i}; a_{n;i+1}; : : : ; a_{n;n-1};A_n)$$

are at a distance atleast Ai apart. Such a basis of L is called a reduced basis [23].

Conjecture III (Woods): If $A_1A_{2...}A_n$=1 and $A_i \le A_1$ for each i then any closed sphere in Rn of radius $\sqrt{n}/2$ contains a point of L.

Woods [10] proved this conjecture for $4 \le n \le 6$ Hans-Gill et al. [12] gave a unified proof of Woods' Conjecture for $n \le 6$ Hans-Gill et al. [12,14] proved Woods' Conjecture for n=7 and n=8 and thus completed the proof of Minkowski's Conjecture for n=7 and 8 Woods [10,24] proved Conjecture and hence Minkowski's Conjecture for n=9. With the assumptions as in Conjecture III, a weaker result would be that

If $w_n \ge n$ any closed sphere in R^n of radius $\sqrt{w_n}/2$ contains a point of L [25,26].

Hans-Gill et al. [12,14] obtained the estimates w_n on Woods' Conjecture for $n^3 \ge 9$ As w_9=9 has been established by the authors [17] recently, we study w_n for $n^3 \ge 10$ in a series of three papers. In this paper we obtain improved estimates w_n on Woods' Conjecture for n=10; 11 and 12. In next papers [18,19], we shall derive improved estimates w_n on Woods' Conjecture for n=13; 14; 15 and for $16 \le n \le 33$ respectively. Together with the following result of Hans-Gill et al. [12], we get improvements of w_n for $n^3 \ge 34$ also.

Proposition 3. Let L be a lattice in R^n with $A_1A_2...A_n$=1 and $A_i \le A_1$ for each i. Let $0 < l_n \le A_n^2 \le m_n$ where l_n and m_n are real numbers. Then L is a covering lattice for the sphere $|x| \le \sqrt{w_n}/2$ where Wn is defined inductively by

$$w_n = \max\{w_{n-1} l_n^{-1/l_{n-1}} + l_n, w_{n-1} m_n^{-1/m_{n-1}} + m_n\}$$

Here we prove

Theorem 1. Let n=10; 11; 12. If d(L)=A1 : : :An=1 and $A_i \le A_1$ for i=2;...; n, then any closed sphere in Rn of radius $\sqrt{w_n}/2$ contains a point of L, where $w_{10} = 10.3, w_{11} = 11.62$ and $w_{12} = 13$.

The earlier best known values were w_{10}=10:5605061, w_{11}=11:9061976 and w_{12}=13:4499927.

To deduce the results on the estimates of Minkowski's Conjecture we also need the following generalization of Proposition 1

Proposition 4. Suppose that we know

$$M_j \le \frac{1}{v_j 2^{j/2} |\Delta|} for 1 \le j \le n-1$$

Let v_n<min $V_{k1} V_{k2}...V_{ks}$, where the minimum is taken over all (k_1; k_2; ; k_s) such that n=k1+k2+ : : :+ks, ki positive integers for all i and $s^3 \geq 2$. Then for all lattices in Rn with homogeneous minimum MH(<)=0, the estimate V_n holds for Minkowski's Conjecture.

Since by arithmetic-geometric inequality the sphere $\{X \in R^n : |X| \leq \frac{\sqrt{w_n}}{2}\}$ is a subset of $\{X : |x_1 x_2 ... x_n| \leq \frac{1}{2^{n/2}} (\frac{w_n}{2_n})^{n/2}\}$ Propositions 2 and 4 immediately imply

Theorem 2: The values of Vn for the estimates of Minkowski's Conjecture can be taken as $(\frac{2n}{w_n})^{n/2}$

For $10 \leq n \leq 33$ these values are listed in Table 1. In Section 2 we state some preliminary results and in Sections 3-5 we prove Theorem 1 for n=10; 11 and 12.

Preliminary Results and Plan of the Proof

Let L be a lattice in Rn reduced in the sense of Korkine and Zolotare. Let (Sn) denotes the critical determinant of the unit sphere Δ Sn with centre O in R^n i.e.

$\Delta(S_n) = Inf\{d(\wedge) : \wedge$ has no point other than O in the interior of $S_n\}$

Let γ_n be the Hermite's constant i.e. γ_n is the smallest real number such that for any positive de nite quadratic form Q in n variables of determinant D, there exist integers u_1; u_2;...; u_n not all zero satisfying

$Q(u_1, u_2, ... u_n) \leq \gamma_n D^{1/n}$

It is well known that We write A_i^2=Bi.

We state below some preliminary lemmas. Lemmas 1 and 2 are due to Woods [25], Lemma 3 is due to Korkine and Zolotare [21] and Lemma 4 is due to Pendavingh and Van Zwam [24]. In Lemma 5, the cases n=2 and 3 are classical results of Lagrange and Gauss; n=4 and 5 are due to Korkine and Zolotare [21] while n=6; 7 and 8 are due to Blichfeldt [3].

Lemma 1. If $2\Delta(S_{n+1}) A_1^n \geq d(l)$ then any closed sphere of radius

$R = A_1(1 - \{A_1^n \Delta(S_{n+1}) / d(L)\}^2)^{1/2}$

in R^n contains a point of L.

Lemma 2. For a Fixed integer i with $1 \leq i \leq n-1$ denote by L_1 the lattice in R^i with reduced basis

$(A_1, 0, ..., 0), (a_{2,1}, A_2, 0, ..., 0), ..., (a_{i,1}, a_{i,2}, ..., a_{i,i-1}, A_i)$

and denote by L2 the lattice in R^{n-i} with reduced basis

$(A_{i+1}$; 0; ; 0); $(a_{i+2;i+1;} A_{i+2}$; 0; ; 0); ; $(a_{n;i+1}$; $a_{n;i+2}$; ; $a_{n;n-1}$;A_n).

If any closed sphere in R_i of radius r1 contains a point of L_1 and if any closed sphere in R_{n-i} of radius r_2 contains a point of L_2 then any closed sphere in Rn of radius $(r_1^2 + r_2^2)^{1/2}$ contains a point of L:

Lemma 3. For all relevant i,

$$B_{i+1} \geq \frac{3}{4} B_i \; and \; B_{i+2} \geq \frac{2}{3} B_i \quad (2.1)$$

Lemma 4. For all relevant i,

$$B_{i+4} \geq (0.46873) B_i \quad (2.2)$$

Throughout the paper we shall denote 0.46873 by ε .

Lemma 5. $\Delta(S_n) = \sqrt{3}/2, 1/\sqrt{2}, 1/2\sqrt{2}, \sqrt{3}/8, 1/8 \; and 1/16$ for n=2; 3; 4; 5; 6; 7 and 8 respectively:

Lemma 6. For any integer s; $1 \leq s \leq n-1$

$B_1 B_2 ... B_{s-1} B_s^{n-s+1} \leq \gamma_{n-s+1}^{n-s+1}$ and

$$B_1 B_2 ... B_s \leq (\gamma_n^{\frac{1}{n-1}} \gamma_{n-1}^{\frac{1}{n-2}} ... \gamma_{n-s+1}^{\frac{1}{n-s}})^{n-s} \quad (2.4)$$

This is Lemma 4 of Hans-Gill et al. [12].

Lemma 7.

$$\{(8.5337)^{\frac{1}{5}} \gamma_n^{\frac{1}{n-1}} \gamma_{n-1}^{\frac{1}{n-2}} ... \gamma_6^{\frac{1}{5}}\}^{-1} \leq B_n \leq \gamma_{n-1}^{\frac{n-1}{n}} \quad (2.5)$$

This is Lemma 6 of Hans-Gill et al. [14].

Remark 1. Let

δ_n =the best centre density of packings of unit spheres in R^n;

δ_n^* =the best centre density of lattice packings of unit spheres in R^n:

Then it is known that

$$\gamma_n = 4(\delta_n^*)^{\frac{2}{n}} \leq 4(\delta_n)^{\frac{2}{n}} \quad (2.6)$$

δ_n^* and hence δ_n is known for $n \leq 8$ Also γ_{24} =4 has been proved by Cohn and Kumar [6]. For $9 \leq n \leq 12$ using the bounds on δ_n given by Cohn and Elkies [5] and inequality (2.6) we find that $\gamma_9 \leq 2.1326324$, $\gamma_{10} \leq 2.2636302$, $\gamma_{11} \leq 2.3933470$, $\gamma_{12} \leq 2.5217871$

We assume that Theorem 1 is false and derive a contradiction. Let L be a lattice satisfying the hypothesis of the conjecture. Suppose that there exists a closed sphere of radius $\sqrt{w_n}/2$ in R^n that contains no point of L in R^n.

Since $B_i=A_i^2$ and d(L)=1; we have $B_1 B_2$: : :B_n=1:

We give some examples of inequalities that arise. Let L1 be a lattice in R4 with basis (A_1; 0; 0; 0), ($a_{2;1}$;A_2; 0; 0); ($a_{3;1}$; $a_{3;2}$;A_3; 0); ($a_{4;1}$; $a_{4;2}$; $a_{4;3}$;A_4); and L_i for $2 \leq i \leq n$ be lattices in R1 with basis (Ai+3). Applying Lemma 2 repeatedly and using Lemma 1 we see that if $2\Delta(S_5) A_1^4 \geq A_1 A_2 A_3 A_4$ then any closed sphere of radius

$$(A_1^2 - \frac{A_1^{10} \Delta(S_5)^2}{A_1^2 A_2^2 A_3^2 A_4^2} + \frac{1}{4} A_5^2 + ... + \frac{1}{4} A_n^2)^{1/2}$$

contains a point of L: By the initial hypothesis this radius exceeds $\sqrt{w_n}/2$ Since $\Delta(S_5) = 1/2\sqrt{2}$ and $B_1 B_2 ..... B_n = 1$ this results in the conditional inequality : if $B_1^4 B_5 B_6 ... B_n \geq 2$ then

$$4B_1 - \frac{1}{2} B_1^5 B_5 B_6 ... B_n + B_5 + B_6 + ... + B_n > w_n \quad (2.7)$$

We call this inequality (4; 1;...; 1); since it corresponds to the ordered partition (4; 1;...; 1) of n for the purpose of applying Lemma 2. Similarly the conditional inequality (1;...; 1; 2; 1;...; 1) corresponding to the ordered partition (1;...; 1; 2; 1;...; 1) is : if $2B_i \geq B_{i+1}$ then

$$B_1 + ... + B_{i-1} + 4B_i - \frac{2B_i^2}{B_{i+1}} + B_{i+2} + ... + B_n > w_n \quad (2.8)$$

Since $4B_i - \frac{2B_i^2}{B_{i+1}} \leq 2B_{i+1}$, (2.8) gives

$B_1 + ... + B_{i-1} + 2B_{i+1} + B_{i+2} + ... + B_n > W_n$:

One may remark here that the condition $2B_i \geq B_{i+1}$ is necessary only if we want to use inequality (2.8), but it is not necessary if we want to use the weaker inequality (2.9). This is so because if $2B_i<B_{i+1}$, using the partition (1; 1) in place of (2) for the relevant part, we get the upper bound 2Bi + B_{i+1} which is clearly less than $2B_{i+1}$. We shall call inequalities of type (2.9) as weak inequalities and denote it by $(1;\ldots;1;2;1;\ldots;1)_w$.

If $(\lambda_1,\lambda_2,\ldots,\lambda_s)$ is an ordered partition of n, then the conditional inequality arising from it, by using Lemmas 1 and 2, is also denoted by $(\lambda_1,\lambda_2,\ldots,\lambda_s)$ If the conditions in an inequality $(\lambda_1,\lambda_2,\ldots,\lambda_s)$ are satisfied then we say that $(\lambda_1,\lambda_2,\ldots,\lambda_s)$ holds. Sometimes, instead of Lemma 2, we are able to use induction. The use of this is indicated by putting (*) on the corresponding part of the partition. For example, if for n=10, B5 is larger than each of B6;B7;…;B10, and if $\frac{B_1^3}{B_1B_3B_4}>2$ the inequality (4; 6*) gives

$$4B_1-\frac{1}{2}\frac{B_1^3}{B_1B_3B_4}+6(B_1B_2B_3B_4)^{-1/6}>w_{10} \tag{2.10}$$

In particular the inequality ((n-1)*; 1) always holds. This can be written as

$$w_{n-1}(B_n)^{\frac{-1}{(n-1)}}+B_n>W_n \tag{2.11}$$

Also we have $B_1 \geq 1$ because if $B_1<1$, then $B_i \leq B_1 <1$ for each I contradicting B1B2:::Bn=1.

Using the upper bounds on and the inequality (2.5), we obtain numerical lower and upper bounds on Bn, which we denote by ln and mn respectively. We use the approach of Hans-Gill et al. [14], but our method of dealing with

Is somewhat different. In Sections 3-5 we give proof of Theorem 1 for n=10; 11 and 12 respectively. The proof of these cases is based on the truncation of the interval [ln;mn] from both the sides.

In this paper we need to maximize or minimize frequently functions of several variables. When we say that a given function of several variables in x; y; is an increasing/decreasing function of x; y;…., it means that the concerned property holds when function is considered as a function of one variable at a time, all other variables being fixed.

Proof of Theorem 1 for n=10

Here we have W_{10}=10:3, $B_1<\gamma_{10}$<2:2636302. Using (2.5), we have l10=0:4007<B10<1:9770808=m_{10}.

The inequality (9*; 1) gives $9(B10)^{\frac{-1}{9}}$ + B10<10:3. But for 0:4398 B10 1:9378, this inequality is not true. Hence we must have either B10<0:4398 or B10>1:9378. We will deal with the two cases 0:4007< B10<0:4398 and 1:9378<B10<1:9770808 separately:

0:4007<B_{10}<0:4398

Using the Lemmas 3 & 4 we have:

$$\begin{cases} B_9 \leq \frac{4}{3}B_{10} < 0.5864 & B_8 \leq \frac{3}{2}B_{10} < 0.6597 & B_7 \leq 2B_{10} < 0.8796 \\ B_6 \leq \frac{B_{10}}{\varepsilon} < 0.9383 & B_5 \leq \frac{4}{3}\frac{B_{10}}{\varepsilon} < 1.2511 & B_4 \leq \frac{3}{2}\frac{B_{10}}{\varepsilon} < 1.4075 \\ B_3 \leq \frac{2B_{10}}{\varepsilon} < 1.8766 & B_2 \leq \frac{B_{10}}{(\varepsilon)^2} < 2.0018 & \end{cases}$$

Claim(i) B_2>1:7046

The inequality (2; 2; 2; 2; 2)w gives $2B_2+2B_4+2B_6+2B_8+2B_{10}$>10:3. Using (3.1), we find that this inequality is not true for $B_2 \leq$ 1:7046. Hence we must have B_2>1:7046.

Claim(ii) B_2<1:8815

Suppose $B_2 \geq 1.8815$ then using (3.1) and that $B_6 \geq \varepsilon B_2$ we find that $\frac{B_2^3}{B_3B_4B_5}>2$ and $\frac{B_6^3}{B_7B_8B_9}>2$ So the inequality (1,4,4,1) holds, i.e. $B_1+4B_2-\frac{1}{2}\frac{B_2^4}{B_3B_4B_5}+4B_6-\frac{1}{2}\frac{B_2^4}{B_7B_8B_9}+B_{10}>10.3$ Applying AM-GM inequality we get

$$B_1+4B_2+4B_6+B_{10}-\sqrt{B_2^5B_6^5B_1B_{10}}>10.3$$

Now since $\varepsilon^2B_2 \leq B_{10}<0.4398$ $B_6 \geq \varepsilon B_2, B_1 \geq B_2$ and $B_2 \geq 1.8815$ we find that the left side is a decreasing function of B_{10} and B_6. So replacing B_{10} by ε^2B_2 and by εB_2 we get $\varnothing_1=B_1+(4+4\varepsilon+\varepsilon^2)B_2-\sqrt{(\varepsilon)^7B_2^{11}B_1}>10.3$ Now the left side is a decreasing function of B2, so replacing B2 by 1.8815 we find that $\varnothing_1<10.3$ for 1<B_1<2:2636302, a contradiction. Hence we must have B_2<1:8815.

Claim (iii) B_3<1:5652

Suppose $B_3 \geq 1.5652$ From (3.1) we have $B_4B_5B_6$<1:6524 and $B_8B_9B_{10}$ <0:1702, so we find that $\frac{B_3^3}{B_4B_5B_6}>2$ and $\frac{B_7^3}{B_8B_9B_{10}} \geq \frac{(\varepsilon B_3)^3}{B_8B_9B_{10}}>2$ for B_3>1:49.

Applying AM-GM to inequality (2,4,4) we get $4B_1-\frac{2B_1^2}{B_2}+4B_3+4B_7-\sqrt{B_3^5B_7^5B_1B_2}>10.3$ Since $B_1 \geq B_2>1.7046, B_7 \geq \varepsilon B_3$ and $B_3 \geq 1.5652$ we find that left side is a decreasing function of B_1 and B_7. So we replace B_1 by B_2, B_7 by εB_3 and get that $\varnothing_2=2B_2+4(1+\varepsilon)B_3-\sqrt{(\varepsilon)^5B_3^{10}B_2^2}>10.3$.

But left side is a decreasing function of B3, so replacing B3 by 1.5652 we find that $\varnothing_2<10.3$ for 1:7046<B_2<1:8815, a contradiction. Hence we must have B_3<1:5652.

Claim (iv) B1>1:9378

Suppose $B_1 \leq 1.9378$ Using (3.1) and that B3<1:5652, B2>1:7046, we find that B_2 is larger than each of B_3; B4;…;B10. So the inequality (1; 9,*) holds. This gives $B_1+9(B_1)^{-1/9}>10.3$ which is not true for $B_1 \leq 1.9378$ So we must have B_1>1:9378.

Claim (v) B3<1:5485

Suppose $B_3 \geq 1.5485$ We proceed as in Claim(iii) and replace B_1 by 1.9378 and B_7 by εB_3 to get that

$$\varnothing_3=4(1.9378)-\frac{2(1.9378)^2}{B_2}+4(1+\varepsilon)B_3-\sqrt{(\varepsilon)^5(1.9378)B_3^{10}B_2}>10.3$$

One easily checks that $\varnothing_3<10.3$ for $1.5485 \leq$ B_3<1:5652 and 1:7046< B_2<1:8815. Hence we have B_3<1:5485.

Claim (vi) B_1<2:0187

Suppose $B_1 \geq 2.0187$ Using (3.1) and Claims (ii), (v) we have $B_2B_3B_4$<4:11. Therefore $\frac{B_1^3}{B_2B_3B_4}>2$ As $B_5 \geq \varepsilon B_1>0.9462$ we see using (3.1) that B_5 is larger than each of $B_6;B_7,\ldots;B_{10}$. Hence the inequality (4; 6,*) holds. This gives $\varnothing_4 = 4B_1 - \frac{1}{2}\frac{B_1^4}{B_2B_3B_4} + 6(B_1B_2B_3B_4)^{-1/6} > 10.3$

Left side is an increasing function of $B_2B_3B_4$ and decreasing function of B1. So we can replace $B_2B_3B_4$ by 4:11 and B1 by 2.0187 to find $\varnothing_4 < 10.3$ a contradiction. Hence we have B_1<2:0187.

Claim (vii) B_4<1:337

Suppose $B_4 \geq 1.337$ then using (3.1) we get $\frac{B_4^3}{B_5B_6B_7} > 2$ Applying AMGM to inequality (1,2,4,2,1) we have

$$B_1 + 4B_2 - \frac{2B_2^2}{B_3} + 4B_4 + 4B_8 + B_{10} - 2\sqrt{B_4^5B_8^5B_1B_2B_3B_{10}} > 10.3$$

Since B_2>1:7046 $B_3 \geq \frac{3}{4}B_2, B_4 \geq 1.337B_8 \geq \varepsilon B_4$ and $B_{10} \geq \frac{2\varepsilon}{3}B_4$ we find that left side is a decreasing function of B_2, B_8 and B_{10}. So we can replace B_2 by 1.7046; B_8 by εB_4 and B_{10} by $\frac{2\varepsilon}{3}B_4$ to get

$$\varnothing_5 = B_1 + 4(1.7046) - \frac{2(1.7046)^2}{B_3} + (4 + 4\varepsilon + \frac{2\varepsilon}{3})B_4 - 2\sqrt{\frac{2}{3}(\varepsilon)^4(1.7046)B_4^9B_1B_3} > 10.3$$

Now left side is a decreasing function of B4, replacing B4 by 1:337, we find that $\varnothing_5 < 10.3$ for 1<B_1<2:0187 and 1<B_3<1:5485, a contradiction. Hence we have B_4<1:337.

Claim (viii) B_5<1:1492

Suppose $B_5 \geq 1.1492$ Using (3.1), we get $B_6B_7B_8$<0:5445: Therefore $\frac{B_2^3}{B_6B_7B_8} > 2$ Also using Lemma 3 & 4, 2 $B_9 \geq 2(\varepsilon B_5)$ >1:077>B_{10}. So the inequality (4*; 4; 2) holds, i.e. 4 $(\frac{1}{B_5B_6B_7B_8B_9B_{10}})^{1/4} + 4B_5 - \frac{1}{2}\frac{B_5^4}{B_6B_7B_8} + 4B_9 - \frac{2B_9^2}{B_{10}} > 10.3$ Now left side is a decreasing function of B_5 and B_9. So we replace B_5 by 1.1492 and B_9 by 1.1492ε and get that $\varnothing_6(x, B_{10}) = 4(\frac{1}{(\varepsilon)(1.1492)^2 X b_{10}})^{1/4} + 4(1+\varepsilon)$ $(1.1492) - \frac{1}{2}\frac{(1.1492)^4}{x} - \frac{2(1.1492\varepsilon)^2}{B_{10}} > 10.3$ where x=$B_6B_7B_8$. Using Lemma 3 & 4 we have x=$B_6B_7B_8 \geq \frac{B_5^3}{4} \geq \frac{(1.1492)^3}{4}$ and $B_{10} \geq \frac{3\varepsilon}{4}B_5 \geq \frac{3\varepsilon}{4}(1.1492)$ It can be verified that $\varnothing_6(x, B_{10}) < 10.3$ for $\frac{(1.1492)^3}{4} \leq x < 0.5445$ and $\frac{3\varepsilon}{4}(1.1492) \leq B_{10} < 0.4398$ giving thereby a contradiction. Hence we must have B_5<1:1492.

Claim (ix) B_2<1:766.

Suppose $B_2 \geq 1.766$ We have $B_3B_4B_5$<2:3793. So $\frac{B_2^3}{B_3B_4B_5} > 2$ Also $B_6 \geq \varepsilon B_2 > 0.8277$ Therefore B6 is larger than each of B_7, B_8, B_9B_{10} Hence the inequality (1; 4; 5,*) holds. This gives $B_1 + 4B_2 - \frac{1}{2}\frac{B_2^4}{B_3B_4B_5} + 5(\frac{1}{B_1B_2B_3B_4B_5})^{\frac{1}{5}} > 10.3$ Left side is an increasing function of $B_3B_4B_5$, a decreasing function of B_2 and an increasing function of B_1. One easily checks that this inequality is not true for B_1<2:0187;

$B_2 \geq 1.766$ and $B_3B_4B_5$<2:3793: Hence we have B_2<1:766.

Final contradiction

As 2(B_2+B_4+B_6+B_8+B_{10})<2(1:766+1:337+0:9383+0:6597+0:4398)<10:3, the weak inequality (2; 2; 2; 2; 2)w gives a contradiction.

9378<B_{10}<1:9770808

Here $B_1 \geq B_{10} > 1.9378$ and B_2=$(B_1B_3\ldots B_{10})^{-1}$

$\leq (B_1B_2B_4\ldots B_{10})^{-1} \leq (\frac{3}{32}\varepsilon^3 B_3^6B_1^2B_{10})^{-1} = (\frac{1}{16}\varepsilon^4 B_2^7B_1B_{10})^{-1}$

Which implies $(B_2)^8 \leq (\frac{1}{16}\varepsilon^4(1.9378)^2)^{-1}$ i.e. B2<1:75076.

Similarly

$B_3 = (B_1B_2B_4\ldots B_{10})^{-1} \leq (\frac{3}{32}\varepsilon^3 B_3^6B_1^2B_{10})^{-1}$

$B_4 = (B_1B_2B_3B_5\ldots B_{10})^{-1} \leq (\frac{3}{32}\varepsilon^2 B_4^5B_1^3B_{10})^{-1}$

$B_6 = (B_1\ldots B_5B_7B_8B_9B_{10})^{-1} \leq (\frac{1}{16}\varepsilon B_6^3B_1^3B_{10})^{-1}$

$B_8 = (B_1\ldots B_7B_9B_{10})^{-1} \leq (\frac{3}{32}\varepsilon^3 B_8B_1^7B_{10})^{-1}$

These respectively give B_3<1:46138, B_4<1:22883, B_6<0:896058 and B_8<0:721763. So we have $B_1^4B_5B_6B_7B_8B_9B_{10} = \frac{B_1^3}{B_2B_3B_4} > 2$ Also $2B_5 \geq 2(\varepsilon B_1) > 1.8166 > B_6$ and $2B_7 \geq 2(\frac{2\varepsilon}{3}B_1) > B_8$ Applying AM-GM to inequality (4,2,2,1,1) we have 4B_1 + 4B_5 + 4B_7 + B_9 +B_{10} -3 $(2B_1^5B_5^3B_7^3B_9B_{10})^{\frac{1}{3}} > 10.3$ We find that left side is a decreasing function of B_7 and B_5, so can replace B_7 by $\frac{2}{3}\varepsilon B_1$ and B_5 by εB_1 then it is a decreasing function of B_1, so replacing B_1 by B_{10} we have 4 $(1 + \varepsilon + \frac{2}{3}\varepsilon)B_{10} + B_9 + B_{10} - 2^{\frac{4}{3}}(\varepsilon)^2(B_{10})4(B_9)^{\frac{1}{3}} > 10.3$ which is not true for $(1.9378)\varepsilon^2 < B_9 \leq B_1 < 2.2636302$ and 1:9378<B10<1:9770808.

Hence we get a contradiction.

Proof of Theorem 1 for n=11

Here we have w_{11}=11.62, $B_1 \leq \gamma_{11} < 2.393347$ Using (2.5), we have 1_{11}=0:3673<B_{11}<2:1016019=m_{11}.

The inequality (10*; 1) gives 10:3 $(B_{11})^{\frac{-1}{10}} + B_{11} > 11.62$ But for $0.4409 \leq B_{11} \leq 2.018$ this inequality is not true. So we must have either $B_{11} < 0.4409$ or B_{11}>2:018.

0:3673<B11<0:4409

Claim (i) B10<0:4692

Suppose $B_{10} \geq 0.4692$ then 2B_{10}>B_{11}, so (9*; 2) holds, i.e. 9 $(\frac{1}{B_{10}B_{11}})^{\frac{1}{9}} + 4B_{10} - \frac{2B_{10}^2}{B_{11}} > 11.62$ As left side is a decreasing function of B_{10}, we can replace B_{10} by 0.4692 and find that it is not true for 0:3673<B_{11}<0:4409.

Hence we must have $B_{10<}$0:4692.

Using Lemmas 3 and 4 we have:

$B_9 \leq \frac{4}{3}B_{10} < 0.6256, B_8 \leq \frac{3}{2}B_{10} < 0.7038, B_7 \leq \frac{B_{11}}{\varepsilon} < 0.94063$

$B_6 \leq \frac{B_{10}}{\varepsilon} < 1.00.., B_5 \leq \frac{4}{3}\frac{B_{10}}{\varepsilon} < 1.3347, B_4 \leq \frac{3}{2}\frac{B_{10}}{\varepsilon} < 1.50151$

$$B_3 \le \frac{B_{11}}{\varepsilon^2} < 2.0068, B_2 \le \frac{B_{10}}{\varepsilon^2} < 2.13557 \quad (4.1)$$

Claim (ii) B2>1:913

The inequality $(2; 2; 2; 2; 2; 1)_w$ gives $2B_2+2B_4+2B_6+2B_8+2B_{10}+B_{11} >$ 11:62. Using (4.1) we find that this inequality is not true for $B_2 \le 1.913$ so we must have $B_2>1:913$.

Claim(iii) B3<1:761

Suppose $B_3 \ge 1.761$ then we have $\frac{B_3^3}{B_4B_5B_6} > 2$ and $\frac{B_7^3}{B_8B_9B_{10}} > \frac{(\varepsilon B_3)^3}{B_8B_9B_{10}} >$ 2. Applying AM-GM to the inequality (2,4,4,1) we get $4B_1 - \frac{2B_1^2}{B_2} + 4B_3 + 4B_7 + B_{11} - \sqrt{B_3^5B_7^5B_1B_2B_{11}} > 11.62$ One easily finds that it is not true for $B_1 \ge B_2 > 1.913, B_3 \ge 1.761, B_7 \ge \varepsilon B_3, B_{11} \ge \varepsilon^2 B_3, 1.913 < B_2 < 2.13557$ and $1.761 \le B_3 < 2.0068$ Hence we must have $B_3<1:761$:

Claim (iv) B1<2:2436

Suppose $B_1 \ge 2.2436$ As $B_2B_3B_{4<}2:13557\times1:761\times1:50151<5:6468$, we have $\frac{B_1^3}{B_2B_3B_4} > 2$ Also $B_5 \ge \varepsilon B_1 > 1.051$ so B5 is larger than each of B6;B7...;B11. Hence the inequality (4; 7,*) holds. This gives $4B_1 - \frac{1}{2}\frac{B_1^4}{B_2B_3B_4} + 7(\frac{1}{B_1B_2B_3})^{\frac{1}{7}} > 11.62$ Left side is an increasing function of $B_2B_3B_4$ and decreasing function of B_1. One easily checks that the inequality is not true for $B_2B_3B_4<5:6468$ and B1 ≥ 2:2436. Hence we have $B_1<2:2436$.

Claim (v) B4<1.4465 and B2>1:9686

Suppose $B_4 \ge$ 1.4465 We have B5B6B7<1:2569 and $B_9B_{10}B_{11}<0:1295$. Therefore for $B_4>1:36$, we have $\frac{B_4^3}{B_5B_6B_7} > 2$ and $\frac{B_8^3}{B_9B_{10}B_{11}} > \frac{(\varepsilon B_4)^3}{B_9B_{10}B_{11}} > 2$

So the inequality (1,2,4,4) holds. Applying AM-GM to inequality(1,2,4,4), we get $B_1 + 4B_2 - \frac{2B_2^2}{B_3} + 4B_8 - \sqrt{B_4^5B_8^5B_1B_2B_3} > 11.62$

A simple calculation shows that this is not true for $B_1 \ge B_2 > 1.913$, $B_4 \ge 1.4465$, $B_8 \ge \varepsilon B_4 \ge 1.4465, B_1 < 2.2436$ and $B_3 < 1.761$ Hence we have B4<1:4465.

Further if $B_2 \le 1.9686$ then $2B_2+2B_4+2B_6+2B_8+2B_{10}+B_{11}<11:62$. So the inequality $(2; 2; 2; 2; 2; 1)_w$ gives a contradiction.

Claim (vi) B4<1:4265 and B2>1:9888

Suppose $B_4 \ge 1.4265$ We proceed as in Claim (v) and get a contradiction with improved bounds on B_2 and B_4.

Claim (vii) B1<2:2056

Suppose $B_1 \ge 2.2056$ As B3B4B5<1:761 × 1:4265 × 1:3347<3:3529, we have $\frac{B_2^3}{B_3B_4B_5} > 2$ Also $B_6 \ge \varepsilon B_2 > 0.9491$ so B6 is larger than each of B7;B8,...,B11. Hence the inequality (1; 4; 6*) holds, i.e. B1 + 4B2 - $\frac{1}{2}\frac{B_2^4}{B_3B_4B_5} + 6(\frac{1}{B_1B_2B_3B_4B_5})^{\frac{1}{6}} > 11.62$

Claim (ix) B1<2:1669

Suppose $B_1 \ge 2.1669$ We proceed as in Claim(iv) and get a contradiction with improved bounds on B_1, B_2 and B_4.

Claim (x) B4<1:403 and B2>2:012

Suppose $B_4 \ge 1.403$ We proceed as in Claim(v) and get a contradiction with improved bounds on B2 and B4.

Final Contradiction:

As now B3B4B5<1:761×1:403 1:3347<3:2977, we have $\frac{B_2^3}{B_3B_4B_5} > 2$ for B2>2:012. Also $B_6 \ge \varepsilon B_2 > 0.943 >$ each of B7; B8; B11. Hence the inequality (1; 4; 6) holds. Proceeding as in Claim (viii) we find that this inequality is not true for $B_1<2:1669$; $B_2>2:012$ and $B_3B_4B_5<3:2977$; giving thereby a contradiction.

2:018<B11<2:1016019

Here $B_1 \ge B_{11} > 2.018$ Therefore using Lemmas 3 & 4 we have $B10=(B1\ \ B9B11)^{-1}$

$$\le (B_1 . \frac{3}{4}B_1 . \frac{2}{3}B_1 \frac{1}{2}B_1 \varepsilon B_1 \frac{3}{4}\varepsilon B_1 \frac{2}{3}\varepsilon B_1 \frac{1}{2}\varepsilon B_1 \varepsilon^2 B_1 . B_1)^{-1}$$

$$= (\frac{1}{16}\varepsilon^6 B_1^9 B_{11})^{-1} < (\frac{1}{16}\varepsilon^6 (2.018)^{10})^{-1} < 1.34702$$

Similarly

$B_4 = (B_1B_2B_3B_4 ... B_{11})^{-1} \le (\frac{1}{16}\varepsilon^3 B_1^3 B_{11})^{-1}$ which gives B4<1:37661.

Claim (i) B10<0:4402

The inequality (9*; 1; 1) gives $9(\frac{1}{B_{10}B_{11}})^{\frac{1}{9}} + B_{10} + B_{11} > 11.62$ But this inequality is not true for $0.4402 \le B_{10}$ <1:34702 and 2:018<B11<2:1016019. Hence we must have B10<0:4402.

Now we have $B_9 \le \frac{4}{3}B_{10} <$ 0:58694, $B_8 \le \frac{3}{2}B_{10} < 0.6603, B_7 \le 2B_{10} < 0.8804$ and $B_6 \le \frac{B_{10}}{\varepsilon} < 0.93914$

Claim (ii) B7<0:768

Suppose $B_7 \ge 0.768$ Then $\frac{B_7^3}{B_8B_9B_{10}} > 2$ so (6*; 4; 1) holds. This gives $\varnothing_7(x) = 6(x)^{1/6} + 4B_7 - \frac{1}{2}B_7^5B_{11}x + B_{11} > 11.62$ where $x=B_1B_2$: : :B_6. The function $\varnothing_7(x)$ has its maximum value at $x = (\frac{2}{B_7^5B_{11}})^{6/5}$ Therefore $\varnothing_7(x) \le \varnothing_7((\frac{2}{B_7^5B_{11}})^{6/5})$ which is less than 11:62 for $0.768 \le B_7 < 0.8804$ 2:018<B11<2:1016019. This gives a contradiction.

Now $B_5 \le \frac{3}{2}B_7 < 1.1521$ and $B_3 \le \frac{B_7}{\varepsilon} < 1.6385$

Claim (iii) B2<1:795

Suppose $B_2 \ge 1.795$ then $\frac{B_2^3}{B_3B_4B_5} > 2$ and $\frac{B_6^3}{B_7B_8B_9} > 2$ Applying AMGM to the inequality (1,4,4,1,1) p we get B1 + 4B2 + 4B6 + B10 + B11 - $\sqrt{B_2^5B_6^5B_1B_{10}B_{11}} > 11.62$ We find that left side is a decreasing function of B_6, so we first replace B_6 by εB_2 then it is a decreasing function of B_2, so we replace B_2 by 1.795 and get that

$$\varnothing_8(B_{11}) = B_1 + 4(1+\varepsilon)(1.795) + B_{10} + B_{11} - \sqrt{(\varepsilon)^5(1.795)^{10}B_1B_{10}B_{11}} > 11.62$$

Now $\varnothing_8(B_{11})>0$ so $\varnothing_8(B_{11})<\max\{\varnothing_8(2.018),\varnothing_8(2.1016019)\}$ which can be verified to be at most 11.62 for $(\varepsilon)^2(1.795)\le B_{10}<0.4402$ and 2:018<B_1<2:393347, giving thereby a contradiction.

Claim (iv) B_5<0:98392

Suppose $B_5\ge 0.98392$ We have $\frac{B_1^3}{B_2B_3B_4}>2$ and $\frac{B_5^3}{B_6B_7B_8}>2$ Also $2B_9\ge 2(\varepsilon B_5)>B_{10}$ Applying AM-GM to the inequality (4; 4; 2; 1) we get $4B_1+4B_5+4B_9-\frac{2B_9^2}{B_{10}}+B_{11}-\sqrt{B_1^5B_5^5B_9B_{10}B_{11}}>11.62$ One can easily check that left side is a decreasing function of B_9 and B_1 so we can replace B_9 by εB_5 and B1 by B11 to get $\varnothing_9=5B_{11}+4(1+\varepsilon)B_5-\frac{2(\varepsilon B_5)^2}{B_{10}}-\sqrt{\varepsilon B_{11}^6B_5^6B_{10}}>11.62$

Now the left side is a decreasing function of B5, so replacing B5 by 0.98392 we see that $\varnothing_9<11.62$ for $\frac{3\varepsilon}{4}(0.98392)<B_{10}<0.4409$ and 2:018<B11 < 2:1016019, a contradiction.

Final Contradiction:

As in Claim(iv), we have $\frac{B_1^3}{B_2B_3B_4}>2$ Also $B_5\ge\varepsilon B_1>0.9458$ each of B_6;B_7,...,B_{10}. Therefore the inequality (4; 6*; 1) holds, i.e. $\varnothing_{10}=4B_1\frac{1}{2}\frac{B_1^4}{B_2B_3B_4}+6(\frac{1}{B_1B_2B_3B_4B_{11}})^{\frac{1}{6}}+B_{11}>11.62$ Left side is an increasing function of $B_2B_3B_4$ and B_{11} and decreasing function of B_1. Using B_5<0:98392, we have $B_3\le\frac{3}{2}B_5<1.47588$ and $B_4\le\frac{4}{3}B_5<1.311894$ One easily checks that $\varnothing_{10}<11.62$ for $B_2B_3B_4$<1:795 × 1:47588 × 1:311894, B11<2:1016019 and $B_1\ge 2.018$ Hence we have a contradiction.

Proof of Theorem 1 for n=12

Here we have w_{12}=13, $B_1\le\gamma_{12}<2.5217871$ Using (2.5), we have l12 =0:3376<B_{12}<2:2254706=m_{12} and using (2.3) we have $B_1B_2^{11}\le\gamma_{11}^{11}$ i.e. $B_2\le\gamma_{11}^{\frac{11}{12}}<2.2254706$

The inequality (11*; 1) gives 11:62$(B_{12})^{-1/11}$ +B12>13. But this is not true for $0.4165\le B_{12}\le 2.17$ So we must have either B12<0:4165 or B12>2:17.

0:3376<B12<0:4165

Claim (i) B11<0:459

Suppose $B_{11}\ge 0.459$ then $B_{12}\ge\frac{3}{4}B_{11}>0.34425$ and 2B11>B_{12}, so (10*; 2) holds, i.e. $\phi_{11}=10.3(\frac{1}{B_{11}B_{12}})^{\frac{1}{10}}+4B_{11}-\frac{2B_{11}^2}{B_{12}}>13$ Left side is a decreasing function of B11, so we can replace B11 by .459 to find that ϕ_{11} <13 for 0:34425<B12<0:4165, a contradiction. Hence we have B_{11}<0:459.

Claim (ii) B10<0:5432

Suppose $B_{10}\ge 0.5432$ From Lemma 3, $B_{11}B_{12}\ge\frac{1}{2}B_{10}^2$ and $B_{10}\le\frac{3}{2}$ B_{12}. Therefore $\frac{1}{2}(0.5432)^2\le B_{11}B_{12}<0.1912$ and $B_{10}^2>B_{11}B_{12}$ so the inequality (9*; 3) holds, i.e. $9(\frac{1}{B_{10}B_{11}B_{12}})^{\frac{1}{9}}+4B_{10}-\frac{B_{10}^3}{B_{11}B_{12}}>13$ One easily checks that it is not true noting that left side is a decreasing function of B_{10}. Hence we must have B_{10}<0:5432.

Claim (iii) B9<0:6655

Suppose $B_9\ge 0.6655$ then $\frac{B_9^3}{B_{10}B_{11}B_{12}}>2$ So the inequality (8*; 4) holds. This gives $\phi_{12}(x)^{1/8}+4B_9-\frac{1}{2}B_9^5x>13$ where x=B_1B_2 ... B_8. The function $\phi_{12}(x)$ has its maximum value at $x=(\frac{2}{B_9^5})^{\frac{8}{7}}$ so $\phi_{12}(x)<\phi_{12}((\frac{2}{B_9^5})^{\frac{8}{7}})<13$ for $0.6655\le B_9-\frac{1}{2}B_9^5x>13$ where x=B_1B_2... B8. The function $\phi_{12}(x)$ has its maximum value at $x=(\frac{2}{B_9^5})^{\frac{8}{7}}$ so $\phi_{12}(x)<x=(\frac{2}{B_9^5})^{\frac{8}{7}}$ <13 for 0:6655 $\le B_9\le\frac{3}{2}B_{11}<0.6885$ This gives a contradiction.

Using Lemmas 3 & 4 we have:

$$B_8\le\frac{3}{2}B_{10}<0.8148, B_7\le\frac{B_{11}}{\varepsilon}<0.9793, B_6\le\frac{B_{10}}{\varepsilon}<1.1589$$

$$B_5\le\frac{B_9}{\varepsilon}<1.4198, B_4\le\frac{3}{2}\frac{B_{10}}{\varepsilon}<1.7384, B_3\le\frac{B_{11}^{\varepsilon}}{\varepsilon^2}<2.0892$$

Claim (iv) B2>1:828, B4>1:426, B6>1:019 and B8>0:715

Suppose $B_2\le 1.828$ Then 2(B2+B4+B6+B8+B10+B12)<2(1:828+1:7384+1:1589+0:8148+0:5432+0:4165)<13, giving thereby a contradiction to the weak inequality (2; 2; 2; 2; 2; 2) w.

Similarly we obtain lower bounds on B_4;B_6 and B_8 using (2; 2; 2; 2; 2; 2)w.

Claim(v) B2>2:0299

Suppose $B_2\le 2.0299$ Consider following two cases:

Case (i) B3>B4

We have B3>B4>1:426>each of B5,....,B12. So the inequality (2; 10*) holds, i.e. $4B_1-\frac{2B_1^2}{B_2}+10.3(\frac{1}{B_1B_2})^{\frac{1}{10}}>13$ The left side is a decreasing function of B_1, so replacing B_1 by B_2 we get 2B2 + 10:3 $(\frac{1}{B_2^2})^{\frac{1}{10}}>13$ which is not true for $B_2\le 2.0299$

Case (ii) $B_3\le B_4$

As B4>1:426>each of B5,....,B12, the inequality (3; 9*) holds, i.e. $\phi_{13}(X)=4B_1-\frac{B_1^3}{x}+9(\frac{1}{B_1x})^{\frac{1}{9}}>13$ where X=B2B3<min $\{B_1^2,(2.0299)(1.7384)\}=\alpha$ say. Now $\phi_{13}(X)$ is an increasing function of X for $B_1\ge B_2>1.828$ and So $\phi_{13}(x)<\phi_{13}(X)$ which can be seen to be less than 13. Hence we have B2>2:0299.

Claim (vi) B1>2:17 and B3<1:9517

Using (2.3) we have $B_3\le(\frac{\gamma_{10}^{10}}{B_1B_2})^{\frac{1}{10}}<1.9648$ Therefore B_2>2:0299>each of B_3,...,B_{12}. So the inequality (1; 11*) holds, i.e. B_1 + 11:62 $(\frac{1}{B_1})^{\frac{1}{11}}>13$ But this is not true for $B_1\le 2.17$ So we must have B1>2:17: Again using (2.3) we have $B_3<(\frac{2.2636302}{2.17\times 2.0299})^{\frac{1}{10}}<1.9517$

Claim (vii) B4<1:646

Suppose $B_4\ge 1.646$ From (5.1) and Claims (i)-(iii), we have $\frac{B_4^3}{B_5B_6B_7}>2$ and $\frac{B_8^3}{B_9B_{10}B_{11}}>\frac{(\varepsilon B_4)^3}{B_9B_{10}B_{11}}>2$ Applying AM-GM to the inequality (1,2,4,4,1) we get $\phi_{14}=B_1+4B_2-\frac{2B_2^2}{B_3}+4B_4+4B_8+B_{12}-\sqrt{B_4^5B_8^5B_1B_2B_3B_{12}}>13$

We find that left side is a decreasing function of B2, B8 and B12. So we can replace B2 by 2:0299, B8 by "B4 and B12 by $\varepsilon^2 B_4$. Then it turns a decreasing function of $\varepsilon^2 B_4$, so can replace B4 by 1.646 to find that ϕ_{14}<13 for B_1<2:52178703 and B_3<1:9517, a contradiction. Hence we have B_4<1:646.

Claim (viii) B1<2:4273

Suppose B1 $\geq$ 2:4273. Consider following two cases:

Case (i) B_5>B_6

Here B_5>each of $B_{6,\ldots,}B_{12}$ as $B_5 \geq \varepsilon B_1$>1.137 >each of B7,…, B12. Also B2B3B4<2:2254706×1:9517×1:646<7:15. So $\frac{B_1^3}{B_2B_3B_4}>2$ Hence the inequality (4; 8*) holds. This gives $4B_1-\frac{1}{2}\frac{B_1^4}{B_2B_3B_4}+8(B_1B_2B_3B_4)^{-1/8}>13$

Left side is an increasing function of $B_2B_3B_4$ and decreasing function of B_1. So we can replace $B_2B_3B_4$ by 7.15 and B_1 by 2.4273 to get a contradiction.

Case (ii) $B_5 \leq B_6$

Using (5.1) we have $B_5 \leq B_6$<1:1589 and so $B_4 \leq \frac{4}{3}B_5 < 1.5452$ Therefore $\frac{B_2^3}{B_3B_4B_5}>2$ as B2>2:0299 and B_3<1:9517. Also from Claim (iv), B_6>1:019>each of $B_7,\ldots,B_{12}$. Hence the inequality (1; 4; 7*) holds. This gives $B_1+4B_2-\frac{1}{2}\frac{B_2^4}{B_4B_5B_6}+7$ $7(B_1B_2B_3B_4B_5)^{-1/7}$>13: Left side is an increasing function of $B_3B_4B_5$ and B_1 and a decreasing function of B_2. One can check that inequality is not true for $B_3B_4B_5$<1:9517×1:5452 ×1:1589, B_1<2:5217871 and for B_2>2:0299: Hence we must have B_1<2:4273:

Claim (ix) B5<1:396

Suppose $B_5 \geq 1.396$ From (5.1), $B_6B_7B_8$<0:925 and $B_{10}B_{11}B_{12}$<0:104, so we have $\frac{B_5^3}{B_6B_7B_8}>2$ and $\frac{B_9^3}{B_{10}B_{11}B_{12}}>\frac{(\varepsilon B_5)^3}{B_{10}B_{11}B_{12}}>2$ Applying AMGM to the inequality (1,2,1,4,4) we get $B_1 + 4B_2 - \frac{2B_2^2}{B_3} + B_4 + 4B_5 + 4B_9 - \sqrt{B_5^5B_9^5B_1B_2B_3B_4}>13$ We find that left side is a decreasing function of B2 and B9. So we replace B2 by 2:0299 and B_9 by εB_5. Now it becomes a decreasing function of B_5 and an increasing function of B_1 so replacing

B_5 by 1.396 and B_1 by 2.4273, we find that above inequality is not true for 1:522<B_3<1:9517 and 1:426<B_4<1:646, giving thereby a contradiction. Hence we must have B_5<1:396.

Claim (x) B3>1:7855

Suppose $B_3 \leq 1.7855$ We have B4>1:426>each of $B_5;B_6,\ldots,B_{12}$, hence the inequality (1; 2; 9*) holds. It gives $\phi_{15}=B_1+4B_2-\frac{2B_2^2}{B_3}+9(\frac{1}{B_1B_2B_3})^{\frac{1}{9}}>13$ It is easy to check that left side of above inequality is a decreasing function of B2 and an increasing function of B1 and B3. So replacing B_1 by 2.4273, B_3 by 1.7855 and B_2 by 2.0299 we get -15<13; a contradiction. Hence we have B_3>1:7855.

Claim (xi) $B_2 > 2.0733$

Suppose $B_2 \leq 2.0733$ We have B3>1:7855>each of $B_4;B_5,\ldots,B_{12}$, hence the inequality (2; 10*) holds. It gives $\phi_{16}=4B_1-\frac{2B_1^2}{B_2}+10.3(\frac{1}{B_1B_2})^{\frac{1}{10}}>13$ The left side is a decreasing function of B_1 and an increasing function of B_2, so replacing B_1 by 2:17 and B_2 by 2.0733 we get $\phi_{16}<13$ a contradiction.

Claim (xii) B7<0:92 and B5<1:38

Suppose $B_7 \geq 0.92$ Here we have $B_4B_5B_6$<2:67 and $B_8B_9B_{10}$<0:295, so $\frac{B_3^3}{B_4B_5B_6}>2$ and $\frac{B_7^3}{B_8B_9B_{10}}>2$ Also $2B_{11} \geq 2\varepsilon B_7 > B_{12}$ Applying AM-GM to the inequality (2,4,4,2) we get $\phi_{17}=4B_1-\frac{2B_1^2}{B_2}+4B_3+4B_7-\sqrt{B_3^5B_7^5B_1B_2B_{11}B_{12}}+4B_{11}-\frac{2B_{11}^2}{B_{12}}>13$ We find that left side is a decreasing function of B1 and B11. So we can replace B1 by 2:17 and B11 by εB_7. Then left side becomes a decreasing function of B_7 and an increasing function of B_2, so can replace B_7 by 0.92 and B_2 by 2.2254706 to see that ϕ_{17}<13 for 1:7855<B_3<1:9517 and 0:3376<B_{12}<0:4156, a contradiction. Hence B_7<0:92. Further $B_5 \leq \frac{3}{2}B_7$ gives B_5<1:38.

Claim (xiii) B6<1:097

Suppose $B_6 \geq 1.097$ Here we have B3B4B5<4:44 and B7B8B9<0:5, so $\frac{B_2^3}{B_3B_4B_5}>\frac{(2.0733)^3}{4.44}>2$ and $\frac{B_6^3}{B_7B_8B_9}>2$ Also $2B_{10} \geq 2\varepsilon B_6 > B_{11}$ Applying AM-GM to the inequality (1,4,4,2,1) we get $\phi_{18}=B_1+4B_2+4B_6-\sqrt{B_2^5B_6^5B_1B_{10}B_{11}B_{12}}+4B_{10}-\frac{2B_{10}^2}{B_{11}}+B_{12}>13$ We find that left side is a decreasing function of B_{10}, B_{12} and B_{11}. So we can replace B_{10} by εB_6 and B_{12} by 0.3376 and B_{11} by $\frac{3\varepsilon}{4}B_6$. Then left side becomes a decreasing function of B_6, so we can replace B_6 by 1.097 to find that ϕ_{18}<13, for 2:17<B_1<2:4273 and 2:0733<B_2<2:2254706, a contradiction. Hence we must have B_6<1:097.

Claim (xiv) B_5>B_6 and $\frac{B_1^3}{B_2B_3B_4}<2$

First suppose B5 $\leq$ B6, then B4B5B6<1:646 × 1:0972<1:981 and $\frac{B_3^3}{B_4B_5B_6}>2$ Also $B_7 \geq \varepsilon B_3 > 0.83 >$ eachofB$_8,\ldots,$B$_{12}$.Hencetheinequality (2; 4; 6*) holds, i.e. $4B_1-\frac{2B_1^2}{B_2}+4B_3-\frac{1}{2}\frac{B_3^4}{B_4B_5B_6}+6(\frac{1}{BlB_2B_3B_4B_5B_6})^{\frac{1}{6}}>13$

Now the left side is a decreasing function of B1 and B3 as well; also it is an increasing function of B_2 and $B_4B_5B_6$. But one can check that this inequality is not true for B_1>2:17, B_3>1:7855, B_2<2:2254706 and $B_4B_5B_6$<1:981, giving thereby a contradiction. Further suppose $\frac{B_1^3}{B_2B_3B_4} \geq 2$ then as B5>B6>1:019>each of B7,…, B12, the inequality (4; 8*) holds. Now working as in Case (i) of Claim (viii) we get contradiction for B1>2:17 and B2B3B4<2:2254706 × 1:9517 × 1:646<7:14934.

Claim (xv) B3<1:9 and B1<2:4056

Suppose $B_3 \geq 1.9$, then for $B_4B_5B_6$<1:646×1:38×1:097<2:492, $\frac{B_3^3}{B_4B_5B_6}>2$ Also B7$\geq \varepsilon$ B3>0:89>each of $B_8,\ldots,B_{12}$. Hence the inequality (2; 4; 6*) holds. Now working as in Claim (xiv) we get contradiction for B_1>2:17, B_2<2:2254706, B_3>1:9 and $B_4B_5B_6$<2:492. So B_3<1:9. Further if $B_1 \geq$ 2:4056, then $\frac{B_1^3}{B_2B_3B_4}>\frac{(2.4056)^3}{2.2254706\times1.9\times1.646}>2$ contradicting Claim (xiv).

Claim (xvi) B_4<1:58 and B_1<2:373

Suppose $B_4 \geq 1.58$ thenforB_5B_6B<1:38×1:097×0:92<1:393, $\frac{B_4^3}{B_5B_6B_7}>2$ Also $B_8 \geq \varepsilon B_4$>0:74>each of $B_9,\ldots,B_{12}$. Hence the inequality (1; 2; 4; 5*)

holds, i.e. $-19 = B_1 + 4B_2 - \frac{2B_2^2}{B_3} + 4B_4 - \frac{1}{2}\frac{B_4^4}{B_5B_6B_7} + 5(\frac{1}{B_1B_2B_3B_4B_5B_6B_7})^{\frac{1}{5}} > 13$

Left side is a decreasing function of B_2 and B_4.

So we replace B_2 by 2.0733 and B_4 by 1.58. Then it becomes an increasing function of B_1, B_3 and $B_5B_6B_7$. So we replace B_1 by 2.4056, B_3 by 1.9 and $B_5B_6B_7$ by 1.393 to find that -19<13, a contradiction. Further if $B_1 \geq 2{:}373$, then $\frac{B_1^3}{B_2B_3B_4} > 2$ contradicting Claim (xiv).

Final Contradiction:

We have $B_3B_4B<1{:}9\times1{:}58\times1{:}38<4{:}15$. Therefore $\frac{B_2^3}{B_3B_4B_5} > 2$ Also B6>1:019>each of $B_7,\ldots,B_{12}$. Hence the inequality (1; 4; 7*) holds. Now we get contradiction working as in Case (ii) of Claim (viii).

5.2 2:17<B12<2:2254706

Here $B_1 \geq B_{12} > 2.17$ Using Lemma 3 and 4, we have

$B^{11} = (B_1B_2\ldots B_{10}B_{12})^{-1} < (\frac{3}{64}\varepsilon^8 B_1^{10}B_{12})^{-1} < 1.8223$

Claim (i) Either B11<0:4307 or B11>1:818

Suppose $0{:}4307 \leq B_{11} \leq 1.818$ The inequality (10*; 1; 1) gives 10:3 $(\frac{1}{B_{11}B_{12}})^{\frac{1}{10}} + B_{11} + B_{12} > 13$ which is not true for $0{:}4307 \leq B_{11} \leq 1.818$ and 2:17<B12<2:2254706. So we must have either $B_{11}<0{:}4307$ or $B_{11}>1{:}818$.

Claim (ii) B11<0:4307

Suppose $B_{11} \geq 0.4307$ then using Claim(i) we have B11>1:818. Now we have using Lemmas 3 & 4,

$B_2 = (B_1B_2\ldots B_{12})^{-1} < (\frac{1}{16}\varepsilon^6 B_2^8B_1B_{11}B_{12})^{-1}$ This gives B2<1:777.

$B_3 = (B_1B_2B_4\ldots B_{12})^{-1} < (\frac{3}{64}\varepsilon^4 B_3^7B_1^2B_{11}B_{12})^{-1}$ This gives B3<1:487

$B_4 = (B_1B_2B_4\ldots B_{12})^{-1} < (\frac{1}{16}\varepsilon^3 B_4^6B_1^3B_{11}B_{12})^{-1}$ This gives B4<1:213.

$B_6 = (B_1..B_5B_7\ldots B_{12})^{-1} < (\frac{1}{16}\varepsilon^2 B_6^4B_1^5B_{11}B_{12})^{-1}$ This gives B6<0:826.

$B_7 = (B_1..B_6B_8\ldots B_{12})^{-1} < (\frac{3}{64}\varepsilon^2 B_7^3B_1^6B_{11}B_{12})^{-1}$ This gives B7<0:697.

$B_8 = (B_1..B_7B_9\ldots B_{12})^{-1} < (\frac{1}{16}\varepsilon^3 B_8^2B_1^7B_{11}B_{12})^{-1}$ This gives B8<0:559.

$B_9 = (B_1..B_8B_{10}B_{11}B_{12})^{-1} < (\frac{3}{64}\varepsilon^3 B_9B_1^7B_{11}B_{12})^{-1}$ This gives B9<0:478.

$B_{10} = (B_1..B_9B_{11}B_{12})^{-1} < (\frac{1}{16}\varepsilon^6 B_1^9B_{11}B_{12})^{-1} < 0.359$

Therefore we have $\frac{B_1^3}{B_2B_3B_4} > 2$ and $B_5 \geq \varepsilon B_1 > 1.01 >$ each of $B_6,\ldots,B_{10}$. So the inequality (4; 6*; 1; 1) holds, i.e. $4B_1 - \frac{1}{2}\frac{B_1^4}{B_2B_3B_4} + 6(B_1B_2B_3B_4B_{11}B_{12})^{-1/6} + B_{11} + B_{12} > 13$ Now the left side is an increasing function of B2B3B4, B11 and of B12 as well. Also it is a decreasing function of B1. So we replace $B_2B_3B_4$ by $1{:}777 \times 1{:}487 \times 1{:}213$, B_{11} by 1.8223, B_{12} by 2.2254706 and B_1 by 2.17 to arrive at a contradiction. Hence we must have $B_{11}<0{:}4307$.

Claim (iii) B10<0:445

Suppose $B_{10} \geq 0.445$ then $2B_{10}>B_{11}$. So the inequality (9*; 2; 1) holds, i.e. $\phi_{20} = 9(\frac{1}{B_{10}B_{11}B_{12}})^{\frac{1}{9}} + 4B_{10} - \frac{2B_{10}^2}{B_{11}} + B_{12} > 13$ $B_{11} \geq \frac{3}{4}$ B10 and B12>2:2254706, the left side is an increasing function of B12 and a decreasing function of B_{10}, so replacing B_{12} by 2.2254706 and B_{10} by 0.445 we find that $\phi_{20}<13$, for $3\ 4(0{:}445)<B_{11}<0{:}4307$, a contradiction. Hence we must have $B_{10}<0{:}445$.

Using Lemmas 3 and 4 we have:

$$B_9 \leq \frac{4}{3}B_{10} < 0.594, B_8 \leq \frac{3}{2}B_{10} < 0.67, B_7 \leq 0.89$$

$$B_6 \leq \frac{B_{10}}{\varepsilon} < 0.9494, B_5 \leq \frac{4}{3}\frac{B_{10}}{\varepsilon} < 1.266, B_4 \leq \frac{3}{2}\frac{B_{10}}{\varepsilon} < 1.4242$$

$$B_3 \leq \frac{2B_{10}}{\varepsilon} < 1.899, B_2 \leq \frac{B_{10}}{(\varepsilon)^2} < 2.0255$$

Claim (iv) B3<1:62

Suppose $B_3 \geq 1.62$ From (5.2), we have B4B5B6<1:712 and B8B9B10<0:178, so $\frac{B_3^3}{B_4B_5B_6} > 2$ and $\frac{B_7^3}{B_8B_9B_{10}} > 2$ Applying AM-GM to the inequality (2,4,4,1,1) we get $\phi_{21} = 4B_1 - \frac{2B_1^2}{B_2} + 4B_3 + 4B_7 - \sqrt{B_3^5B_7^5B_1B_2B_{11}B_{12}} + B_{11} + B_{12} > 13$ We find that left side is a decreasing function of B_1, B_7 and B_{11}. So we can replace B_1 by B_{12}, B_7 by εB_3 and B_{11} by $\varepsilon^2 B_3$. Then it becomes a decreasing function of B_3, so replacing B_3 by 1.62 we find that $\phi_{21}<13$; for $1{:}6275<B_2<2{:}0255$ and $2{:}17<B_{12}<2{:}2254706$, a contradiction. Hence we must have $B_3<1{:}62$.

Claim (v) B12>2:196

Suppose $B_{12} \leq 2.196$ From (5.2), we have $B_2B_3B_4<4{:}674$ and $\frac{B_1^3}{B_2B_3B_4} > 2$ Also $B_5 \geq \varepsilon B_1 > 1.01 >$ each of B6,..., B11. Therefore the inequality (4; 7*; 1) holds, i.e. ϕ_{22} $\phi_{22} = 4B_1 - \frac{1}{2}\frac{B_1^4}{B_2B_3B_4} + 7(B_1B_2B_3B_4B_{12})^{-1/7} + B_{12} > 13$ Left side is an increasing function of B2B3B4 and of B_{12} as well. Also it is a decreasing function of B_1. So we can replace $B_2B_3B_4$ by 4.674, B_{12} by 2.196 and B_1 by 2.17 to get $\phi_{22}<13$, a contradiction. Hence we must have $B_{12}>2{:}196$.

Final Contradiction

Now we have $B_1 \geq B_{12}>2{:}196$. We proceed as in Claim(v) and use (4; 7*; 1). Here we replace $B_2B_3B_4$ by 4.674, B_{12} by 2.2254706 and B_1 by 2.196 to get $\phi_{22}<13$, a contradiction.

References

1. Bambah RP, Dumir VC, Hans-Gill RJ (2000) Non-homogeneous prob-lems: Conjectures of Minkowski and Watson, Number Theory, Trends in Mathematics, Birkhauser Verlag, Basel 15-41.
2. Birch BJ, Swinnerton-Dyer HPF (1956) On the inhomogeneous min-imum of the product of n linear forms, Mathematika 3: 25-39.
3. Blichfeldt HF (1934) The minimum values of positive quadratic forms in six, seven and eight variables, Math Z 39: 1-15.
4. Cebotarev N (1940) Beweis des Minkowski'schen Satzes uber lineare inhomo-gene Formen, Vierteljschr. Naturforsch. Ges. Zurich, 85 Beiblatt 27-30.
5. Cohn H, Elkies N (2003) New upper bounds on sphere packings, I. Ann of Math 157: 689-714.
6. Cohn H, Kumar A (2004) The densest lattice in twenty-four dimensions, Electron. Res Announc Amer Math Soc 10: 58-67.

7. Conway JH, Sloane NJA (1993) Sphere packings, Lattices and groups, Springer-Verlag, Second edition, New York.

8. Gruber P (2007) Convex and discrete geometry, Springer Grundlehren Series 336.

9. Gruber P, Lekkerkerker CG (1987) Geometry of Numbers, Second Edi-tion, North Holland, 37.

10. Hans-Gill RJ, Raka M, Sehmi R, Sucheta (2009) A uni ed simple proof of Woods' conjecture for n 6, J Number Theory 129: 1000-1010.

11. Hans-Gill RJ, Raka M, Sehmi R (2009) On conjectures of Minkowski and Woods for n=7, J Number Theory 129: 1011-1033.

12. Hans-Gill RJ, Raka M, Sehmi R (2010) Estimates On Con-jectures of Minkowski and Woods, Indian Jl Pure Appl Math 41: 595-606.

13. Hans-Gill RJ, Raka M, Sehmi R (2011) On Conjectures of Minkowski and Woods for n=8, Acta Arithmetica 147: 337-385.

14. Hans-Gill RJ, Raka M, Sehmi R (2011) Estimates On Con-jectures of Minkowski and Woods II, Indian Jl. Pure Appl. Math. 42: 307-333.

15. Il'in IV (1986) A remark on an estimate in the inhomogeneous Minkowski conjecture for small dimensions,) 90, Petrozavodsk. Gos. Univ., Petrozavodsk 24-30.

16. IV (1991) Chebotarev estimates in the inhomogeneous Minkowski con-jecture for small dimensions, Algebraic systems, Ivanov. Gos. Univ, Ivanovo 115-125.

17. Kathuria L, Raka M (2014) On Conjectures of Minkowski and Woods for n=9.

18. Kathuria L, Raka M (2011) Refined Estimates on Conjectures of Woods and Minkowski-II, To be Submitted.

19. Kathuria L, Raka M (2014) Generalization of a result of Birch and Swinnerton-Dyer, To be Submitted.

20. Korkine A, Zolotare G (1877) Sur les formes quadratiques, Math. Ann. 366-389; Sur les formes quadratiques positives. Math Ann 11: 242-292.

21. McMullen CT (2005) Minkowski's conjecture, well rounded lattices and topo-logical dimension. J Amer Math Soc 18: 711-734.

22. Mordell LJ (1960) Tschebotare's Theorem on the product of Non-homogeneous Linear Forms (II). J London Math Soc 35: 91-97.

23. Pendavingh RA, Van Zwam SHM (2007) New Korkine-Zolotarev in-equalities. SIAM J Optim 18: 364-378.

24. Woods AC (1965) The densest double lattice packing of four spheres. Math-ematika 12: 138-142.

25. Woods AC (1965) Lattice coverings of ve space by spheres. Mathematika 12: 143-150.

26. Woods AC (1972) Covering six space with spheres. J Number Theory 4: 157-180.

Existence of Multiple Solutions for P-Laplacian Problems Involving Critical Exponents and Singular Cylindrical Potential

Mohammed El Mokhtar Ould El Mokhtar

Qassim University, Department of Mathematics, Qassim University, Buraidah, Kingdom of Saudi Arabia

Abstract

In this paper, we establish the existence of multiple solutions for p-Laplacian problems involving critical exponents and singular cylindrical potential, by using Ekeland's variational principle and mountain pass theorem without Palais-Smale conditions.

Keywords: P-Laplacian; Critical exponents; Cylindrical potential; Dimensional

Introduction

The aim of this paper is to establish the existence and multiplicity of solutions to the following quasilinear elliptic problem

$$(\mathcal{P}_{\lambda,\mu})\begin{cases}-\Delta_p u-\mu|y|^{-p}|u|^{p-2}u=h(y)|y|^{-s}|u|^{q-2}u+\lambda g(x) \text{ in } \mathbb{R}^N,\ y\neq 0\\ u\in\mathcal{D}_1^p(\mathbb{R}^N),\end{cases}$$

Where $\Delta_p u=div\left(|\nabla u|^{p-2}\nabla u\right)$,1<p<k,k and N are integers with N>p, $p\le k\le N$, $\mathbb{R}^N=\mathbb{R}^k\times\mathbb{R}^{N-k}$\$., the point $x\in\mathbb{R}^N$ can be written as $x=(y,z)\in\mathbb{R}^k\times\mathbb{R}^{N-k}, -\infty<\mu<\bar{\mu}_{k,p}:=((k-p)/p)^p, 0\le s<p, q:=p^*(s)=p(N-s)/(N-p)$ is the critical Sobolev-Hardy exponent, λ and μ are positive parameters which we will specify later, g is a continuous function on $\mathbb{R}^N$ and h is a bounded positive function on $\mathbb{R}^k$.

Let $\mathcal{H}_\mu=\mathcal{D}_1^p(\mathbb{R}^N)$ be the space defined as the completion of $C_c^\infty(\mathbb{R}^N)$ with respect to the norm $\|\nabla u\|_p=\left(\int_{\mathbb{R}^N}|\nabla u|^p dx\right)^{\frac{1}{p}}$.

When $\mu<\bar{\mu}_{k,p}$, Hardy type inequality implies that the norm

$$\|u\|=\|u\|_{\mu,p}=\left(\int_{\mathbb{R}^N}\left(|\nabla u|^p-\mu|y|^{-p}|u|^p\right)dx\right)^{1/p},$$

is will defined in $\mathcal{H}_\mu$ and $\|.\|$ is equivalent to $\|\nabla.\|_p$; since the following inequalities hold: $\left(1-\left(\max(\mu,0)/\bar{\mu}_{k,p}\right)\right)^{1/p}\|\nabla u\|_p\le\|u\|\le\left(1-\left(\min(\mu,0)/\bar{\mu}_{k,p}\right)\right)^{1/p}\|\nabla u\|_p$, for all $u\in\mathcal{H}_\mu$

We define the weighted Sobolev space $\mathcal{D}:=\mathcal{H}_\mu\cap L^p(\mathbb{R}^N,|y|^{-s}dx)$ which is a Banach space with respect to the norm defined by $\mathcal{N}(u):=\|u\|_\mu+(\int_{\mathbb{R}^N}|y|^{-s}|u|^q dx)^{1/q}$.

Several existence results are available in the case p = 2 and k = N; we quote for example [1-3]; and the references therein. For more details, when $h\equiv 1$, $\mu=0$ and $q=2^*$, the regular problem $(\mathcal{P}_{1,0})$ has been considered, on the bounded domain Ω, by Tarantello [4]. She proved that for $g\in\left(H_0^1(\Omega)\right)'$ not identically zero and satisfying a suitable condition, the problem considered admits two solutions. Also, they are two nontrivial non-negative solutions when g is nonnegative. The problem $(\mathcal{P}_{\lambda,\mu})$ has been studied by Bouchekif and Matallah in [2], by using Ekeland.s variational principle and mountain pass theorem, they established the existence of two nontrivial solutions when $0<\mu\le\bar{\mu}_N, \lambda\in(0,\Lambda_*)$, where Λ_* is a positive constant and under sufficient conditions on functions g and h.

For the case p=2 and k<N, there are much less studies in the literature at our knowledge. We cite for example [4-6], and the references therein. As noticed in [6] considered the minimization problem

$$S(p)=S(N,p,k,s)$$
$$=\inf\left\{\int_{\mathbb{R}^N}|\nabla u|^p, u\in c_c^\infty((\mathbb{R}^k\setminus\{0\}\times\mathbb{R}^{N-k})\ and \int_{\mathbb{R}^N}|y|^{-s}|u|^q dx=1\right\}$$

and in [6], solutions which are radially symmetric in the x-variable receive impor-tance with regard to certain elliptic equations on the n=N-k+1 dimensional hyperbolic space $\mathbb{H}^n$. In particular, Musina in [6] has considered the problem $(\mathcal{P}_{0,\mu})$ with $h\equiv 1$. She established the existence of ground state solution when $0<\mu<\bar{\mu}_k$ and $2<k\le N$ and the support of the ground state solution is a half-space when k = 1 and $N\ge 4$

In case p>2 and 1<k<N, equations with cylindrical potentials were also studied by many people [1,4,7-10]. For instance, in [11], Xuan studied the multiple weak solutions for p-Laplace equation with singularity and cylindrical symmetry in bounded domains. However, they only considered the equation with sole critical Hardy-Sobolev term.

Since our approach is variational, we de.ne the functional $I_{\lambda,\mu}$ on $\mathcal{D}$ by

$$I_{\lambda,\mu}(u):(1/p)\|u\|^p-(1/q)\int_{\mathbb{R}^N}h(y)|y|^{-s}|u|^q dx-\lambda\int_{\mathbb{R}^N}g(x)udx.$$

Throughout this work, we consider the following assumptions

(G) g $\in\mathcal{H}_\mu'$(dual of $\mathcal{H}_\mu$),

(H) $\lim_{|y|\to 0}h(y)=\lim_{|y|\to\infty}h(y)=h_0>0,\ h(y)\ge h_0, y\in\mathbb{R}^k$.

In our work, we prove the existence of at least two distinct critical points of $I_{\lambda,\mu}$.

One by the Ekeland variational principle in [10] with negative energy, and the other by mountain pass theorem in [7] without Palais-

***Corresponding author:** El Mokhtar Ould El Mokhtar Mohammed, Qassim University, College of Science, Department of Mathematics, BO 6644, Buraidah: 51452, Kingdom of Saudi Arabia, E-mail: med.mokhtar66@yahoo.fr

Smale conditions with positive energy.

Our main result is given as follows

Theorem 1: Suppose that $p<k<N, 0\le s<p, \mu<\bar{\mu}_{k,p}$, hypothesis (H) holds, $g\in\mathcal{H}_{\mu}'\cap C(\mathbb{R}^N)$ and $g\neq 0$. Then there exists $\Lambda_*>0$ such that the problem $(\mathcal{P}_{\lambda,\mu})$ has at least two solutions for any $\lambda\in(0,\Lambda_*)$.

This paper is organized as follows. In Section 2, we give some preliminaries.

Section 3 is devoted to the proof of Theorem 1.

Preliminaries

We start by recalling the following definition and properties from the paper [6].

The first inequality that we need is the Hardy inequality

$$\int_{\mathcal{R}^N}|\nabla u|^p dx \ge \bar{\mu}\%_{k,p}\int_{\mathcal{R}^N}|y|^{-p}|u|^p dx, \text{ for all } u\in\mathcal{D}_1^p(\mathbb{R}^N), \tag{4.1}$$

the constant $\bar{\mu}_{k,p} := ((k-p)/p)^p$ is sharp but not achieved [2].

Definition 1: An entire solution v to $(\mathcal{P}_{\lambda,\mu})$ is a ground state solution if it achieves the best constant

$$S_{\mu,p}=S_{\mu,p}(k,N):=\lim_{v\in\mathcal{H}_\mu((\mathbb{R}^k\setminus\{0\})\times\mathbb{R}^{N-k})}\frac{\int_{\mathbb{R}^N}\left(\int_{\mathcal{R}^N}(|\nabla u|^p-\mu|y|^{-p}|u|^p)dx\right)dx}{\left(\int_{\mathbb{R}^N}|y|^{-s}|v|^q dx\right)^{p/q}}, \tag{4.2}$$

Lemma 1: Assume [6] that $p<k<N, 0\le s<p$ and $\mu<\bar{\mu}_{k,p}$. Then, the in fimum $S_{\mu,p}$ is achieved on $\mathcal{H}_\mu((\mathbb{R}^k\setminus\{0\}\setminus\{0\}\times\mathbb{R}^{N-k})$.

Lemma 2: Let $(u_n)\subset\mathcal{D}$ be a Palais-Smale sequence $\left[(PS)_c \text{ in short}\right]$ of $I_{\lambda,\mu}$, i.e.,

$I_{\lambda,\mu}(u_n)\to c$ and $I'_{\lambda,\mu},(u_n)\to 0$ in $\mathcal{D}'(duqlof\mathcal{D})$ as $n\to\infty$ for some $c\in\mathbb{R}$. (4.3)

Then, $u_n\rightharpoonup u$ in $\mathcal{D}$ and $I'_{\lambda,\mu}(u)=0$.

Proof: From (4.3);

We have

$$(1/p)\|u_n\|^p-(1/q)\int_{\mathbb{R}^N}h(y)|y|^{-s}|u_n|^q dx-\lambda\int_{\mathbb{R}^N}g(x)u_n dx=c+o_n(1)$$

and

$$\|u_n\|^p-\int_{\mathbb{R}^N}h(y)|y|^{-s}|u_n|^q dx-\lambda\int_{\mathbb{R}^N}g(x)u_n dx=o_n(1), \text{ for } n \text{ large},$$

where $o_n(1)$ denotes $o_n(1)\to 0$ as $n\to\infty$. Then,

$$c+o_n(1)=I_{\lambda,\mu}(u_n)-(1/q)\langle I'_{\lambda,\mu}(u_n),u_n\rangle$$

$$\ge ((q-p)/pq)\|u_n\|^p-\lambda((q-1)/q)\|g\|\mathcal{H}'_\mu\|u_n\|,$$

(u_n) is bounded in $\mathcal{D}$. Up to a subsequence if necessary, we obtain that

$u_n \rightharpoonup u$ in $\mathcal{D}$

$u_n \rightharpoonup u$ in $L_q(\mathbb{R}^N;|y|^{-s})$

$u_n \to u$ a.e in $\mathbb{R}^N$.

Consequently, we get

$I'_{\lambda,\mu}(u)=0.$

Lemma 3: Let $(u_n)\subset\mathcal{D}$ be a Palais-Smale sequence $(PS)_c\, I_{\lambda,\mu}$ for some $c\in\mathbb{R}$.

Then, $u_n\rightharpoonup u$ in $\mathcal{D}$ and either $u_n\to u$ or $c\ge I_{\lambda,\mu}(u)+((q-p)/pq)\left(h_0^{-p/q}S_{\mu,q}\right)^{q/(q-p)}$ for all $q\in\left(p,p^*(0)\right]$

Proof: We know that (u_n) is bounded in $\mathcal{D}$ Up to a subsequence if necessary, we have that

$u_n \rightharpoonup u$ in $\mathcal{D}$

$u_n \to u$ a.e in $\mathbb{R}^N$.

Denote $v_n=u_n-u$, then $v_n\rightharpoonup 0$. As in Brézis and Lieb [2]; we have $|v_n|_q^p=|u_n|_q^p-|u|_q^p$

and

$$\lim_{n\to\infty}\int_{\mathbb{R}^N}h(y)\left(|y|^{-s}|u_n|^q-|y|^{-s}|u_n-u|^q\right)dx=\int_{\mathbb{R}^N}h(y)|y|^{-s}|u|^q dx.$$

On the other hand, by using the assumption (H), we obtain

$$\lim_{n\to\infty}\int_{\mathbb{R}^N}h(y)|y|^{-s}|u_n|^q dx=h_0\lim_{n\to\infty}\int_{\mathbb{R}^N}|y|^{-s}|u_n|^q dx.$$

Then, we get

$$I_{\lambda,\mu}(u_n)=I_{\lambda,\mu}(u)+(1/p)\|v_n\|^p-(h_0/q)\int_{\mathbb{R}^N}|y|^{-s}|v_n|^q+o_n(1)$$

and

$$\langle I'_{\lambda,\mu}(u_n),u_n\rangle=\|v_n\|^p-h_0\int_{\mathbb{R}^N}|y|^{-b}|u_n|^q+o_n(1).$$

Then we can assume that

$$\lim_{n\to\infty}\|v_n\|^p=h_0\lim_{n\to\infty}\int_{\mathbb{R}^N}|y|^{-s}|u_n|^q=l\ge 0.$$

Assume $l>0$, we have by definition of $S_{\mu,q}$

$$l\ge S_{\mu,q}\left(lh_0^{-1}\right)^{p/q},$$

and so that

$$l\ge\left(h_0^{-p/q}S_{\mu,q}\right)^{q/(q-p)}.$$

Thus we get

$$c=I_{\lambda,\mu}(u)+((q-p)/pq)l$$

$$\ge I_{\lambda,\mu}(u)+((q-p)/pq)\left(h_0^{-p/q}S_{\mu,q}\right)^{q/(q-p)}.$$

Proof of Theorem 1

The proof of Theorem 1 is given in two parts.

Existence of a local minimizer

We prove that there exists $\lambda_*>0$ such that for any $\lambda\in(0,\lambda_*)$, $I_{\lambda,\mu}$ can achieve a local minimizer. First, we establish the following result.

Proposition 1: Suppose that $p<k<N, 0\le s<p, \mu<\bar{\mu}_{k,p}$, hypothesis (H) holds, $g\in\mathcal{H}'_\mu\cap C(\mathbb{R}^N)$ and $g\neq 0$. Then there exists λ_*, ϱ and δ such that for all $\lambda\in(0,\lambda_*)$ we have

$$I_{\lambda,\mu}(u)\ge\delta>0 \text{ for } \|u\|=\varrho \tag{5.1}$$

Proof: By the Holder inequality and the definition of we get for all $u\in\mathcal{D}\setminus\{0\}$ and $\varepsilon>0$

$$I_{\lambda,\mu}(u) := (1/p)\ \|u\|^p - (1/q)\int_{\mathbb{R}^N} h(y)|y|^{-s}|u|^q\,dx - \lambda\int_{\mathbb{R}^N} g(x)u\,dx,$$

$$\geq\ (1/p)\|u\|^p - (h_\infty/q)S_{\mu,q}\|u\|^q - \lambda\|g\|\mathcal{H}_\mu^{'}\|u\|,$$

$$\geq (1/p-\varepsilon)\|u\|^p - (|h|_\infty/p)S_{\mu,q}\|u\|^q - -C_\varepsilon\|\lambda g\|H_\mu^{'}.$$

Taking $\varepsilon < 1/p$ and $\varrho = \|u\|_\mu$, then there exist $\varrho > 0$ small enough and a positive constant λ_* such that

$$I_{\lambda,\mu}(u) \geq \delta > 0 \text{ for } \|u\|_\mu = \varrho \text{ and } \lambda \in (0,\lambda_*). \tag{5.2}$$

Since g is a continuous function on $\mathbb{R}^N$, not identically zero, we can choose $\phi \in C_0^\infty(\mathbb{R}^N\setminus\{0\})$ such that $\int_{\mathbb{R}^N} g(x)\phi dx > 0$. It follows that for $t>0$ small,

$$I_{\lambda,\mu}(t\phi) := (t^p/p)\|\phi\|^p - (t^q/q)\int_{\mathbb{R}^N} h(y)|y|^{-s}|\phi|^q\,dx - \lambda t\int_{\mathbb{R}^N} g(x)\phi dx < 0. \tag{5.3}$$

We also assume that t is so small enough such that $\|t\phi\|_\mu < \varrho$. Thus, we have $c_1 = \inf\{I_{\lambda,\mu}(u) : u \in B\varrho\} < 0$, where $B_\varrho = \{u \in \mathcal{D}, \mathcal{N}(u) \leq \varrho\}$. (5.4)

Using the Ekeland's variational principle, for the complete metric space $\overline{B}_\varrho$ with respect to the norm of $\mathcal{D}$, we can prove that there exists a $(PC)_{c_1}$ sequence $(u_n) \subset \overline{B}_\varrho$ such that $u_n \rightharpoonup u_1$ for some u_1 with $\mathcal{N}(u_1) \leq \varrho$.

Now, we claim that $u_n \to u_1$. If not, by Lemma??; we have

$$c_1 \geq I_{\lambda,\mu}(u_1) + ((q-p)/pq)\left(h_0^{-p/q}S_{\mu,q}\right)^{q/(q-p)}$$

$$\geq c_1 + ((q-p)/pq)\left(h_0^{-p/q}S_{\mu,q}\right)^{q/(q-p)}$$

$$> c_1,$$

which is a contradiction.

Then we obtain a critical point u_1 of $I_{\lambda,\mu}$ for all $\lambda \in (0,\lambda_*)$ satisfying $c_1 = I_{\lambda,\mu}(u_1) < 0$.

On the other hand we have

$$c_1 = ((q-p)/pq)\|u_1\|^p - ((q-1)/q)\int_{\mathbb{R}^N} \lambda g(x)u_1 dx$$

$$\geq -(1/pq)(q-1)^p(q-p)^{-1}\lambda^p\|g\|^p_{\mathcal{H}_\mu^{'}}. \tag{5.5}$$

Thus u1 is a nontrivial solution of our problem with negative energy.

Existence of mountain pass type solution

We use the mountain pass theorem without Palais-Smale conditions to prove the existence of a nontrivial solution with positive energy. For this, we need the following Lemma.

Lemma 4: Let $\lambda^* > 0$ such that

$c^*_{\lambda,p} > 0$ for all $\lambda \in (0,\lambda^*)$.

Then, there exist $\Lambda \in (0,\lambda^*)$ and $\varphi_\varepsilon \in \mathcal{D}$ for $\varepsilon > 0$ such that $\sup_{t\geq 0} I_{\lambda,\mu}(t\varphi_\varepsilon) < c^*_{\lambda,p}$, for all $\lambda \in (0,\Lambda)$.

Proof: Let

$$\varphi_\varepsilon(x) = \begin{cases} \omega_\varepsilon(x) & \text{if } g(x) \geq 0 \text{ for all } x \in \mathbb{R}^N \\ \omega_\varepsilon(x-x_0) & \text{if } g(x_0) > 0 \text{ for } x_0 \in \mathbb{R}^N \\ -\omega_\varepsilon(x) & \textit{if } g(x) \leq 0 \textit{ for all } x \in \mathbb{R}^N \end{cases}, \tag{5.6}$$

Where ω_ε veri.es (2:2)

Then, we claim that there is an ε_0 such that

$$\int_{\mathbb{R}^N} g(x)\varphi_\varepsilon(x) > 0, \text{for any } \varepsilon \in (0,\varepsilon_0). \tag{5.7}$$

In fact, $g(x) \geq 0$ or $g(x) \leq 0$ for all $x \in \mathbb{R}^N$, and (5.7) holds obviously. If there exists an $x_0 \in \mathbb{R}^N$ such that $g(x_0) > 0$, by the continuity of $g(x)$ there is an $\eta > 0$ such that $g(x) > 0$ for all $x \in B_\eta(x_0)$. Then, by the definition of $\omega_\varepsilon(x-x_0)$ it is easy to see that there exists an ε_0 small enough such that

$$\int_{\mathbb{R}^N} g(x)\omega_\varepsilon(x-x_0) > 0,\ \textit{for any } \varepsilon \in (0,\varepsilon_0). \tag{5.8}$$

Now, we consider the following functions

$$f(t) = I_{\lambda,\mu}(t\varphi_\varepsilon)$$

and

$\tilde{f}(t) = (t^p/p)\ \|\varphi_\varepsilon(x)\|^p - (t^q/q)h_0\int_{\mathbb{R}^N}|y|^{-s}|\varphi_\varepsilon(x)|^q\,dx.$ Then, we get for all $\lambda \in (0,\lambda^*)$

$$0 = f(0) < c^*_{\lambda,p}.$$

By the continuity of f(t), there exists t_1 a sufficiently small positive quantity such that $f(t) < c^*_{\lambda,p}$, for all $t \in (0,t_1)$. On the other hand, we have

$$\max_{t\geq 0}\tilde{f}(t) = ((q-p)/pq)\left(h_0^{-p/q}S_{\mu,q}\right)^{q/(q-p)},$$

then, we obtain

$$\sup_{t\geq 0} I_{\lambda,\mu}(t\varphi_\varepsilon) < ((q-p)/pq)\left(h_0^{-p/q}S_{\mu,q}\right)^{q/(q-p)} - \lambda t_1\int_{\mathbb{R}^N}|y|^{-s}g(x)\varphi_\varepsilon dx.$$

Taking $\lambda > 0$ such that

$$\lambda t_1\int_{\mathbb{R}^N} g(y)\varphi_\varepsilon dx > (1/pq)(q-1)(q-p)^{-1/p}\lambda^p\|g\|^p_{\mathcal{H}_\mu^{'}}.$$

By (5.7) we get

$$0 < \lambda < Q.$$

Where

$$Q := (pq(q-p)^{1/p}(q-1)^{-1})t_1\left(\int_{\mathbb{R}^N} g(x)\varphi_\varepsilon dx\right)\|g\|^{-p}_{\mathcal{H}_\mu^{'}}.$$

Set

$$\Lambda = \min\{\lambda^*, Q\}.$$

We deduce that

$$\sup_{t\geq 0} I_{\lambda,\mu}(t\varphi_\varepsilon) < c^*_{\lambda,p},\ \text{for all } \lambda \in (0,\Lambda).$$

Since $\lim_{t\to\infty} I_{\lambda,\mu}(t\varphi_\varepsilon) = -\infty$, we can choose $T>0$ large enough such that $I_{\lambda,\mu}(T\varphi_\varepsilon) < 0$. From Proposition 1, we have $I_{\lambda,\mu|\partial B_\varrho} \geq \delta > 0$ for all $\lambda \in (0,\lambda_*)$. By mountain pass theorem without the Palais-Smale condition, there exists a $(PC)_{c_2}$ sequence (u_n) in $\mathcal{D}$ which is characterized by

$$c_2 = \inf_{\gamma\in\Gamma}\max_{t\in[0,1]} I_{\lambda,\mu}(\gamma(t)),$$

with

$$\Gamma = \{\gamma \in C([0,1],\mathcal{D}), \gamma(0) = 0, \gamma(1) = T\varphi_\varepsilon\}.$$

Then, (u_n) has a subsequence, still denoted by (u_n) such that $u_n \rightharpoonup u_2$ in $\mathcal{D}$.

By Lemma 3, if u_n doesn.t converge to u_2; we get

$$c_2 \geq I_{\lambda,\mu}(u_2) + ((q-p)/pq)\left(h_0^{-p/q} S_{\mu,q}\right)^{q/(q-p)} \geq c^*_{\lambda,p},$$

what contradicts the fact that, by Lemma 4, we have $\sup_{t\geq 0} I_{\lambda,\mu}(t\varphi_\varepsilon) < c^*_{\lambda,p}$, for all $\lambda \in (0,\Lambda)$. Then $u_n \to u_2$ in $\mathcal{D}$.

Thus, we obtain a critical point u_2 of $I_{\lambda,\mu}$ for all $\lambda \in (0,\lambda_*)$ with $\Lambda_* := \min\{\lambda_*, \Lambda\}$ satisfying $I_{\lambda,\mu}(u_2) > 0$.

References

1. Badiale M, Tarantello G (2002) A Sobolev-Hardy inequality with applications to a nonlinear elliptic equation arising in astrophysics. Arch Ration Mech Anal 163: 252-293.
2. Brézis H, Lieb E (1983) A Relation between point convergence of functions and convergence of functional. Proc Amer Math Soc 88: 486-490.
3. Gazzini M, Musina R (2009) On the Hardy-Sobolev-Maz'ja inequalities: symmetry and breaking symmetry of extremal functions. Commun Contemp Math 11: 993-1007.
4. Tarantello G (1992) On nonhomogeneous elliptic equations involving critical Sobolev exponent. Ann Inst Henri Poincare 9: 281-304.
5. Maz'ja VG (1980) Sobolev Spaces. Springer-Verlag, Berlin.
6. Musina R (2008) Ground state solutions of a critical problem involving cylindrical weights. Non-linear Anal 68: 3972-3986.
7. Badiale M, Bergio V, Rolando S (2007) A nonlinear elliptic equation with singular potential and applications to nonlinear field equations. J Eur Math Soc 9: 355-381.
8. Badiale M, Guida M, Rolando S (2007) Elliptic equations with decaying cylindrical potentials and power-type nonlinearities. Adv Differential Equations 12: 1321-1362.
9. Bhakta M, Sandeep K (2009) Hardy-Sobolev-Maz'ya type equations in bounded domains. J Differ Equ 247: 119-139.
10. Gazzini M, Mussina R (2009) On a Sobolev type inequality related to the weighted p-Laplace operator. J Math Anal Appl 352: 99-111.
11. Xuan B (2003) Multiple solutions to p-Laplacian equation with singularity and cylindrical symmetry. Nonlinear Analysis 55: 217-232.

Breakable Solid Transportation Problem with Hybrid and Fuzzy Safety Factors using LINGO and Genetic Algorithm

Abhijit Baidya*, Uttam Kumar Bera and Manoranjan Maiti

Department of Mathematics, National Institute of Technology, Agartala, Jirania 799055, West Tripura, India

Abstract

In this paper we present a solution of solid transportation problem (STP) for breakable items with different environments. If we carrying the produce from sources to destination by the means of unlike conveyances then due to insurgency, land slide and bad road, there are some risks or difficulties to transport the items. By this motive we initiate "Safety Factors" in transportation problem. Due to this reason desired total safety factor is being introduced. Also our objective is to evaluate the solution of STP using expected value model. Here we develop six models where first three models are formulated taking crisp unit transportation cost but the remaining three models are formulated taking hybrid unit transportation cost. To build up the different models we consider breakability and safety factor which is taken as crisp, fuzzy and hybrid for assorted models. All the fuzzy and hybrid models are reduced into its crisp equivalent using expected value modeling. Finally by Generalized Reduced Gradient (GRG) method using LINGO.13 optimization software and Genetic Algorithm we solve the mathematical models and put a enlarge discussion on it.

Keywords: Solid transportation problem; Safety factor; Hybrid variables; Triangular fuzzy number; breakability; Expected value model; Genetic algorithm

Introduction

The transportation problem (TP) was developed by Hitchcock [1]. The classical transportation problem deals with transportation goods from some sources to some destinations. The solid transportation problem (STP) is a generalization of the well-known transportation problem (TP) in which three-dimensional properties is taken into account in the objective and constraint set instead of source and destination. The STP was first stated by Shell [2]. In many industrial problems, a homogeneous product is delivered from an origin to a destination by means of different modes of transport called conveyances, such as trucks, cargo flights, goods trains, ships, etc. These conveyances are taken as the third dimension. A solid transportation problem can be converted to a classical transportation problem by considering only a single type of conveyance. Transportation problems normally are formulated as arrangement problems in which carrying are made from sources to destinations as well as from destinations to destinations also. Sometimes there are restrictions on the flow of transportation. In the projected traditional transportation problem, transportation may be made from all sources to all destinations, if required and no transportation from destinations to destinations or from sources to sources is permissible. Also, the distances between the origins and destinations are not here taken into account as in the network problems. Actuality most of transportation problems are unbalanced for breakable items as the supplied amount by the suppliers (i.e., origins) is not equal to the received amount by the retailers (i.e., destinations). Few of these items are glass-goods, toys, ceramic goods, etc. Till now the materials of these type has not been considered for transportation models. Zadeh [3] first introduced the concept of fuzzy set theory. Later several authors such as Zadeh, Kaufmann, Zimmermann, Liu, Dubois and Prade [4-8] developed and applied fuzzy set theory. Chanas and Kuchta [9] studied transportation problem with fuzzy cost coefficients. [10] considered two types of uncertain STP, one with interval numbers and other with fuzzy numbers. Liu and Liu [11] presented expected value model for fuzzy programming. Yang and Liu [12] applied expected value model, chance-constrained programming model and dependent-chance programming in fixed charge solid transportation problem in fuzzy environment. Applied possibility programming approach to a material requirement planning problem with fuzzy constraints and fuzzy coefficients, using the definition of possibility measure of fuzzy number. Hybrid variable was first stated by Liu in 2006 but in 2009 Li and Liu proposed the expected value of hybrid variable. In some realistic transportation systems, transported amount from a source inversely depends on the level of unit transportation cost. When the unit transportation cost in a particular route is low, then a decision maker (DM) tries to transport the maximum amount of the item through that route, i.e., the transported amount is high. Again on the contrary, if the unit transportation cost is high, then less resource is transported. For medium unit transport cost, the transported amount is also medium. Chanas and Kuchta and Omar and Samir [13,14] discussed the solution algorithm for solving the transportation problem in fuzzy environment. Grzegorzewski and Chanas [15,16] approximated the fuzzy number to its nearest interval. Grzegorzewski and Mrowka [17] approximated a general fuzzy number to trapezoidal or triangular fuzzy numbers. Due to insufficient information, lack of confirmation and vary financial market, the available data of a transportation system such as resources, demands and conveyance capacities are not always crisp or precise but are fuzzy or stochastic or both. So the fuzziness and randomness can be present in the objective function as well as in the constraints of a STP. Dealing with different types of uncertainty in many practical problems is still an emerging problem. Recently Kundu et al. [18] solve a multi-objective solid transportation problem with

***Corresponding author:** Abhijit Baidya, Department of Mathematics, National Institute of Technology, Agartala, Jirania 799055, West Tripura, India
E-mail: abhijitnita@yahoo.in

hybrid penalty cost. Also Baidya et al. [19] works on safety factor and uncertainty. Our aspire in this paper is to formulate and solve single-objective solid transportation problem (SOSTP) with safety constraints with special types of uncertain (fuzzy and hybrid) parameters. The fixed charge problem was initialized by Hirsch in 1968. Up to now, it has been widely applied in many decision-making and optimization problems. Interested readers may refer to Kennington and Unger, Sun et al., Gottlieb and Paulmann [20-23], and so on. In this paper, we shall consider the fuzzy fixed charge STP. In spite of so many developments in literature, there are some lacunas in solid transportation problem and these are:

(i) Some researcher such as Chanas and Kuchta [13] studied transportation problem with fuzzy cost coefficients. Jimenez and Verdegay in 1999 [10] considered two types of uncertainty (interval and fuzzy) in solid transportation problem but nobody can solve any STP by taking hybrid uncertainty.

(ii) Very few STPs are available for breakable items; no STP model is formulated for hybrid and fuzzy safety factor and breakability.

In this paper, an item with breakability rate is transported from origins to destinations through dissimilar conveyances. We formulate six models without and with safety factor and breakability where these safety factor and breakability are crisp, fuzzy and hybrid. To organize this manuscript we employ the unit transportation cost as crisp number, fuzzy number and hybrid number and also to convert the uncertainty models into its crisp corresponding we apply expected value modeling. In some model decision maker (DM) likes to minimize the transportation cost choosing the particular routes and modes of transportation for particular so that total safety for the system is greater or equal to a predefined safety value. In rising countries, due to insurgency, all routes for transportation are not equally safe. Furthermore there are some risks of running some conveyances (modes) in some particular routes. In STP, none has taken this safety factor into account for development, through it is very much prevalent in different parts of India, including North-East region and Maoist dominated areas.

Preliminaries

Definitions

Fuzzy number: A fuzzy subset $\tilde{A}$ of real number with membership function is said to be a fuzzy number if

$\mu_{\tilde{A}}(x)$ is upper semi-continuous membership function;

$\tilde{A}$ is normal, i.e., there exists an element x_0 such that $\mu_{\tilde{A}}(x_0)$;

$\tilde{A}$ is fuzzy convex, i.e. $\mu_{\tilde{A}}(\lambda x_1 + (1-\lambda)x_2 \geq \mu_{\tilde{A}}(x_1) \wedge \mu_{\tilde{A}}(x_2) \forall x_1, x_2 \in R$ and $\lambda \in [0,1]$

Support of $\tilde{A} = \{x \in R : \mu_{\tilde{A}}(x) > 0\}$ is bounded.

Fuzzy numbers are represented by two types of membership functions: (a) Linear membership functions e.g. triangular fuzzy number (TFN), Trapezoidal fuzzy number, Piecewise Linear fuzzy number etc. (b) Non-linear membership functions e.g. Parabolic fuzzy number (PFN), Exponential fuzzy number and other non-linear fuzzy number. We used the following fuzzy numbers:

Triangular Fuzzy Number (TFN): Triangular Fuzzy Number (TFN) is the fuzzy number $\tilde{A} = (a_1, a_2, a_3)$ with the membership function $\mu_{\tilde{A}}(x)$

a continuous mapping: $\mu_{\tilde{A}} : R \to [0,1]$

$$\mu_{\tilde{A}}(x) = \begin{cases} 0 & for\ -\infty < x < a_1 \\ \dfrac{x-a_1}{a_2-a_1} & for\ a_1 \leq x < a_2 \\ \dfrac{a_3-x}{a_3-a_2} & for\ a_2 \leq x \leq a_3 \\ 0 & for\ a_2 < x < \infty \end{cases}$$

General Fuzzy Number (GFN): It is known that for any fuzzy number $\tilde{A}$ there exist four numbers $a_1, a_2, a_3, a_4 \in R$ and two functions $f(x), g(x) : R \to [0,1]$ where f(x) is non-decreasing and g(x) is non-increasing, such that we can describe a membership function $\mu_{\tilde{A}}(x)$ in a following manner (Figures 1 and 2)

$$\mu_{\tilde{A}}(x) = \begin{cases} 0 & for\ x < a_1 \\ f(x) & for\ a_1 \leq x < a_2 \\ 1 & for\ a_2 \leq x \leq a_3 \\ g(x) & for\ a_3 \leq x \leq a_4 \\ 0 & for\ a_4 \leq x \leq \infty \end{cases}$$

Credibility measure: Credibility measure was presented by Liu and Liu (2002). For a fuzzy variable ξ with membership function $\mu_\xi(x)$ and for any set B of real numbers, credibility measure of fuzzy event $\{\xi \in B\}$ is defined as

$$Cr\{\xi \in B\} = \frac{1}{2}(\text{Pos}\{\xi \in B\} + Nec\{\xi \in B\}),$$

where possibility and necessity measures of $\{\xi \in B\}$ are respectively defined as

$$\text{Pos}\{\xi \in B\} = \sup_{X \in B^c} \mu_\xi(x)$$

And

$$Nec\{\xi \in B\} = 1 - Sup_{X \in B^c} \mu_\xi(x)$$

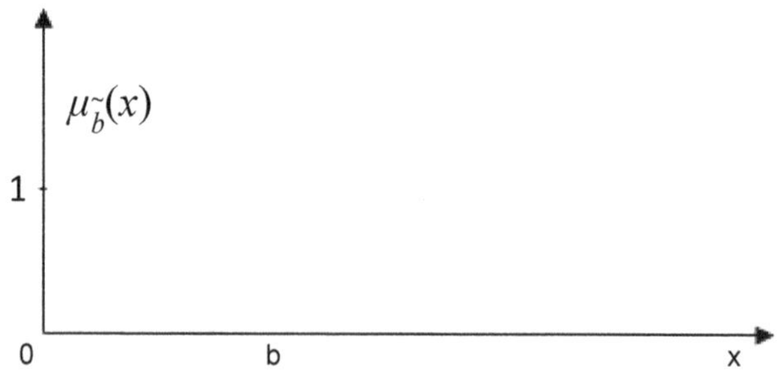

Figure 1: Membership function of a crisp number b.

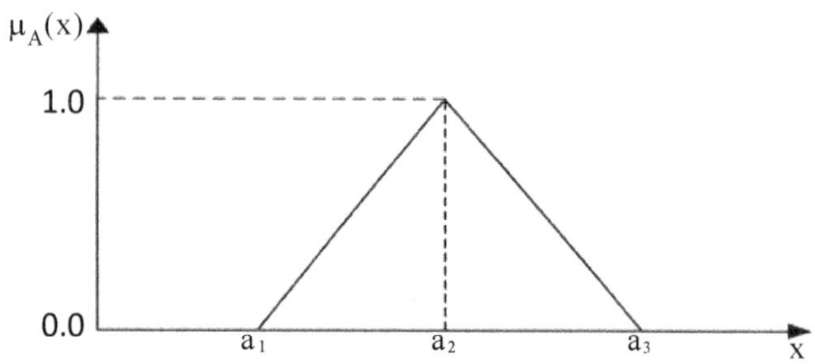

Figure 2: Triangular Fuzzy Number.

Expected value of fuzzy variable:

Definition: Let ξ be a fuzzy variable. Then the expected value of ξ is defined by $E(\xi) = \int_0^{+\infty} Cr\{\xi \geq r\}dr - \int_{-\infty}^{0} Cr\{\xi \leq r\}dr$

Provided that at least one of the two integrals is finite. The equipossible fuzzy variable on [a, b] has an expected value $\frac{(a+b)}{2}$ The triangular fuzzy variable (a, b, c) has an expected value $\frac{(a+2b+c)}{4}$ The trapezoidal fuzzy variable (a, b, c, d) has an expected value $\frac{(a+b+c+d)}{4}$

Let ξ be the continuous nonnegative fuzzy variable with membership function μ if μ is decreasing on

$$E[\xi] = \frac{1}{2}\int_0^{\infty} \mu(x)\,dx$$

Example-1: A fuzzy variable ξ is called exponentially distributed if has an exponential membership function

$$\mu(x) = 2(1+\exp(\frac{\pi x}{\sqrt{6}m}))^{-1}, x \geq 0, m > 0$$

Then the expected value is $\frac{\sqrt{6}m\, in 2}{\pi}$

Theorem-1: Let ξ be a continuous fuzzy variable with membership function μ If its expected value exists, and there is a point x0 such that $\mu(x)$ is increasing on $(-\infty, x_0)$ and decreasing on $(x_0, +\infty)$ then

$$E[\xi] = x_0 + \frac{1}{2}\int_{x_0}^{+\infty} \mu(x)\,dx - \frac{1}{2}\int_{-\infty}^{x_0} \mu(x)\,dx$$

Proof:

If $x_o \geq 0$, then

$$Cr\{\xi \leq r\} = \begin{cases} \frac{1}{2}[1+1-\mu(x)], \text{if } 0 < r \leq x_0 \\ \frac{1}{2}\mu(x), \text{if } r > x_0 \end{cases}$$

And $Cr\{\xi \leq r\} = \{\frac{1}{2}\mu(x)$

$$E[\xi] = \int_0^{x_0} [1 - \frac{1}{2}\mu(x)]dx + \int_{x_0}^{+\infty} [\frac{1}{2}\mu(x)]dx - \int_{-\infty}^{0} [\frac{1}{2}\mu(x)]dx = x_0 + \frac{1}{2}\int_{x_0}^{+\infty} [\frac{1}{2}\mu(x)]dx - \frac{1}{2}\int_{-\infty}^{x_0} [\frac{1}{2}\mu(x)]dx$$

If $x_o < 0$, a similar way may prove the equation (5). The theorem is proved. Especially, let $\mu(x)$ be symmetric function on the line $x = x_o, then\, E[\xi] = x_o$

Example-2: Let ξ be triangular fuzzy variable (a, b, c), then it has an expected value

$$E[\xi] = b + \frac{1}{2}\int_b^c \frac{x-c}{b-c}dx - \frac{1}{2}\int_b^c \frac{x-a}{b-a}dx = b + \frac{c-b}{4} + \frac{a-b}{4} = \frac{a+2b+c}{4}$$

Example-3: Let ξ be equipossible fuzzy variable on [a, b], then it has an expected value

$$E[\xi] = \frac{a+b}{2} + \frac{1}{2}\int_{\frac{a+b}{2}}^{b} 1\,dx - \frac{1}{2}\int_a^{\frac{a+b}{2}} 1\,dx = \frac{a+b}{2}$$

Example-4: Let ξ be trapezoidal fuzzy variable (a, b, c, d), then it has an expected value

$$E[\xi] = \frac{b+c}{2} + \frac{1}{2}\int_{\frac{b+c}{2}}^{c} 1dx + \frac{1}{2}\int_c^d \frac{x-d}{c-d}dx - \frac{1}{2}\int_a^b \frac{x-a}{b-a}dx - \frac{1}{2}\int_b^{\frac{b+c}{2}} 1\,dx = \frac{a+b+c+d}{4}$$

Especially, let ξ be triangular fuzzy variable (a, b, c) and b-a=c-b, then it has an expected value $E[\xi] = b$. Let ξ be trapezoidal fuzzy variable (a, b, c, d) and b-a=d-c, then it has an expected value $E[\xi] = \frac{b+c}{2}$

A fuzzy variable ξ is called normally distributed if it has a normal membership function

$$\mu(x) = 2(1+\exp(\frac{\pi\,|\,x-e\,|}{\sqrt{6}\sigma}))^{-1}, x \in R, \sigma > 1$$

By theorem 1, the expected value is e.

The definition of expected value operation is also applicable to discrete case. Assume that ξ be a simple fuzzy variable whose membership function is given by

$$\mu(x) = \begin{cases} \mu_1, if\ x = a_1 \\ \mu_2, if\ x = a_2 \\ \mu_m, if\ x = a_m \end{cases}$$

Where $a_1, a_2, \ldots a_m$ are distinct numbers. Note that Then $\mu_1 V \mu_2 V \mu_3 V \ldots\ldots V \mu_m = 1$ the expected value of is

$$E[\xi] = \sum_{i=1}^{m} w_i a_i$$

Where the weights are given by

$$w_i = \frac{1}{2}(\max_{1\leq j\leq m}\{\mu_j \mid a_j \geq a_i\} - \max_{1\leq j\leq m}\{\mu_j \mid a_j < a_i\} + \max_{1\leq j\leq m}\{\mu_j \mid a_j \geq a_i\} - \max_{1\leq j\leq m}\{\mu_j \mid a_j > a_i\})$$

For i=1, 2, …..,m. It is easy to verity that all and the sum of all $w_i \geq 0$ weights is just 1.

Optimistic and pessimistic value: Let ξ be a fuzzy variable and $\alpha \in [0,1]$. Then $\xi_{sup}(\alpha) = \text{SUP}\{\xi \geq r\} \geq \alpha\}$ is called $\alpha-$ optimistic value to ξ;

And

$\xi_{inf}(\alpha) = \text{SUP}\{r : cr\{\xi \geq r\} \geq \alpha\}$ is called $\alpha-$ pessimistic value to ξ;

Random variable: For the probability space (Ω, S, P) where Ω is a set of elementary events, S is a set of all events $(a\,\sigma - \text{field of events})$ and $P : S \to [0,1]$ is a probability function, the mapping $\widehat{X} : (\Omega, S, P) \to R$ is called a random variable. By the probability distribution function of the random variable $\widehat{X}$ we mean a function $F(x) = P\{w : \widehat{X}(w) \leq x\}$ for all $x \in N$ with $F(-\infty) = 0\, and\, F(\infty) = 1$

Hybrid variable

Definition: Suppose (Θ, P, Cr) is a credibility space (Ω, S, Pr) and be a probability space. The product $(\Theta, P, Cr) \times (\Omega, S, Pr)$ is a chance space.

Definition: Then a chance measure of an event $\wedge$ is defined as

$$ch(\wedge) = \begin{cases} \sup_{\theta\in\Theta}(Cr\{\theta\} \wedge \Pr\{\wedge(\theta)\}), \text{if } \sup_{\theta\in\Theta}(\mathrm{Cr}\{\wedge(\theta)\}) < 0.5; \\ 1-\sup_{\theta\in\Theta}(Cr\{\theta\} \wedge \Pr\{\wedge(\theta)\}), \text{if } \sup_{\theta\in\Theta}(\mathrm{Cr}\{\theta\} \wedge \Pr\{\wedge(\theta)\}) \geq 0.5; \end{cases}$$

$Then\, ch(\phi) = 0, \mathrm{ch}\{\Theta \times \Omega\} = 1, 0 \leq \mathrm{ch}\{\wedge\} \leq$

1 For any event $\wedge$

Definition: A hybrid variable is a measurable function from a chance space $(\Theta, P, \mathrm{Cr}) \times (\Omega, \mathrm{S}, \mathrm{Pr})$ to real numbers, i.e. for any Borel set B of real numbers, the set $\{\xi \in \mathrm{B}\} = \{(\theta, \mathrm{w}) \in \Theta \times \Omega : \xi(\theta, \mathrm{w}) \in \mathrm{B}\}$ is an event.

Remark: A hybrid variable degenerates to a fuzzy variable if the value of

$\xi(\theta, \mathrm{w})$ does not vary with w. For example, $\xi(\theta, w) = \theta, \xi(\theta, w) = \theta 2 + 1, \xi(\theta, w) = \sin\theta$

Remark: A hybrid variable degenerate to a random variable if the value of $\xi(\theta, \mathrm{w})$ does not vary with θ. For example, $\xi(\theta, w) = w, \xi(\theta, w) = w^2 + 1, \xi(\theta, w) = \sin w$

Remark: A hybrid variable $\xi(\theta, \mathrm{w})$ may also be regarded as a function from a credibility space $(\Theta, \mathrm{P}, \mathrm{Cr})$ to the set $\{\xi(\theta, .) \mid \theta \in \Theta\}$ random variable. Thus ξ is a random fuzzy variable defined by Liu (2006).

Remark: A hybrid variable $\xi(\theta, \mathrm{w})$ may also be regarded as a function from a probability space $(\Omega, \mathrm{A}, \mathrm{Pr})$ to the set $\{\xi(., \mathrm{w}) \mid \mathrm{w} \in \Omega\}$ of fuzzy variables. If $Cr\{\xi(., \mathrm{w}) \in \mathrm{B}\}$ is a measurable function of w for any Borel set B of real number, then ξ is a fuzzy random in the sense of Liu and Liu (2009) (Figure 3).

Example: If $\tilde{a}$ is a fuzzy variable and $\hat{n}$ is a random variable, then the sum $\tilde{\xi} = \tilde{a} + \hat{n}$ is a hybrid variable, i.e. if $f : \Re^2 \rightarrow \Re$ is a measurable function, then $\tilde{\xi} = f(\tilde{a}, \hat{n})$ is a hybrid variable. Now suppose that $\tilde{a}$ has a membership function $\mu\, and\, n$ has a probability density function ϕ Then for any Borel set B of real numbers, we have

$$ch\{f(\tilde{a}, \hat{n}) \in \mathrm{B}\} = \begin{cases} \sup_z[\frac{\mu(\mathrm{x})}{2} \wedge \int_{f(\mathrm{x},\mathrm{y})\in\mathrm{B}} \phi(\mathrm{y})\,\mathrm{dy}],\ if \sup_z[\frac{\mu(\mathrm{x})}{2} \wedge \int_{f(\mathrm{x},\mathrm{y})\in\mathrm{B}} \phi(\mathrm{y})\,\mathrm{dy}] < 0.5; \\ 1-\sup_z \sup_z[\frac{\mu(\mathrm{x})}{2} \wedge \int_{f(\mathrm{x},\mathrm{y})\in\mathrm{B}} \phi(\mathrm{y})\,\mathrm{dy}],\ if \sup_z[\frac{\mu(\mathrm{x})}{2} \wedge \int_{f(\mathrm{x},\mathrm{y})\in\mathrm{B}} \phi(\mathrm{y})\,\mathrm{dy}] \geq 0.5; \end{cases}$$

Definition: Let $\approx\!\xi$ be a hybrid variable. Then the expected value of ξ is defined by

$E[\xi] = \int_0^{+\infty} ch\{\xi \geq r\}dr - \int_{-\infty}^0 ch\{\xi \leq r\}dr$ provided that at least one of the integrals is finite.

Example: For the hybrid variable $\xi = \tilde{a} + \hat{n}$, expected value of $\xi\, is\, E(\xi) = \mathrm{E}(\tilde{a}) + E(\hat{n})$ For example, if $\tilde{a} = (\mathrm{t}_1, \mathrm{t}_2, \mathrm{t}_3)$ is

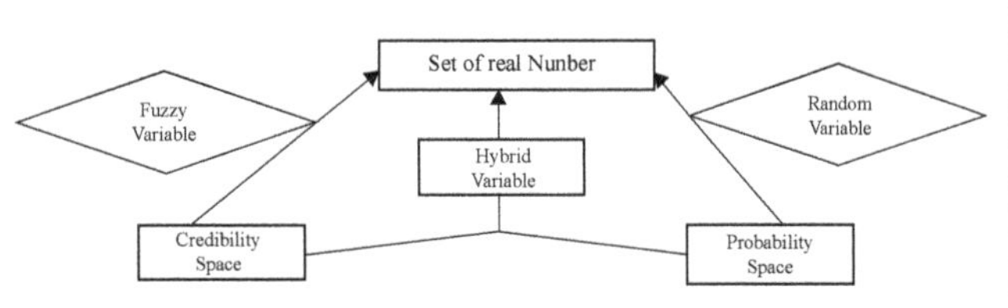

Figure 3: Geographical representation of Hybrid variable.

triangular fuzzy number and $\hat{n} = N(\mu, \sigma^2)$ is normally distributed random variable. Then $E(\xi) = \frac{t_1 + t_2 + t_3}{4} + \mu$.

Genetic Algorithm

The Genetic algorithm (GA) was first proposed by Holland [23]. Genetic algorithm is a well-known computerized stochastic search method based on the evolutionary theory of Charles Darwin survival of the fittest and natural genetics (Goldberg, 1989). GA has successfully been applied to optimization problems in different fields, like engineering design, optimal control, transportation and assignment problems, job scheduling, inventory control and other real-life decision-making problems. The most fundamental idea of Genetic Algorithm is to imitate the natural evolution process artificially in which populations undergo continuous changes through genetic operators, like crossover, mutation and selection. Genetic algorithm can easily be implemented with the help of computer programming. In particular, it is very useful for solving complicated optimization problems which cannot be solved easily by direct or gradient based mathematical techniques. It is very effective to handle large-scale, real-life, discrete and continuous optimization problems without making unrealistic assumptions and approximations. Keeping the imitation of natural evolution as the foundation, genetic algorithm can be designed appropriately and modified to exploit special features of the problem to solve. This algorithm starts with an initial population of possible solutions (called individuals) to a given problem where each individual is represented using some form of encoding as a chromosome. These chromosomes are evaluated for their fitness. Based on their fitness, chromosomes in the population are to be selected for reproduction and selected individuals are manipulated by two known genetic operations, like crossover and mutation. The crossover operation is applied to create offspring from a pair of selected chromosomes. The mutation operation is used for a little modification/change to reproduce offspring. The repeated applications of genetic operators to the relatively fit chromosomes result in an increase in the average fitness of the population over generation and identification of improved solutions to the problem under investigation. This process is applied iteratively until the termination criterion is satisfied. The following functions and values are adopted in the proposed GA to solve the problem.

Chromosome representation

The concept of chromosome is normally used in the GA to stand for a feasible solution to the problem. A chromosome has the form of a string of genes that can take on some value from a specified search space. The specific chromosome representation varies based on the particular problem properties and requirements. Normally, there are two types of chromosome representation – (i) the binary vector representation based on bits and (ii) the real number representation. In this research work, the real number representation scheme is used.

Here, a "K dimensional real vector" $X = (\mathrm{x}_1, \mathrm{x}_2, \ldots, \mathrm{X}_\mathrm{k})$ is used to represent a solution, where $\mathrm{x}_1, \mathrm{x}_2, \ldots, \mathrm{X}_\mathrm{k}$ represent different decision variables of the problem.

Initialization

A set of solutions (chromosomes) is called a population. N such solutions $\mathrm{x}_1, \mathrm{x}_2, \ldots, \mathrm{X}_N$ are randomly generated from search space by random number generator such that each X_i satisfies the constraints of the problem. This solution set is taken as initial population and is the starting point for a GA to evolve to desired solutions. At this step, probability of crossover P_c and probability of mutation P_m are also initialized. These two parameters are used to select chromosomes from mating pool for genetic operations- crossover and mutation respectively.

Fitness value

All the chromosomes in the population are evaluated using a fitness function. This fitness value is a measure of whether the chromosome is suited for the environment under consideration. Chromosomes with higher fitness will receive larger probabilities of inheritance in subsequent generations, while chromosomes with low fitness will more likely be eliminated. The selection of a good and accurate fitness function is thus a key to the success of solving any problem quickly. In this thesis, value of a objective function due to the solution X, is taken as fitness of X. Let it be f(X).

Selection process to create mating pool

Selection in the GA is a scheme used to select some solutions from the population for mating pool. From this mating pool, pairs of individuals in the current generation are selected as parents to reproduce off spring. There are several selection schemes, such as roulette wheel selection, ranking selection, stochastic universal sampling selection, local selection, truncation selection, tournament selection, etc. Here, Roulette wheel selection process is used in different cases. This process consist of following steps-

(i) Find total fitness of the population $F = \sum_{i=1}^{N} f(x_i)$

(ii) Calculate the probability of section pr of each solution by the formula $pr_i = \dfrac{f(x_i)}{F}$.

(iii) Calculate the cumulative probability qr_i for each solution X_i by the formula $qr_i = \sum_{j=0}^{i} pr_j$

(iv) Generate a random number "r" from the range [0,1].

(v) If $r < qr_1$ then select X_1 otherwise select X_i $(2 \le i \le N)$ where $qr_{i-1} \le r \le qr_i$.

(vi) Repeat step (iv) and (v) N times to select N solutions from current population. Clearly one solution may be selected more than once.

(vii) Let us denote this selected solution set by $P^1(T)$.

Crossover

Crossover is a key operator in the GA and is used to exchange the main characteristics of parent individuals and pass them on the children. It consists of two steps:

(i) Selection for crossover: For each solution $P^1(T)$ generate a random number r from the range [0,1]. If $r < p_c$ then the solution is taken for crossover, where P_c is the probability of crossover.

(ii) Crossover process: Crossover taken place on the selected solutions. For each pair of coupled solutions Y_1, Y_2 a random number c is generated from the range [0,1] and Y_1, Y_2 are replaced by their offspring's Y_{11}, Y_{21} respectively where $Y_{11} = cY_1 + (1 = c)\ Y_2, Y21 = cY2 + (1 - c)\ Y_1$ provided Y_{11}, Y_{21} satisfied the constraints of the problem.

Mutation

The mutation operation is needed after the crossover operation to maintain population diversity and recover possible loss of some good characteristics. It is also consist of two steps:

(i) Selection for mutation: For each solution of $P^1(T)$ generate a random number r from the range [0,1]. If $r < p_m$ then the solution is taken for mutation, where p_m is the probability of mutation.

(ii) Mutation process: To mutate a solution $X = (X_1, X_2, \dots, X_K)$ select a random integer r in the range [1..K]. Then replace x_r by randomly generated value within the boundary of rth component of X. Following selection, crossover and mutation, the new population is ready for its next iteration, i.e., $P^1(T)$ is taken as population of new generation. With these genetic operations a simple genetic algorithm takes the following form. In the algorithm T is iteration counter, P(T) is the population of potential solutions for iteration T, evaluate fitness of each members of $P(T)$

GA Algorithm

1. Set iteration counter T=0.

2. Initialize probability of crossover p_c and probability of mutation p_m

3. Initialize P(T).

4. Evaluate P(T).

5. Repeat

a. Select N solutions from P(T), for mating pool using Roulette-wheel selection process. Let this set be P(T)1.

b. Select solutions from P(T)1, for crossover depending on p_c

c. Made crossover on selected solutions for crossover to get population P(T)2.

d. Select solutions from P(T)2, for mutation depending on p_m

e. Made mutation on selected solutions for mutation to get population P(T+1).

f. $T \leftarrow T+1$

g. Evaluate P(T).

6. Until (Termination condition does not hold).

7. Output: Fittest solution (chromosome) of P(T).

Constraints Handling in GA

The main idea of handling constraints is to design chromosomes carefully by genetic operators to keep all these within the feasible solution set. To ensure that the chromosomes (solutions) are feasible, we have to check all the new chromosomes (x) generated by genetic operators. We suggest that a function is designed for each target optimization problem, the output value 1 means that the chromosome is feasible, 0 for infeasible. The algorithm for finding the feasibility of an individual (solution) (x) for the optimization problem (2.60) is as follows:

for j = 1 to l do

if $(g_j(X) \le 0)$ continue; else return 0; endif endfor

for k=1 to m do

if $(h_k(x) = 0)$ continue; else return 0; endif endfor return 1

Description of the Problem

STP is a problem of transporting goods from some sources to some customers through some conveyances (modes of transportation) and the main objective is to find optimal transportation plan so that the total transportation cost is minimum. Also the goods transported through each source cannot exceed its supply capacity, the requirements of each destination must be satisfied and the total transported amount must not exceed the capacity of vehicles. Again if we carry breakable item on some route then it is required to inflict the rate of breakability on that route, for this motive in this manuscript we inflict the percentage of breakability. In our respective problems safety constraint are imposed for some risk or difficulties in the road. The determination of future transportation planning is generally based on the past record. But the available data from previous experiments are not always precise, often those are imprecise due to uncertainty in ruling, fluctuate financial market, linguistic information, imperfect statistical analysis, insufficient information, etc. For example, transportation cost depends upon fuel price, labor charges, tax charges, etc., each of which are fluctuate time to time. Similarly supply of a source can't be always exact, because it depends upon the availability of manpower, raw-materials, market competition, product demands, etc. Fuzzy set theory and random set theory are most widely used and successfully applied tools to deal with uncertainty. In the next section we formulate four STPs with safety constraints and different uncertain (crisp, fuzzy, hybrid) parameters.

Assumption and Notation

(i) M: Number of plants in Solid transportation problem.

(ii) N: Number of destinations in Solid transportation problem.

(iii) K: Number of conveyances in Solid transportation problem.

(iv) ξ_{ijk} = Unit transportation cost to transport the commodity from i–th source to j–th destination by k–th conveyances.

(v) ξ_{ijk} = Unknown quantity which is to be transported from i–th source to j–th destination by k–th conveyance(decision variable).

(vi) a_i = amount of homogeneous product available at the i–th plant.

(vii) b_j = demand at the j–th destination.

(viii) e_k = amount of product which can be carried by the k–th conveyance.

(ix) α_{ijk} = Rate of breaking item of the Solid transportation problem from i-th plant to j-th by k–th conveyance.

(x) ξ_{ijk} = Hybrid unit transportation cost to transport the commodity from i–th source to j–th destination by k–th conveyances.

(xi) $\tilde{a}_i$ = Fuzzy amount of homogeneous product available at the i–th plant.

(xii) $\tilde{b}_j$ = Fuzzy demand at the j–th destination.

(xiii) $\tilde{e}_k$ = Fuzzy amount of product which can be carried by the k–th conveyance.

(xiv) S_{ijk} = the safety factor when an item is transported from the i–th plant to j–th destination by k–th conveyance.

(xv) $\tilde{S}_{ijk}$ = Fuzzy safety factor when an item is transported from the i–th plant to j–th destination by k–th conveyance.

(xvi) $\hat{\tilde{S}}_{ijk}$ = Hybrid safety factor when an item is transported from the i–th plant to j–th destination by k–th conveyance.

(xvii) B= Desired safety measure (DSM) for whole transportation system.

(xviii) $\tilde{\alpha}_{ijk}$ Fuzzy rate of breaking item of the Solid transportation problem from i-th plant to j-th by k–th conveyance.

(xix) $\tilde{\alpha}_{ijk}$ Hybrid rate of breaking item of the Solid transportation problem from i-th plant to j-th by k–th conveyance.

(xx) If an item is transported from source i-th to destination j-th by k-th conveyance and then the safety factor S_{ijk} is considered. This implies that if $x_{ijk} > 0$, and then we consider the safety factor for this route as a part of the safety constraint. Thus for the convenience of modeling, the following notation is introduced:

$$Y_{ijk} = \begin{cases} 1 \\ for x_{ijk} > 0\, otherwise \\ 0 \end{cases}$$

(xxi) f_{ijk} = fixed charge to transport the commodities from i-th plant to j-th destination by k-th conveyance.

Model Formulation

According to the above assumption and notation we formulate the following models on transportation problems:

Model-1: Formulation of fixed charge STP with crisp penalties, resources, demands, conveyance capacities and without safety factors, breakability:

After taking all the system parameters and the resources, etc., are deterministic and precisely we formulate the following model as without safety factor and without breakability:

$$Min(Z) = \sum_{i=1}^{M}\sum_{j=1}^{N}\sum_{k=1}^{K}\xi_{ijk}\, x_{ijk} + \sum_{i=1}^{M}\sum_{j=1}^{N}\sum_{k=1}^{K} f_{ijk}\, y_{ijk} \quad (1)$$

Subject to the constraints

$$\sum_{j=1}^{N}\sum_{k=1}^{K} x_{ijk} \le a_i,\ i = 1,2,3,\dots\dots\dots, \mathrm{M} \quad (2)$$

$$\sum_{i=1}^{M}\sum_{k=1}^{K} x_{ijk} \ge b_i, j = 1,2,3,\dots\dots\dots, N \quad (3)$$

$$\sum_{j=1}^{n}\sum_{k=1}^{k} x_{ijk} \le e_k, k = 1,2,3,\dots\dots\dots, K \quad (4)$$

$$x_{ijk} \ge 0,\ \forall i, j, k$$

The problem is unbalanced since $\sum_{i=1}^{M} a_i \ne \sum_{i=1}^{N} b_i \ne \sum_{k=1}^{K} e_k$ and has feasible solution if $\sum_{i=1}^{M} a_i \ge \sum_{j=1}^{N} b_j\ and\ \sum_{k=1}^{K} e_k \ge \sum_{j=1}^{N} b_j$

Model-2: Formulation of STP with crisp penalties, resources, demands, conveyance capacities, breakability and without safety factors:

Considering the entire scheme parameters and the resources, etc., are deterministic and precisely we formulate the following model without safety factor and with crisp breakability:

$$Min(Z)=\sum_{i=1}^{M}\sum_{j=1}^{N}\sum_{k=1}^{K}\xi_{ijk}\,x_{ijk}+\sum_{i=1}^{M}\sum_{j=1}^{N}\sum_{k=1}^{K}f_{ijk}\,y_{ijk}$$

Subject to the constraints

$$\sum_{j=1}^{N}\sum_{k=1}^{K}x_{ijk}\le a_i, i=1,2,3,\ldots\ldots\ldots,M, \quad (5)$$

$$\sum_{i=1}^{N}\sum_{k=1}^{K}x_{ijk}(1-\alpha_{ijk})\ge b_i,\ j=1,2,\ldots,N, \quad (6)$$

$$\sum_{j=1}^{n}\sum_{k=1}^{k},x_{ijk}\le e_k,\ k=1,2,\ldots\ldots,K, \quad (7)$$

$x_{ijk}\ge 0, \forall\, i,j,k.$

Model-3: Formulation of STP with crisp penalties, resources, demands, conveyance capacities, safety factors and breakability:

In view of the unit transportation cost, supplies, demands, conveyances capacities, safety factors and breakability as a crisp number we formulate the model as follows:

$$Min(Z)=\sum_{i=1}^{m}\sum_{j=1}^{K}\sum_{k=1}^{K}\xi_{ijk}\,x_{ijk}+\sum_{i=1}^{M}\sum_{j=1}^{N}\sum_{k=1}^{K}f_{ijk}\,y_{ijk}$$

Subject to the constraints (5), (7) and

$$\sum_{i=1}^{m}\sum_{j=1}^{n}\sum_{k=1}^{K}S_{ijk}\,y_{ijk}\ge B \quad (8)$$

$$\sum_{i=1}^{M}\sum_{k=1}^{K}x_{ijk}\ (1-\alpha_{ijk})\ge \mathrm{b_j}, j=1,2,\ldots,\mathrm{N} \quad (9)$$

$x_{ijk}\ge 0,\ \forall i,j,k.$

Model 4: Formulation of STP with hybrid penalties, fuzzy resources, demands, conveyance capacities, breakability and without safety factors.

So in this model formulation we consider the unit transportation cost, resources, demands, conveyance capacities and breakability as fuzzy:

$$Min(Z)=\sum_{i=1}^{m}\sum_{j=1}^{n}\sum_{k=1}^{K}\tilde{\xi}_{ijk}\,x_{ijk}+\sum_{i=1}^{M}\sum_{j=1}^{N}\sum_{k=1}^{K}\tilde{f}_{ijk}\,y_{ijk} \quad (10)$$

Subject to the constraints,

$$\sum_{j=1}^{N}\sum_{k=1}^{K}x_{ijk}\le\tilde{a}_i,\ i=1,2,\ldots,\mathrm{M} \quad (11)$$

$$\sum_{j=1}^{M}\sum_{k=1}^{K}x_{ijk}(1-\tilde{a}_{ijk})\ge\tilde{b}_j\ \ j=1,2,\ldots,N \quad (12)$$

$$\sum_{j=1}^{n}\sum_{k=1}^{K}x_{ijk}\le\tilde{e}_k, k=1,2,\ldots,K \quad (13)$$

$x_{ijk}\ge 0,\forall i,j,k.$

Model-5: Formulation of STP with hybrid penalties, fuzzy resources, demands, conveyance capacities, safety factors and breakability:

In this model formulation we consider unit transportation cost, resources, demands, conveyance capacities, safety factors and breakability as fuzzy number:

$$Min(Z)=\sum_{i=1}^{m}\sum_{j=1}^{n}\sum_{k=1}^{K}\tilde{\xi}_{ijk}\,x_{ijk}+\sum_{i=1}^{M}\sum_{j=1}^{N}\sum_{k=1}^{K}\tilde{f}_{ijk}\,y_{ijk}$$

Subject to the constraints (11), (13) and

$$\sum_{i=1}^{m}\sum_{j=1}^{n}\sum_{k=1}^{K}\tilde{S}_{ijk}\,x_{ijk}\ge\tilde{B}, \quad (14)$$

$$\sum_{i=1}^{M}\sum_{k=1}^{K}x_{ijk}(1-\tilde{\alpha}_{ijk})\ge\tilde{\mathrm{b}}_j, j=1,2,\ldots N \quad (15)$$

$x_{ijk}\ge 0,\forall i,j,k.$

Model-6: Formulation of STP with hybrid penalties, resources, demands, conveyance capacities, safety factors and breakability:

It may happen that the demand or any factor of a commodity in the society is uncertain, not precisely known, but some past data about it is available. From the available records, the probability distribution of demand or any other factor of the commodity can be determined. In a hybrid number we have the combination of fuzziness and randomness. Here we consider the unit transportation cost, resources, demands, conveyance capacities, safety factors and breakability as hybrid number:

$$Min(Z)=\sum_{i=1}^{m}\sum_{j=1}^{n}\sum_{k=1}^{K}\tilde{\xi}_{ijk}\,x_{ijk}+\sum_{i=1}^{M}\sum_{j=1}^{N}\sum_{k=1}^{K}\tilde{f}_{ijk}\,y_{ijk}$$

Subject to the constraint

$$\sum_{j=1}^{n}\sum_{k=1}^{k}x_{ijk}\le\hat{\tilde{a}}_i, i=1,2,\ldots,M \quad (16)$$

$$\sum_{i=1}^{M}\sum_{k=1}^{K}x_{ijk}(1-\hat{\tilde{a}}_{ijk})\ge\hat{\tilde{b}}_j,\ j=1,2,\ldots,N \quad (17)$$

$$\sum_{j=1}^{n}\sum_{k=1}^{k}x_{ijk}\le\hat{\tilde{e}}_k,\ k=1,2,\ldots,K \quad (18)$$

$$\sum_{i=1}^{m}\sum_{j=1}^{n}\sum_{k=1}^{k}\hat{\tilde{S}}_{ijk}\,Y_{ijk}\ge\hat{\tilde{B}} \quad (19)$$

$x_{ijk}\ge 0,\forall\, i,j,k.$

Solution Methodology

Liu and Liu in 2002 introduced a spectrum of expected value model of fuzzy programming to obtain optimum expected value of objective function under some expected constraints. Considering hybrid penalty $\hat{\tilde{\xi}}_{ijk}$ are in the form, $\xi_{ijk}=\tilde{\xi}_{ijk}+\hat{\xi}_{ijk};\tilde{\xi}_{ijk}$ where ξ denote the triangular fuzzy number (TFN) and $\hat{\xi}_{ijk}$ is normally distributed random variable with known mean $\bar{\xi}$ and variance Var $(\hat{\xi}_{ijk})$ and constructing expected value model (Liu and Liu 2002; Yang and Liu 2007; Yang and Feng 2007) and using expected value model we have the crisp equivalent of the respective models. The reduced crisp models are solved using constraint optimization software LINGO 13.0 and Genetic Algorithm (GA) (soft computing technique).

Crisp conversion of the Model-4, 5, 6:

For model-4, we have the corresponding crisp form as

$$\sum_{i=1}^{m}\sum_{j=1}^{n}\sum_{k=1}^{k}E(\hat{\tilde{\xi}}_{ijk})x_{ijk}+\sum_{i=1}^{m}\sum_{j=1}^{n}\sum_{k=1}^{k}E(\hat{\tilde{f}}_{ijk})y_{ijk} \quad (20)$$

Subject to the constraints,

$$\sum_{j=1}^{N}\sum_{k=1}^{k}x_{ijk}<=E(\tilde{\mathrm{a}}_\mathrm{i}),\mathrm{i}=1,2,\ldots,\mathrm{M} \quad (21)$$

$$\sum_{i=1}^{M}\sum_{k=1}^{k}x_{ijk}(1-E(\tilde{\alpha}_{\mathrm{ijk}}))\ge E(\tilde{\mathrm{b}}_\mathrm{j}),\mathrm{j}=1,\ldots,\mathrm{N} \quad (22)$$

$$\sum_{j=1}^{n}\sum_{k=1}^{k}x_{ijk}\le \mathrm{E}(\tilde{\mathrm{e}}_\mathrm{k}),\mathrm{k}=1,2,\ldots,\mathrm{K} \quad (23)$$

$\mathrm{x}_{\mathrm{ijk}}\ge 0,\forall i,j,k.$

For Model-5, we have the corresponding crisp form as

(20), (21), (22), (23) and

$$\sum_{i=1}^{m}\sum_{j=1}^{n}\sum_{k=1}^{k} E(\tilde{S}_{ijk})y_{ijk} > E(\tilde{B}) \quad (24)$$

$x_{ijk} \geq 0, \forall i, j, k.$

For Model-6, we have the corresponding crisp form as, (20) and subject to the constraint

$$\sum_{j=1}^{N}\sum_{k=1}^{k} x_{ijk} \leq E(\overset{\approx}{a}_i), i = 1,2,\ldots, M \quad (25)$$

$$\sum_{i=1}^{M}\sum_{k=1}^{k} x_{ijk}(1 - E(\overset{\approx}{\alpha}_{ijk})) \geq E(\overset{\approx}{b}_j), j = 1,2,\ldots, N \quad (26)$$

$$\sum_{j=1}^{n}\sum_{k=1}^{k} x_{ijk} \leq E(\overset{\approx}{e}_k), k = 1,2,\ldots, K \quad (27)$$

$$\sum_{i=1}^{m}\sum_{j=1}^{n}\sum_{k=1}^{k} E(\overset{\approx}{S}_{ijk})y_{ijk} > E(\tilde{B}) \quad (28)$$

$x_{ijk} \geq 0, \forall i, j, k.$

Numerical Experiments

An item is transported from three plants (at Kolkata, Delhi and Mumbai) to three destinations (at Agartala, Agra and Assam) by three different modes of transport as cargo flight, train and truck. Due to breakability few item as damage.

Sometime the percentage of breakability in the transportation problem, the unit transportation costs, resources, demands at destinations, capacities of the conveyances and safety factor are not known precisely but some past data is available. So it is possible to assume the above as hybrid or fuzzy.

Therefore for model-1, 2, 3 the percentage of breakability in the transportation is 3 and for model-4, 5 and model-6 the above correspondence are (2, 3, 5) and (1, 2.3, 2.5) + (1.5, .99) respectively. The unit transportation costs, resources, demands at destinations, capacities of the conveyances and safety factor (for models 1 to 6) are given below:

Input data

Crisp Unit transportation costs (in $):

$\xi_{111} = 18, \xi_{211} = 20, \xi_{311} = 25, \xi_{121} = 26,$

$\xi_{221} = 29, \xi_{321} = 24, \xi_{113} = 30, \xi_{213} = 21,$

$\xi_{313} = 24, \xi_{131} = 25, \xi_{231} = 23, \xi_{331} = 30,$

$\xi_{112} = 22.5, \xi_{212} = 21, \xi_{312} = 19, \xi_{123} = 35,$

$\xi_{223} = 38, \xi_{323} = 21.7, \xi_{122} = 23.9, \xi_{222} = 26.8,$

$\xi_{322} = 32.1, \xi_{132} = 23.5, \xi_{232} = 22.4, \xi_{332} = 37.6,$

$\xi_{133} = 27, \xi_{233} = 25.1, \xi_{333} = 30.2,$

Hybrid Unit transportation costs (in $):

$\overset{\approx}{\xi}_{111} = (10,12,13) + (7,0.5), \overset{\approx}{\xi}_{211} = (8,11,14) + (9,1.7),$

$\overset{\approx}{\xi}_{311} = (11,15,17) + (10,2), \overset{\approx}{\xi}_{121} = (12,14,17) + (11,1.8),$

$\overset{\approx}{\xi}_{221} = (13,18,19) + (13,3), \overset{\approx}{\xi}_{321} = (7,8,11) + (15,1.2),$

$\overset{\approx}{\xi}_{113} = (6,7,11) + (22,3), \overset{\approx}{\xi}_{213} = (7,12,18) + (8,1.35),$

$\overset{\approx}{\xi}_{313} = (12,18,22) + (6,1.23), \overset{\approx}{\xi}_{131} = (16,17,18) + (8,1.44),$

$\overset{\approx}{\xi}_{231} = (13,17,18) + (7,3), \overset{\approx}{\xi}_{331} = (8,17,19) + (15,1.22),$

$\overset{\approx}{\xi}_{112} = (11,12,14) + (11,1.2), \overset{\approx}{\xi}_{212} = (11,12,15) + (9,2),$

$\overset{\approx}{\xi}_{312} = (12,13,15) + (6,1), \overset{\approx}{\xi}_{123} = (14,19,20) + (19,0.85),$

$\overset{\approx}{\xi}_{223} = (13,15,19) + (23,2.22), \overset{\approx}{\xi}_{323} = (15,16,20) + (5,0.27),$

$\overset{\approx}{\xi}_{122} = (9,10,11) + (14,2.2), \overset{\approx}{\xi}_{222} = (10,11,12) + (17,2.3),$

$\overset{\approx}{\xi}_{322} = (16,17,19) + (16,1.44), \overset{\approx}{\xi}_{132} = (14,17,20) + (7,1.9),$

$\overset{\approx}{\xi}_{232} = (9,13,18) + (10,0.65), \overset{\approx}{\xi}_{332} = (17,19,21) + (19,1.75),$

$\overset{\approx}{\xi}_{133} = (9,11,12) + (17,1.33), \overset{\approx}{\xi}_{233} = (10,15,16) + (12,0.95), \overset{\approx}{\xi}_{333} = (15,19,23) + (11,0.77)$

Crisp Safety Factors:

$S_{111} = 0.7, S_{211} = 0.75, S_{311} = 0.85, S_{121} = 0.61, S_{221} = 0.5,$

$S_{321} = 0.57, S_{131} = 0.82, S_{231} = 0.75, S_{331} = 0.82,$

$S_{112} = 0.71, S_{212} = 0.91, S_{312} = 0.52, S_{122} = 0.73,$

$S_{222} = 0.79, S_{322} = 0.9, S_{132} = 0.83, S_{232} = 0.73,$

$S_{332} = 0.65, S_{113} = 0.55, S_{213} = 0.74, S_{313} = 0.86,$

$S_{123} = 0.96, S_{223} = 0.68, S_{323} = 0.64, S_{133} = 0.77,$

$S_{233} = 0.78, S_{333} = 0.59.$

Safety Factors as Triangular Fuzzy Number:

$\tilde{S}_{111} = (.7,.71,.72), \tilde{S}_{211} = (.71,.73,.75),$

$\tilde{S}_{311} = (.81,.84,.86), \tilde{S}_{121} = (.55,.59,.63),$

$\tilde{S}_{221} = (.46,.50,.51), \tilde{S}_{321} = (.52,.56,.58),$

$\tilde{S}_{131} = (.80,.82,.83), \tilde{S}_{231} = (.72,.75,.78),$

$\tilde{S}_{331} = (.81,.82,.83), \tilde{S}_{112} = (.70,.71,.73),$

$\tilde{S}_{212} = (.89,.91,.94), \tilde{S}_{312} = (.47,.52,.53),$

$\tilde{S}_{122} = (.71,.73,.76), \tilde{S}_{222} = (.76,.79,.82),$

$\tilde{S}_{322} = (.87,.92,.93), \tilde{S}_{132} = (.80,.83,.84),$

$\tilde{S}_{232} = (.7,.73,.74), \tilde{S}_{332} = (.62,.64,.67),$

$\tilde{S}_{113} = (.53,.55,.58), \tilde{S}_{213} = (.73,.74,.78),$

$\tilde{S}_{313} = (.83,.86,.88), \tilde{S}_{123} = (.92,.96,.98),$

$\tilde{S}_{223} = (.64,.68,.69), \tilde{S}_{323} = (.63,.64,.68),$

$\tilde{S}_{133} = (.74,.77,.78), \tilde{S}_{233} = (.75,.78,.80),$

$S_{333} = (.55,.59,.62).$

Safety Factors as Hybrid variables:

$\overset{\approx}{S}_{111} = (.45,.46,.47) + (.20,.15),$

$\overset{\approx}{S}_{211} = (.45,.47,.5) + (.30,.43),$

$\overset{\approx}{S}_{311} = (.45,.47,.48) + (.40,.25),$

$\overset{\approx}{S}_{121} = (.32,.34,.38) + (.20,.41),$

$\overset{\approx}{S}_{221} = (.35,.40,.44) + (.1,.25),$

$\overset{\approx}{S}_{321} = (.40,.43,.45) + (.17,.32),$

$\overset{\approx}{S}_{131} = (.55,.80,.83) + (.22,.32),$

$\overset{\approx}{S}_{231} = (.50,.55,.58) + (.25,.34),$

$\overset{\approx}{S}_{331} = (.52,.53,.54) + (.35,.3),$

$\overset{\approx}{S}_{112} = (.40,.41,.42) + (.3,.40),$

$\overset{\approx}{S}_{212} = (.53,.54,.55) + (.37,.41),$

$\overset{\approx}{S}_{312} = (.29,.30,.33) + (.20,.13),$

$\overset{\approx}{S}_{122} = (.41,.43,.45) + (.30,.14),$

$\overset{\approx}{S}_{222} = (.37,.39,.4) + (.40,.15),$

$\overset{\approx}{S}_{322} = (.51,.53,.54) + (.35,.24),$

$\overset{\approx}{S}_{132} = (.45,.46,.47) + (.35,.37),$

$\overset{\approx}{S}_{232} = (.3,.33,.34) + (.39,.26),$

$\overset{\approx}{S}_{232} = (.3,.33,.34) + (.39,.26),$

$\overset{\approx}{S}_{113} = (.22,.25,.28) + (.31,.24),$

$\overset{\approx}{S}_{213} = (.34,.35,.38) + (.35,.21),$

$\overset{\approx}{S}_{313} = (.52,.56,.59) + (.31,.22),$

$\overset{\approx}{S}_{123} = (.61,.66,.68) + (.29,.23),$

$\overset{\approx}{S}_{223} = (.31,.32,.35) + (.34,.31),$

$\overset{\approx}{S}_{323} = (.41,.44,.45) + (.18,.32),$

$\overset{\approx}{S}_{133} = (.34,.37,.38) + (.41,.19),$

$\overset{\approx}{S}_{233} = (.45,.48,.52) + (.30,.12),$

$\overset{\approx}{S}_{333} = (.33,.39,.40) + (.20,.46),$

Crisp fixed charge:

$f_{111} = 11.8, f_{211} = 12, f_{311} = 12.6, f_{221} = 12.9,$

$f_{321} = 12.4, f_{131} = 12.5, f_{231} = 12.3, f_{331} = 13.0,$

$f_{112} = 10.5, f_{212} = 12.1, f_{312} = 11.9, f_{122} = 12.34,$

$f_{222} = 12.68, f_{322} = 13.21, f_{132} = 12.35, f_{232} = 12.24,$

$f_{332} = 13.76, f_{113} = 13.0, f_{213} = 12.1, f_{313} = 13.0,$

$f_{123} = 13.5, f_{223} = 13.8, f_{323} = 12.17, f_{133} = 12.7,$

$f_{233} = 12.51, f_{333} = 13.02.$

Hybrid fixed charge:

$\overset{\approx}{f}_{111} = (3,4,5) + (7,0.3), \overset{\approx}{f}211 = (4,6,7) + (6,0.5),$

$\overset{\approx}{f}_{311} = (5,5.5,7) + (5.8,2), \overset{\approx}{f}_{121} = (8,9,10) + (4,0.9),$

$\overset{\approx}{f}_{221} = (8,9,11) + (2,0.5), \overset{\approx}{f}_{321} = (10,11,12) + (1,0.5),$

$\overset{\approx}{f}_{131} = (4,6,8) + (7,7.7), \overset{\approx}{f}_{231} = (8,9,10) + (3.2,2),$

$\overset{\approx}{f}_{331} = (7,7.8) + (5,0.6), \overset{\approx}{f}_{112} = (6,7,8) + (3.25,3),$

$\overset{\approx}{f}_{212} = (3,5,7) + (7.1,3), \overset{\approx}{f}_{312} = (5,6,7) + (5.9,2),$

$\overset{\approx}{f}_{122} = (3,4.5,5.5) + (7.34,1), \overset{\approx}{f}_{222} = (6,7,8.1) + (5.68),$

$\overset{\approx}{f}_{322} = (6.1,7.9,8) + (6.68,3), \overset{\approx}{f}_{132} = (12.35), \overset{\approx}{f}_{232} = (6,7.8,8.2) + (4.24,.8), \overset{\approx}{f}_{332} = (8,8.9,9.5) + (5.78,2.1),$

$\overset{\approx}{f}_{113} = (6,8,10.4) + (5.6,2),$

$\overset{\approx}{f}_{213} = (2.5,5.7,8.9) + (6,1.2), \overset{\approx}{f}_{213} = (10,11,12) + (2,3),$

$\overset{\approx}{f}_{213} = (6.7,7.8,9.8) + (5.6,3), \overset{\approx}{f}_{213} = (2.5,5.7,8.9) + (6,1.2),$

$\overset{\approx}{f}_{313} = (10,11,12) + (2,3),$

$\overset{\approx}{f}_{123} = (6.7,7.8,9.8) + (5.6,3),$

$\overset{\approx}{f}_{223} = (10,11.2,12.3) + (2.8,2), \overset{\approx}{f}_{323} = (10,10.2,11.3) + (3.8,2.3)12.17,$

$\overset{\approx}{f}_{133} = (4,6,8) + (5.7,4), \overset{\approx}{f}_{233} = (8,8.1,8.9) + (3,2.4),$

$\overset{\approx}{f}_{333} = (10,11.4,11.9) + (3.03,1).$

Crisp Resources, Demands, and Conveyance capacities:

$a_1 = 64, a_2 = 64, a_3 = 70, b_1 = 41, b_2 = 43, b_3 = 41, e_1 = 45, e_2 = 43, e_3 = 41$

Resources, Demands and Conveyance capacities as a TFN:

$\tilde{a}_1 = (61,64,67), \tilde{a}_2 = (62,66,68), \tilde{a}_3 = (70,71,72),$

$\tilde{b}_1 = (40,42,43), \tilde{b}_2 = (42,43,45), \tilde{b}_3 = (43,45,46),$

$\tilde{e}_1 = (47,49,50), \tilde{e}_2 = (43,46,48), \tilde{e}_3 = (41,44,47.)$

Resources, Demands and Conveyance capacities as a Hybrid variable:

$\overset{\approx}{a}_1 = (51,54,57) + (9,2), \overset{\approx}{a}_2 = (52,53,56) + (12,2.3), \overset{\approx}{a}_3 = (60,63,64) + (8,5),$

$\overset{\approx}{b}_1 = (35,36,37) + (8,2), \overset{\approx}{b}_2 = (26,29,30) + (16,6), \overset{\approx}{b}_3 = (25,28,29) + (13,2),$

$\overset{\approx}{e}_1 = (27,30,31) + \overset{\approx}{e}_2 = (31,32,34) + (12,7)\overset{\approx}{e}_3 = (28,29,30) + (13.9,8).$

Desired minimum total safety measure for the system:

$$B = 7.72, \widetilde{B} = (7.4,7.5,7.9), \widetilde{\mathrm{B}} = (6.5,6.9,7) + (0.85,1).$$

Other parametric values

Firstly, we set the different parameters on which this GA depends. These are the number of generation (MAXGEN), population size (POPSIZE), probability of crossover (PXOVER), probability of mutation (PMU). There is no clear indication as to how large a population should be. If the population is too large, there may be difficulty in storing the data, but if the population is too small, there may not be enough string for good crossovers. For the present problems, POPSIZE=100, PXOVER=0.7, PMU=0.3 and MAXGEN=4000.

Results

To solve the crisp models we use the LINGO.13 optimization software and GA and the optimal results of the mentioned models are given (Table 1).

Overview of the Results of the Four Models

The hybrid parameters were first introduced in the literature in 1978, consideration of this type of parameters in the decision making problems is in the developing stage. But, in some real-life problems, available data are hybrid. Hence, decision making problems with hybrid data are of great importance though there are very few such models in the literature. Here for the first time, some constrained unbalanced STPs are formulated with different types of hybrid costs and resources and reduced to corresponding crisp ones using appropriate method. These problems are solved using GA and GRG technique and numerically illustrated. models-3, 5 and 6 are greater than the transportation cost of the models-2 and 3 since the models 3, 5 and 6 are solved with safety factor as well as breakability but models-2 and 4 are solved with breakability and without safety factor. The transportation cost of the models with breakability is more than the cost of the other models without breakability because due to breakability the requirement of customer at the end are not fulfill so supplier deliver more quantity as required by customer. Again the transportation cost obtained by generalized reduced gradient technique (LINGO 13.0 Software) is greater than the cost obtained by using Genetic Algorithm. Thus we conclude that to solve any solid transportation problem with and without breakability and safety factor GA is very useful than LINGO (Table 2).

For further research one may apply these techniques for solving interval valued optimization

After solving the respective we observe that the total transportation cost of the model-1 is less than the total transpiration cost of the remaining models since the models-2 and 4 is solved using breakability and models-3, 5, 6 are solved with breakability and safety factors. Again the total transportation cost of models-3, 5 and 6 are greater than the transportation cost of the models-2 and 3 since the models 3, 5 and 6 are solved with safety factor as well as breakability but models-2 and 4 are solved with breakability and without safety factor. The transportation cost of the models with breakability is more than the cost of the other models without breakability because due to breakability the requirement of customer at the end are not fulfill so supplier deliver more quantity as required by customer. Again the transportation cost obtained by generalized reduced gradient technique (LINGO 13.0 Software) is greater than the cost obtained by using Genetic Algorithm. Thus we conclude that to solve any solid transportation problem with

Optimal	Model-1	Model-2	Model-3	Model-4	Model-5	Model-6
Solution						
Min(Z)	2848.65	2936.37	2957.04	2817.27	2867.02	2893.1
x_{111}	0	0	0	0	26.15	1.88
x_{211}	0	0	0	29.6	0	0
x_{311}	0	2.6	0.5	0	0	0
x_{121}	0	0	0	0	4.84	8.41
x_{221}	0	0	0.62	0	0	0
x_{321}	0	0	1.47	3.44	8.01	0.7
x_{131}	0	0	31.97		6.1	0
x_{231}	41	42.27	10.3	12.16	0	34.8
x_{331}	0	0	0	0	0	0
x_{112}	32.66	17.94	18.45	0	0.62	2.47
x_{212}	8.34	21.73	19.4	0	0	13.2
x_{312}	0	0	0	12.78	0	17.88
x_{122}	2	3.33	5.15	0	7.95	1.918
x_{222}	0	0	0	0	0	0
x_{322}	0	0	0	0	0	0
x_{132}	0	0	0	7.97	0	5.839
x_{232}	0	0	0	22.24	34.43	1.45
x_{332}	0	0	0	0	0	0
x_{113}	0	0	0	0	0	0
x_{213}	0	3.91	0	0	15.608	4.7
x_{313}	0	0	0	0	0	2.36
x_{123}	0	0	0	0	0	0
x_{223}	0	0	0	0	0	0
x_{323}	41	41	37.08	41	23.64	33.537
x_{133}	0	0	0	0	0.56	0
x_{233}	0	0	0	0	0.27	0.4
x_{333}	0	0	0	0	0.92	0

Table 1: Optimal Results of different Models using LINGO-13.0.

Optimal Solution	Model–1	Model–2	Model–3	Model–4	Model–5	Model–6
Min(Z)	2339.2	2286.2	2343.5	2319.3	2353.8	2155.5
x_{111}	0	0	0	0	0	0
x_{211}	0	0	0	0	0	0
x_{311}	33	11	0	1	2	11
x_{121}	0	0	0	0	0	0
x_{221}	0	0	0	0	0	0
x_{321}	0	0	0	31	0	0
x_{131}	8	0	0	0	0	0
x_{231}	0	0	0	0	39	21
x_{331}	0	21	32	0	0	0
x_{112}	0	0	0	0	6	0
x_{212}	17	0	0	0	0	0
x_{312}	0	0	0	0	0	0
x_{122}	0	0	0	23	14	0
x_{222}	0	0	0	0	0	0
x_{322}	0	0	0	0	0	0
x_{132}	0	23	23	0	17	23
x_{232}	0	0	0	0	0	0
x_{332}	0	0	0	0	0	0
x_{113}	0	0	0	0	0	0
x_{213}	0	39	39	0	0	39
x_{313}	0	0	1	0	0	0
x_{123}	0	0	0	0	0	0
x_{223}	0	0	0	0	0	1
x_{323}	0	1	0	1	31	0
x_{133}	15	0	0	0	0	0
x_{233}	22	0	0	39	0	0
x_{333}	0	0	0	0	0	0

Table 2: Optimal Results of different Models using Genetic Algorithm.

and without breakability and safety factor GA is very useful than LINGO.

For further research one may apply these techniques for solving interval valued optimization problems in the areas of engineering disciplines and management science.

Conclusion

To prepare this manuscript we consider the unit transportation cost, demand, supplies, conveyances capacity, breakability and safety factor as crisp, fuzzy and hybrid variable. In our manuscript we solve the models-1 without breakability and safety factors, models-2 and 3 are solve without safety factor and with breakability and models-3, 5 and 6 are solved with safety factor and breakability. Sometimes impreciseness occurs in transportation for this reason we formulate model-4, 5 and 6 in imprecise environment (fuzzy and hybrid). The result of the respective models as per our expectation i.e., the optimal cost of model-2 and 4 is greater than model-1 due to breakability and the optimal cost of models-3, 4 and 5 is greater than models-2 and 4 due to safety factors. The methods, used for solution here are quite general in nature and these can be applied to other similar uncertain/imprecise models in other areas such as inventory control, ecology, sustainable farm management, etc. In our approach we introduce hybrid transportation cost in solid transportation problem. Finally the entire mathematical models are solved by using LINGO 13.0 software and GA. So our technique is decidedly productive in the wisdom of real life problems of practical importance. Practical numerical examples are provided to demonstrate the feasibility of all decision variables of the proposed methods.

References

1. Hitchcock FL (1941) The distribution of a product from several sources to numerous localities. Journal of Mathematical Physics 20: 224-230.
2. Shell E (1955) Distribution of a product by several properties. Directorate of Management Analysis. Proceedings of the 2nd Symposium in Linear Programming, DCS/Comptroller HQUSAF Washington, DC, 2: 615-642.
3. Zadeh LA (1965) Fuzzy sets. Information and Control 8: 338-353.
4. Zadeh LA (1978) Fuzzy sets as a basis for a theory of possibility. Fuzzy Sets and Systems 1: 3-28.
5. Kaufmann A (1975) Introduction to the theory of fuzzy subsets, New York: Academic Press.
6. Zimmermann HJ (1996) Fuzzy set theory and its applications. Boston: Kluwer Academic Publishers (4th edition).
7. Liu B (2006) A survey of credibility theory. Fuzzy Optimization and Decision Making 5: 387-408.
8. Dubois D, Prade H (1998) Possibility Theory: An Approach to Computerized Processing of Uncertainty. New York: Plenum.
9. Chanas S, Kuchta D (1996) A concept of the optimal solution of the transportation problem with fuzzy cost. Coefficients. Fuzzy Sets and System 82: 299-305.
10. Jimenez F, Verdegay JL (1999) Solving fuzzy Solid transportation problems by an evolutionary algorithm based parametric approach. European Journal of Operational Research 117: 485-510.
11. Liu B, Liu YK (2002) Expected value of fuzzy variable and fuzzy expected value models. IEEE Transactions on Fuzzy Systems 10: 445-450.
12. Yang L, Liu L (2007) Fuzzy fixed charge solid transportation problem and algorithm. Applied Soft Computing 7: 879-889.
13. Chanas S, Kuchta D (1996) Multi-objective programming in optimization of interval objective functions-A generalized approach. European Journal of Operational Research 94: 594-598.

14. Saad OM, Abass SA (2002) A Parametric study on transportation problem under fuzzy environment. Engineering Journal of the University of Qatar 15: 165-176.

15. Grzegorzewski P, Mrowka E (2007) Trapezoidal approximation of fuzzy number-revisited. Fuzzy Sets and Systems 158: 757-768.

16. Chanas S (2001) On the interval approximation of a fuzzy number. Fuzzy Sets and System 122: 353-356.

17. Grzegorzewski P (2002) Nearest interval approximation of a fuzzy number. Fuzzy Sets and Systems 130: 321-330.

18. Kundu P, Kar S, Maiti M (2013) Multi-objective solid transportation problem with budget constraint in uncertain environment. International Journal of System Science.

19. Baidya A, Bera UK, Maiti M (2013) Multi-item interval valued solid transportation problem with safety measure under fuzzy-stochastic environment. Journal of Transportation Security 6: 151-174.

20. Kennington JL, Unger VE (1976) A new branch and bound algorithm for the fixed charge transportation problem. Management Science 22: 1116-1126

21. Sun M, Aronson JE, Mckeown PG, Dennis D (1998) A tabu search heuristic procedure for fixed charge transportation problem. European Journal of Operational Research 106: 411-456.

22. Gottlieb J, PaulmannL (1998) Genetic algorithms for the fixed charge transportation problems. Proceedings of the IEEE Conference on Evolutionary Computation. ICEC 330-335.

23. Holland HJ (1975) Adaptation in natural and artificial systems. University of Michigan.

Bicompletable Standard Fuzzy Quasi-Metric Space

Jehad R Kider*

Department of Applied Science, University of Technology, Iraq

Abstract

In this paper we introduce the definition of standard fuzzy quasi-metric space then we discuss several properties after we give an example to illustrate this notion. Then we showed the existence of a standard fuzzy quasi-metric space which is not bicompletable. Here we prove that every bicompletable standard qussi-metric space admits a unique [up to F-isometic] bicompletion.

Keywords: Standard fuzzy metric space, Standard fuzzy quasi-metric space, Bicompletable standard qussi-metric space

Introduction

In [1] Kider started the study of a notion of standard fuzzy metric space that constitutes an interesting modification of the notion of metric fuzziness due to George and Veeramani [2]. In this paper we extend the notion standard fuzzy metric space to a standard fuzzy quasi-metric space [3]. On the other hand, it was presented in [4] an example of a standard fuzzy metric space that is not completable, also it has been obtained an internal characterization of completable standard fuzzy metric spaces. taking these results into account and the fact that the concept of bicompletion provides a theory of completion to quasi-metric spaces in the classical sense [5]. It seems natural and interesting to discuss the problem of characterizing standard fuzzy quasi-metric spaces that are bicompletable. The main purpose of this paper is to solve this problem. Following the modern terminology of [5] by a quasi-metric on a set X we mean a function $d: XX \to [0,\infty)$ such that for all x, y, z ∈ X.

(i) d(x,y)=d(y,x)=0 if and only if x=y

(ii) $d(x,y) \le d(x,z) + (z,y)$

Each quasi-metric d on X generates a T0-topology which has a base the family of Υ_d Open balls $\{B_\varepsilon(x): X, \varepsilon > 0\}$ where $B_\varepsilon(x) = \{y \in X : d(x,y) < \varepsilon\}$.

Standard Fuzzy Metric Space

Definition 1.1: A binary operation : [0,1] × [0,1] → [0,1] is a continuous t-norm if * satisfies the following conditions [1]:

1- * is associative and commutative.

2- * is continuous.

3- a * 1=a for all a ∈ [0,1].

4- a * b ≤ c * d whenever a ≤ c and b ≤ d where a,b, c,d ∈ [0,1].

Remark 1.2: For any r1 > r2 we can find r3 such that r1* r3 ≥ r2 and for any r4 we can find an r5 such that r5* r5 ≥ r4 where r1,r2,r3,r4,r5 ∈ (0,1) [2].

We introduce the following definition.

Definition 1.3: A triple (X,M,*) is said to be standard fuzzy metric space if X is an arbitrary set, is a continuous t- norm and M is a fuzzy set * on X^2 satisfying the following conditions [1]:

(FM1) M(x,y) > 0 for all x, y ∈ X

(FM2) M(x,y)=1 if and only if x=y

(FM3) M(x,y)=M(y,x) for all x, y ∈ X

(FM4) M(x,z) ≥ M(x,y) M(y,z) for all x, y and z ∈ X

(FM5) M(x,y) is a continuous fuzzy set

Example 1.4: Let X= N, and let a * b=a.b for all a, b ∈ [0,1] [1].

$$\text{Define } M(x,y) = \begin{cases} \frac{x}{y} \text{ if } x \le y \\ \frac{y}{x} \text{ if } y \le x \end{cases}$$

for all x, y N .

Then (N, M,) is a standard fuzzy metric space.

Example 1.5: Let X= R and let a b=a.b for all a, b ∈ [0,1] [1].

Define $M(x,y) = \dfrac{1}{e|x-y|}$ for all x, y ∈ R

Then (R, M,) is a standard fuzzy metric space.

Definition 1.6: Let (X,M) be a standard fuzzy metric space then M is continuous if whenever xn→x and yn→y in X then M(xn,yn) →M(x,y) that is $\lim_{n\to\infty} M(x_n, y_n) = M(x,y)$ [1].

Definition 1.7: Let (X,M) be a standard fuzzy metric space .Then B(x,r) ={y∈X: M(x,y)>1-r} is an open ball with center x∈X and radius r, 0<r<1 [1].

Proposition 1.8: Let B(x,r1) and B(x,r2) be two open balls with same center x in a standard fuzzy metric space (X,M,).Then either [1]

B(x,r1) ⊆ B(x,r2) or B(x,r2) ⊆ B(x,r1) where r1,r2 ∈ (0,1).

Definition 1.9: A subset A of a standard fuzzy metric space (X,M,) is said to be open if given any point a in A there exists r, 0<r<1 such that B(a,r) ⊆ A. A subset B is said to be closed if B^c is open [1].

***Corresponding author:** Jehad R Kider, Department of Applied Science, University of Technology, Iraq, E-mail: jehadkider@gmail.com

Definition 1.10: Let (X,M,*) be a standard fuzzy metric space and let $A \subset X$ then the closure of A is denoted by $\bar{A}$ A or CL(A) and is defined to be the smallest closed set contains A [1].

Definition 1.11: A subset A of a standard fuzzy metric space (X,M,*) is said to be dense in X if $\bar{A} = X$ [1].

Theorem 1.12: Every open ball in a standard fuzzy metric space (X,M,*) is an open set [1].

Theorem 1.13: Let (X,M,) is a standard fuzzy metric space. Define $\Gamma_M = \{A \subset X : x \in A$ if and only if there exists $0<r<1$ such that $B(x,r) \subset A\}$ then Γ_M is a topology on X.

Theorem 1.14: Every standard fuzzy metric space is a Hausdorff space.

Definition 1.15: A sequence (xn) in a standard fuzzy metric space (X,M,*) is said to be converge to a point x in X if for each r ,0<r<1 there exists a positive number N such that M(xn,x)>(1-r) , for each $n \geq N$.

Theorem 1.16: Let (X,M,) be a standard fuzzy metric space then for a sequence (xn) in X converge to x if and only if $\lim_{n\to\infty} M(x_n, x) = 1$

Definition 1.17: A sequence (xn) in a standard fuzzy metric space (X,M,) is Cauchy if for each r, 0 < r < 1, there exists a positive number N such that M(xn,xm) > (1-r), for each m, n ≥ N.

Proposition 1.18: Let (X,d) be an ordinary metric space and let a b=a.b for all $a, b \in [0,1]$ Define $M_d(x, y) = \frac{1}{1+d(x, y)}$ then $(X,M_d,*)$ is a standard fuzzy metric space and it is called the standard fuzzy metric induced by the metric d [1].

Proposition 1.19: Let (X,d) be a metric space and let $(X,M_d,*)$ be the standard fuzzy metric space induced by d. Let (X_n) be a sequence in X. Then (X_n) converges to $x \in X$ in (X,d) if and only if (X_n) converges to x in $(X,M_d,*)$.

Proposition 1.20: Let (X,d) be a metric space and let $M_d(X,Y) = \frac{1}{1+d(x,y)}$. Then (X_n) is a Cauchy sequence in (X,d) if and only if (X_n) is a Cauchy sequence in $(X, M_d,*)$ [1].

Definition 1.21: Let (X,M,*) be a standard fuzzy metric space. A subset A of X is said to be F-bounded if there exists 0<r<1 such that, M(x,y)> 1- r, for all $x, y \in A$ [1].

Proposition 1.22: Let (X,d) be a metric space and let $M_d(X,Y) = \frac{1}{1+d(x,y)}$ then a subset A of X is F-bounded if and only if it is bounded [1].

Definition 1.23: A standard fuzzy metric space (X,M,*) is complete if every Cauchy sequence in X converges to a point in X [1].

Definition 1.24: Let $(X,M_x,*)$ and $(Y,M_Y,*)$ be standard fuzzy metric spaces and $A \subseteq X$. A function f:A→Y is said to be continuous at $a \in A$, if for every $0< \varepsilon <1$, there exist some $0< \delta <1$, such that $M_Y(f(x),f(a))> (1- \varepsilon)$ whenever $x \in A$ and $M_X(x,a)> (1- \delta)$. If f is continuous at every point of A, then it is said to be continuous on A.

Theorem 1.25: Let $(X, M_X,*)$ and $(Y,M_Y,*)$ be standard fuzzy metric spaces and $A \subseteq X$. A function f:A→Y is continuous at $a \in A$ if and only if whenever a sequence (X_n) in A converge to a, the sequence $(f(X_n))$ converges to f(a).

Theorem 1.26: A function f:X→Y is continuous on X if and only if $f^{-1}(G)$ is open in X for all open subset G of Y.

Theorem 1.27: A mapping f:X→Y is continuous on X if and only if $f^{-1}(F)$ is closed in X for all closed subset F of Y [1].

Lemma 1.28: Let A be a subset of a standard fuzzy metric space (X,M,*) then $a \in \bar{A}$ if and only if there is a sequence (a_n) in A such that $a_n \to a$ [2].

Theorem 1.29: Let A be a subset of a standard fuzzy metric (X,M,) then A is dense in X if and only if for every $x \in X$ there is $a \in A$ such that M(x,a) $1- \varepsilon$ for some $0< \varepsilon <1$ [3].

Definition 1.30: Let $(X,M_x,*)$ and $(Y,M_Y,*)$ be any two standard fuzzy metric spaces. A mapping f:X→Y which is both one-to-one and onto is said to be a homeomorphism if and only if the mapping f and f^{-1} are continuous on X and Y, respectively. Two standard fuzzy metric spaces X and Y are said to be homeomorphic if and only if there exists a homeomorphism of X onto Y, and in this case, Y is called a homeomorphic image of X [4].

Remark 1.31: If X and Y are homeomorphic, the homeomorphism puts their points in one-to-one correspondence in such a way that their open sets also correspond to one another. For standard fuzzy metric space X and Y, let $X \sim Y$ means that X and Y are homeomorphic. It is easily verified that the relation is reflexive, symmetric and transitive.

Definition 1.32: A mapping f from a standard fuzzy metric space $(X, M_X,*)$ into a standard fuzzy metric space $(Y,M_Y,*)$ is an F-isometry if

$M_Y(f(x),f(y))=M_X(x,y)$ for all $x, y \in X$.

It is obvious that an F-isometry is one-to-one and uniformly continuous. X and Y are said to be F-isometric if there exists an F-isometry between them that is onto. An F-isometry is necessarily a homeomorphism but the converse is not true [4].

Proposition1.33: Let (X,d) be a metric space and let $(X,M_d,*)$ be the induced standard fuzzy metric space . Let (Y) be another metric space and let $(Y,N_d,*)$ be the induced standard fuzzy metric space. Let f:X→Y be a mapping then f is isometry if and only if f is F-isometry [4].

Theorem 1.34: Let A be a dense subset of a standard fuzzy metric space (X,M,*). If every Cauchy sequence of point of A converges in X then (X,M,*) is complete.

Proof: Let (X_n) be a Cauchy sequence in X, since A is dense then for every $Xn \in X$

there is $a_n \in A$ such that $M(X_n,a_n)>(1- s)$ for some 0<s<1 by

Theorem 1.29 Then by Remark 1.2 there is $(1- \varepsilon) \in (0,1)$ such that

$(1- s) * (1- s)> (1- \varepsilon)$.

Since (X_n) is Cauchy so (a_n) is Cauchy so $a_n \to x$ by assumption

Now $M(x_n,x) \geq M(x_n,x) * M(a_n,x) \geq (1- s) * (1- s) > (1- \varepsilon)$

Hence $x_n \to x$

Definition 1.35: Let (X,M,*) be a standard fuzzy metric space. A completion of (X,M,*) is a complete standard fuzzy metric space (Y,N,*) such that (X,M,*) is F-isometric to a dense subset of Y.

In [4] it was presented the following example of a standard fuzzy metric space

That is not completable.

Example 1.36:

Let a b=max {0, a + b -1} for all a, b $\in$ [0,1] . Now let { X_n: n=3, 4, 5,... , ∞} and {y_n : n=3,4,..., ∞} be two sequences of distinct points such that $A \cap B = \phi$ where A= {X_n : n $\geq$ 3} and B={ y_n: n $\geq$ 3}. Put X=A B, define M : X x X→ [0,1] as follows :

$$M(X_n,X_m)=M(y_n , y_m)=1- [\frac{1}{n \wedge m} - \frac{1}{n \vee m}]$$

Where $n \wedge m$ =min{n,m} and $n \vee m$ =max{n,m}.

$$M(X_n , Y_m)=M(y_m ,X_n)=\frac{1}{n}+\frac{1}{m}$$

It was shown that (X,M,*) is a standard fuzzy metric space and (X,M,*) is not completable.

Definition 1.37: A standard fuzzy metric space (X,M,*) is called completable if it admits a completion.

Theorem 1.38: Every completable standard fuzzy metric space admits a completion.

Standard Fuzzy Quas-Metric Space

Definition 2.1: The triple (X,M,*) is called a standard fuzzy quasi-metric space where X is a nonempty set, * is a continuous t-norm and M is a fuzzy set on X x X satisfying the following conditions:

(1) For all x, y$\in$X, M(x,y) > 0

(2) M(x,y)=M(y,x)=1 if and only if x=y

(3) M(x,y) * M(y,z) $\leq$ M(x,z) for all x, y, z$\in$X

(4) M is a continuous fuzzy set

Propostion 2.2:

If (X,M, *) is a standard fuzzy quasi-metric space then define

M^{-1}: XxX→[0,1] by : M^{-1} (x,y)=M(y,x) for all x, y$\in$X. Then

(X,M^{-1},*) is a standard fuzzy quasi-metric space.

Proof:

(1) M^{-1} (x,y)> 0 since M(y,x)> 0 for all x, y$\in$X

(2) M^{-1} (x,y)=1 if and only if M(y,x)=1=M(x,y) $\Leftrightarrow$ y=x

(3) M^{-1} (x,y)* M^{-1} (y,z)=M(y,x) * M(z,y)

= M(z,y) * M(y,x)

$\leq$M(z,x)=M^{-1} (x,z)

(4) M^{-1} is continuous since M is continuous.

Therefore (X, M^{-1}, *) is a standard fuzzy quasi-metric space

Proposition 2.3:

Let (X,M,*) be a standard fuzzy quasi-metric space. Define G:XxX→[0,1] by: G(x,y)=min{M(x,y), M^{-1} (x,y)}. Then (X,G,*) is a standard fuzzy space. We shall refer to (X,G, *) as the standard fuzzy metric induced by (X,M,*).

Proof:

It is sufficient to show that G(x,y)=G(y,x) for each x, y$\in$X.

If G(x,y)=M(x,y) then G(y,x) must equal to M^{-1} (y,x) but

M^{-1} (y,x)=M(x,y) that is G(x,y)=M(x,y).

Hence G(x,y)=G(y,x)

Similarly if G(x,y)=M^{-1} (x,y) then G(x,y)=G(y,x)

Therefore (X,G, *) is a standard fuzzy metric space

Proposition 2.4:

Let (X,M,*) be a standard fuzzy quasi-metric space. Then Γ_m = {A $\subset$ X : a$\in$A $\Leftrightarrow$ $\exists$ r, 0< r <1, such that B(a,r) $\subset$ A} is a topology on X.

Proof: The proof is similar to the proof of Theorem 1.13, hence is omitted [5].

Example 2.5:

Let (X,d) be an ordinary quasi-metric space and let M_d be the function defined on X X to [0,1] by: M_d (x,y)=$\frac{1}{1+d(x,y)}$

Then for each continuous t-norm *, (X, M_d,*) is a standard fuzzy quasi-metric space,which is called the standard fuzzy induced by the quasi-metric d. Furthermore, it is easy to check that $(M_d)^{-1}$=$M_{d^{-1}}$ and G_d=M_ds where d^{-1} (x,y)=d(y,x), d^s (x,y)=max {d(x,y), d^{-1} (x,y)} G_d (x,y)=min{M_d (x,y),M_d^{-1} (x,y)}

Definition 2.6: A standard fuzzy quasi-metric space (X,M,*) is called bicomplete if (X,G,*) is a complete standard fuzzy metric space.

Definition 2.7: Let (X,M,*) be a standard fuzzy quasi-metric space. A bicompletion of (X,M,*) is a bicomplete standard fuzzy quasi-metric space (Y,N,*) such that (X,M,*) is F-isometric to a dense subset of Y.

Lemma 2.8: Let (X,M,*) be a standard fuzzy quasi-metric space. Denote by S the collection of all Cauchy sequence in (X,G,*). Define a relation $\sim$ on S by (x_n) $\sim$ (x'_n) if and only if lim G(x_n, x'_n)=1, where by lim G(x_n, x'_n) we denote the lower limit of the sequence (G(x_n, x'_n)) i.e G(x_n, x'_n)=$\sup_k \inf_{n \geq k}$ G(x_n, x'_n) Then $\sim$ is an equivalence relation on S.

Proof:

1- $\sim$ is reflexive because G(X_n ,X_n)=1 for all n$\in$N so (X_n) $\sim$ (X_n)

2- If (X_n) (y_n), it immediately follows that (y_n) (X_n) because

G(X_n , y_n)=G(y_n, X_n) for all n$\in$N , So that

Lim G(y_n, X_n) Lim=G(X_n , y_n)=1

3- is transitive, suppose that (X_n) $\sim$ (y_n) and (y_n) $\sim$ (Z_n) . We shall prove Lim G(X_n ,Z_n)=1. Since (X_n) $\sim$ (Y_n) then G(X_n , y_n)=1.

Also (Y_n) $\sim$ (Z_n) so Lim G(y_n, X_n)=1 for all n$\in$N.

Now G(X_n ,Z_n) $\geq$ G(X_n ,Y_n) *G(y_n, Z_n)

Hence Lim G(X_n ,Z_n)=1

Lemma 2.9: Define M_s ((X_n),(y_n))=lim M(X_n,y_n) for all (X_n), (y_n)$\in$S where : SxS→[0,1]. Then M_s satisfies 1, 3 and 4 of Definition 2.1.

Proof:

1- M_s ((X_n), (y_n)) > 0 because M(X_n, y_n) >0 so, lim M(X_n, y_n) >0.

3- Let (X_n), (y_n),(Z_n)$\in$S and put α = M_s ((X_n),(y_n)),β =M_s ((y_n),(Z_n))

and $\gamma = M_s((X_n),(Z_n))$. We shall show that $\alpha * \beta \leq \gamma$

If=0 or=0 the conclusion is obvious. So we assume that 0

and β =0. Choose an arbitrary $\varepsilon \in (0, \min \frac{\{\alpha\beta\}}{2})$. Then

$\alpha - \varepsilon < M_s((X_n),(y_n))$ and $\beta - \varepsilon < M_s((y_n),(z_n))$

Furthermore, there exists such that for all k $k \geq N_\varepsilon$

$M_s((x_n),(y_n)) - \varepsilon < M(x_k, y_k)$ And $M_s((y_n),(z_n)) - \varepsilon < M(y_k, z_k)$

Then $(\alpha - 2\varepsilon) * (\beta - 2\varepsilon) \leq [M_s((x_n),(y_n)) - \varepsilon] * [M_s((y_n),(z_n)) - \varepsilon]$

$\leq M(x_k, y_k) * M(y_k, z_k)$

$\leq M(x_k, z_k)$ for all $K \geq N_\varepsilon$

Therefore $(\alpha - 2\varepsilon) * (\beta - 2\varepsilon) \leq \inf_{k \geq N_\varepsilon} M(X_k, z_k)$

$\leq \lim M(X_n, Z_n) = \gamma$

By continuity of *, it follows that $\alpha * \beta \leq \gamma$

4- M_s is continuous because M is continuous

Notation 2.10:

We denote the quotient $s/\sim$ by $\tilde{X}$ and $[(X_n)]$ the class of the element (X_n) of S.

Lemma 2.11:

If $(X_n) \sim Y$ **and** $(y_n) \sim (b_n)$ Then $M_s((X_n),(y_n)) = M_s((a_n),(b_n))$

Proof:

$$M_s((X_n),(y_n)) \geq M_s((x_n),(a_n)) * M_s((a_n),(b_n) * M_s((b_n),(y_n))$$
$$= M_s((a_n),(b_n))$$

Thus $M_s((X_n),(y_n)) \geq M_s((a_n),(b_n))$ **Now**

$$M_s((a_n),(b_n)) \geq M_s((a_n),(b_n)) * M_s((x_n),(y_n) * M_s((y_n),(b_n))$$
$$= M_s((x_n),(y_n))$$

So, $M_s((a_n),(b_n)) \geq M_s((x_n),(y_n))$

Therefore $M_s((x_n),(y_n)) \geq M_s((a_n),(b_n))$

Definition 2.12:

For each $[(x_n)]$, $[(y_n)] \in \tilde{X}$ define $\tilde{M}([(x_n)],[(y_n)]) = M_s(x_n),(y_n))$. Then $\tilde{M}$ is a function from $\tilde{X}x\tilde{X}$ to [0,1] and it is well defined by Lemma 2.11. Also we define T:X→ $\tilde{X}$ such that for each x ∈ X, T(x) is the class of constant sequence x, x, ...

Now, from the above construction we obtain the main result in this section.

Theorem 2.13:

Let (X,M,*) be a standard fuzzy quasi-metric space.

(a) (X,M,*) is a standard fuzzy quasi-metric space

(b) T(X) is dense in (X,M,*)

(c) (X,M,) is F-isometry to (T(X), M,*)

(d) (X,M,*) is bicomplete

Proof (a):

M satisfies conditions 1, 3 and 4 of Definition 2.1 as an immediate consequence of Lemma 2.9. Now, let (Xn), (Yn) ∈ S such that

$\tilde{M}([x_n],[(y_n)])$ if $(z_n) \in [(y_n)]$ it follows that from Lemma 2.11 that $M_s((z_n),(y_n))$ 1 The same argument shows that $(z_n) \in [(x_n)]$ implies that $M_s((z_n),(x_n)) = 1$ We conclude that $\tilde{M}([x_n],[(y_n)]) = 1$ if and only if $[(x_n)] = [(y_n)]$ Hence $(\tilde{X}, \tilde{M}, *)$ is a standard fuzzy quasi-metric space.

Proof (b):

Let $(x_n) \in S$ and 0< < 1. Since (x_n) is Cauchy sequence in (X,M,*) then there is

N_ε such that M $(x_k, x_{N_\varepsilon}) > (1 - \frac{\varepsilon}{2})$ for all $k \geq N_\varepsilon$

Thus $\tilde{M}([X_n], T(x_{n_\delta})) = M_s((x_n), T()x_{N_\delta})$

$= \sup_n \inf_{k>n} M(x_k, x_{N_\varepsilon})$

$\geq 1 - \frac{\varepsilon}{2} > 1 - \varepsilon$

We have shown that T(X) is dense in $(\tilde{X}, \tilde{M}, *)$

Proof (c):

This is almost obvious because for each x, y ∈ X, we have $\tilde{M}(Tx, Ty) = M(x, y)$

Proof (d):

Let $\tilde{G}([x_n].[x_n]) = \min\{\tilde{M}([(x_n),[(x_n)]), \tilde{M}^{-1}([(x_n),[(x_n)])\}$

Let $(\tilde{X}_n)$ be a Cauchy sequence in $(\tilde{X}, \tilde{G}, *)$ then there is an increasing sequence (n_k) in N such that $\tilde{G}(\tilde{X}_n, \tilde{X}_m) > 1 - 2^{-k}$ for all n, m $\geq n_k$ Since T(X) is dense in $(\tilde{X}, \tilde{G}, *)$ then for each $k \in N$ there is $y_k \in X$ such that $\tilde{G}(\tilde{X}_{nk}, T(y_k)) > 1 - 2^{-k}$ for all $k \in N$ We show that (y_k) is a Cauchy sequence in (X,G,*). To this end, choose $0 < \varepsilon < 1$ Take $j \in N$ such that $(1-2^{-j})*(1-2^{-j})*(1-2^{-j})>(1-\varepsilon)$ Then for each k, m $\geq j$ we have

$$M(y_k y_m) = \tilde{M}(T(y_k), T(y_m))$$
$$\geq \tilde{M}(T(y_k), \tilde{X}_{nk}) * \tilde{M}(\tilde{x}_{nm}, T(y_m))$$
$$\geq (1 - 2^{-k}) * (1 - 2^{-(k \wedge m)}) * (1 - 2^{-m})$$
$$\geq (1 - 2^{-j}) * (1 - 2^{-j}) * (1 - 2^{-j}) > (1 - \varepsilon)$$

And consequently (y_k) is a Cauchy sequence in (X,G,*). Therefore $\tilde{y} \in \tilde{X}$ where $\tilde{y} = [(y_k)]$ Finally, we prove that $(\tilde{X}_n)$ converges to $\tilde{y}$ in $(\tilde{X}, \tilde{G}, *)$

Indeed, as in part (c) choose $0 < \varepsilon < 1$ Take $j \in N$

$(1 - 2^{-j}) * (1 - 2^{-j}) * (1 - 2^{-j}) > (1 - \varepsilon)$

Since (y_k) is a Cauchy sequence in $(\tilde{X}, \tilde{G}, *)$ the proof of part (b) shows that there is $k \geq j$ such that $\tilde{G}(\tilde{y}, T(y_k)) > 1 - 2^{-j}$

Then for $n \geq n_k$ we obtain

$\tilde{G}(\tilde{y}, \tilde{x}_n) \geq \tilde{G}(\tilde{y}, T(y_k)) * \tilde{G}(T(y_k), \tilde{X}_{nk}) * \tilde{G}(\tilde{X}_{nk}, \tilde{X}_n)$

$\geq (1 - 2^{-j}) * (1 - 2^{-k}) * (1 - 2^{-k}) *$

$\geq (1 - 2^{-j}) * (1 - 2^{-j}) * (1 - 2^{-j}) > (1 - \varepsilon)$

We conclude that $(\tilde{X}, \tilde{M}, *)$ is bicomplete

Definition 2.14: A standard fuzzy quasi-metric space (X,M,*) is called bicompletable if it admits a bicompletion

Theorem 2.15: Let (X,M,*) be a standard fuzzy quasi-metric space and let (Y,N,*) be a bicomplete standard fuzzy quasi-metric space. If there is an F-isometry mapping f from a dense subset A of X to Y then f has a unique extension f^*: X→Y.

Proof: We consider any $x \in X$ but $X=\overline{A}$ so $x \in \overline{A}$ then there is a sequence (X_n) in A such that (X_n) converges to x by Lemma 1.28. Then (X_n) is Cauchy.

Since f is F-isometry (f(X_n)) is Cauchy in Y but Y is complete hence there is $y \in Y$ such that (f(X_n)) converges to y. Now we define f (x)=y.

We now show that this definition is independent of the particular choice of the sequence in A converging to x .Suppose that (X_n) in A converges to x and (z_n) in A converges to x. Then (v_m) converges to x where (v_m)=(x_1 ,z_1 x_2,z_2 , ,...). Hence (f(v_m)) converges and the two subsequence (f(x_n)) and (f(z_n)) of (f(v_m)) must have the same limit. This prove is uniquely defined at every $x \in X$. Clearly (x)=f(x) for every $x \in$ A so that is an extension of f.

Theorem 2.16: Let (X,M,*) be a standard fuzzy quasi-metric space and let (Y,N,*) be a bicomplete standard fuzzy quasi-metric space .If f is an F-isometry mapping from a dense subset A of X to Y then the unique extension f:X→Y is an F-isometry.

Proof:

Let x, $y \in X$ then there exists two sequences (x_n) and (y_n) in A such that x_n→ x and y_n→ y. Choose an arbitrary 0< ε < 1. Now:

$\varepsilon + M(x,y) > M(x,y)$. Furthermore, it follows that (x_n) and (y_n) are Cauchy sequences in A so (f(x_n)) and (f(y_n)) are Cauchy sequences in Y. But Y is complete hence (f(y_n)) converges to (y) and (f(x_n)) converges to f(x). Then there is a positive integer N such that

$M(x,x_n) > (1-\varepsilon), M(y_n,y) > (1-\varepsilon)$

$N(f^*(x_n), f^*(x)) > (1-\varepsilon)$ and $N(f^*(y_n),(y)) > (1-\varepsilon)$ for all $n \geq N$

Thus we have

$$\begin{aligned} \varepsilon + M(x,y) &> M(x,y) \\ &\geq M(x,x_n) * M(x_n,y_n) * M(y_n,y) \\ &\geq (1-\varepsilon) * N(f^*(x_n), f^*(y_n)) * (1-\varepsilon) \end{aligned}$$

But

$$\begin{aligned} N(f^*(x_n), f^*(y_n)) &\geq N(f^*(x_n), f^*(x)) * N(f^*(x), f^*(y)) * N(f^*(y_n), f^*(y)) \\ &\geq (1-\varepsilon) * N(f^*(x), f^*(y)) * (1-\varepsilon) \text{ for all } n \geq N \end{aligned}$$

Therefore $\varepsilon + M(x,y) > (1-\varepsilon) * [(1-\varepsilon) * N(f^*(x), f^*(y)) * (1-\varepsilon)] * (1-\varepsilon)$

By continuity of * and * it follows that M(x,y) ≥ N((x), (y))

A similar argument shows that N((x), (y)) ≥ M(x,y) For all x, $y \in X$

We conclude that f* is an F-isometry from (X,M,*) to (Y,N,*)

Theorem 2.17: Every bicompletable standard fuzzy quasi-metric space admits a unique [up to F-isometry] bicompletion.

Proof: Let (Y,M_1,*) and (Z,M_2,0) be two bicompletions of (X,M,*) then we will prove that (Y,M_1,*) and (Z,M_2,0) are F- isometric. Since (Y, M_1,*) is a bicompletion of (X,M,*) then there is an F-isometry f from (X,M,*) to a dense subset of (Y, M_1,*). By Theorem 2.15 and Theorem 2.16 f admits a unique extension f* onto (Y, M_1,*) which is also an F-isometry. Similarly is an isometry extension (X,M,*) onto (Z, M_2,0). To prove that and are F-isometric it remains to see that and are onto we will show that is onto. Indeed given $y \in Y$ there is a sequence (x_n) in X such that (x_n) →y. Since is an F-isometry (x_n) is a Cauchy sequence, so it converges to some point $x \in X$. Consequently f* (x)=y. Similarly we can prove that is onto. Hence f* and f are F-isometric.

Now (Y, M_1,*) is F-isometric to (X,M,*) and (X,M,*) is F-isometric to (Z, M_2,0). Hence (Y, M_1,*) is F-isometric to (Z, M_2,0).

References

1. Kider RJ (2014) Compact Standard Fuzzy Metric Space, ijma 5: 129-136.
2. George A, Veeramani P (1994) On Some Results in Fuzzy metric Spaces, Fuzzy Sets and Systems 64: 395-399.
3. George A, Veeramani P (2001) "On Some Results of Analysis for Fuzzy metric Spaces" Fuzzy Sets and Systems 90: 365-368.
4. Ameer ZA (2013) On Some Results of Analysis in a Standard Fuzzy Metric Spaces, M. Sc. Thesis. University of Technology, Iraq.
5. Fletcher P, Lindgren W (1982) Quasi-Uniform Spaces, Marcel Dekker, New York.

Generalizing of Differential Transform Method for Solving Nonlinear Differential Equations

Fabien Kenmogne*

Laboratory of Modelling and Simulation in Engineering and Biological Physics, Faculty of Science, University of Yaound´e I, Cameroon

Abstract

In this paper, the differential transform method (DTM) is proved to be an excellent tool to investigate analytical and numerical solutions of nonlinear ordinary differential equations. The concept of differential transformation is briefly presented and some well-known properties of this DTM are rewritten in a more generalized forms. Several classes of nonlinear differential equations are solved analytically and numerically, in order to show the efficiency of the DTM, and the fundamental algorithms are presented in Matlab language.

Keywords: Differential transform method; Nonlinear ordinary differential equation; Exact solution; Numerical investigation

Introduction

Ordinary or partial differential equations are commonly encountered in several branches of Sciences including Biology, Physics, Chemistry and Mathematics [1-3]. They are particularly used in physical branches to describe qualitatively the dynamical behaviors of physical systems including the population growth, the motion of particles in potential fields, the signal voltage in electrical circuits, the dynamical behaviors of trees under the effects of wind [4-7]. These differential equations obtained via physical laws (for e.g the Kirchhoff and Maxwell laws) are either linear or nonlinear. Although linear differential equations are easily solvable, nonlinear differential equations are difficult to solve and several of them do not admit analytical solutions and their solutions need to be approximated.

Many classical methods have been proposed to find exact or approximated solutions of these nonlinear differential equations, including the Lyapunov approach, the direct integration, the multi-scale expansion technics, the harmonic balance method [1,8-10], the fractional homotopy analysis transform method based on the innovative adjustment in Laplace transform algorithm [11-13] and the differential transform method (DTM) [14,15]. The DTM has been used in recent years for solving nonlinear oscillator dynamics problems and this semi-exact method, based on the *kth* order derivative of state variables around the initial time *t*0 or space *x*0, does not need linearization [16].

It is important to mention that in the previous works in which the DTM is used, the initial time or space are chosen to be the origin (*t*0=0 or *x*0=0) and the obtained solutions which is in the Taylor series expansion form are compared to the known analytical solutions of the models equations, which are interesting but remain nevertheless unsatisfactory to study the influence of initial conditions on dynamical behaviors of nonlinear differential equations.

The motivation of this paper is to give an answer at the above problem. For this aim, the rest of the paper is organized as follows: Section (2) describes the DTM in which some modifications have been carried out, in order to take into account the initial conditions. In section (3), some analytical and numerical examples are presented to illustrate the efficiency of the DTM and the obtained numerical results are compared to the well-known four order Runge Kutta numerical scheme. Finally, we give the conclusion in Section (4).

Basic Concept of Differential Transform Method

The differential transformation method is one of the semi-analytical method commonly used for solving ordinary and partial differential equations in the forms of polynomials as approximations of the exact solutions. The basic definition and the fundamental theorems of the DTM and its applicability for different kinds of differential equations are given in [14,16]. However in these works, all expansions are around the origin *x*0=0. Let us first remember that the differential transformation of the *kth* order derivative of any one dimensional variable y is given as follows:

$$Y(k) = \lim_{x\to x_0} \frac{1}{k!}\left(\frac{d^k y(x)}{dx^k}\right) \tag{1}$$

If all values of the the transformation *Y* (*k*) are computed, one can find the expression of the variable *y*(*x*) by using the inverse transformation defined as follows

$$Y(x) = \sum_{k=0}^{\infty} Y(k)(x-x_0)^k \tag{2}$$

It is obvious that the DTM satisfies the following set of conditions:

i) If $y(x) = ay1(x) \pm by2(x)$, then $Y(k) = aY1(k) \pm bY2(k)$, where *a* and *b* are any arbitrary constants.

ii) $\text{if } y(x) = \frac{d^n}{dx^n} y_1(x), \text{then } Y(k) = \frac{(k+n)!}{k!} y_1(k=n)$

iii) $\text{if } y(x) = \int y_1(x)\,dx, \text{then } Y(k) = \frac{1}{k} y_1(k-1)$

iv) $\text{if } y(x) = y_1(x)\,y_2(x), \text{ then } Y(k) = \sum_{k1=0}^{k} Y_1(k_1)\,Y_2(k-k_1)$

***Corresponding author:** Fabien Kenmogne, Laboratory of Modelling and Simulation in Engineering and Biological Physics, Faculty of Science, University of Yaound´e I, Cameroon, E-mail: kenfabien@yahoo.fr

v) if $y(x)=y_1(x)y_2(x)..y_n(x)$, then $Y(k)=\sum_{kn-1=0}^{k}\sum_{kn-2=0}^{kn-1}\sum_{k2=0}^{k3}\sum_{k1=0}^{k2}Y_1(k_1)Y_2(k_2-k_1)..Y_{n-1}(k_{n-1}-k_{n-2})Y_n(k-k_{n-1})$

All these five first theorems are well known and given in [14,16]. In addition to the above theorems, the last other theorems of the previous works are generalized for any given initial conditions as follows:

Vi) if $y(x)=x^n$, then $Y(k)=\frac{n!}{k!(n-k)}\Theta(n-k)!\Theta(n-k)x_0^{(n-k)}$

where Θ is the step function,

$$\Theta(u)=\begin{cases}1 & if\ \ u\geq 0\\ 0 & else\end{cases} \quad (3)$$

and $x0$ the initial value of x. It is obvious that when $x0=0$, one recover the result obtained in [14,16] (that is $Y(k)=\delta(n-k)$), where $\delta(u)$ is the Dirac delta function.

$$\delta(u)=\begin{cases}1 & if\ \ u=0\\ 0 & else\end{cases} \quad (4)$$

Vii) If $y(x)=\exp(\lambda x)$, then $Y(k)=\frac{\lambda k}{k!}\exp(\lambda x0)$, where λ is any arbitrary constant.

Viii) If $y(x)=\sin(wx+\alpha)$, then $Y(k)=\frac{w^k}{k!}\sin(\frac{k\Pi}{2}+wx_0+\alpha)$ where w and α are constants.

iX) if $y(x)=\cos(wx+\alpha,)$ then $Y(k)=\frac{w^k}{k!}\cos(\frac{k\Pi}{2}+wx_0+\alpha)$

X) If $y(x)=a$, then $Y(k)=a\,\delta(k)$, where a is a constant.

Xi) if $y(x)=\exp(\lambda y_1(x))$, then $Y(k)=\frac{\lambda k}{k!}\exp(\lambda y_1(x_0))[\sum_{kn-2=0}^{kn-1}\sum_{k2=0}^{k3}\sum_{k1=0}^{k2}N_k!\lambda-N_2Y_2(k_2-k_1)..Y_2(k_{n-1}-kn-2)Y_2(k-k_{n-1})]$, with $N_k=\delta(k_1)+\delta(k_2-k_1)+..+\delta(k_{n-1}-k_{n-2})+\delta(k-k_{n-1})$

$$Y_2(u)=\begin{cases}1 & if\ \ u=0\\ Y_1(u) & else\end{cases} \quad (5)$$

The five firsts terms of $Y(k)$ of this theorem are given in the extended form in the Appendix (5.1).

Applications of DTM to the Analytical and Numerical Results

In this section, some examples are given to illustrate the procedure outlined above.

The DTM in analytical results

The exact analytical solution of any set of solvable ordinary differential equations can be obtained by using properties i, ii, iii,..,xi and by choosing the initial differentiation point to be the origin $x0=0$. Let us first consider for illustrating the efficiency of the DTM, the following set of differential equations:

$$\begin{cases} y'_1(x)\ \ 2\exp(4x)y_4(x)\\ y'_2(x)=y_1(x)-y_3(x)+\cos(x)-\exp(2x)\\ y'_3(x)=y_2(x)-y_4(x)-\sin(x)+\exp(-x)\\ y'_4(x)=-\exp(-5x)y_1(x)\end{cases} \quad (6)$$

and subjected to the initial conditions $y1(0)=y2(0)=y4(0)=1$, $y3(0)=0$, where the prime stands for the derivation with respect to variable x. Note that these equations under these initial conditions have exact analytical solutions in the form:

$$\begin{cases} y_1(x)=\exp(2x)\\ y_2(x)=\sin(x)+\cos(x)\\ y_3(x)=\sin(x)\\ y4(x)=\exp(-x)\end{cases} \quad (7)$$

By using the above properties around the expansion point $x0=0$, Eq.(6) is transformed in differential form as follows:

$$\begin{cases} Y_1(k+1)=\frac{2}{k+1}\left[\sum_{k2=0}^{k}\sum_{k1=0}^{k2}\frac{4^{k_1}}{k_1^!}Y_4(k_2-k_1)Y_4(k-k_2)\right]\\ Y_2(k+1)=\frac{2}{k+1}\left[Y_1(k)-Y_3(k)+\frac{1}{k!}\cos\left(\frac{k\pi}{2}\right)-\frac{2^k}{k!}\right]\\ Y_3(k+1)=\frac{2}{k+1}\left[Y_2(k)-Y_4(k)+\frac{1}{k!}\sin\left(\frac{k\pi}{2}\right)+\frac{(-1)^k}{k!}\right]\\ Y_4(k+1)=-\frac{1}{k+1}\left[\sum_{k2=0}^{k}\sum_{k1=0}^{k2}\frac{(-5)^{k_1}}{k_1^!}Y_1(k_2-k_1)Y_1(k-k_2)\right]\end{cases} \quad (8)$$

and subjected to the initial condition:

$$\begin{cases} Y_1(0)=\lim_{x\to 0}y_1(x)=1\\ Y_2(0)=\lim_{x\to 0}y_2(x)=1\\ Y_3(0)=\lim_{x\to 0}y_3(x)=0\\ Y_4(0)=\lim_{x\to 0}y_4(x)=1\end{cases} \quad (9)$$

From the above initial condition and accounting to the set of Eqs. (8), the values of $Yi(k)$, with $i=1;\ 2;\ 3;\ 4$ are progressively computed, leading to: Substituting the results of the above Table 1 into Eq.(2) leads to the following solution of the set of Eqs.(6)

$$\begin{cases} y_1(x)=1+2x+2x^2+\frac{8}{3^!}x^3+\frac{16}{4^!}x^4+\frac{32}{5^!}x^5+....\\ y_2(x)=\left(x-\frac{1}{3^!}x^3+\frac{1}{5^!}x^5+...\right)+\left(1-\frac{1}{2^!}x^2+\frac{1}{4^!}x^4+...\right)\\ y_3(x)=x-\frac{1}{3^!}x^3+\frac{1}{5^!}x^5+...\\ y_4(x)=\left(1-x+\frac{1}{2^!}x^2-\frac{1}{3^!}x^3+\frac{1}{4^!}x^4-\frac{1}{5^!}x^5...\right)\end{cases} \quad (10)$$

which coincide with solution (7) expanded as a power series near the origin $x0=0$.

The DTM in numerical investigations

As proved above, by considering the differential transformation near the constant and fixed initial value $x0$ (Figure 1), one obtains just the solution as a power series of the exact solution near this initial value. In order to seek the solution of the above problem with the DTM and valid for all values of x, we divide the space in small subspace, with space step h. To find the value of $y(x+h)$ knowing the value of $y(x)$, this method requires first the calculation of Y(k) at position x using the

Order of the differentiation	Y1(k)	Y2(k)	Y3(k)	Y4(k)
k=1	2	1	1	-1
k=2	2	-1/2	0	1/2
k=3	8/3	-1/3	-1/3	-1/3
k=4	16/4	-1/4	0	1/4
k=5	32/5	-1/5	-1/5	-1/5

Table 1: Table into Eq.(2).

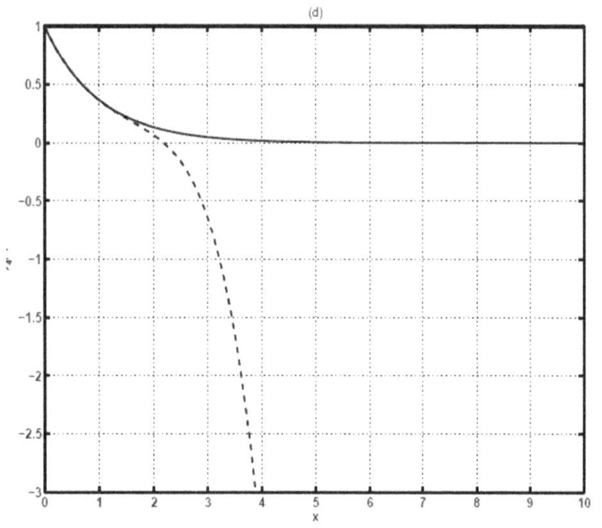

Figure 1: Solution of Eq.(6). Solid line, exact solution of the system given by (7). Dash line, approximated solution (9) at the origin *x*=0. As one can see, they coincide only near the origin *x*=0.

transformation (1) and the iterative process needs the value of *y*(*x*) at the initial space *x*=*x*0 and next, values of

$$y(x+h) = \sum_{k=0}^{\infty} h^k Y(k) \tag{11}$$

As the differentiation is obtained here at *kth* order, with $k \to \infty$, this method is accurate to $(k + 1)th$ order in step size *h*, which tend to zero, leading this method more stable in numerical investigations as compared to other scheme. The proposed numerical technique that we can use to calculate Eq.(11) is summarized in the flow chart shown in Figure 2. In what follows, the above example and two others are chosen to illustrate the procedure outlined in this subsection.

Example 1: Let us first consider the set of Eq.(6), with the arbitrary initial conditions (*y*1(*x*0); *y*2(*x*0); *y*3(*x*0); *y*3(*x*0)), leading to the fact that the solution of this equation is not necessary given by (10). By using properties i, ii, iii,..,xi with now *x*0=0, instead of Eq.(8), Equation(6) is now transformed in the following differential form:

$$\begin{cases} Y_1(k+1) = \dfrac{2}{k+1}\left[\sum_{k2=0}^{k}\sum_{k1=0}^{k2}\dfrac{4^{k_1}\exp(4x_0)}{k_1^!}Y_4(k_2-k_1)Y_4(k-k_2)\right] \\ Y_2(k+1) = \dfrac{2}{k+1}\left[Y_1(k)-Y_3(k)+\dfrac{1}{k!}\cos\left(\dfrac{k\pi}{2}+x_0\right)-\dfrac{2^k\exp(2x_0)}{k!}\right] \\ Y_3(k+1) = \dfrac{2}{k+1}\left[Y_2(k)-Y_4(k)+\dfrac{1}{k!}\sin\left(\dfrac{k\pi}{2}+x_0\right)+\dfrac{(-1)^k\exp(-x_0)}{k!}\right] \\ Y_4(k+1) = -\dfrac{1}{k+1}\left[\sum_{k2=0}^{k}\sum_{k1=0}^{k2}\dfrac{(-5)^{k_1}\exp(-5x_0)}{k_1^!}Y_1(k_2-k_1)Y_1(k-k_2)\right] \end{cases} \tag{12}$$

which is well function of initial condition *x*0. This equation is subjected to the initial condition:

$$\begin{cases} Y_1(0) = \lim_{x\to x_0} y_1(x) = y_1(x_0) \\ Y_2(0) = \lim_{x\to x_0} y_2(x) = y_2(x_0) \\ Y_3(0) = \lim_{x\to x_0} y_3(x) = y_3(x_0) \\ Y_4(0) = \lim_{x\to x_0} y_4(x) = y_4(x_0) \end{cases} \tag{13}$$

By using the procedure (12) and the inverse DTM (11), the values of *yi*(*x*) can be progressively computed to give the profile of *yi*(*x*), which coincide with the solid line of Figure 3, which is the exact solutions given by Eq. (7) if the initial conditions (9) are considered. In Matlab the algorithm may be given by the function DTMequation1 as given in Appendix(5.2):

Example 2: The forced Vander Pol Duffing oscillator: Our second example is the forced Vander Pol Duffing oscillator

$$\ddot{y} - \varepsilon(1-y^2)\dot{y} + y + \alpha y^3 = F\cos(\omega t) \tag{14}$$

where the dot stands for the derivation with respect to time *t*. This equation can be rewritten in the set of first order differential equations as follows:

$$\begin{cases} \dot{y}_1 = y_2 \\ \dot{y}_2 = \varepsilon y_2 - y_1 - (\varepsilon y_2 + \alpha y_1)y_1^2 + F\cos(\omega t) \end{cases} \tag{15}$$

with $y1 \equiv y$ and $y2=y^{\cdot}1$, and subjected to the initial condition *y*1(*t*0) and *y*2(*t*0). Using again the DTM, it is easy to show that the set of Eq.(15) can be rewritten in the differential transformation form as:

$$\begin{cases} Y_1(k+1) = \dfrac{1}{k+1}Y_2(k) \\ Y_2(k+1) = \dfrac{1}{k+1}\left[\varepsilon Y_2(k) - Y_1(k) - \sum_{k_2=0}^{k}\sum_{k_1=0}^{k_2}\left(\varepsilon Y_2(k_1)+\alpha Y_1(k_1)\right)Y_1(k_2-k_1) + F\dfrac{\omega^k}{k^!}\cos\left(\dfrac{k\pi}{2}+\omega t_0\right)\right] \end{cases} \tag{16}$$

As in the previous example, the values of *y*1(*t*) and *y*2(*t*) are progressively computed using both the procedure (16) and the inverse DTM (11), and the obtained results are plotted in Figure 4, which coincide exactly with results obtained by using the well-known four order Runge Kutta scheme. The Matlab algorithm used to compute *yi*(*t*) is now given by the function DTMequation2 as shown in Appendix(5.3):

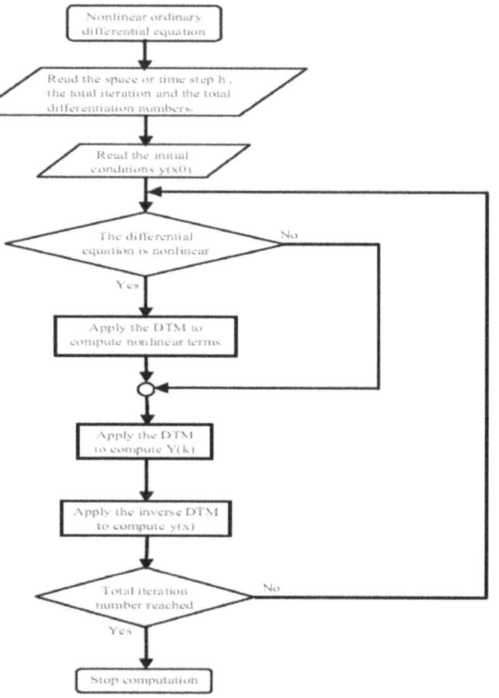

Figure 2: Flow chart for solving nonlinear ordinary differential equations viaDTM.

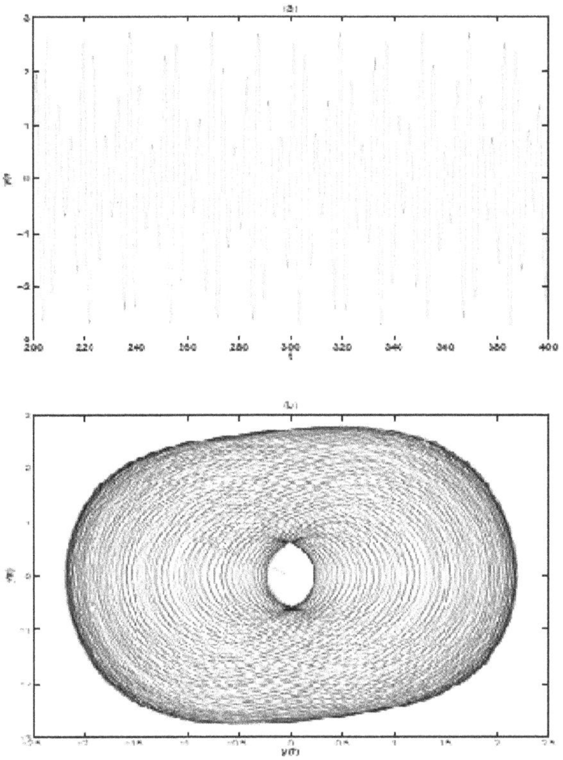

Figure 3: Solution of Eq.(14) (a): Evolution of *y*(*t*), (b): phase space plot forthe following choice of parameters: *α*= 0:4, "=0:1, *F*=0:8. and *w*=1.

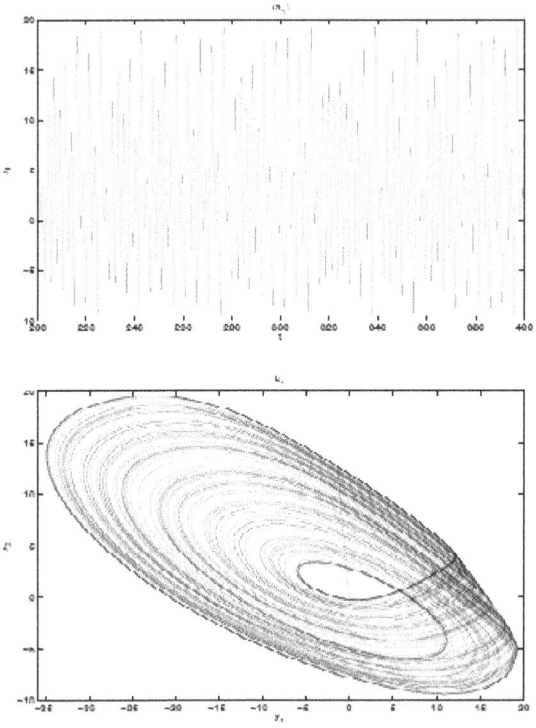

Figure 4: Phase space plot of the solution of Eq.(14), obtained for the choiceof parameters: *σ*1=1:49, *σ*2 =0:872,=51:6, *σ*=255=256.

Example 3: Our third example is the following set of ordinary differential equation:

$$\begin{cases} \overline{y}_1 = -\sigma_1(y_1 + y_2) + y_4 - \gamma(\exp(y_1 + y_3) - 1), \\ \overline{y}_2 = -\varepsilon_1\sigma_1(y_1 + y_2) + \varepsilon_1 y_4 \\ \overline{y}_3 = \varepsilon_2[y_4 + \gamma(\alpha - 1)(\exp(y_1 + y_2) - 1)], \\ \overline{y}_4 = -(y_1 + y_2 + y_3 + \sigma_2 y_4) \end{cases} \quad (17)$$

subjected again to the initial condition *y*1(*t*0), *y*2(*t*0), *y*3(*t*0) and *y*4(*t*0), which can model several systems oscillators in which nonlinear diodes or nonlinear bipolar junction transistors are present, like the Colpitts oscillators [17,18]. This equation can be rewritten in the differential form as:

$$\begin{cases} y_1(k+1) = \frac{1}{k+1}[-\sigma_1(y_1(k) + y_2(k)) + y_4(k) - \gamma(v(k) - \delta(k))] \\ y_2(k+1) = \frac{\varepsilon_1}{k+1}[-\sigma_1(y_1(k) + y_2(k)) + y_4(k)] \\ y_3(k+1) = \frac{\varepsilon_1}{k+1}[-\sigma_1(y_4(k) + \gamma(\alpha - 1)(v(k) - \delta(k))] \\ y_4(k+1) = -\frac{1}{k+1}(y_1(k) + y_2(k) + y_3(k) + \sigma_2 y_4(k)] \end{cases} \quad (18)$$

where δ is again the delta function and where

$$v(k) = \frac{1}{k!}\exp(y_1(t_0) + y_3(t_0) + y_3(t_0))[\sum_{k_{n-1=0}}^{k} \sum_{k_{n-2=0}}^{k_{n-1}} \sum_{k2=0}^{k_3} \sum_{k1=0}^{k} N_k!U(k_1)U(k_2 - k_1) \\ ..U(K_{n-1} - k_{n-2})U(k - k_{n-1})] \quad (19)$$

With $N_k = \delta(k_1) + \delta(k_2 - k_1) + .. + \delta(k_{n-1} - k_{n-2}) + \delta(k - k_{n-1})$ and

$$\begin{cases} U_2(\Delta k) = 1 \text{ if } \Delta k = 0 \\ U_2(\Delta k) = Y_1(\Delta k) + Y_3(\Delta k) \text{ else} \end{cases} \quad (20)$$

In this work we have demonstrated the applicability of the differential transform method for solving analytically and numerically the set of nonlinear ordinary differential equations, with the help of some concretes examples. Firstly, we have ameliorated the existing properties of the DTM found in works preceding this work and we have next shown that the DTM is very powerful and efficient tool in finding exact solutions for a wide class of problems. The results of numerical simulation applied to certain classes of nonlinear differential equations show that the DTM is more accurate and the obtained solutions are very rapidly convergent in comparison with the traditional numerical schemes.

References

1. Strogatz SH (1994) Nonlinear dynamics and chaos with applications to Physics, Biology, Chemistry and Engineering. Perseus books, Massachusetts.
2. Kumar Sunil (2013) A Numerical Study for Solution of Time Fractional Nonlinear Shallow-Water Equation in Oceans. Zeitschrift fur Naturforschung A 68: 1-7.
3. Kumar Sunil (2013) Numerical Computation of Time-Fractional Fokker-Planck Equation Arising in Solid State Physics and Circuit theory. Zeitschrift fur Naturforschung A 68: 1-8.
4. Nana Nbendjo BR, Yamapi R (2007) Active control of extended Van der Pol equation. Commun Nonlinear SciNumerSimul 12: 1550
5. Cuomo KM, Oppenhein AV (1993) Circuit implementation of synchronized chaos with applications to communications. Phys Rev Lett 71: 65.
6. Mattheck C, Breloer H (1994) The Body Language Of Trees. HMSO, London, UK.

7. Wood CJ (1995) Understanding wind forces on trees. Cambridge University Press, M.P. Coutts and J. Grace, eds, Cambridge, UK.

8. Biazar J, Ghazvini H (2007) He's variational iteration method for solving linear and non-linear systems of ordinary differential equations. Appl Math Comput 191: 287-297.

9. Butcher JC (2003) Numerical methods for ordinary differential equations. John Wiley and Sons.

10. Jafari H, Daftardar-Gejji V (2006) Revised Adomian decomposition method for solving systems of ordinary and fractional differential equations. Appl Math Comput 181: 598 - 608.

11. Kumar Sunil (2014) A new analytical modelling for fractional telegraph equation via Laplace transform. Applied Mathematical Modelling 38: 31543163.

12. Khader MM, Kumar Sunil, Abbasbandy S (2013) Newhomotopy analysis transform method for solving the discontinued problems arising in nanotechnology Chinese Physics B 22(11): 1-5.

13. Kumar Sunil, Mehdi Rashidi Mohammad (2014) New Analytical Method for Gas Dynamics Equation Arising in Shock Fronts. Computer Physics Communications 185: 1947-1954.

14. Mirzaee F (2011) Differential transform method for solving linear and nonlinear systems of ordinary differential equations. Applied Mathematical Sciences 5: 3465-3472.

15. Arikoglu A, Ozkol I (2005) Differential transform method for solving Volterra integral equation with separable kernels. Appl Math Comput 168: 1145-1158.

16. Zhou JK (1986) Differential transformation and Application for electrical circuits, Huazhong University Press, Wuhan, China.

17. Kengne J, Chedjou JC, Kenne G, Kyamakya K (2012) Dynamical properties and chaos synchronization of improved Colpitts oscillators. Commun Nonlinear SciNumerSimulat 17: 2914 2923.

18. Maggio GM, De Feo O, Kennedy MP(1999)Nonlinear analysis of the Colpitts oscillator with applications to design. IEEE Trans. on circ. and Syst. I: Fundamental theory and application 46.

On Hydromagnetic Channel Flow of an Oldroyd-B Fluid Induced by Tooth Pulses in a Rotating System

Ghosh S*

Department of Computer Science, BIT Mesra, Kolkata Extension Centre, 1582 Rajdanga Main Road, Kolkata-700107, India

Abstract

An initial value problem concerning the motion of an incompressible electrically conducting viscoelastic Oldroyd-B fluid in a channel bounded by two infinite rigid non-conducting plates in presence of an external transverse magnetic field when both the fluid and the plates are in a state of solid body rotation with constant angular velocity about an axis normal to the plates is solved. The unsteady motion is generated impulsively from rest in such a fluid when the upper plate is subjected to velocity tooth pulses with the lower plate held fixed. It is assumed that no electric current exists in the basic state and the magnetic Reynolds number is very small. Exact solutions of the problem are obtained by utilizing two methods, of them, one is the method of Fourier analysis and the other is the method of Laplace transforms. The enquiries are made about the velocity field and the skin-friction on the walls. It is shown that both the methods give the same exact solution of the problem. The influence of rotation, the magnetic field and the elasticity of the fluid on the components of fluid velocity and the wall skin-frictions are examined quantitatively. Some known results are found to emerge as special cases of the present analysis.

Keywords: Hydromagnetic; Pulsatile flow; Oldroyd-B fluid; Rotating system

Introduction

The fluid flow generated by pulsatile motion of the boundary is found to have immense importance in aerospace science, nuclear fusion, astrophysics, geophysics and cosmical gas dynamics. The investigation in this direction was carried out by Chakraborty and Ray [1] who examined the magneto-hydrodynamic Couette flow of a viscous fluid between two parallel plates when one of the plates is set in motion by random pulses. Makar [2] presented the solution of magneto-hydrodynamic viscous flow between two parallel plates when one of the plates is subjected to velocity tooth pulses and the induced magnetic field is neglected. Bestman and Njoku [3] constructed the solution of hydromagnetic channel flow of an incompressible, electrically conducting viscous fluid produced by tooth pulses including the effect of induced magnetic field, ignored by the author [2], and using the methodology of Fourier analysis instead of applying the commonly used technique of Laplace transforms which involve complicated inversions. Ghosh and Debnath [4] considered the hydromagnetic channel flow of a two- phase fluid-particle system induced by tooth pulses and obtained solution using the method of Laplace transforms. Datta and Dalal [5,6] discussed the heat transfer to pulsatile flow of a dusty fluid in pipes and channels following the method of perturbation. On the other hand, Hayat et al. [7] studied some simple flows of an Oldroyd-B fluid using the method of Fourier transforms. Asgar et al. [8] also utilized the same methodology as that of authors [7] to solve the problem concerning Hall effects on the unsteady hydromagnetic flow of an Oldroyd-B fluid while Hayat et al. [9] constructed the solution of hydromagnetic Couette flow of an Oldroyd-B fluid in a rotating system following the method of perturbation. In the present paper, we intend to construct solution of hydromagnetic channel flow of a rotating Oldroyd-B fluid induced by tooth pulses with a view to its application in hydromagnetic spin-up in a contained fluid [10], the motion of the earth's liquid core [11], the development of sunspot, the solar cycle and the structure of the magnetic stars [12] and in the determination of the effects of the external magnetic field and rotation on the flow of blood in the cardiovascular system, particularly at low rates of shear [13].

The problem is devoted to the study of unsteady hydromagnetic flow of an incompressible, electrically conducting Oldryod-B fluid confined in a channel bounded by two infinite rigid non-conducting parallel plates separated by a distant h when both the plates and the fluid are in a state of solid body rotation with a constant angular velocity Ω about the z-axis normal to the plates. The unsteady motion is supposed to generate impulsively from rest in the fluid due to velocity tooth pulses applied on the upper plate with the lower plate held fixed. Exact expression for the fluid velocity is obtained using the methods of Fourier analysis and Laplace transforms separately. The results for the skin-friction on the walls are also obtained in both the cases. It is shown that both the methods give the same exact solution of the problem. The effects of rotation, the magnetic field and the fluid elasticity on the components of fluid velocity and the wall skin-frictions are examined quantitatively. It is observed that the viscoelastic fluids grow and decay less faster than the ordinary viscous fluids. The magnetic field exerts a damping effect on such flows. The increasing and decreasing effects of rotation alone on the components of fluid velocity and skin-frictions during increasing and decreasing motions of the fluid are shown separately through pictorial representation.

Basic Equations

Following Oldroyd [14] the constitutive equations for an oldroyd-B fluid can be written as

$$T=-p\,I+S, \tag{2.1}$$

$$s+\lambda_1\frac{DS}{Dt}=\mu[1+\lambda_2\frac{D}{Dt}]A_1 \tag{2.2}$$

***Corresponding author:** Ghosh S, Department of Computer Science, BIT Mesra, Kolkata Extension Centre, 1582 Rajdanga Main Road, Kolkata-700107, India
E-mail: bijarn@yahoo.com

Where T=Cauchy stress tensor, p=fluid pressure, I=identity tensor, S=extra stress tensor, $\mu, \lambda_1, \lambda_2 =$ viscosity coefficient, relaxation time, retardation time (assumed constants).

The tensor A_1 is defined as

$$A_1=\nabla V+(\nabla V)^T: \tag{2:3}$$

In a cartesian system, $\frac{D}{Dt}$ (upper convected time derivative) operating on any tensor B_1

$$\text{Is } \frac{DB_1}{Dt}=\frac{\partial B_1}{\partial t}+(v.\nabla)B_1-(\nabla v)B_1-B_1(\nabla V)^T \tag{2.4}$$

It is to be mentioned here that this model includes the viscous fluid as a particular case for $\lambda_1=\lambda_2$; the Maxwell fluid when l_2=0 and an oldroyd-B fluid when $0<l_2<l_1<1$.

The stress equations of motion for an incompressible electrically conducting oldroyd-B fluid in a rotating system in presence of an external magnetic fluid are

$$\nabla.v=0 \tag{2.5}$$

$$\rho[\frac{\partial V}{\partial t}+(v.\nabla)V+2\Omega\times V+\Omega\times(\Omega\times r)]=\nabla.T+J\times B \tag{2.6}$$

$$\nabla.B=0,\ \nabla\times B=\mu_0 J \tag{2.7ab}$$

$$\nabla\times E^*=-\frac{\partial B}{\partial t}, J=\sigma_0[E^*+V\times B] \tag{2.8ab}$$

where V=(u,v,w)=fluid velocity, $r^2=(x^2+y^2)$, =fluid density, J=current density, B=magnetic flux density, E^*=electric field, μ_0=magnetic permeability (assumed constant), s_0=electrical conductivity (assumed finite) and Ω=angular velocity of solid body rotation.

Formulation of the Problem

In view of the physical nature of the problem, as shown in Figure 1, we take

$$V=(u(z;t); v(z;t); 0); S=S(z;t); \tag{3.1a}$$

$$B=(0; 0; B0); \Omega=(0; 0; \Omega) \tag{3.1b}$$

Where z-axis is normal to the plate, x-axis is along the plate and y-axis perpendicular to it with origin at the plate.

The equation (2.2) then yields

$$S_{xx}+\lambda_1[\frac{\partial}{\partial t}S_{xx}-2S_{xz}\frac{\partial u}{\partial z}]=-2\mu\lambda_2(\frac{\partial u}{\partial z})^2 \tag{3.2a}$$

$$S_{xy}+\lambda_1[\frac{\partial}{\partial t}S_{xy}-S_{yz}\frac{\partial u}{\partial z}-S_{xz}\frac{\partial u}{\partial z}]=-2\mu\lambda_2(\frac{\partial u}{\partial z})(\frac{\partial u}{\partial z}) \tag{3.2b}$$

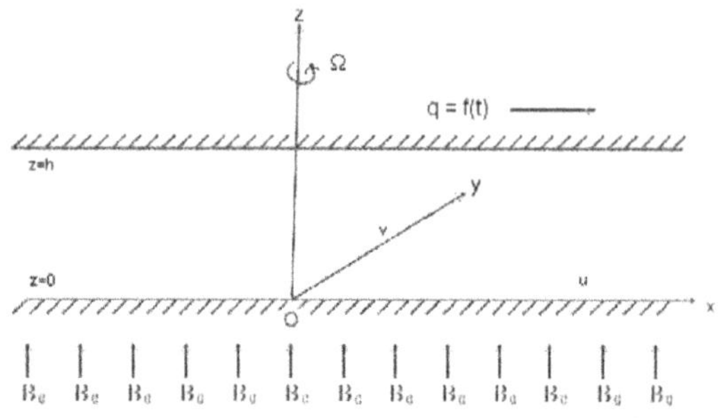

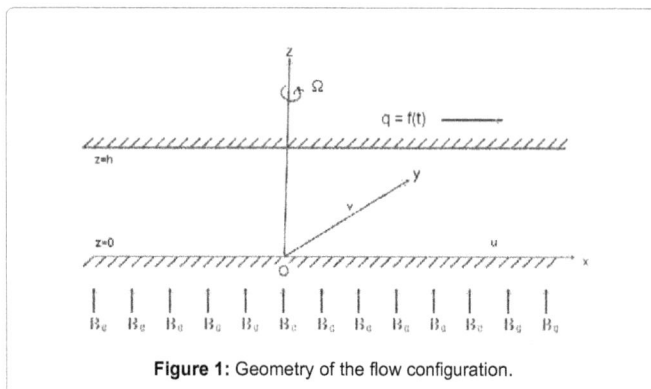

Figure 1: Geometry of the flow configuration.

$$S_{xz}+\lambda_1[\frac{\partial}{\partial t}S_{xz}-S_{zz}\frac{\partial u}{\partial z}]=\mu(\frac{\partial u}{\partial z})+\mu\lambda_2(\frac{\partial^2 u}{\partial z\,\partial t}) \tag{3.2c}$$

$$S_{yy}+\lambda_1[\frac{\partial}{\partial t}S_{yy}-2S_{yz}\frac{\partial u}{\partial z}]=-2\mu\lambda_2(\frac{\partial u}{\partial z})^2 \tag{3.2d}$$

$$S_{yz}+\lambda_1[\frac{\partial}{\partial t}S_{yz}-2S_{zz}\frac{\partial u}{\partial z}]=-\mu(\frac{\partial u}{\partial z})+\mu\lambda_2\frac{\partial^2 v}{\partial_z\partial_t} \tag{3.2e}$$

$$S_{zz}+\lambda_1[\frac{\partial}{\partial t}S_{zz}=0 \tag{3.2f}$$

Integrating (3.2f), we get $S_{zz}F(z)\exp[-(t/\lambda_1)]$ (3.3)

Where F(z) is an arbitrary function of z. We shall investigate the possibility of the solution in which F(z)=0 i.e., S_{zz}=0 [9].

We now assume that no applied or polarization voltage exists i.e, E*=0 so that no energy is added or extracted from the fluid by the electric field. We further assume that the magnetic Reynolds number is very small which is plausible for most electrically conducting fluids. This implies that the current is mainly due to induced electric field so that j=σ_0(u×B) and the applied magnetic field remains essentially unaltered by the electric current flowing in the fluid. We also assume that the induced magnetic field produced by the motion of the fluid is negligible compared to the applied magnetic field so that the Lorentz force term in (2.6) becomes $-\sigma_0 B_0^2 V$:

The equations of motion in (2.6) then reduces to

$$\rho[\frac{\partial u}{\partial t}-2\Omega v]=-\frac{\partial p}{\partial x}+\frac{\partial S_{xz}}{\partial_z}-\sigma_0 B_0^2 u \tag{3.4a}$$

$$\rho[\frac{\partial u}{\partial t}+2\Omega u]=-\frac{\partial p}{\partial y}+\frac{\partial S_{yz}}{\partial_z}-\sigma_0 B_0^2 u \tag{3.4b}$$

$$-\frac{\partial p}{\partial z}=0 \tag{3.4c}$$

Where p is the modified pressure including the centrifugal force term. The Equation (3.4c) shows that $\frac{\partial p}{\partial x}, \frac{\partial p}{\partial y}$ have the same value as in the free stream. Accordingly, we assume that both the quantities are zero.

We now combine (3.4a) and (3.4b) with the help of (3.2c) and (3.2e) to obtain

where H(t - T) = 0, t < T and H(t - T) = 1, t >T. (3.5)

Where q=u + i v is the complex velocity of the fluid.

We now introduce the non-dimensional quantities

$$\bar{z}=\frac{z}{\sqrt{v\lambda_1}}, d=\frac{h}{\sqrt{v\lambda_1}}, \bar{t}=\frac{t}{\lambda_1}, \bar{q}=\frac{q}{U_0}, M^2=\frac{\sigma_0 B_0^2}{\rho}\lambda_1,$$

$$E=2\Omega\lambda_1,\ k=\frac{\lambda_2}{\lambda_1},\ 0\leq k\leq 1.$$

in equation (3.5) and on dropping bars we get

$$(1+\frac{\partial}{\partial t})[\frac{\partial q}{\partial t}+iEq]=(1+k\frac{\partial}{\partial t})\frac{\partial^2 q}{\partial z^2}-M^2(1+\frac{\partial}{\partial t})q \tag{3.6}$$

The problem now reduces to solving (3.6) subject to the boundary and initial conditions given by

q(z, t) = 0 at z = 0, t > 0; q(z, t) = f(t) at z = d, t > 0 (3.7 a, b)

and

$$[q(z,t), \frac{\partial q}{\partial t}(z,t)] = [0,0] \text{ at } t \leq \textit{for all } z \qquad (3.7c)$$

where f(t) representing the tooth pulses is an even periodic function with period 2T and strength E_0 T.

Solution of the Problem

Method of fourier analysis

According to the nature of f(t) mentioned above, the mathematical form of u(d, t) may be written as

$$q(d,t) = \frac{E_0}{T}\left\{t\,H(t) + 2\sum_{n=1}^{\infty}(-1)^n (t-nT)H(t-nT)\right\} \qquad (4.1)$$

where H(t - T) = 0, t < T and H(t - T) = 1, t >T.

Using half-range Fourier series, the condition (4.1) can also be expressed in the form

$$u(d,t) = \frac{E_0}{2} - \frac{4E_0}{\pi^2}\sum_{m=0}^{\infty}\frac{1}{(2m+1)^2}\cos\left\{(2m+1)\frac{\pi t}{T}\right\} \qquad (4.2)$$

By virtue of (4.2) we assume the solution of (3.6) as

$$q(z,t) = q_s + \frac{1}{2}\sum_{m=0}^{\infty}[q_{2m+1}(z)e^{(2m+1)\frac{i\pi t}{T}} + \overline{q}_{2m+1}(z)e^{-(2m+1)\frac{i\pi t}{T}}] + \sum_{n=1}^{\infty}W_n(t)\sin(\frac{n\pi z}{d}) \qquad (4.3)$$

Where $\overline{q}$ is the conjugate of q. The first two terms in (4.3) are chosen so as to satisfy (4.2)

While the last term accommodates the initial condition.

Substituting (4.3) in (3.6) and then using (3.7a) and (4.2), we have the following equations

With appropriate conditions as

$$\frac{d^2q_s}{dz^2} - L^2 q_s = 0 \qquad (4.4)$$

With q_s=0 on z=0; $q_s = \frac{E_0}{2}$ on z=d $\qquad \frac{d^2 q_{2m+1}}{dz^2} - L_m^2 q2_{m+1} = 0 \qquad (4.5)$

With

$$q_{2m+1} = 0 \; on \; z = 0, q_{2m+1} = -|\frac{4E_0}{\pi^2}\frac{1}{(2m+1)^2} \; on \; z = d$$

And

$$\frac{d^2W_n}{dt} + (1 + M^2 + \frac{n^2\pi^2 k}{d^2} + iE)\frac{dW_n}{dt} + (M^2 + \frac{n^2\pi^2}{d^2} + iE)W_n = 0 \qquad (4.6)$$

With

$$W_n(t) = W_n(0), W_n'(t) = W_n'(0) \text{ at } t = 0$$

Where $W_n(0)$ and $W_n'(0)$ are to be determined.

In the above,

$$L^2 = (M^2 + iE), L_m^2 = \frac{[M^2 + i(E + \beta_m)](1 + i\beta_m)}{1 + i\beta_m k} \quad and \quad \beta_m = \frac{(2m+1)\pi}{T}$$

The solutions of equations (4.4)-(4.6) are

$$q_s(Z) = \frac{E_0}{2}\frac{\sinh Lz}{\sinh Ld} \qquad (4.7)$$

$$q_{2m+1}(z) = -\frac{4E_0}{\pi^2}\frac{1}{(2m+1)^2}\frac{\sinh L_m^z}{\sinh L_m^d} \qquad (4.8)$$

$$W_n(t) = W_n'(0)\frac{e^{m1t} - e^{m2t}}{m_1 - m_2} + W_n(0)\frac{m_1 e^{m2t} - m_2 e^{m1t}}{m_1 - m_2} \qquad (4.9)$$

Where

$$2m_1, 2m_2 = -\left[(1 + M^2 + \frac{n^2\pi^2 k}{d^2} + iE) \mp \left\{\left(1 + M^2 + \frac{n^2\pi^2 k}{d^2} + iE\right)^2 - 4\left(M^2 + \frac{n^2\pi^2}{d^2} + iE\right)\right\}^{\frac{1}{2}}\right] \qquad (4.10)$$

The initial conditions in (3.7b) provide

$$\frac{Wn(0)}{E_0} = n\pi(-1)^n\left[\frac{1}{n^2\pi^2 + L^2 d^2} - \frac{8}{\pi^2}\text{Re}\sum_{m=0}^{\infty}\frac{1}{(2m+1)^2(n^2\pi^2 + L_m^2 d^2)}\right] \qquad (4.11)$$

$$\frac{W'n(0)}{E_0} = \frac{8n(-1)^n}{\pi}\text{Im}\sum_{m=0}^{\infty}\frac{\beta_m}{(2m+1)^2(n^2\pi^2 + L_m^2 d^2)} \qquad (4.12)$$

The m-series in (4.11) and (4.12) are of orders b_m^{-3} and b_m^{-2} when $m \to \infty$. The n-series is also convergent since m_1; m_2 are of order $-N_1^2$ and m_1-m_2 has the order N_1^2 as $n \to \infty$ where $N_1^2 = M^2 + \frac{n^2\pi^2 k}{d^2}$.

Finally, the fluid velocity takes the form

$$\frac{q(z,t)}{E_0} = \frac{\sinh Lz}{2\sinh Ld} - \frac{4}{\pi^2}\text{Re}\sum_{m=0}^{\infty}\frac{e^{i\beta mt}}{(2m+1)^2}\frac{\sinh L_m z}{\sinh L_m d} + \sum_{n=1}^{\infty}\left\{\frac{W_n'(0)}{E_0}\frac{e^{m_1 t} - e^{m_2 t}}{m_1 - m_2} + \frac{W_n(0)}{E_0}\frac{m_1 e^{m_2 t} - m_2 e^{m_1 t}}{m_1 - m_2}\right\} \sin\frac{n\pi z}{d} \qquad (4.13)$$

which in the limit t → ∞ provides the steady velocity field

$$\frac{S_0}{E_0} = \frac{L}{2\sinh Ld} - \frac{4}{\pi^2}\text{Re}\sum_{m=0}^{\infty}\frac{e^{i\beta mt}}{(2m+1)^2}\frac{L_m}{\sinh L_m d} + \sum_{n=1}^{\infty}\frac{n^2\pi^2(-1)^n}{d} \times [\frac{1}{L^2 d^2 + n^2\pi^2} - \frac{8}{\pi^2}\text{Re}\sum_{m=0}^{\infty}\frac{1}{(2m+1)}\frac{1}{L_m^* d^2 + n^2\pi^2}]e - (M^2 + \frac{n^2\pi^2}{d^2} + iE)t \qquad (4.14)$$

where the harmonic part contains the effect of fluid elasticity due to the presence of pulsation.

On the other hand, the solution corresponding to classical viscous fluid in a rotating system can be obtained from (4.13) in the limit k → 1. This solution is given by

$$\frac{q(z,t)}{E_0} = \frac{\sinh Lz}{2\sinh Ld} - \frac{4}{\pi^2}\text{Re}\sum_{m=0}^{\infty}\frac{e^{i\beta mt}}{(2m+1)^2}\frac{\sinh L_m^* z}{\sinh L_m^* d} + \sum_{n=1}^{\infty}n\pi(-1)^n[\frac{1}{L^2 D^2 + n^2\pi^2} - \frac{8}{\pi^2}\sum_{m=0}^{\infty}\frac{1}{(2m+1)^2}\frac{M^2 d^2 + n^2\pi^2}{(M^2 d^2 + n^2\pi^2)^2 + (E + \beta_m)^2 d^4}] \times e^{-(M^2 + n^2\pi^2/d^2 + iE)t}\sin\frac{n\pi z}{d} \qquad (4.15)$$

Where $L_m^* = \sqrt{M^2 + i(E + \beta_m)}$

The result (4.15) in a non-rotating system (E=0) yields

$$\frac{u(y,t)}{E_0} = \frac{\sinh My}{2\sinh Md} - \frac{4}{\pi^2}\text{Re}\sum_{m=0}^{\infty}\frac{e^{i\beta_m t}}{(2m+1)^2}\frac{\sinh L_1 y}{2\sinh L_1 d} + \sum_{n=1}^{\infty}n\pi(-1)^n[\frac{1}{M^2 d^2 + n^2\pi^2} - \frac{8}{\pi^2}\sum_{m=0}^{\infty}\frac{1}{(2m+1)^2}\frac{M^2 d^2 + n^2\pi^2}{(M^2 d^2 + n^2\pi^2)^2 + \beta_m^2 d^4}] \times e^{-(M^2 + n^2\pi^2/d^2 + iE)t}\sin\frac{n\pi z}{d} \qquad (4.16)$$

Where $L_1 = \sqrt{M^2 + i\beta_m}$

The solution (4.16) is identical to that of Bestman and Njoku [3]. In particular, when T → 0 and E_0=2 the general result (4.13) reduces to

$$q(z,t) = \frac{\sinh Ly}{\sinh Ld} + 2\pi\sum_{n=1}^{\infty}\frac{n(-1)^n}{L^2 d^2 + n^2\pi^2}\frac{m_1 e^{m2t} - m_2 e^{m1t}}{m_1 - m_2}\sin\frac{n\pi z}{d} \qquad (4.17)$$

which describes the hydromagnetic channel flow of a rotating Oldroyd-B fluid generated by impulsive motion of the upper plate with a constant velocity. The result (4.17) in dimensionless form and in the limit M → 0; E → 0 coincides exactly with that of authors [7] while the result (4.16) in the limit T → 0 and E_0=2V agrees completely with that of Soundalgekar [15].

The skin-friction on the plates z=0 and z=d are given by

$$\frac{S_0}{E_0}=\frac{L(1-e^{-t})}{2\sinh Ld}-\frac{4}{\pi^2}\mathrm{Re}\sum_{m=0}^{\infty}\frac{L_m}{(2m+1)^2}\frac{1+ik\beta_m}{1+i\beta_m}\frac{e^{i\beta mt}-e^{-t}}{\sinh L_m d}+$$
$$\sum_{n=1}^{\infty}\frac{n\pi}{d}\{\frac{W_n'(0)}{E_0}\frac{1}{m_1-m_2}[\frac{1+km_1}{1+m_1}(e^{m1t}-e^{-t})-\frac{1+km_2}{1+m_2}(e^{m2t}-e^{-t})]+\frac{W_n(0)}{E_0}\frac{1}{M_1-m_2} \tag{4.18}$$
$$[\frac{(1+km_2)m_1}{1+m_2}(e^{m2t}-e^{-t})-\frac{(1+km_1)m_2}{1+m_1}(e^{m1t}-e^{-t})]\}$$

$$\frac{S_d}{E_0}=\frac{L\cosh Ld}{2\sinh Ld}(1-e^{-t})-\frac{4}{\pi^2}\mathrm{Re}\sum_{m=0}^{\infty}\frac{L_m}{(2m+1)^2}\frac{1+ik\beta_m}{1+i\beta_m}(e^{i\beta mt}-e^{-t})$$
$$\frac{\cosh L_m d}{\sinh L_m d}+\sum_{n=1}^{\infty}\frac{n\pi}{d}(-1)^n\{\frac{W_n(0)}{E_0}\frac{1}{m_1-m_2}[\frac{1+km_1}{1+m_1}(e^{m1t}-e^{-t})-\frac{1+km_2}{1+m_2} \tag{4.19}$$
$$(e^{m2t}-e^{-t})]+\frac{W_n(0)}{E_0}\frac{1}{m_1-m_2}[\frac{(1+km_2)m_1}{1+m_2}(e^{m1t}-e^{-t})-\frac{(1+km_2)m_2}{1+m_2}(e^{m1t}-e^{-t})]\}$$

Above results in the limit $k\to 1$ (viscous fluid) reduces to

$$\frac{S_0}{E_0}=\frac{L}{2\sinh Ld}-\frac{4}{\pi^2}\mathrm{Re}\sum_{m=0}^{\infty}\frac{e^{i\beta mt}}{(2m+1)^2}\frac{L_m}{\sinh L_m d}+\sum_{n=1}^{\infty}\frac{n^2\pi^2(-1)^n}{d}\times$$
$$[\frac{1}{L^2d^2+n^2\pi^2}-\frac{8}{\pi^2}\mathrm{Re}\sum_{m=0}^{\infty}\frac{1}{(2m+1)}\frac{1}{L^*_m d^2+n^2\pi^2}]e-(M^2+\frac{n^2\pi^2}{d^2}+iE)t \tag{4.20}$$

And

$$\frac{S_d}{E_0}=\frac{L\cosh Ld}{2\sinh Ld}-\frac{4}{\pi^2}\mathrm{Re}\sum_{m=0}^{\infty}\frac{e^{i\beta mt}}{(2m+1)^2}\frac{L^*_m\cosh L^*_m d}{\sinh L^*_m d}+$$
$$\sum_{n=1}^{\infty}\frac{n^2\pi^2}{d}[\frac{1}{L^2d^2+n^2\pi^2}-\frac{8}{\pi^2}\mathrm{Re}\sum_{m=0}^{\infty}\frac{1}{(2m+1)^2}\frac{1}{L^*_m d^2+n^2\pi^2}]\times \tag{4.21}$$
$$e^{-(M2+\frac{n^2\pi^2}{d^2}+iE)t}$$

However, when T→0 and E0=2, (4.20) and (4.21) provide the hydromagnetic solutions in a rotating system which are given by

$$S_0=\frac{L}{\sinh Ld}+\frac{2\pi^2}{d}\sum_{n=1}^{\infty}\frac{n^2(-1)^n}{L^2d^2+n^2\pi^2}e-(M^2+\frac{n^2\pi^2}{d^2}+iE)t \tag{4.22}$$

And

$$S_d=\frac{L\cosh Ld}{\sinh Ld}+\frac{2\pi^2}{d}\sum_{n=1}^{\infty}\frac{n^2}{L^2d^2+n^2\pi^2}e-(M2+\frac{n^2\pi^2}{d^2}+iE)t \tag{4.23}$$

Finally, when $E\to 0$, (4.22) and (4.23) provide the classical results as

$$S_0=\frac{M}{\sinh Md}+\frac{2\pi^2}{d}\sum_{n=1}^{\infty}\frac{n^2(-1)^n}{M^2d^2+n^2\pi^2}e-(M^2+\frac{n^2\pi^2}{d^2})t$$

And

$$S_d=\frac{M\cosh Md}{\sinh Md}+\frac{2\pi^2}{d}\sum_{n=1}^{\infty}\frac{n^2}{M^2d^2+n^2\pi^2}e^{-(M2+\frac{n^2\pi^2}{d^2}+iE)t} \tag{4.24}$$

II. Method of Laplace transforms

The problem, when solved by the method of Laplace transform technique, reduces to solving the transformed equation

$$\frac{d^2\bar{q}}{dz^2}-\frac{(1+s)(s+m^2+iE)}{1+ks}\bar{q}=0 \tag{4.25}$$

subject to the conditions

$$\bar{q}=0\,at\,z=0,\bar{q}=\frac{E_0}{Ts^2}\tanh(\frac{sT}{2})\,\mathrm{at}\,z=d \tag{4.26}$$

The expression for the fluid velocity q(z,t) then obtained as

$$\frac{q(z,t)}{E_0}=\frac{\sinh L_z}{2\sinh Ld}-\frac{4}{\pi^2}\mathrm{Re}\sum_{m=0}^{\infty}\frac{ei\beta mt}{(2m+1)^2}\frac{\sinh L_m z}{\sinh L_m d}$$
$$-\frac{2\pi}{Td^2}\sum_{n=1}^{\infty}n(-1)^n G\sin\frac{n\pi z}{d} \tag{4.27}$$

Where

$$\beta_m=\frac{(2m+1)\pi}{T}L=\sqrt{M^2+iE},L_m=[\frac{(1+i\beta_m)(M^2+i(E+\beta_m))}{1+i\beta_m k}]^{1/2}$$

$G=G_1+G_2$

$$G_j=\frac{e^{s}j^{t}\tanh\frac{s_jT}{2}}{s_j^2\{a_1+\frac{b_1}{(1+ks_j)^2}\}},j=1,2$$

$$a_1=\frac{1}{k},b_1=(M^2+iE-\frac{1}{k})(1-k)$$

However, in the limit $k\to 1$, the result (4.27) reduces to

$$\frac{q(z,t)}{E_0}=\frac{\sinh Lz}{2\sinh Ld}-\frac{4}{\pi^2}\mathrm{Re}\sum_{m=0}^{\infty}\frac{e^{i\beta mt}}{(2m+1)^2}\frac{\sinh L_1 z}{\sinh L_1 d}-$$
$$\frac{2\pi}{Td^2}\sum_{n=1}^{\infty}n(-1)\frac{Exp(-L_2t)\tanh(\frac{L_2T}{2})}{L_2^2}\sin\frac{n\pi z}{d} \tag{4.28}$$

Where

$$L=\sqrt{M^2+iE},L_1=\sqrt{M^2+i(E+\beta_m)}L_2=M^2+\frac{n^2\pi^2}{d^2}+iE$$

The result (4.28) is in excellent agreement with those of author [2] and authors [4] in the limit $E\to 0$.

The corresponding expressions for skin-friction at the plates are

$$\frac{S_0}{E_0}=\frac{L(1-e^{-t})}{2\sinh Ld}-\frac{4}{\pi^2}\mathrm{Re}\sum_{m=0}^{\infty}\frac{L_m}{(2m+1)^2}\frac{1+ik\beta_m}{1+i\beta_m}\times$$
$$\frac{e^{i\beta mt}-e^{-t}}{\sinh_m d}-\frac{2\pi^2}{Td^3}\sum_{n=1}^{\infty}n^2(-1)^n H \tag{4.29}$$

$$\frac{S_d}{E_0}=\frac{L\cosh Ld}{2\sinh Ld}(1-e^{-t})-\frac{4}{\pi^2}\mathrm{Re}\sum_{m=0}^{\infty}\frac{L_m}{(2m+1)^2}\frac{1+ik\beta_m}{1+i\beta_m}\times$$
$$\times(e^{i\beta mt}-e^{-t})\frac{\cosh L_m d}{\sinh L_m d}-\frac{2\pi^2}{Td^3}\sum_{n=1}^{\infty}n^2 H \tag{4.30}$$

where $L=\sqrt{M^2+iE}$, H=H1 + H2;

$$H_j=\frac{1+ks_j}{1+s_j}\frac{(e^{s_jt}-e^{-t})\tanh\frac{s_jT}{2}}{S_j^2\left\{a_1+\frac{b_1}{(1+ks_j)^2}\right\}},j=1,2 \tag{4.31}$$

The result (4.13) obtained by the method of Fourier analysis when compared with (4.27) obtained by the method of Laplace transforms reveals that the two results are exactly identical in respect of their steady and harmonic parts but the transient parts of them are of different forms. In order to show that these two results for the fluid velocity represent the same exact solution of the problem we incorporate the analysis given in the appendix.

Numerical Results and Discussions

The nature of pulses subjected on the upper plate produces both the developing (increasing) and the retarding (decreasing) flows in the fluid. To investigate the effects of various flow parameters on the fluid velocity corresponding to developing flow at t=0.5 and the retarding flow at t=1.75, the exact solution (4.13) is evaluated for the cases E=0, E=0.1 and E=1.0 when T=1. The non-zero values of flow parameters are chosen arbitrarily within their range of validity. The changing nature of the velocity components are incorporated in Figures 2 and 4(a,b) to 6(a,b) for different values of the fluid elasticity k, the magnetic field M and the rotation E.

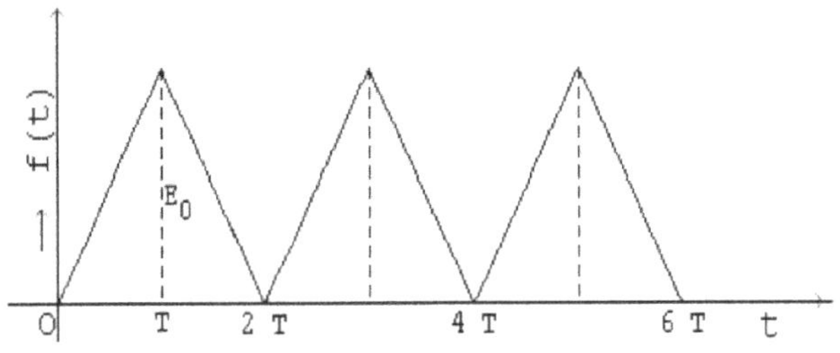

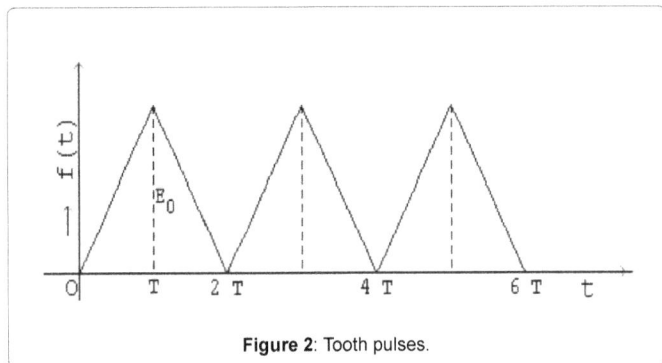

Figure 2: Tooth pulses.

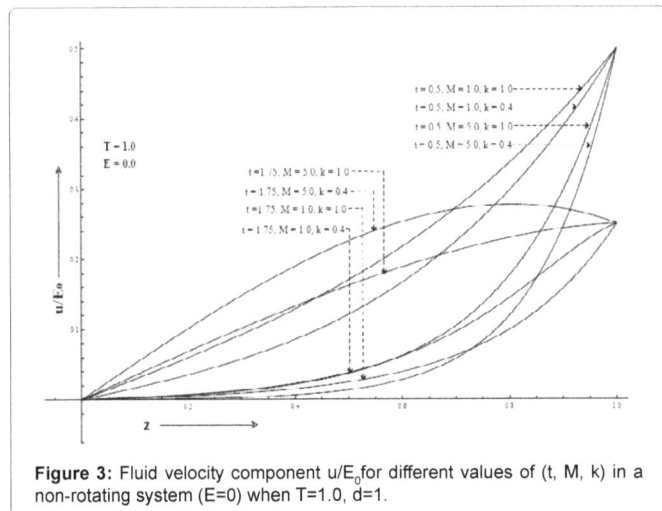

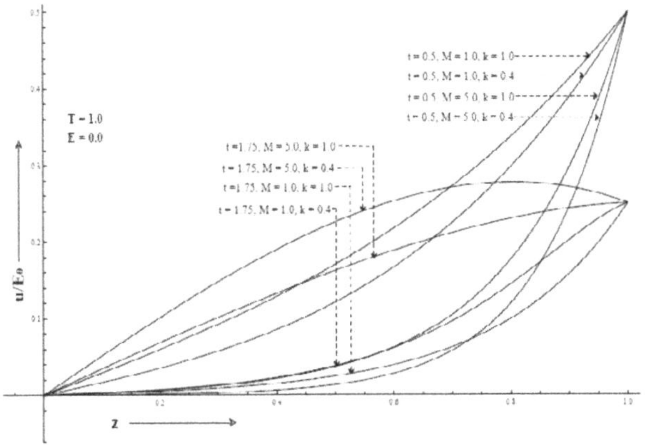

Figure 3: Fluid velocity component $u/E_0$for different values of (t, M, k) in a non-rotating system (E=0) when T=1.0, d=1.

It is observed from Figure 3 that in absence of rotation (E=0) and for fixed values of the magnetic field M, the decrease in elastic parameter k decreases the fluid velocity $u=E_0$ when the flow is developing and increases the same when the flow is retarding. Such a result is expected because the fluid is more and more viscoelastic for smaller and smaller values of k with k=1 representing the clean viscous fluid and the viscoelastic fluids neither grow nor decay as quickly as clean viscous fluids. It is further noticed that the effect of k on $u=E_0$ whether decreasing or increasing enhances with the increase of the magnetic field although the magnetic field has an overall damping effect on the flow for fixed values of k. As a result the effect of k persists even when the steady state is attained (eqn.(4.14)).

However, in presence of rotation, the fluid velocity component $u=E_0$ varies in a manner similar to that of non-rotating case excepting a significant diminution in its magnitude with the increase of rotation for all values of k and M when the flow is developing. On the contrary, exactly reverse effect is found when the flow is retarding. On the other hand, the increase in magnetic field continues to produce its damping effect on all kinds of flow in presence of rotation and the elasticity k of the fluid. The above findings are exhibited in Figures 4a and 5a. More specifically, for fixed values of M,k,T, the decreasing and increasing effects of rotation E on the component of fluid velocity $u=E_0$ respectively for the increasing and decreasing motion is shown in Figure 6a.

It is noticed that the lateral component of fluid velocity $v=E_0$ appears in a contained fluid only in presence of rotation. In the developing flow, the magnitude of $v=E_0$ increases with the increase of rotation E and the elasticity k but decreases with the increase of the magnetic field M. A reverse effect with respect to elastic parameter k is found when the flow is retarding. The variation of fluid velocity component $v=E_0$ for different values of flow parameters and time is illustrated in Figures 4b and 5b while the increasing effects of rotation E for fixed values of (M,k,T) on the magnitude of $v=E_0$ both for the increasing and decreasing motions are presented in Figure 6b.

To investigate the effects of various flow parameters on the components of skin-friction on the plates, the results (4.18) and (4.19)

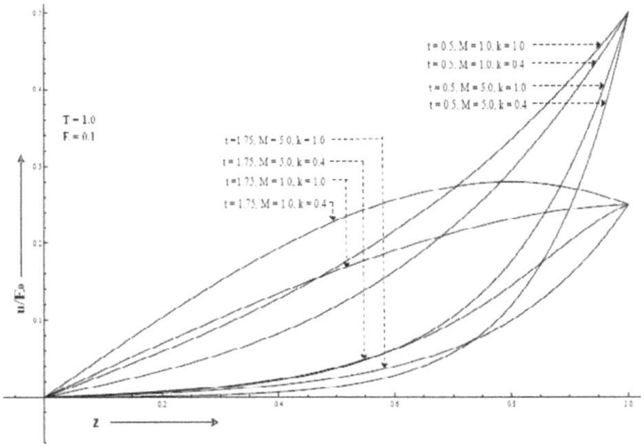

Figure 4a: Fluid velocity component $u/E_0$for different values of (t, M, k) in a rotating system (E=0.1) when T =1.0, d=1.

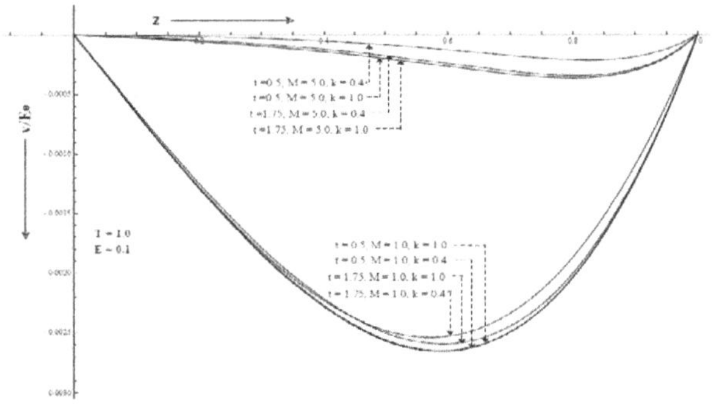

Figure 4b: Fluid velocity component $u/E_0$for different values of (t, M, k) in a rotating system (E=0.1) when T =1.0, d=1.

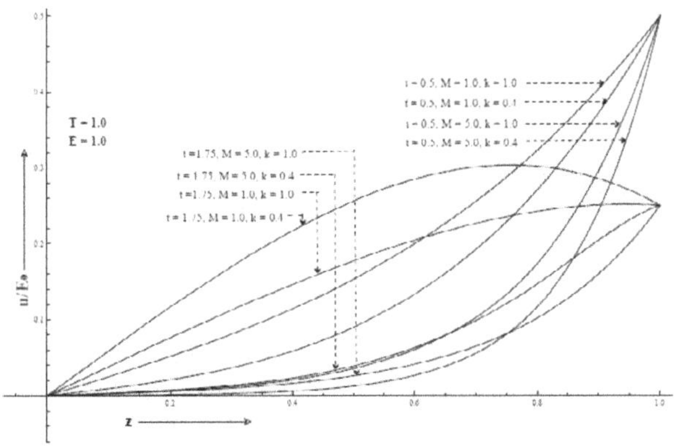

Figure 5a: Fluid velocity component $u/E_0$for different values of (t, M, k) in a rotating system (E=1.0) when T =1.0, d=1.

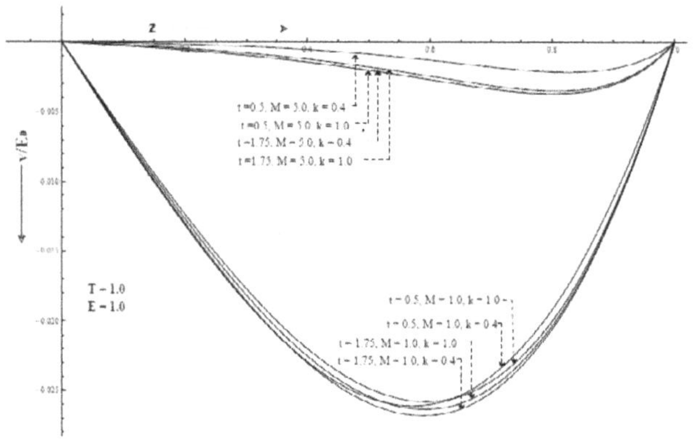

Figure 5b: Fluid velocity component u/E_0 for different values of (t, M, k) in a rotating system (E=1.0) when T =1.0, d=1.

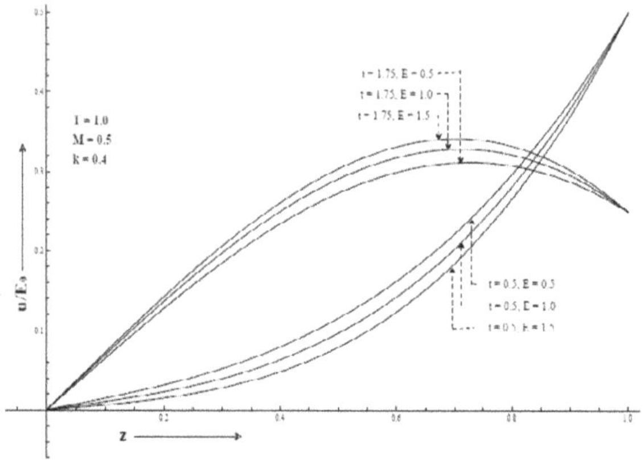

Figure 6a: Fluid velocity component u/E_0 for different values of (E,t) when T=1.0, M=0.5, k=0.4, d=1.

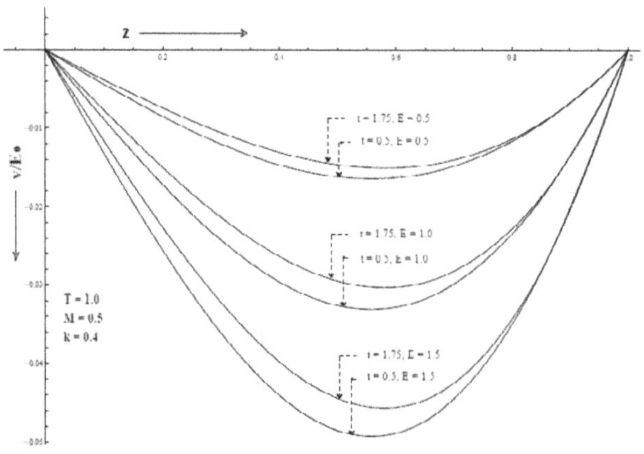

Figure 6b: Fluid velocity component u/E_0 for different values of (E,t) when T =1.0, M= 0.5, k= 0.4, d=1.

are evaluated numerically for the case E=1.0 when T=1.0 and d=1. The results are presented in Figures 7a to 10b. It is observed from Figures 7a and 8a that the longitudinal component of skin-friction on both the plates fluctuate in a manner similar to that of pulses imparted in the fluid. On the lower plate, the longitudinal component of skin-friction decreases with the decrease of the elasticity k for all values of

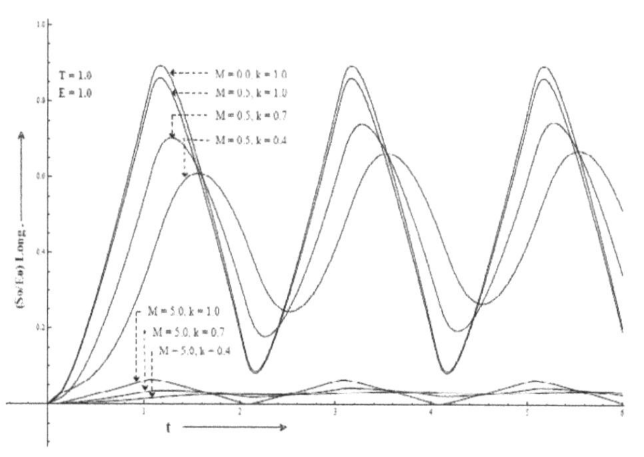

Figure 7a: Longitudinal component of skin friction S_0/E_0 at the lower plate for different values of (M, k) when T =1.0, d=1.

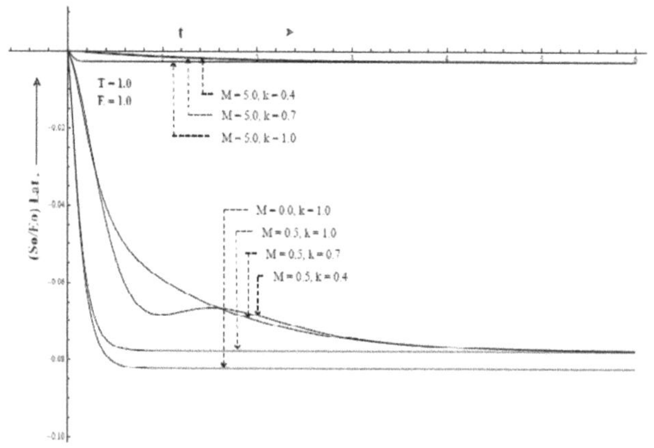

Figure 7b: Lateral component of skin friction S_0/E_0 at the lower plate for different values of (M, k) when T =1.0, d=1.

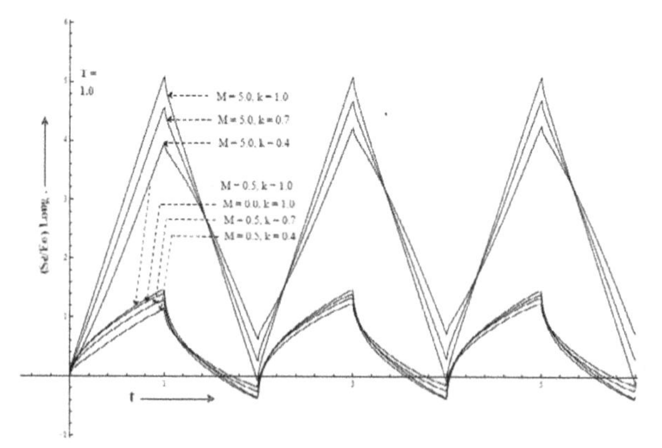

Figure 8a: Longitudinal component of skin friction S_d/E_0 at the upper plate for different values of (M, k) when T =1.0, d=1.

the magnetic field M when the flow is developing but a reverse effect is found when the flow is retarding. On the contrary, the magnitude of the longitudinal component of the skin-friction at the upper plate although decreases with decrease of the elasticity k but increases with the increase of the magnetic field M when the flow is developing and a

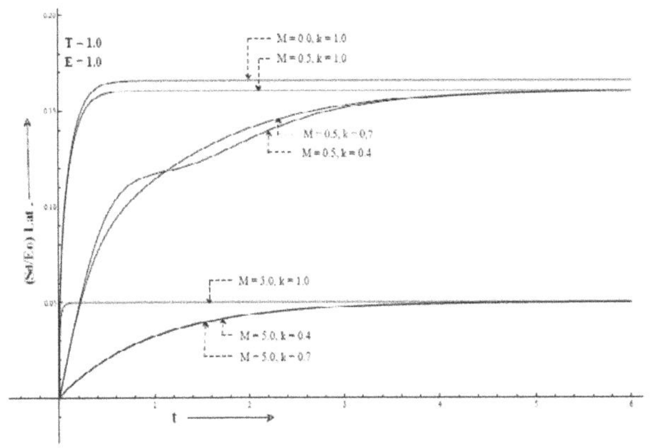

Figure 8b: Lateral component of skin friction S_d/E_0 at the upper plate for different values of (M, k) when T =1.0, d=1.

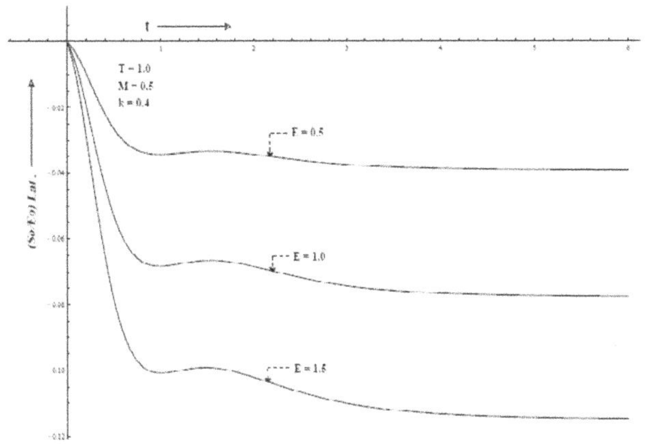

Figure 9b: Variation of Lateral component of skin friction S_d/E_0at the lower plate with rotation (E) for fixed values of (T, M, k) and d=1.

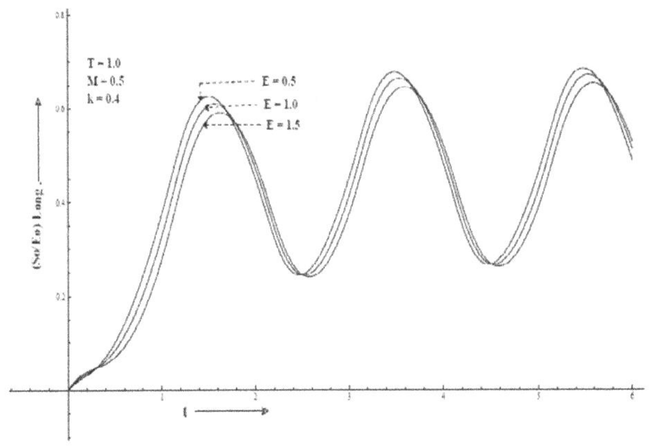

Figure 9a: Variation of Longitudinal component of skin friction S_d/E_0 at the lower plate with rotation (E) for fixed values of (T, M, k) and d=1.

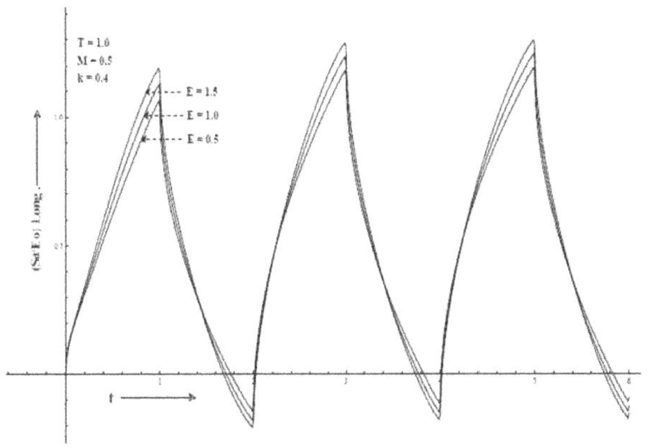

Figure 10a: Variation of Longitudinal component of skin friction S_d/E_0at the upper plate with rotation (E) for fixed values of (T, M, k) and d=1.

reverse effect with respect to the elasticity k is found for all values of the magnetic field M when the flow is retarding. The curves corresponding to M=0.0 and k=1.0 in Figures 7a and 8a represent the quantitative response of the longitudinal component of skin-friction respectively on the lower and the upper plate for the case of classical viscous fluid.

It is further noticed from Figure 9a that for fixed values of M,k,T, the longitudinal components of the skin-friction at the lower plate decrease with the increase of rotation E when the flow is developing but increase with E when the flow is retarding. However, a reverse effect is found at the upper plate which is evident from Figure 10a. On the other hand, the magnitude of lateral components of skin-frictions on both the plates increase with the increase of the elastic parameter k for all values of the magnetic field M as evidenced from Figures 7b and 8b. However, the lateral component of skin-frictions are negative at the lower plate and positive at the upper plate for all values of M and k. Finally, for fixed values of (M,k,T) the magnitude of the lateral component skin-frictions on both the plates increases with the increase of rotation E as exhibited in Figures 9b and 10b.

Conclusion

An analysis concerning the hydromagnetic channel flow of an Olroyd-B fluid induced by tooth pulses in a rotating system has been

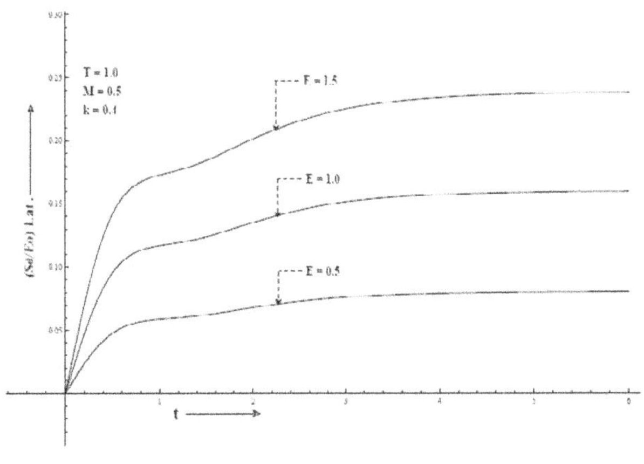

Figure 10b: Variation of Lateral component of skin friction S_d/E_0at the upper plate with rotation (E) for fixed values of (T, M, k) and d=1.

presented in this paper. Both the method of Fourier analysis and the method of Laplace transforms are adopted to arrive at the solution of the problem separately. It is shown that both the methods provide the same exact solution of the problem. The quantitative analysis of the results are incorporated following the solutions obtained by the method of Fourier analysis not due to compulsion. One can also use the solutions obtained by the method of Laplace transforms to explain the results quantitatively. The present analysis seems to be useful in various branches of science and technology with particular reference to geophysical and physiological fluid flow situations.

References

1. Chakraborty A, Ray J (1980) Unsteady magnetohydrodynamic Couette flow between two plates when one of the plates is subjected to random pulses. J Phys Soc Jpn 48: 1361-1364.
2. Makar MN (1987) Magnetohydrodynamic flow between two plates when one ˜of the plates is sujected to tooth pulses. Acta Phys Pol A 71: 995.
3. Bestman AR, Njoku FI (1988) On hydromagnetic channel flow induced by tooth pulses, Preprint IC/88/10, Miramare-Trieste.
4. Ghosh AK, Debnath L (1996) On hydromagnetic pulsatile flow of a two-phase fluid. ZAMM 76: 121-123.
5. Datta N, Dalal DC, Misra SK (1993) Unstesdy heat transfer to pulsatile flow of a dusty viscous incompressible fluid in a channel. Int J Heat Mass Transfer 36: 1783-1788.
6. Datta N, Dalal DC (1995) Pulsatile flow and heat transfer of a dusty fluid through an infinitely long annular pipe. Int. Multiphase Flow 21: 515-528.
7. Hayat T, Siddiqui AM, Asgar S (2001) Some simple flows of an Oldroyd-B fluid. Int. J Engg Sci 39: 135-147.
8. Asgar S, Parveen S, Hanif S, Siddiqui AM, Hayat T (2003) Hall effects on the unsteady hydromagnetic flows of an Oldroyd-B fluid. Int J Engg Sci 41: 609-619.
9. Hayat T, Nadeem S, Asghar S (2004) Hydromagnetic Couette flow of an Oldroyd-B fluid in a rotating system. Int J Engg Sci 42: 65-78.
10. Loper ED, Benton RE (1970) On the spin-up of an electrically conducting fluid Part 2, Hydromagnetic spin-up between infinite flat insulating plates. J Fluid Mech 43: 785-799.
11. Hide R, Roberts PH (1961) Origin of the mean geomagnetic field in Physics and Chemistry of the Earth. (Ed 4), Pergamon Press, New York.
12. Dieke HR (1970) Internal rotation of the Sun. In: Goldberg L(ed) Annual Reviews of Astronomy and Astrophysics. (Ed 8), Annual Reviews Inc.
13. EL-Shehawey EF, Elbarbary EME, Afifi NAS, Elshahed M (2000) MHD flow of an Elastico-Viscous fluid under periodic body acceleration. Int J Math and Math Sci 23: 795-799.
14. Oldroyd JG (1950) On the formulation of the rheological equations of state. Proc Roy Soc London A200: 523.
15. Soundalgekar VM (1967) On the flow of an electrically conducting, incompressible fluid near an accelerated plate in the presence of a parallel plate, under transverse magnetic field. Proc. Indian Acad Sci A 65: 179-187.

The Studying of Random Cauchy Convection Diffusion Models under Mean Square and Mean Fourth Calculus

Sohalya MA*, Yassena MT and Elbaza IM

Faculty of Science, Department of Mathematics, Mansoura University, Egypt

Abstract

The random partial differential equations have a wide range of physical, chemical, and biological applications. The finite difference method offers an attractively simple approximations for these equations. In this paper, the finite difference technique is performed in order to find an approximation solutions for the linear one dimensional convection-diffusion equation with random variable coefficient. We study the consistency and stability of the finite difference scheme under mean square sense. A statistical measure such as mean for the numerical approximation, and the exact solution based on different statistical distributions is computed.

Keywords: Random convection-diffusion equation; Finite difference method; Mean square calculus; Mean fourth calculus

Introduction

The convection-diffusion equation is a parabolic partial differential equation combining the diffusion equation and the advection equation, which describes physical phenomena where particles or energy (or other physical quantities) are transferred inside a physical system due to two processes: diffusion and convection. In its simplest form (when the diffusion coefficient and the convection velocity are constant and there are no sources or sinks).

We can see that the convection-diffusion model in a membrane containing pores or channels lined with positive fixed charges acts as a barrier between intracellular and extracellular compartments filled with electrolyte solutions. The external salt concentration is greater than the internal concentration, thus making it possible to associate the action of the salt solution with sodium. The reason for choosing positive fixed charges in the channels is that this assumption leads to a conductance increase with membrane depolarization. A potential difference E applied across the membrane creates a convectional flow (i.e., bulk flow) with the linear velocity (volume flow per unit of membrane area) by the process of electro osmosis, in the presence of fixed charges, whose density is considered to be relatively low. A pressure difference *P* across the membrane may also be present to influence the volume flow [1-3].

The pollutants solute transport from a source through a medium of air or water is described by a partial differential equation of parabolic type derived on the principle of conservation of mass, and is known as advection-diffusion equation (ADE). In one-dimension it contains two coefficients, one represents the diffusion parameter and the second represents the velocity of the advection of the medium like air or water. In case of porous medium, like aquifer, velocity satisfies the Darcy law and in non-porous medium, like air it satisfies the laminar conditions. The dispersive property differs from pollutant to pollutant.

In water, a pollutant may enter the groundwater zone directly to a landfill site from an industrial site such as nuclear power plants, chemical industries, construction industries etc., and mathematical modelling of the transport of salinity, pollutants and suspended matter in shallow waters involves the numerical solution of a convection-diffusion equation when a pollutant on the surface of a narrow channel. The main purpose in the Pollutants Transport Model is to describe the evolution of the concentration of the pollutant. There are three types for pollutant water surface waters, groundwater, point-source pollution, non point-source pollution and transponder pollution.

Many forms of atmospheric pollution affect human health and the environment at levels from local to global. These contaminants are emitted from diverse sources, and some of them react together to form new compounds in the air. Industrialized nations have made important progress toward controlling some pollutants in recent decades, but air quality is much worse in many developing countries, and global circulation patterns can transport some types of pollution rapidly around the world [4-8]. To complete this model we need to assign the physically relevant boundary and the initial condition. There are three types of boundary conditions:

- The Dirichlet type in which the concentration of the pollutant is prescribed on the boundary.
- The Neumann type flux condition in which the concentration flux normal to the boundary is prescribed.
- The mixed type in which the concentration between the boundary and outside medium.

The initial conditions to be prescribed are generally expressed in terms of background concentration. Although precise background concentration is normally not available, one can consider arbitrary functional form in terms of spatial coordinates.

The performance of this paper is studying the mean square consistency, stability and convergence for one scheme of finite difference method in solving the following linear random convection diffusion equation

$$\begin{cases} u_t + \beta u_x = \alpha u_{xx}, & t \in [0,\infty), x \in R, \\ u(x,0) = u_0(x). \end{cases} \tag{1}$$

***Corresponding author:** Sohalya MA, Faculty of Science, Department of Mathematics, Mansoura University, Egypt
E-mail: m_stat2000@yahoo.com; mtyassen@yahoo.com
islamelbaz88@gmail.com

where α is a constant, β is a random variable, t is the time variable, x is the space coordinate and u_t, u_x denote the derivatives with respect to t and x, respectively. Additionally, $u_0(x)$ is a deterministic initial data function. The convection-diffusion equation is one of the most popular equations in physics. Usually, it arises often e.g., in pollutant transport, to describe time and space variations in the motion of particle. Natural phenomena laws are usually enough to derive this equation. Random models can be also used to derive this equation, with the worthy advantage that they provide us with informations about the statistical properties for activity of particle [9,10].

The deterministic convection-diffusion equation reflects two transport mechanisms, the convection and the diffusion [11]. The convection is a kind of heat transfer, it takes place in liquids and gases only because liquids and gases have a physical moving. Additionally, the convection transfers the large mass of particles from a hot part of a fluid rises to a cooler part sinks [12,13]. The diffusion happens when,a single particle of a fluid moving from a higher concentration area to a lower concentration area [14]. For example, in Thongmoon and McKibbin [15], the author has dealt with the deterministic case of this problem numerically by using cubic splines and two standard finite difference schemes. In the same context, the exponential B-spline functions are used for the Galerkin numerical solution of the advection-diffusion equation, also, the extended B-spline are used for the Collection numerical solution of the advection-diffusion equation [16,17], respectively. On the other hand, the random convection diffusion equation happens when the concentration field be under uncertain inputs arises from random flow (velocity) transport or with source (forcing) term. Bishehniasar and Soheili [18], a compact finite difference approximation and semi-Millstream scheme are used in solving the one dimensional advection-diffusion equation with white noise term. Stochastic explicit finite difference methods are discussed for the one dimensional advection-diffusion equation of Ito type [19]. Also, it is clear in Wan et al. [20], the two dimensional advection-diffusion equation of Ito type is solved by using spectral element method.

Our problem (1) states that at a particular location the rate of change of fish numbers with respect to time is determined by the fish's population dynamics and the fluxes of fish to or from that location by means of advection (the velocity terms) and turbulent diffusion (the diffusivity terms). This equation usually has been applied to the transport of non living entities such as pollutants or salt, although its application to the transport of biota such as plankton is becoming more common.

The relative contribution of each term in the problem (1) must be known if the drift migration itself is to be understood, and each term may have biological and physical components. For the physical process of transport the advective terms represent displacement due to the average currents while the diffusivity terms express the dispersal of fish by ocean turbulence. The computations are relative to the velocity at some selected depth, and if the velocity at that depth is not zero, the calculated current velocity is not a true "over-the ground" speed. Geotropic computations neglect currents driven by wind and other frictional forces, yet these currents may be responsible for the majority of the transport, if the larvae occupy the surface layer, so we can take the velocity as a random variable. Even if the currents are well known, larvae undergoing what are nominally drift migrations may nonetheless act in some way to modify their transport and add a biological component to the velocity vectors in the advection-diffusion equation. Because currents may vary substantially with depth, especially if they are produced by the wind or tides, an important way the fish may modify its drift is by controlling its vertical position [21-23]. All provided examples in which fish larvae that were concentrated in surface waters were transported by (wind-induced) currents.

We can develop this model as a random Cauchy problem for advection diffusion model, if we talk about the drift migration of Fish from position to another by random velocity and random diffusivity along a farm in x direction with unbounded spatial domain also, without forces.

In our work, we focus on the convection-diffusion equation including one random variables since, for instance, $u(x,t)$ denotes the concentration of pollutant at x point and t time. β Refers to the advection velocity (wind speed is random variable) in x direction and α denotes the diffusivity coefficient (diffusion of particle is a constant). It is worth to point out in this paper that, the difficulties is in proving the consistency and stability for the scheme we use under mean square sense, where the solution of the problem depends on the involved random variable in the equation.

Our paper has been partitioned as follows, the next section presents some concepts of mean square sense and functional analysis in $L_2(\Omega)$ space. Section 3, studies the finite difference method for the random convection-diffusion equation. Moreover, discussing how to prove the consistency and stability in mean square for our scheme. Section 4, studies numerical approximation with its statistical mean and standard deviation by introducing a numerical example. Section 5,6 are devoted to conclusion and references, respectively.

Preliminaries

In this section, we present some definitions and some important inequalities that we will use in this paper. A real random variable X defined on the probability space $(\Omega,\mathcal{F},P)$ and satisfying the property that $E\left[|X|^p\right]<\infty$, is called p-order random variable ($p-r.v$) where, $p \geq 1$ nd $E[\ ]$ denotes the expected value operator. If $X\in L_p(\Omega)$, then the L_p norm is defined as $\|X\|_p=\left[E\left[|X|^p\right]\right]^{\frac{1}{p}}$.

Proposition 1: A sequence $\{X_n,n>0\}\in L_2(\Omega)$ is mean square convergent to a random variable $X\in L_p(\Omega)$ if $\lim_{n\to\infty} E\left[|X_n-X|^2\right]=0$.

Some Important Inequalities

• Schwarz's Inequality.

$$E[|XY|]\leqslant[E[X^2]E[Y^2]]^{\frac{1}{2}}\ [24].$$

• Hölder's Inequality

$$E[|XY|]\leqslant[E[X^n]]^{\frac{1}{n}}[E[Y^m]]^{\frac{1}{m}},$$

where $n,m>1,\frac{1}{n}+\frac{1}{m}=1$ [25].

If $X,Y\in L_p(\Omega)$, $q\leq 1$ we have

$$\|XY\|_q\leqslant\|X\|_{2q}\|Y\|_{2q}.$$

• Minkowski's Inequality If $1\leq p<\infty$ and $X,Y\in L_p(\Omega)$, then $X,Y\in L_p(\Omega)$, and $[E[|X+Y|^p]]^{\frac{1}{p}}\leqslant[E[|X|^p]]^{\frac{1}{p}}+[E[|Y|^p]]^{\frac{1}{p}}$.

• Lyapunov's inequality For $1\leq r<s<\infty$, then we have $[E[|X|^r]]^{\frac{1}{r}}\leqslant[E[|X|^s]]^{\frac{1}{s}}$.

Random Finite Difference Scheme

Firstly, in order to apply the finite difference technique to find the approximation solutions for our problem (1), we will discretize the space and the time by finite increasing sequences as follows, the grid points for the space as, $a=x_0<x_1<x_2<x_3<\ldots<x_k=b$. Also, the time points as, $0=t_0<t_1<t_2<t_3<\ldots<t_n=\infty$. Let we define the grid cells for the space to be $\Delta t=(x_k-x_{k-1})$ for $k\geq 1$ and also, define the time steps to be $\Delta t=(t_n-t_{n-1})$ for $t\geq 1$. Consider $u_k^n=u(k\Delta x,n\Delta t)$ approximates the exact solution for the problem (1) as, $u(x,t)$ at the point $(k\Delta x,n\Delta t)$. To formulate the difference scheme according to the problem (1), we must replace the first and second derivative in (1) by difference formulas as follows:

- First-order forward finite difference approximation to u_t

$$u_t(k\Delta x,n\Delta t)\approx\frac{u_k^{n+1}-u_k^n}{\Delta t}$$

- First-order forward finite difference approximation to u_x

$$u_x(k\Delta x,n\Delta t)\approx\frac{u_{k+1}^n-u_k^n}{\Delta x}$$

- Second-order centered finite difference approximation to u_{xx}

$$u_{xx}(k\Delta x,n\Delta t)\approx\frac{u_{k+1}^n-2u_k^n+u_{k-1}^n}{(\Delta x)^2}$$

By substituting in (1), we get the random difference scheme

$$\begin{cases} u_k^{n+1}=(1+r\beta\Delta x-2r\alpha)u_k^n+(r\alpha-r\beta\Delta x)u_{k+1}^n+r\alpha u_{k-1}^n \\ u_k^0=u_0(k\Delta x)=u_0(x_k) \\ r=\dfrac{\Delta t}{(\Delta x)^2},t_n=n\Delta tandx_k=k\Delta x \end{cases} \tag{2}$$

Consistency of RFDS (2)

A random finite difference scheme (RFDS) $\mathrm{L}_k^n\mathrm{u}_k^n=\mathrm{G}_k^n$ that approximating the random partial differential equation (RPDE) $Lv=G$ is consistent in mean square at time $t=(n+1)\Delta t$, if for any smooth function $\Phi=\Phi(x,t)$, we have in mean square

$$E\left[\left|(L\Phi-G)_k^n-(L_k^n\Phi(k\Delta x,n\Delta t)-G_k^n)\right|^2\right]\to 0, \tag{3}$$

$\Delta t\to 0$, $\Delta x\to 0$ *and* $(k\Delta x,n\Delta t)\to(x,t)$.

Theorem 1: The RFDS (2) that according to the problem (1) is mean square consistent such that $t\to 0$, $\Delta x\to 0$ and $(k\Delta x,n\Delta t)\to(x,t)$.

Proof

$$L(\Phi)_k^n=\frac{\Phi(k\Delta x,(n+1)\Delta t)-\Phi(k\Delta x,n\Delta t)}{\Delta t}+\beta\frac{\Phi((k+1)\Delta x,n\Delta t)-\Phi(k\Delta x,n\Delta t)}{\Delta x}-\alpha\int_{n\Delta t}^{(n+1)\Delta t}\Phi_{xx}(k\Delta x,s)ds,$$

$$L_k^n\Phi(k\Delta x,n\Delta t)=\frac{\Phi(k\Delta x,(n+1)\Delta t)-\Phi(k\Delta x,n\Delta t)}{\Delta t}+\beta\frac{\Phi((k+1)\Delta x,n\Delta t)-\Phi(k\Delta x,n\Delta t)}{\Delta x}-\alpha\frac{\Phi((k+1)\Delta x,n\Delta t)-2\Phi(k\Delta x,n\Delta t)+\Phi((k-1)\Delta x,n\Delta t)}{(\Delta x)^2}.$$

Then,

$$\frac{\Phi((k+1)\Delta x,n\Delta t)-2\Phi(k\Delta x,n\Delta t)+\Phi((k-1)\Delta x,n\Delta t)}{(\Delta x)^2}=\frac{\partial^2\Phi(k\Delta x,n\Delta t)}{\partial x^2}+\mathcal{O}((\Delta x)^2).$$

From Taylor expansion, the second derivative

$$\frac{\Phi((k+1)\Delta x,n\Delta t)-2\Phi(k\Delta x,n\Delta t)+\Phi((k-1)\Delta x,n\Delta t)}{(\Delta x)^2}=\frac{\partial^2\Phi(k\Delta x,n\Delta t)}{\partial x^2}+\mathcal{O}((\Delta x)^2).$$

Then, we have

$$E\left[|L(\Phi)_k^n-L_k^n(\Phi)|^2\right]=E[|\alpha\frac{\partial^2\Phi(k\Delta x,n\Delta t)}{\partial x^2}+\mathcal{O}((\Delta x)^2)-\alpha\int_{n\Delta t}^{(n+1)\Delta t}\Phi_{xx}(k\Delta x,s)ds|^2].$$

As $\Delta t\to 0$, $\Delta x\to 0$ and $(k\Delta x,n\Delta t)\to(x,t)$,

$$E\left[\left|(L\Phi-G)_k^n-(L_k^n\Phi(k\Delta x,n\Delta t)-G_k^n)\right|^2\right]\to 0.$$

Hence, the RFDS (2) is mean square consistent as Δx, $\Delta t\to 0$ and $(k\Delta x,n\Delta t)\to(x,t)$

Stability of RFDS (2)

A random difference scheme $\mathrm{L}_k^n\mathrm{u}_k^n=\mathrm{G}_k^n$ that approximating RPDE $Lv=G$ is mean square stable, if there exist some positive constants ε, δ, non-negative constants η, ξ and u^0 is an initial data such that

$$\mathrm{E}\left[\left|\mathrm{u}^{n+1}\right|^2\right]\leq\eta e^{\xi t}\mathrm{E}\left[\left|\mathrm{u}^0\right|^2\right], \tag{4}$$

for all, $t=(n+1)\Delta t$, $0<\Delta x\leq\varepsilon$, $0<\Delta t\leq$.

Theorem 2 The RFDS equation (2) that according to the problem (1) is mean square stable under the conditions

1. $\Delta t\to 0$, Δx is fixed.
2. β Has positive random distribution.
3. $E[|\beta|^4]$ (4th order random variable).
4. u^0 is a deterministic function.

Proof

Since,

$$u_k^{n+1}=(1+r\beta\Delta x-2r\alpha)u_k^n+(r\alpha-r\beta\Delta x)u_{k+1}^n+r\alpha u_{k-1}^n,$$

$$E[|u_k^{n+1}|^2]=E[|(1+r\beta\Delta x-2r\alpha)u_k^n+(r\alpha-r\beta\Delta x)u_{k+1}^n+r\alpha u_{k-1}^n|^2].$$

Also since,

$$E[|X+Y|^2]\leqslant[\sqrt{E(|X|^2)}+\sqrt{E(|Y|^2)}]^2,$$

Then,

$$\begin{aligned}E[|u_k^{n+1}|^2]\leqslant\ & E[|u_k^n+r\beta(\Delta x)u_k^n-2r\alpha u_k^n|^2]\\&+2E[|r\alpha u_k^nu_{k+1}^n-r\beta(\Delta x)u_k^nu_{k+1}^n+3r^2\alpha\beta(\Delta x)u_k^nu_{k+1}^n-r^2\beta^2(\Delta x)^2u_k^nu_{k+1}^n-2r^2\alpha^2u_k^nu_{k+1}^n|]\\&+2E[|r\alpha u_k^nu_{k-1}^n+r^2\alpha\beta(\Delta x)u_k^nu_{k-1}^n-2r^2\alpha^2u_k^nu_{k-1}^n|]\\&+2E[|r^2\alpha^2u_{k+1}^nu_{k-1}^n-r^2\alpha\beta(\Delta x)u_{k+1}^nu_{k-1}^n|]\\&+E[|r\alpha u_{k+1}^n|^2]+2(E[|r\alpha u_{k+1}^n|^2])^{1/2}(E[|r\beta(\Delta x)u_{k+1}^n|^2])^{1/2}\\&+E[|r\beta(\Delta x)u_{k+1}^n|^2]+E[|r\alpha u_{k-1}^n|^2].\end{aligned}$$

Since,

$$E[|X+Y+Z|]\leqslant E[|X|]+E[|Y|]+E[|Z|],$$

Then,

$$\begin{aligned}E[|u_k^{n+1}|^2]\leqslant\ & E[|u_k^n|^2]+2(E[|u_k^n|^2])^{1/2}(E[|r\beta(\Delta x)u_k^n-2r\alpha u_k^n|^2])^{1/2}+E[|r\beta(\Delta x)u_k^n|^2]\\&+2(E[|r\beta(\Delta x)u_k^n|^2])^{1/2}(E[|2r\alpha u_k^n|^2])^{1/2}+E[|2r\alpha u_k^n|^2]+2E[|r\alpha u_k^nu_{k+1}^n|]\\&+2E[|r\beta(\Delta x)u_k^nu_{k+1}^n|]+6E[|r^2\alpha\beta(\Delta x)u_k^nu_{k+1}^n|]+2E[|r^2\beta^2(\Delta x)^2u_k^nu_{k+1}^n|]+4E[|r^2\alpha^2u_k^nu_{k+1}^n|]\\&+2E[|r\alpha u_k^nu_{k-1}^n|]+2E[|r^2\alpha\beta(\Delta x)u_k^nu_{k-1}^n|]+4E[|r^2\alpha^2u_k^nu_{k-1}^n|]\\&+2E[|r^2\alpha^2u_{k+1}^nu_{k-1}^n|]+2E[|r^2\alpha\beta(\Delta x)u_{k+1}^nu_{k-1}^n|]+E[|r\alpha u_{k+1}^n|^2]\\&+2(E[|r\alpha u_{k+1}^n|^2])^{1/2}(E[|r\beta(\Delta x)u_{k+1}^n|^2])^{1/2}+E[|r\beta(\Delta x)u_{k+1}^n|^2]+E[|r\alpha u_{k-1}^n|^2].\end{aligned}$$

Since,

$$\|X\|_2=[E(X^2)]^{1/2}\quad\forall X\in L_2(\Omega),$$

Then,

$$\begin{aligned}\|u_k^{n+1}\|_2^2\leqslant\ &\|u_k^n\|_2^2+2\|u_k^n\|_2\|(r\beta(\Delta x)-2r\alpha)u_k^n\|_2+\|r\beta(\Delta x)u_k^n\|_2^2+2\|r\beta(\Delta x)u_k^n\|_2\|2r\alpha)u_k^n\|_2\\&+\|2r\alpha)u_k^n\|_2^2+2r\alpha\|u_k^n\|_2\|u_{k+1}^n\|_2+2r(\Delta x)\|\beta\|_2\|u_k^n\|_2\|u_{k+1}^n\|_2+6r^2\alpha(\Delta x)\|\beta\|_2\|u_k^n\|_2\|u_{k+1}^n\|_2\\&+2r^2(\Delta x)^2\|\beta^2\|_2\|u_k^n\|_2\|u_{k+1}^n\|_2+4r^2\alpha^2\|u_k^n\|_2\|u_{k+1}^n\|_2+2r\alpha\|u_k^n\|_2\|u_{k-1}^n\|_2\\&+2r^2\alpha(\Delta x)\|\beta\|_2\|u_k^n\|_2\|u_{k-1}^n\|_2+4r^2\alpha^2\|u_k^n\|_2\|u_{k-1}^n\|_2+2r^2\alpha^2\|u_{k+1}^n\|_2\|u_{k-1}^n\|_2\\&+2r^2\alpha(\Delta x)\|\beta\|_2\|u_{k+1}^n\|_2\|u_{k-1}^n\|_2+r^2\alpha^2\|u_{k+1}^n\|_2^2+2r^2\alpha(\Delta x)\|u_{k+1}^n\|_2\|\beta u_{k+1}^n\|_2\\&+r^2(\Delta x)^2\|\beta u_{k+1}^n\|_2^2+r^2\alpha^2\|u_{k-1}^n\|_2^2.\end{aligned}$$

Since,

$$\|XY\|_2\leqslant\|X\|_4\|Y\|_4\quad\forall X,Y\in L_4(\Omega),$$

Then,

$$\begin{aligned}\| u_k^{n+1} \|_2^2 \leqslant & \| u_k^n \|_4^2 + 2r(\Delta x) \| u_k^n \|_4^2 \| \beta \|_4 + 4r\alpha \| u_k^n \|_4^2 + r^2(\Delta x)^2 \| \beta \|_4^2 \| u_k^n \|_4^2 \\ & + 4r^2\alpha(\Delta x) \| \beta \|_4 \| u_k^n \|_4^2 + r^2\alpha^2 \| u_k^n \|_4^2 + 2r\alpha \| u_k^n \|_4 \| u_k^{n+1} \|_4 + 2r(\Delta x) \| \beta \|_4 \| u_k^n \|_4 \| u_k^{n+1} \|_4 \\ & + 6r^2\alpha(\Delta x) \| \beta \|_4 \| u_k^n \|_4 \| u_{k+1}^n \|_4 + 2r^2(\Delta x)^2 \| \beta^2 \|_4 \| u_k^n \|_4 \| u_{k+1}^n \|_4 + 4r^2\alpha^2 \| u_k^n \|_4 \| u_{k+1}^n \|_4 \\ & + 2r\alpha \| u_k^n \|_4 \| u_{k-1}^n \|_4 + 2r^2\alpha(\Delta x) \| \beta \|_4 \| u_k^n \|_4 \| u_{k-1}^n \|_4 + 4r^2\alpha^2 \| u_k^n \|_4 \| u_{k-1}^n \|_4 \\ & + 2r^2\alpha^2 \| u_{k+1}^n \|_4 \| u_{k-1}^n \|_4 + 2r^2\alpha(\Delta x) \| \beta \|_4 \| u_{k+1}^n \|_4 \| u_{k-1}^n \|_4 + r^2\alpha^2 \| u_{k+1}^n \|_4^2 + 2r^2\alpha(\Delta x) \| \beta \|_4 \| u_{k+1}^n \|_4^2 \\ & + r^2(\Delta x)^2 \| \beta \|_4^2 \| u_{k+1}^n \|_4^2 + r^2\alpha^2 \| u_{k-1}^n \|_4^2 .\end{aligned}$$

Then,

$$\begin{aligned}\sup_k \| u_k^{n+1} \|_2^2 \leqslant & \sup_k \| u_k^n \|_4^2 + 2r(\Delta x) \| \beta \|_4 \sup_k \| u_k^n \|_4^2 + 4r\alpha \sup_k \| u_k^n \|_4^2 + r^2(\Delta x)^2 \| \beta \|_4^2 \sup_k \| u_k^n \|_4^2 \\ & + 4r^2\alpha(\Delta x) \| \beta \|_4 \sup_k \| u_k^n \|_4^2 + 4r^2\alpha^2 \sup_k \| u_k^n \|_4^2 + 2r\alpha \sup_k \| u_k^n \|_4^2 + 2r(\Delta x) \| \beta \|_4 \sup_k \| u_k^n \|_4^2 \\ & + 6r^2\alpha(\Delta x) \| \beta \|_4 \sup_k \| u_k^n \|_4^2 + 2r^2(\Delta x)^2 \| \beta^2 \|_4 \sup_k \| u_k^n \|_4^2 + 4r^2\alpha^2 \sup_k \| u_k^n \|_4^2 \\ & + 2r\alpha \sup_k \| u_k^n \|_4^2 + 2r^2\alpha(\Delta x) \| \beta \|_4 \sup_k \| u_k^n \|_4^2 \| + 4r^2\alpha^2 \sup_k \| u_k^n \|_4^2 \\ & + 2r^2\alpha^2 \sup_k \| u_k^n \|_4^2 + 2r^2\alpha(\Delta x) \| \beta \|_4 \sup_k \| u_k^n \|_4^2 + r^2\alpha^2 \sup_k \| u_k^n \|_4^2 + 2r^2\alpha(\Delta x) \| \beta \|_4 \sup_k \| u_k^n \|_4^2 \\ & + r^2(\Delta x)^2 \| \beta \|_4^2 \sup_k \| u_k^n \|_4^2 + r^2\alpha^2 \sup_k \| u_k^n \|_4^2 .\end{aligned}$$

Then,

$$\begin{aligned}\sup_k \| u_k^{n+1} \|_2^2 & \leqslant [1 + 8r\alpha + 16r^2\alpha^2 + 4r(\Delta x) \| \beta \|_4 + 4r^2(\Delta x)^2 \| \beta \|_4^2 + 16r^2\alpha(\Delta x) \| \beta \|_4] \sup_k \| u_k^n \|_4^2 \\ & \vdots \\ & \leqslant [1 + 8r\alpha + 16r^2\alpha^2 + 4r(\Delta x) \| \beta \|_4 + 4r^2(\Delta x)^2 \| \beta \|_4^2 + 16r^2\alpha(\Delta x) \| \beta \|_4]^{n+1} \sup_k \| u_k^0 \|_4^2 .\end{aligned}$$

Taking

$$8r\alpha + 16r^2\alpha^2 + 4r(\Delta x) \| \beta \|_4 + 4r^2(\Delta x)^2 \| \beta \|_4^2 + 16r^2\alpha(\Delta x) \| \beta \|_4 \leqslant \lambda^2(\Delta t).$$

Then,

$$\sup_k \| u_k^{n+1} \|_2^2 \leqslant (1+\lambda^2 \Delta t)^{n+1} \sup_k \| u_k^0 \|_4^2 .$$

Since, u^0 is a deterministic function,

$$\sup_k \| u_k^{n+1} \|_2^2 \leqslant (1+\lambda^2 \Delta t)^{n+1} \sup_k \| u^0 \|^2 ,$$

and $\Delta t = \frac{t}{n+1}$, we have

$$E[| u_k^{n+1} |^2] \leqslant (1+\frac{\lambda^2 t}{n+1})^{n+1} E[| u^0 |^2] \leqslant e^{\lambda^2 t} E[| u^0 |^2].$$

Hence, the RFDS (2) is mean square stable with $\eta=1$, $\xi=\lambda^2$

Case Studies

We can get random Cauchy problem for the convection diffusion equation in a membrane model if the electro osmosis transportations randomly in the presence of fixed charges acts as a barrier between intracellular component at the initial position and extracellular component at any position along the unbounded domain.

Also, We can get it in a pollutant model if the water streams speed transports randomly, the pollutant diffuse in a deterministic case along the unbounded spatial domain when a pollutant on the surface of an open channel flow, the depth of the water is not constant and also, the turbulent diffusion in the surface of water, channel shapes, channel slop and the nature of the channel material. Turbulence is difficult to define exactly, nevertheless, there are several important characteristics that all turbulent flows possess. These characteristics include unpredictability, rapid diffusivity, high levels of fluctuating velocity, and dissipation of kinetic energy. The velocity at a point in a turbulent flow will appear to an observer to be random or chaotic. The velocity is unpredictable in the sense that knowing the instantaneous velocity at some instant of time is insufficient to predict the velocity a short time later.

In the random case of this model:

The random convection term: The flux is determined by the water stream only but if the velocity of the water stream, the bulk of pollutant that is driven by the stream, without deformation or expansion is random also.

In the deterministic diffusion term:

The pollutant expands from higher concentration regions to lower ones but if the diffusion of pollutant is deterministic, the concentration of pollutant in any region will be deterministic as in this work.

We can also get it in a pollutant model if the wind streams speed transports randomly, the pollutant diffuse in a deterministic case along the unbounded spatial domain when a pollutant of type nitrogen dioxide, ozone, and total suspended particulate matter and carbon monoxide. We can write for example the membrane model as follows.

Let the concentration $u(x,t)$ inside a pore in the membrane, according to our problem, is given by the following partial differential equation

$$\begin{cases} u_t + \beta u_x = \alpha u_{xx}, & t \geq 0, x \in R, \\ u(x,0) = e^{-x^2}, & x \in R \end{cases} \qquad (5)$$

Where X is the unbounded space co-ordinate perpendicular to the membrane surfaces, t is the time, the diffusion coefficient is a constant and the advection velocity is a random variable and with an exponential initial condition.

The exact solution

$$u(x,t) = \frac{1}{\sqrt{1+4\alpha t}} e^{-\frac{(x-\beta t)^2}{1+4\alpha t}} . \qquad (6)$$

The numerical solution

The random finite difference scheme for this problem is

$$u_k^{n+1} = (1 + r\beta\Delta x - 2r\alpha)u_k^n + (r\alpha - r\beta\Delta x)u_{k+1}^n + r\alpha u_{k-1}^n,$$

$$u_k^0 = u_0(k\Delta x) = u_0(x_k) = e^{-(k\Delta x)^2},$$

where $r = \frac{\Delta t}{(\Delta x)^2}$, $t_n = n\Delta t$ and $x_k = k\Delta x$.

From the RFDS (2)

$$u_1^1 = (1 + r\beta\Delta x - 2r\alpha)u_1^0 + (r\alpha - r\beta\Delta x)u_2^0 + r\alpha u_0^0.$$

$$u_1^1 = (1 + r\beta\Delta x - 2r\alpha)e^{-(\Delta x)^2} + (r\alpha - r\beta\Delta x)e^{-(2\Delta x)^2} + r\alpha.$$

$$u_1^2 = (1 + r\beta\Delta x - 2r\alpha)u_1^1 + (r\alpha - r\beta\Delta x)u_2^1 + r\alpha u_0^1.$$

$$\begin{aligned} u_1^2 = & [(1 + r\beta\Delta x - 2r\alpha)^2 + 2r\alpha(r\alpha - r\beta\Delta x) + (r\alpha)^2]e^{-(\Delta x)^2} \\ & + 2[(r\alpha - r\beta\Delta x)(1 + r\beta\Delta x - 2r\alpha)]e^{-(2\Delta x)^2} + (r\alpha - r\beta\Delta x)^2 e^{-(3\Delta x)^2} \\ & + 2r\alpha(1 + r\beta\Delta x - 2r\alpha). \end{aligned}$$

Verification of the convergence of mean

The ß~Binomial (1.0,0.5), α=1 was explained in Table 1 and ß~ Beta distribution (1.0,2.0), α=1 was explained in Table 2.

The ß~Binomial (1.0,0.5), α=1 was explained in Table 3 and ß~ Beta distribution (1.0,2.0), α=1 was explained in Table 4.

λ^2 for stability

The ß~Binomial distribution (1.0,0.5), α=1 was explained in Table 5 and ß~Beta distribution (1.0,2.0) was explained Table 6. The ß~Exponential (0.5), α=1 was explained in Table 7.

Conclusion

In this work, we have shown that the random finite difference method can be used to obtain the approximation solution stochastic process for the random Cauchy advection diffusion model in one

k	n	x_k	t_n	$E(u(x,t)_{x_k,t_n})$	$E\|u_k^n\|$	$\frac{\|E(u(x,t)_{x_k,t_n}) - E\|u_k^n\|\|}{E(u(x,t)_{x_k,t_n})}$
1	1	0.5	0.01	0.7747527612	0.7753211115	0.00073358925
1	2	0.5	0.005	0.7747527612	0.7752913846	0.00069521972

Table 1: ß~Binomial (1.0,0.5), α=1.

k	n	x_k	t_n	$E(u(x,t)_{x_k,t_n})$	$E\|u_k^n\|$	$\frac{\|E(u(x,t)_{x_k,t_n}) - E\|u_k^n\|\|}{E(u(x,t)_{x_k,t_n})}$
1	1	0.5	0.01	0.7735224945	0.7739513739	0.00055444980
1	2	0.5	0.005	0.7735224945	0.7739421289	0.00054249799

Table 2: ß~ Beta distribution (1.0,2.0), α=1.

k	n	x_k	t_n	$E(u(x,t)_{x_k,t_n})$	$E\|u_k^n\|$	$\frac{\|E(u(x,t)_{x_k,t_n}) - E\|u_k^n\|\|}{E(u(x,t)_{x_k,t_n})}$
1	1	0.5	0.01	0.7768138752	0.7770609473	0.00031805829
1	2	0.5	0.005	0.776813852	0.7770535156	0.00030849140

Table 3: ß~Binomial (1.0,0.5), α=1.

k	n	x_k	t_n	$E(u(x,t)_{x_k,t_n})$	$E\|u_k^n\|$	$\frac{\|E(u(x,t)_{x_k,t_n}) - E\|u_k^n\|\|}{E(u(x,t)_{x_k,t_n})}$
1	1	0.5	0.01	0.7761821248	0.7763760781	0.00024988117
1	2	0.5	0.0025	0.7761821248	0.7763737678	0.00024690468

Table 4: ß~Beta distribution (1.0,2.0), α=1.

Δt	0.1	0.05	0.025	0.005	0.0001	0.000001
λ^2	64.168761	19.578549	6.6628165	0.83233003	0.01419545	0.00014145

Table 5: ß~Binomial distribution (1.0,0.5), α=1 and Δx=0.25.

Δt	0.1	0.05	0.025	0.005	0.0001	0.000001
λ^2	59.9415369	18.388637	6.2987860	0.79647193	0.01365934	0.00013613

Table 6: ß~Beta distribution (1.0,2.0), α=1 and Δx=0.25.

Δt	0.1	0.05	0.025	0.005	0.0001	0.000001
λ^2	67.646950	20.554410	6.95993389	0.86122520	0.01462376	0.00014571

Table 7: ß~Exponential (0.5), α=1 and Δx=0.25.

dimension. The study has been conducted through proving the consistency and stability for the random finite difference scheme we used in this paper under mean square calculus. The convection velocity coefficient must be bounded according to the stability condition. The usefulness of applying our technique to deal with this class of problems has been shown through a number of illustrative examples.

References

1. Teorell T (1971) Handbook of Sensory Physiology. Springer-Verlag, Berlin, Heidelberg and New York 1: 10.
2. Prashanth P (2008) Asymptotic and particle methods in nonlinear transport phenomena, membrane separations and drop dynamics. ProQuest.
3. Rubinow SI (1975) Introduction to mathematical biology. Courier Corporation, USA.
4. Ayyoubzadeh SA, Zahiri A (2003) New envelope sections method to study hydraulics of compound varying river channels using a depth-averaged 2d model. Int J Eng Sci 7: 7.
5. Ayyoubzadeh SA (1997) Hydraulic Aspects of Straight-Compound Channel flow and Bed Load Sediment Transport. The University of Birmingham, UK.
6. Blackadar AK (1997) Turbulence and diffusion in the atmosphere: lectures in Environmental Sciences. SpringerVerlag, p: 185.
7. Hanna SR (1982) Applications in air pollution modeling, Atmospheric Turbulence and Air Pollution Modeling. FTM Nieuwstadt and Reidel-Dordrecht H, Cap, p: 7.
8. Salsa S (2015) Partial differential equations in action: from modelling to theory. Springer, USA.
9. Mohammed WW, Sohaly MA, El-Bassiouny AH, Elnagar KA (2014). Mean square convergent finite difference scheme for stochastic parabolic pdes. American Journal of Computational Mathematics 4: 280.
10. Yassen MT, Sohaly MA, Elbaz I (2016) Random Crank-Nicolson Scheme for Random Heat Equation in Mean Square Sense. American Journal of Computational Mathematics 6: 66-73.
11. Ding H, Zhang Y (2011) A new numerical method for solving convection-diffusion equations. Non linear Mathematics for Uncertainty and its Applications 100: 463-470.
12. Carslaw HS (1921) Introduction to the mathematical theory of the conduction of heat in solids. London 2: 268.
13. Lienhard JH (2013) A heat transfer textbook. Courier Corporation.

14. Crank J (1975) Diffusion in a plane sheet. The Mathematics of Diffusion 2: 44-68.

15. Thongmoon M, McKibbin R (2006) A comparison of some numerical methods for the advection diffusion equation. Research Letters in the Information and Mathematical Sciences 10: 49-62.

16. Gorgulu MZ, Dag I (2016) Galerkin method for the numerical solution of the advection-diffusion equation by using exponential b-splines. arXiv Preprint arXiv: 1604.04267.

17. Irk D, Dağ İ, Tombul M (2015) Extended cubic B-spline solution of the advection-diffusion equation. KSCE Journal of Civil Engineering 19: 929-934.

18. Bishehniasar M, Soheili AR (2013) Approximation of stochastic advection-diffusion equation using compact finite difference technique. Iranian Journal of Science and Technology 37: 327-333.

19. Soheili AR, Arezoomandan M (2013) Approximation of stochastic advection diffusion equations with stochastic alternating direction explicit methods. Applications of Mathematics 58: 439-471.

20. Wan X, Xiu D, Karniadakis GE (2004) Stochastic solutions for the two-dimensional advection-diffusion equation. SIAM Journal on Scientific Computing 26: 578-590.

21. Nelson WR, Ingham MC, Schaaf WE (1977) Larval transport and year-class strength of atlantic menhaden, brevoortia tyrannus. US National Marine Fisheries Service Fishery Bulletin 75: 23-41.

22. Powles H (1981) Distribution and movements of neustonic young estuarine dependent (mugil spp., pomotomus saltatrix) and estuarine independent (coryphaena spp.) fishes off the southeastern united states. Rapports et Proces-Verbaux des Reunions Conseil International pour l'Exploration de la Mer 178.

23. Sette OE (1943) Biology of the atlantic mackerel (saomber saambrus) of north america. US Fish and Wildlife Service Fishery Bulletin, p: 50.

24. Soong TT (1973) Random Differential Equations in Science and Engineering. Academic Press, New York.

25. Villafuerte L, Cortés JC (2013) Solving random differential equations by means of differential transform methods. Advances in Dynamical Systems and Applications 2: 413-425.

Simulations of Three-dimensional Second Grade Fluid Flow Analysis in Converging-Diverging Nozzle

Ahmed S* and Bano Z
Department of Mathematics, Sukkur Institute of Business Administration, Sukkur, 62500, Pakistan

Abstract

Very little work have been reported on the computational methods for non-Newtonian fluid turbulent flows. This is due to nonlinear system of elliptic partial differential equations that makes the solution very difficult. Another concern is the meshing of flow domain that accounts for complications in solving most problems. In this work we present standard Galerkin finite element method for the steady incompressible non Newtonian fluid flow in a converging-diverging nozzle. The flow is fully three-dimensional with turbulent characteristics. The main aim is to study the velocity and shear stress profiles. Shock profiles are noted for specific pressure boundary conditions. Moreover the plotted results shows variations of velocity components, pressure and turbulence energy dissipation. It is observed that for non-Newtonian fluid flow the mass flow rate and pressure loss effects are significant in diverging part of nozzle.

Keywords: Computational; Newtonian fluid flows; Unsteady; Three-dimensional

Introduction

Analyzing flow patterns in a converging-diverging nozzle has been one of interesting topic in computational fluid dynamics. There are numerous applications of this flow phenomena in aerospace and engineering sciences. Such process are difficult to handle analytically due to complex mathematical model associated to the flow and ensuing instabilities carried by flow parameters. Looking back to the history Jaffery [1] and Hamel [2], in their studies considered the converging diverging channel steady two dimensional Newtonian fluid flow. They observed quiet interesting results by treating Navier-Stokes equations with similarity transforms. Further developments were presented in Schlichtinh [3] and Batchelor [4] based on the boundary layer approximations. Makinde [5] examined the incompressible Newtonian fluid flow by incorporation of linearly diverging symmetrical channel. Recently, Zarqa et al. [6] performed approximate analytical analysis using Adomian decomposition method for a channel with variable diverging ratio. It is evident by several studies that the mechanism of such flows is characterized due the fact that shocks instabilities are produced within the flow domain. Mehta et al. [7] formulated finite element method for solid rocket motor. He worked on the transient flow with axisymmetric and anisotropic model on throat nozzle. Later on Nikhi et al. [8] presented the computational analysis for flow in da Laval nozzles. Apart from the flow analysis, there is significant contribution by researchers on the shape optimization analysis for nozzle flows [9].

Although many fluids in real world applications carry Newtonian behavior, there is a number of fluids that exhibit non- Newtonian behavior. The applications ranges from industrial materials like clay coatings, doilling muds, suspensions, oils, greases, polymer melts to biological materials. Because of the great diversity in the physical structure of non-Newtonian fluids there are many constitutive models that represent different types of non-Newtonian fluids. Among these, the second grade fluids exhibits a linear relation between the stress and the Rivilin-Ericksen tensor, its square and the second Rivilin- Ericksen tensor [10]. The resulting mathematical model is more complex than that for Newtonian fluid. To respond to this concern we opt for computational techniques to investigate the flow parameters. The theoretical foundations of finite element method for non-Newtonian fluid flow is presented by Adelia [11]. They presented finite element method for elliptic type system of partial differential equations that represents steady state incompressible second grade fluid flow. Numerical testing of the proposed method has not reported yet. It is due to computational complexity of the mathematical model. A large number of solution techniques have been developed to solve such non-linear problems, these include approximate analytical techniques such as perturbation analysis, Homotopy analysis, adomian decomposition and numerical techniques such as finite difference method, finite volume method, finite element methods, operator splitting methods and spectral methods. The finite difference method is strongly influenced by meshing quality problem while finite volume method requires additional treatment for analyzing different flow regimes. The flexibility of meshing and segregated formulation of pressure and velocity in terms of the basic functions are the most desirable features of finite element method that makes it suitable and superior for investigating non-Newtonian fluid flow in converging-diverging channel. Another concern is the convergence and stability of numerical methods. Finite element method is not just suitable in terms of efficiency but also there is a sound stability analysis before implementation. This work addresses the application of finite element method for second grade fluid flow in a converging-diverging channel. We present two different results for two different set of primitive variables in order to see changes in flow regimes.

The remaining article is organized as follows. Section 2 presents basic equations for the flow problem and corresponding weak formulation. The numerical methodology and simulation set up

***Corresponding author:** Ahmed S, Department of Mathematics, Sukkur Institute of Business Administration, Sukkur, 62500, Pakistan
E-mail: dr.sidrah@iba-suk.edu.pk

are included in section 3 along with results and discussions. Finally concluding remarks are summarized in section 4.

Basic Equations

The mathematical model represents the three dimensional flow of incompressible second grade fluid. It is to be noted here that the fluid material is incompressible, i.e. density is kept constant. The body forces and thermal effects are ignored.

Hence the dynamics are represented as:

$$\text{div } V=0 \quad (1)$$

$$\rho \dot{V} = \text{div } T, \quad (2)$$

where ρ is density, V stands for three dimensional velocity vector and T is the stress tensor defines as:

$$T=-pI+\mu A_1+\alpha_2 A_2+\alpha_1 A_1^2 \quad (3)$$

With A_1 and A_2 be the first and second Rivilin-Ericksen tensors, μ is the coefficient of viscosity and α_1 and α_2 are the normal stress moduli. Thus the equations that controls the transport of momentum with the fluid flow reads as:

$$0=-\nabla p+\mu\Delta^2 V+(\alpha_1+\alpha_2)divA_1^2+\alpha_1\left[\Delta^2 V_t+\Delta^2(\nabla\times V)\times V+\left|\nabla\left(V\bullet\nabla^2+\frac{1}{4}|A_1^2|\right)\right)\right] \quad (4)$$

The coupled equations 1 and 4 constitute a nonlinear elliptic system of PDEs. Moreover, boundary conditions are associated in order to have a well-posed problem. A rigorous weak formulation is given next. The flow domain Ω is three- dimensional and ∂Ω represents boundary of the domain. In this case it comprises of an inlet, outlet and curved surface of nozzle. The domain is discretized into non-overlapping cells. In order to derive weak formulation for the system (1-3) together with appropriate boundary condition, multiply the equations by a test function and then integrate over a finite element cell. The test function is taken in a Sobolove space with the property that it vanishes on the prescribed boundary. The solution is sought in terms of trial functions which satisfies the prescribed boundary conditions. For the continuity equation we get:

$$\int_\Omega q.divVd\Omega=0 \quad (5)$$

Defining $\vec{u}=(u_1,u_2,u_3)^T$ with u_1, u_2 and u_3 be the separate test functions for momentum equations, it follows:

$$0=\int_\Omega\left[-\nabla p+\mu\Delta^2 V+(\alpha_1+\alpha_2)divA_1^2\right]\cdot\vec{u}\,d\Omega \quad (6)$$

Consider the second integral on right hand side:

$$\int_\Omega(\mu\Delta^2 V)\cdot\vec{u}\,d\Omega=\int_\Omega\sigma\cdot\nabla\vec{u}\,d\Omega-\int_\Gamma(u_n\sigma^{nm}+u_t\sigma_{nt})d\Gamma \quad (7)$$

Where ΓΩ denotes the boundary of Ω, un and ut are the normal and tangential components of $\vec{u}$ respectively. Next is to define the basic functions $\Psi_i(\vec{x})$ and $\phi_i(\vec{x})$ for pressure and velocity components respectively. The approximation for V and p will be:

$$p_h=\sum_{i=1}^{m}p_j\Psi j(\vec{x}) \quad (8)$$

and

$$\vec{V}_h=\sum_{j=1}^{n}v_{1j}\varphi_{j1}j(\vec{x})+v_{2j}\varphi_{j2}j(\vec{x})+v_{3j}\varphi_{j3}j(\vec{x})=\sum_{j=1}^{\beta n}v_j\varphi_j j(\vec{x}) \quad (9)$$

with

$$\varphi_{j1}=\begin{pmatrix}\varphi_j(\vec{x})\\0\\0\end{pmatrix},\varphi_{j2}=\begin{pmatrix}0\\\varphi_j(\vec{x})\\0\end{pmatrix}\text{ and }\varphi_{j3}=\begin{pmatrix}0\\0\\\varphi_j(\vec{x})\end{pmatrix} \quad (10)$$

Working according to standard Galerkin FEM approach, we obtain:

$$\int_\Omega\Psi_j div\vec{V}_h\,d\Omega=0, j=1:m, \quad (11)$$

$$\int_\Omega[\mu A_1+\alpha_1 A_1+\alpha_2 A_2]\cdot\nabla\varphi_j+\int_\Omega\rho\left(\vec{V}_h\cdot\nabla\varphi_j\right)\bullet\varphi_j d\Omega=0 \quad (12)$$

The tangential and normal components u_t and u_n are taken as $\varphi_j\cdot\vec{t}$ and $\varphi_j\cdot\vec{n}$ respectively using conventional notations for tangential and normal vectors to element. Finally the above discretized set of equations are written as a system of nonlinear algebraic equations as follows:

$$LU=0, \quad (13)$$

$$SU+N(U)-L^Tp=F. \quad (14)$$

Here U stands for the vector of unknowns v_{1j}, v_{2j} and v_{3j}, p denotes the vector of unknowns p_j, N represents the discretized nonlinear convective terms, LU is the discretization of continuity equation and $-L^Tp$ is the discretization of pressure gradient. All the contributions from boundary integrals are transferred to right hand side. In order to handle nonlinear terms we have chosen standard Newton linearization method. So in linearized form the system is written as:

$$SU+N(U^k)U-L^Tp=F, \quad (15)$$

$$LU=0. \quad (16)$$

where U^k refers to the solution from previous iteration.

Numerical Implementation

The implementation for the simulation of model presented in section II are carried within the framework of open source Autodesk. In order to achieve accurate and stable computations of gradients a second order upwinding scheme is used for momentum equations.

Simulation setup

The formulation presented in section II is implemented in a three dimensional converging-diverging nozzle. Figure 1 shows the simulation domain. The nozzle is aligned along x-axis and three dimensional flow is considered. A finite velocity magnitude is provided at the inlet and velocity constant is kept unknown at outlet with a fixed pressure value. No-slip boundary are used for the symmetry walls of channel.

Figure 1: Schematic diagram of converging-diverging nozzle aligned along x-axis.

Figure 2: X-component of velocity vector.

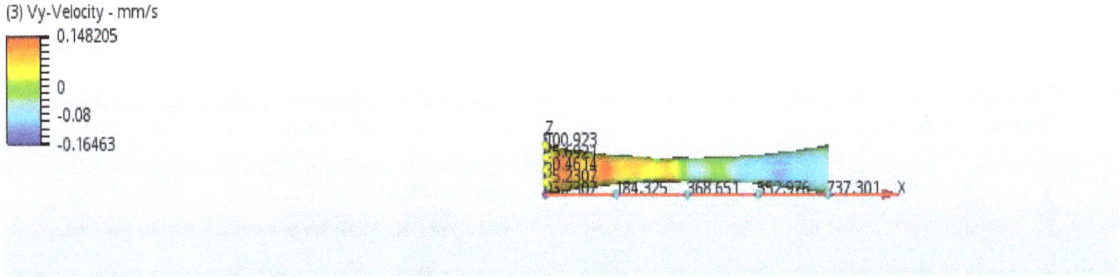

Figure 3: Y-component of velocity vector.

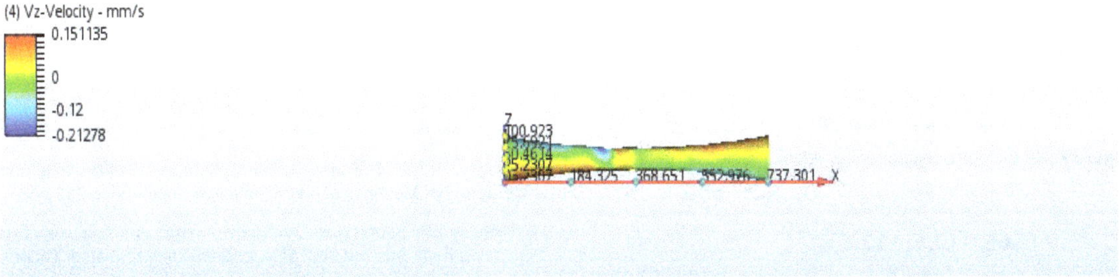

Figure 4: Z-component of velocity vector.

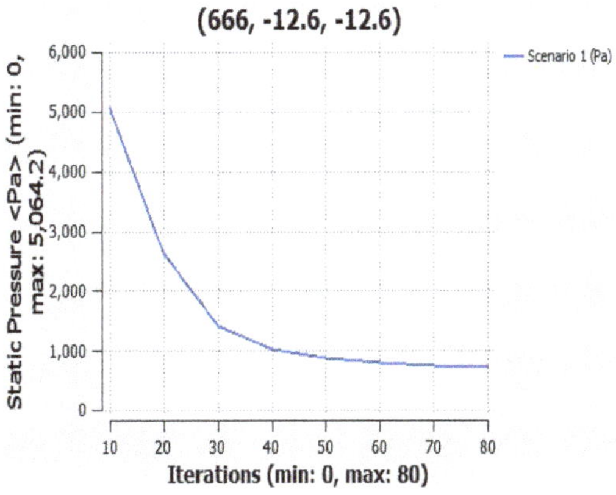

Figure 5: XY Plot for Static Pressure.

Simulation Results

Figures 1-8 give the overview of different flow parameters. These are velocity components in each direction and shear stresses. The power law non-Newtonian model is used with viscosity coefficient equals to 0.0033. The mass flow rate of 5 g/s is applied at the inlet. The flow is turbulent with k-epsilon model and Kappa taken as 0.4.

Conclusion

The steady turbulent second grade fluid flow has been observed by computational method. The flow regime has been bounded by a converging-diverging nozzle. It is observed that flow is fully three-dimensional near converging throat area inlet. But this effect reduces near inlet and outlet boundaries. Moreover shear stresses are maximum

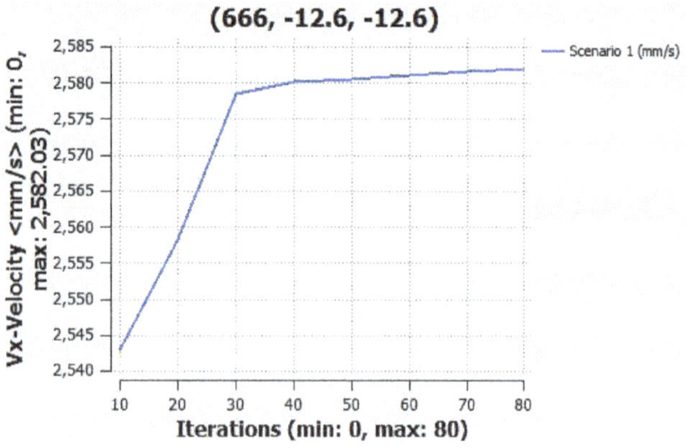

Figure 6: XY plot for x-component of velocity vector.

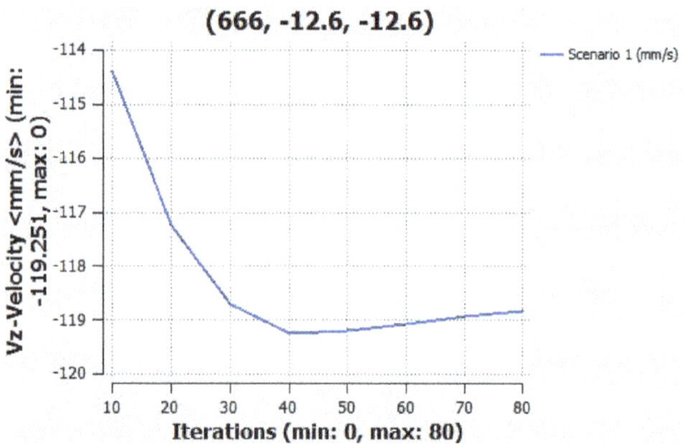

Figure 7: XY plot for z-component of velocity vector.

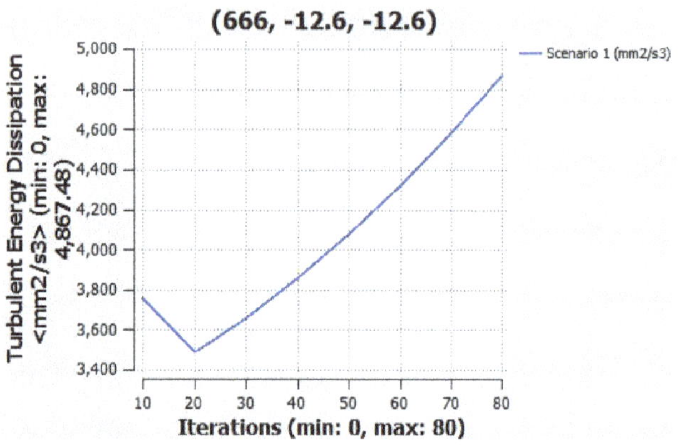

Figure 8: XY plot for turbulence kinetic energy dissipation.

in converging throat area. There is a significant decrease in pressure in diverging regime while turbulence kinetic energy dissipation effects are maximum for this region. Hence it can inferred that there is a relation between pressure loss and mass flow rate.

The future goals are to extend the study for multi physics problems by considering the thermal and magnetic effects. Moreover the nozzle walls can be designed as wavy or irregular to optimize the flow parameters.

Acknowledgments

Sidrah Ahmed and Zarqa Bano acknowledge the financial and technological support from Sukkur Institute of Business Administration for this work.

References

1. Jeffery GB (1915) L The two-dimensional steady motion of a viscous fluid. The London, Edinburgh, and Dublin Philosophical Magazine and Journal of Science 29: 455-465.
2. Hamel G (1917) Spiralförmige Bewegungen zäher Flüssigkeiten. Jahresbericht der Deutschen Mathematiker-Vereinigung 25: 34-60.
3. Schlichting H, Gersten K (2003) Boundary-layer theory. Springer Science and Business Media.
4. Batchelor GK (2000) An introduction to fluid dynamics. Cambridge university press.
5. Makinde OD, Sibanda P (2000) Steady flow in a diverging symmetrical channel: numerical study of bifurcation by analytic continuation. Quaestiones Mathematicae 23: 45-57.
6. Siddiqui AM, Haroon T, Bano Z (2014) Steady 2-D Flow of a Second Grade Fluid in a Symmetrical Diverging Channel of Varying Width. Applied Mathematical Sciences 8: 4675-469.
7. Mehta RC, Suresh K, Iyer RN (1998) Thermal stress analysis of a solid rocket motor nozzle throat insert using finite element method. Indian journal of engineering and materials sciences 5: 271-277.
8. Jagtap R (2014) Theoretical & CFD analysis of de laval nozzle. International Journal of Mechanical and Production Engineering 2: 33-36.
9. Sanjay RP, Subba Rao DV (2015) Design and Analysis of Hot Runner Nozzle Using Fem. International of Mechanical and Industrial Technology 3: 228-241.
10. Walters K (1970) Relation between coleman-noll, rivlin-ericksen, green-rivlin and oldroyd fluids. Zeitschrift für angewandte Mathematik und Physik ZAMP 21: 592-600.
11. Sequeira A, Baía M (1999) A finite element approximation for the steady solution of a second-grade fluid model. Journal of Computational and Applied Mathematics.

The k-out-of-n System Model with Degradation Facility

El-Damcese MA* and Shama MS
Department of Mathematics, Faculty of Science, Tanta University, Tanta, Egypt

Abstract

In this paper, we study the reliability analysis of k-out-of-n system with degradation facility. Let failure rate, degradable rate and repair rate of components are assumed to be exponentially distributed. There are two types of repair. The first is due to failed state. The second is due to degraded state. The expressions of reliability and mean time to system failure are derived with repair and without repair. We used several cases to analyze graphically the effect of various system parameters on the reliability system.

Keywords: Reliability; Mean time to system failure; Simplex system

Notations

n: Number of components in the system.

k: Minimum number of components that must work for the k-out-of-n system to work.

λ_1: The failure rate of the unit.

λ_2: The degradable rate of the unit.

μ_1: The repair rate of failed unit.

μ_2: The repair rate of degraded unit.

$P_{i,j}(t)$: Probability that there are i degradable units and j failed units in the system at time t where i=0,1,2,…, n, j=0,1,2,…, n-k+1.

S: Laplace transform variable.

$P_{i,j}^*(s)$: Laplace transform of $P_{i,j}(t)$.

$P_i(t)$: Probability for i=0, 1, 2 (0→operable (normal) state; 1→ degradable state; 2→ failed state).

$P_i^*(s)$: Laplace transform of $P_i(t)$.

$R_{sim}(t)$: Reliability function of simplex system with degradation.

$MTTF_{sim}$: Mean time to failure of simplex system with degradation.

$R_{tmr}(t)$: Reliability function of TMR system with degradation.

$MTTF_{sim}$: Mean time to failure of TMR system with degradation.

$R_N(K,N)$: Reliability of k-out-of-n system or probability that at least k out of the n components are working (nonrepairable system), where $0 \leq k \leq n$ and both k and n are integers.

$R_N(K,n)$: Reliability of k-out-of-n system or probability that at least k out of the n components are working (repairable system), where $0 \leq k \leq n$ and both k and n are integers.

$MTTF_R(k,n)$: Mean time to failure of a k-out-of-n system.

μ_1: The repair rate of failed unit.

μ_2: The repair rate of degraded unit.

$\mu_{1,n}$: Mean repair rate when there are n failed units in the system.

$\mu_{2,n}$: Mean repair rate when there are n degraded units in the system.

Introduction

The general structure of series and parallel systems: the so-called k-out-of n system. In this type of system, if any combinations of k units out of n independent units work, it guarantees the success of the system. For simplicity, assume that all units are identical. Furthermore, all the units in the system are active. The parallel and series systems are special cases of this system for k=1 and k=n, respectively.

In the past decades; many articles concerning the reliability and availability of standby systems have been published. Among them Galikowsky et al. [1] analyzed the series systems with cold standby components. Wang and Sivazlian [2] examined the reliability characteristics of a multiple-server (M+W) unit system with exponential failure and exponential repair time distributions. Ke et al. [3] studied the reliability measures of a repairable system with warm standby switching failures and reboot delay. Yuge et al. [4] introduced the reliability of a k-out-of-n system with common-cause failures using multivariate exponential distribution. Zhang et al. [5] analyzed the availability and reliability of k-out-of-(M+N): G warm standby systems with two types of failure.

In this paper, we provide a detailed coverage on reliability evaluation of the k-out-of-n systems with degradation. We study the simple model "one unit with one degradable state (simplex)" then we investigate triple modular redundancy (TMR) without repair components. Finally we study the k-out-of-n in details with repair and without repair of components. In addition, we perform numerical results to analyze the effects of the various system parameters on the system reliability.

Non-repairable k-out-of-n System

Simplex system

We get Simplex system when n=k=1

Based on the state transition diagram in Figure 1, we can derive the following differential equations:

***Corresponding author:** El-Damcese MA, Department of Mathematics, Faculty of Science, Tanta University, Tanta, Egypt
E-mail: meldamcese@yahoo.com

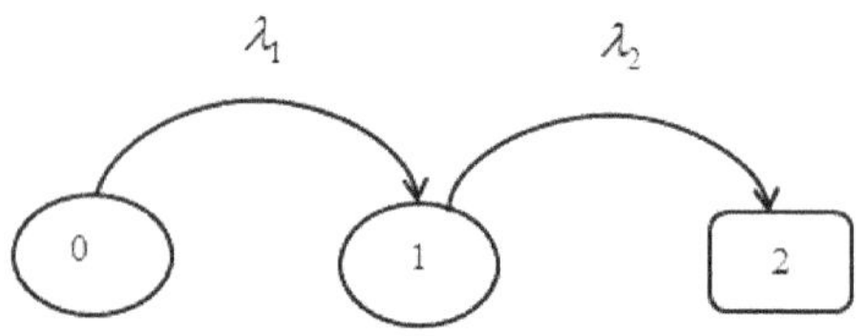

Figure 1: Simplex system with degradation diagram.

$$\frac{dP_0(t)}{dt} = -\lambda_1 P_0(t) \quad (1)$$

$$\frac{dP_1(t)}{dt} = -\lambda_2 P_1(t) + \lambda_1 P_0(t) \quad (2)$$

$$\frac{dP_2(t)}{dt} = -P_2(t) + \lambda_2 P_1(t) \quad (3)$$

Initial conditions:

$$P_i(0) = \begin{cases} 1, & where \quad i = 0 \\ 0, & otherwise \end{cases}$$

Taking the Laplace transform of equations (1-3) and applying initial conditions, we have

(s=λ1) P

$$(s+\lambda_1)\, P_0^*(s) = 1 \quad (4)$$

$$(s+\lambda_2)\, P_1^*(s) - \lambda_1 P_0^*(s) = 0 \quad (5)$$

$$(s)\, P_2^*(s) - \lambda_2 P_1^*(s) = 0 \quad (6)$$

Using the Laplace transform technique, the solutions of $P_i^*(s)$, i=0, 1, 2 are given

$$P_0^*(s) = \frac{1}{(s+\lambda_1)} \quad (7)$$

$$P_1^*(s) = \frac{\lambda_1}{s^2 + s(\lambda_1+\lambda_2) + \lambda_1\lambda_2} \quad (8)$$

$$P_2^*(s) = \frac{\lambda_1\lambda_2}{s\left(s^2 + s(\lambda_1+\lambda_2) + \lambda_1\lambda_2\right)} \quad (9)$$

Inverse Laplace transforms of these equations yield

$$P_0(t) = e^{-\lambda_1 t} \quad (10)$$

$$P_1(t) = \frac{\lambda_1(e^{-\lambda_2 t} - e^{-\lambda_1 t})}{\lambda_1 - \lambda_2} \quad (11)$$

The reliability function of the system can be written as

$$R_{sim}(t) = P_0(t) + P_1(t) = \frac{\lambda_1 e^{-\lambda_2 t} - \lambda_2 e^{-\lambda_1 t}}{\lambda_1 - \lambda_2} \quad (12)$$

The $MTTF$ can be obtained from this equation

$$MTTF_{sim} = \frac{1}{\lambda_1} + \frac{1}{\lambda_2} \quad (13)$$

Triple modular redundancy: TMR

In this scheme, three identical redundant units or modules perform the same task simultaneously with degradable rate. The TMR system only experiences a failure when more than one component fails. In other words, this type of redundancy can tolerate failure of a single component. Figure 2 shows a diagram of the TMR scheme.

From Figure 2, taking LaPlace transforms of the state equations yields:

$$(s+3\lambda_1)P_{0,0}^*(s)=1 \quad (14)$$

$$(s=2\lambda_1)P_{0,1}^*(s) - \lambda_2 P_{1,0}^*(s)=0 \quad (15)$$

$$(s+2\lambda_1+\lambda_2)P_{1,0}^*(s) - 3\lambda_1 P_{0,0}^*(s)=0 \quad (16)$$

$$(s+\lambda_1+\lambda_2)P1,1^*(s) - 2\lambda 1p0,1^*(s)-2\lambda 2P2,0^*(s)=0 \quad (17)$$

$$(s+\lambda_1+2\lambda_2)\, P_{2,0}^*(s) - 2\lambda_1 P_{1,0}^*(s)=0 \quad (18)$$

$$(s+2\lambda_2)P_{2,1}^*(s) - \lambda_1 P_{1,1}^*(s) - 3\lambda_2 P_{3,0}^*(s)=0 \quad (19)$$

$$(s)P_{0,2}^*(s) - \lambda_2 P_{1,1}^*(s)0 \quad (20)$$

$$(s)\, P_{1,2}^*(s) - 2\lambda_2 P_{2,1}^*(s)=0 \quad (21)$$

Solving equations (14-21and taking inverse Laplace transforms of these equations, we get the reliability function of the system

$$R_{tmr}(t) = \sum_{i,j=0}^{i=3,\, j=1} P_{i,j}(t)$$

$$R_{tmr}(t) = \frac{3\left(\lambda_1^2 e^{-2\lambda_2 t} + \lambda_2^2 e^{-2\lambda_1 t} - 2\lambda_1\lambda_2 e^{-(\lambda_1+\lambda_1)t}\right)}{\lambda_1 - \lambda_2}$$

$$-\frac{2\left(\lambda_1^3 e^{-3\lambda_2 t} - \lambda_2^3 e^{-3\lambda_1 t} + 3\lambda_1\lambda_2^2 e^{-(\lambda_1+2\lambda_1)t} - 3\lambda_1^2\lambda_2 e^{-(\lambda_1+2\lambda_2)t}\right)}{\lambda_1 - \lambda_2} \quad (22)$$

As we know the reliability of the triple modular redundancy of a component with one failure rate can be obtained from this equation

$$R_{TMR}(t) = 3R^2 - 2R^3 \quad (23)$$

Where $R=e^{-\lambda t}$ reliability of a component with failure rate λ, from (22) we get

$$R_{tmr}(t) = 3(R_{sim})^2 - 2(R_{sim})^3 \quad (24)$$

$R_{tmr}(t)=3(R_{sim})^2 - 2(R_{sim})^3$

The $MTTF_{tmr}$ can be obtained from this equation

$$MTTF_{tmr} = \int_0^\infty R_{tmr}(t) \quad (25)$$

General system

We will examine a general model for analysis of such systems when they are nonrepairable. When a k-out-of-n system is put into operation, all n components are in good condition. The system is failed when the number of working components goes down below k or the number of

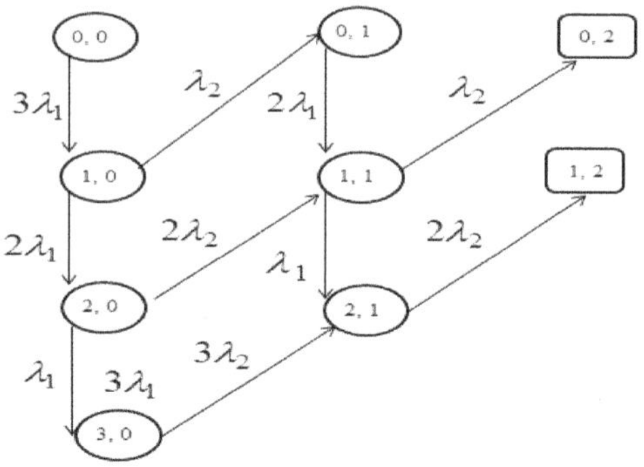

Figure 2: TMR with degradation diagram.

failed components has reached n−k+1. We consider the components in a k-out-of-n system are i.i.d. Let the failure rate and degradable rate occur independently of the states of other units and follow exponential distributions with λ_1, λ_2, respectively.

At time t=0 the system starts operation with no failed units. The Laplace transforms of $P_{i,j}(t)$ are defined by:

$$P^*_{i,j}(s) = \int_0^\infty e^{-st} P_{i,j}(t)\,dt,\ i=0,1,2,\ldots,n \text{ and } j=0,1,2,\ldots,n-k+1.$$

Based on the model descriptions, the system state transition diagram is given in

Figure 3 and it leads to the following Laplace transform expressions for $P_{i,j}^*(s)$:

$$\left(s+n\lambda_1\right)P^*_{0,0}(s)=1 \tag{26}$$

$$\left(s+(n-i)\lambda_1\right)P^*_{0,i}(s)-\lambda_2 P^*_{1,i-1}(s)=0, 1\le i\le n-kc \tag{27}$$

$$\left(s+(n-i)\lambda_1+i\lambda_2\right)P^*_{i,0}(s)-(n-i+1)\lambda_1 P^*_{i-1,0}(s)=0, 1\le i\le n-1 \tag{28}$$

$$\left(s+n\lambda_2\right)P^*_{n,0}(s)-\lambda_1 P^*_{n-1,0}(s)=0 \tag{29}$$

$$\left(s+(n-i)\lambda_2\right)P^*_{n-i,i}(s)-(n-i+1)\lambda_2 P^*_{n-i+1,i-1}(s)-\lambda_1 P^*_{n-i-1,i}(s)=0, 1\le i\le n-k \tag{30}$$

$$\begin{aligned}&\left(s+(n-k-i)\lambda_2\right)P^*_{n-k-i,i}(s)-(n-k-i)\lambda_2 P^*_{n-k-i+1,i-1}(s)\\&-(n-k-1)\lambda_1 P^*_{n-k-i-1,i}(s)=0, 1\le i\le n-k-1\end{aligned} \tag{31}$$

$$\begin{aligned}&\left(s+(n-k-j+i)\lambda_2+(k-i)\lambda_1\right)P^*_{n-k-j+i,j}(s)-(k-i+1)\lambda_1 P^*_{n-k-j+i-1,j}(s)\\&-(n-k-j+i+1)\lambda_2 P^*_{n-k-j+i+1,j-1}(s)=0, 1\le i\le k-1, 1\le j\le n-k\end{aligned} \tag{32}$$

$$\begin{aligned}&\left(s+(i+1)\lambda_2+(n-j-i-1)\lambda_1\right)P^*_{i+1,j}(s)-(i+2)\lambda_2 P^*_{i+2,j-1}(s)\\&-(n-j-i)\lambda_1 P^*_{i,j}(s)=0, 0\le i\le n-k-2, 1\le j\le n-k-i-1\end{aligned} \tag{33}$$

$$\left(s\right)P^*_{i,n-k+1}(s)-(i+1)\lambda_2 P^*_{i+1,n-k}(s)=0, 0\le i\le k-1 \tag{34}$$

From solving equations (26-34) and taking inverse Laplace, we obtain the reliability function as follows:

$$R_N(k,n)=L^{-1}\left(\sum_{\substack{i=n,\\ j=n-k+1}}^{i,j=0} P^*_{i,j}(s)\right)=\left(\sum_{\substack{i=n,\\ j=n-k+1}}^{i,j=0} P_{i,j}(t)\right) \tag{35}$$

$$R_N(k,n)=\sum_{i=K}^{n}\binom{n}{i}(R_{sim})(1-R_{sim})^{n-i} \tag{36}$$

The mean time to failure $MTTF_N(k,n)$ can be obtained from the following relation.

$$MTTF_N(k,n)=\lim_{s\to 0} R_r^*(s)=\lim_{s\to 0}\left\{\sum_{\substack{i=n\\ j=n-k+1}}^{i,j=0} P^*_{i,j}(s)\right\} \tag{37}$$

When we perform a sensitivity analysis for changes in the $RN(k,n)$ resulting from changes in system parameters λ_1 and λ_2 and. By differentiating equation (36) with respect to λ_1 we obtain,

$$\frac{\partial R_N(k,n)}{\partial \lambda_1}=\frac{\partial}{\partial \lambda_1}\left\{\sum_{i+j=0}^{n-k} P_{i,j}(t)\right\}=\left\{\sum_{i+j=0}^{n-k}\frac{\partial}{\partial \lambda_1} P_{i,j}(t)\right\} \tag{38}$$

We use the same procedure to get $\dfrac{\partial R_N(k,n)}{\partial \lambda_2}, \dfrac{\partial R_N(k,n)}{\partial \mu_1}, \dfrac{R_N(k,n)}{\partial \mu_2}$.

We use two cases to study the effect of k and n on system reliability

Case 1: Fix λ_1=0.001, λ_2=0.008, n=3 and choose k=1,2,3.

Case 2: Fix λ_1=0.001, λ_2=0.008, k=2 and choose n=2,3,4.

From Figures 4 and 5 we can be observed that the system reliability increases as k increases or n increases.

Then, we perform a sensitivity analysis with respect to λ_1 and λ_2. In Figure 6 we can easily observe that the biggest impact almost happened at different time and the order of magnitude of the effect is ($\lambda_1>\lambda_1$).

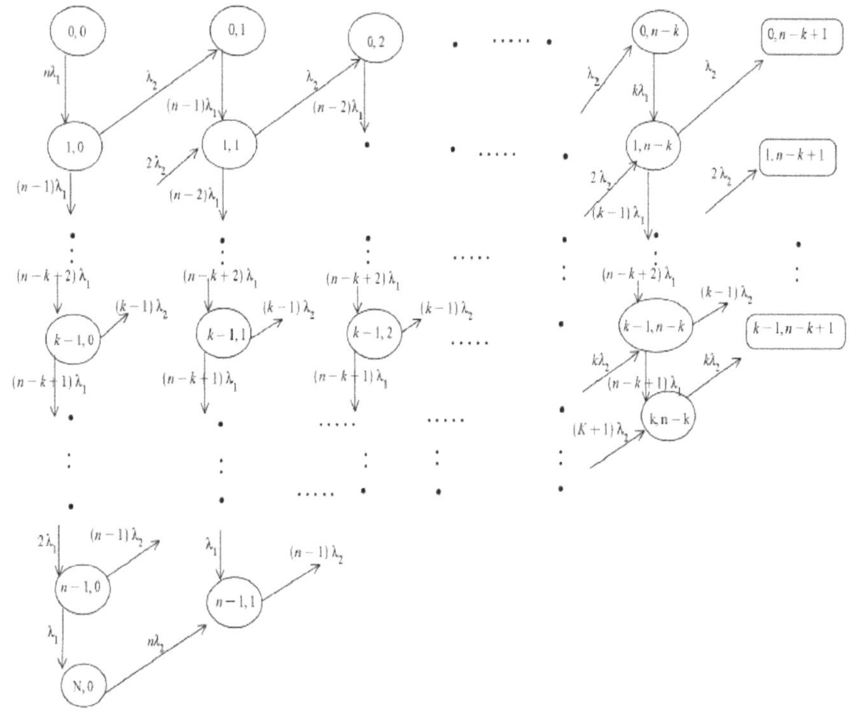

Figure 3: State-transition-rate diagram of non-repairable System with degradation.

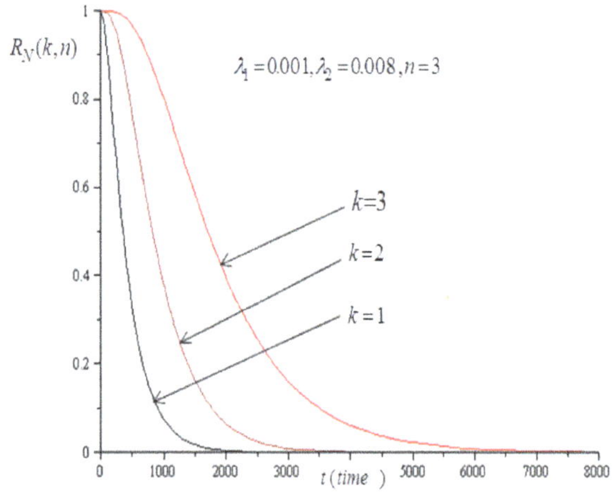

Figure 4: Non-repairable system reliability with degradation for different numbers of unit's k.

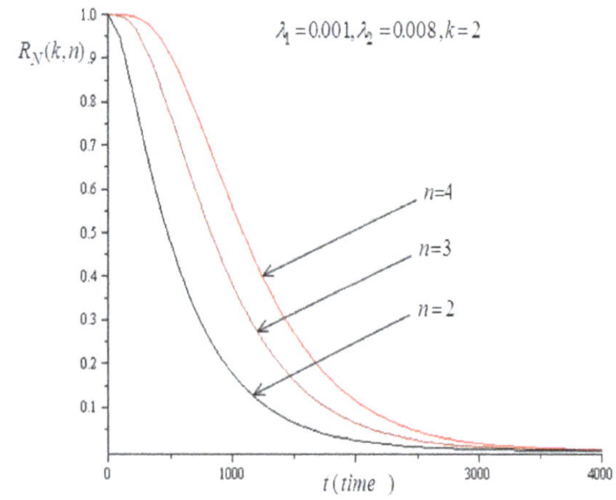

Figure 5: Non-repairable system reliability with degradation for different numbers of unit's n.

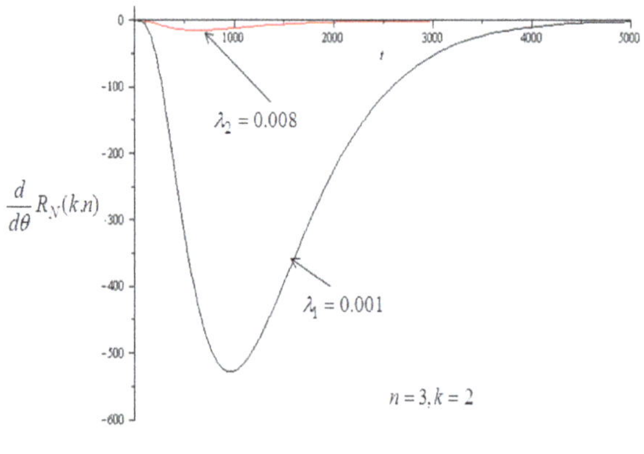

Figure 6: Sensitivity of repairable system reliability with respect to system parameters.

Repairable k-out-of-n system

In this section, we will develop a general model for analysis of such systems when they are repairable. Let failure rate and degradable rate that occur independently of the states of other units and follow exponential distributions with λ_1 and λ_2 respectively in addition, repair rates of failure and degradation are assumed to be exponentially distributed with parameters μ_1 and μ_2 respectively.

The system starts at time t=0 and there is no failed or degradable components. When a unit failed it is immediately sent to the first service line where it is repaired with time-to-repair which is exponentially distributed with parameter μ_1.we have two service lines the first one repair failed units and the second one repair degraded units. When an operating unit degraded it is it is repaired with time-to-repair which is exponentially distributed with parameter μ_2 during it working. We assume that the secession of failure times and repair times are independently distributed random variable. Let us assume that failed units arriving at the repairmen form a single waiting line and are repaired in the order of their breakdowns; i.e. according to the first-come, first-served discipline. Suppose that the repairmen in the two service lines can repair only one failed unit at a time and the repair is independent of the failure of the units. Once a unit is repaired, it is as good as new.

The mean repair rate $\mu_{1,j}$ is given by:

$$\mu_{1,j} = \begin{cases} j\mu_1 & , \quad \text{if } 1 \le \text{ n} \le \min(\text{R}_1,\text{n-k}) \\ R_1\mu_1 & , \quad \text{if } \text{R}_1 \le \text{ j} \le \text{n} - k \\ 0 & \quad otherwise \end{cases}$$

The mean repair rate $\mu_{2,j}$ is given by:

$$\mu_{2,j} = \begin{cases} j\mu_2 & , \quad \text{if } 1 \le \text{ n} \le \min(\text{R}_2,\text{n}) \\ R_2\mu_2 & , \quad \text{if } \text{R}_2 \le \text{ j} \le \text{n} \\ 0 & \quad otherwise \end{cases}$$

Based on the model descriptions, the system state transition diagram is given in

Figure 7 and it leads to the following Laplace transform expressions for $P_{i,j}^{*}(s)$:

$$\left(s+n\lambda_1\right)P_{0,0}^{*}(s)-\mu_{1,1}P_{0,1}^{*}(s)-\mu_{2,1}P_{1,0}^{*}(s)=1 \tag{39}$$

$$\left(s+(n-j)\lambda_1+\mu_{1,j}\right)P_{0,j}^{*}(s)-\mu_{1,j+1}P_{0,j+1}^{*}(s)-\mu_{2,1}P_{1,j}^{*}(s)-\lambda_2P_{1,j-1}^{*}(s)=0 \\ ,1\le j\le n-k-1 \tag{40}$$

$$\left(s+k\lambda_1+\mu_{1,n-k}\right)P_{0,n-k}^{*}(s)-\lambda_2P_{1,n-k-1}^{*}(s)=0 \tag{41}$$

$$\left(s+(n-i)\lambda_1+i\lambda_2+\mu_{2,i}\right)P_{i,0}^{*}(s)-(n-i+1)\lambda_1P_{i-1,0}^{*}(s)-\mu_{1,1}P_{i,1}^{*}(s)-\mu_{2,i+1}P_{i+1,0}^{*}(s)=0 \\ ,1\le i\le n-1 \tag{42}$$

$$\left(s+n\lambda_2+\mu_{2,n}\right)P_{n,0}^{*}(s)-\lambda_1P_{n-1,0}^{*}(s)=0 \tag{43}$$

$$\left(s+(n-i)\lambda_2+\mu_{2,n-i}+\mu_{1,i}\right)P_{n-i,i}^{*}(s)-(n-i+1)\lambda_2P_{n-i+1,i-1}^{*}(s)-\lambda_1P_{n-i-1,i}^{*}(s)=0 \\ ,1\le i\le n-k \tag{44}$$

$$\left(s+i\lambda_2+(k-i)\lambda_1+\mu_{1,n-k}+\mu_{2,i}\right)P_{i,n-k}^{*}(s)-(i+1)\lambda_2P_{i+1,n-k-1}^{*}(s)-\mu_{2,i+1}P_{i+1,n-k}^{*}(s) \\ -(k-i+1)\lambda_1P_{i-1,n-k}^{*}(s)=0,1\le i\le k-1 \tag{45}$$

$$\left(s+(n-k-i)\lambda_2+\mu_{2,n-k-i}+\mu_{1,i}\right)P_{n-k-i,i}^{*}(s)-(n-k-i)\lambda_2P_{n-k-i+1,i-1}^{*}(s) \\ -(n-k-1)\lambda_1P_{n-k-i-1,i}^{*}(s)-\mu_{2,n-k-i+1}P_{n-k-i+1,i}^{*}(s)-\mu_{1,i+1}P_{n-k-i,i+1}^{*}(s)=0 \\ ,1\le i\le n-k-1 \tag{46}$$

$$\left(s+(n-k-j+i)\lambda_2+(k-i)\lambda_1+\mu_{1,j}+\mu_{2,n-k-j+i}\right)P_{n-k-j+i,j}^{*}(s) \\ -\mu_{2,n-k-j+i+1}P_{n-k-j+i+1,j}^{*}(s)-(n-k-j+i+1)\lambda_2P_{n-k-j+i+1,j-1}^{*}(s) \\ -(k-i+1)\lambda_1P_{n-k-j+i-1,j}^{*}(s)-\mu_{1,j+1}P_{n-k-j+i,j+1}^{*}(s)=0,1\le i\le k-1,1\le j\le n-k-1 \tag{47}$$

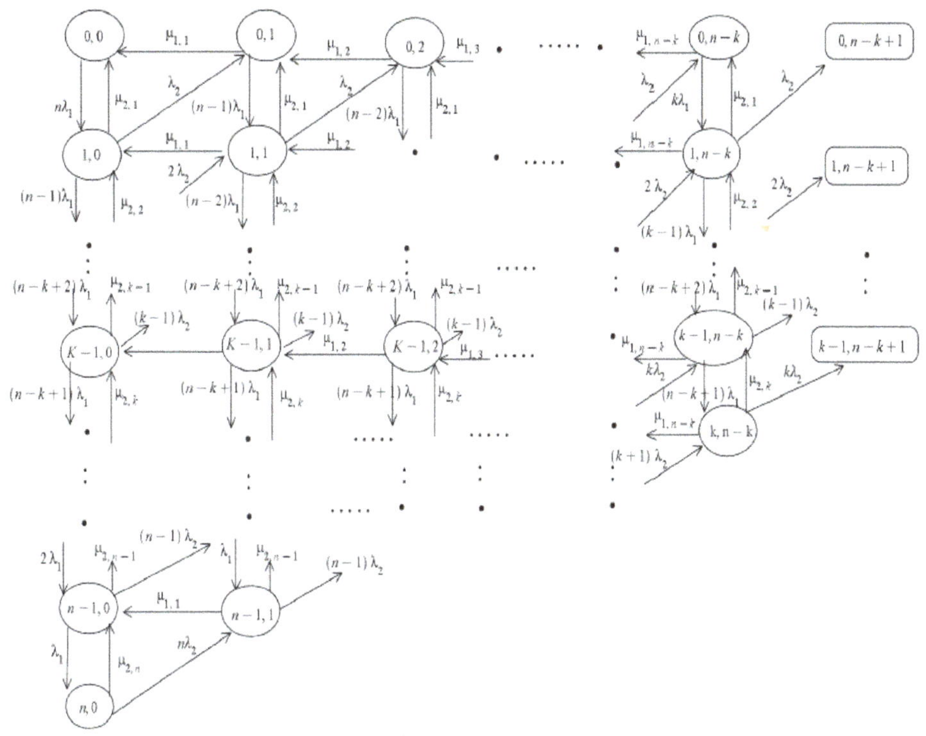

Figure 7: State-transition-rate diagram of repairable System with degradation facility.

$$\left(s+(i+1)\lambda_2+(n-j-i-1)\lambda_1+\mu_{1,j}+\mu_{2,i+1}\right)P^*_{i+1,j}(s)-(n-j-i)\lambda_1 P^*_{i,j}(s)$$
$$-\mu_{2,i+2}P^*_{i+2,j}(s)-(i+2)\lambda_2 P^*_{i+2,j-1}(s)-\mu_{1,j+1}P^*_{i+1,j+1}(s)=0 \tag{48}$$
$$,0\le i\le n-k-2, 1\le j\le n-k-i-1$$

$$(s)P^*_{i,n-k+1}(s)-(i+1)\lambda_2 P^*_{i+1,n-k}(s)=0, 0\le i\le k-1 \tag{49}$$

By solving equations (39-49) and taking inverse Laplace transforms (using maple program). We obtain the reliability function as follows:

$$R_R(k,n)=L^{-1}\left(\sum_{i+j=0}^{n-k}P^*_{i,j}(s)\right)=\left(\sum_{i+j=0}^{n-k}P_{i,j}(t)\right) \tag{50}$$

Where $i,j=0,1,2,\ldots,n-k$.

The mean time to failure $MTTFR(k,n)$ can be obtained from the following relation.

$$MTTF_R(k,n)==\lim_{s\to 0}\left\{\sum_{i+j=0}^{n-k}P^*_{i,j}(s)\right\}=\left\{\sum_{i+j=0}^{n-k}P^*_{i,j}(0)\right\} \tag{51}$$

We perform a sensitivity analysis for changes in the reliability of the system $R_R(k,n)$ from changes in system parameters λ_1, λ_2, μ_1 and μ_2. by differentiating equation (50) with respect to λ_1 we obtain

$$\frac{\partial R_R(k,n)}{\partial\lambda_1}=\frac{\partial}{\partial\lambda_1}\left\{\sum_{i+j=0}^{n-k}P_{i,j}(t)\right\}=\left\{\sum_{i+j=0}^{n-k}\frac{\partial}{\partial\lambda_1}P_{i,j}(t)\right\} \tag{52}$$

We use the same procedure to get $\frac{\partial R_R(k,n)}{\partial\lambda_2},\frac{\partial R_R(k,n)}{\partial\mu_1},\frac{R_R(k,n)}{\partial\mu_2}$.

Numerical results

In this section, we use MAPLE computer program to provide the numerical results of the effects of various parameters on system reliability and system availability. We choose λ_1=0.001, λ_2=0.008 and

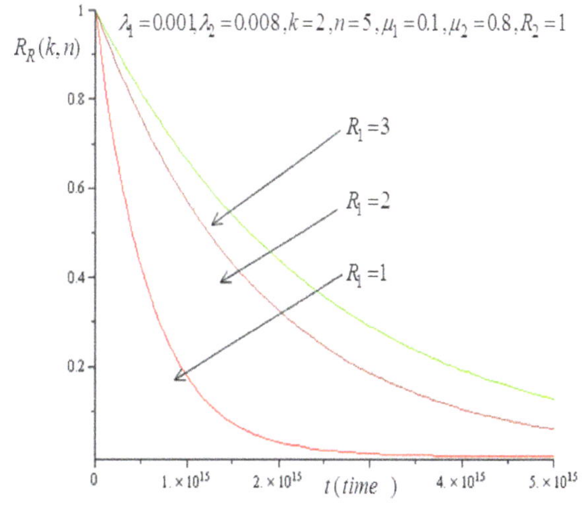

Figure 8: Repairable system reliability for different numbers of repairmen in first service line.

fix μ_1=0.1, μ_2=0.8. The following cases are analyzed graphically to study the effect of various parameters on system reliability.

Case 1: Fix n=5, k=2, R_2=1, and choose R_1=1,2,3.

Case 2: Fix n=5, k=2, R_1=1, and choose R_2=1,2,3.

From Figure 8 we find that R_1 don't effect on system reliability when number of repairman more than one. Figure 9 shows that the repairable system reliability increases as R_2 increases.

Finally we perform sensitivity analysis for system reliability $R_R(k,n)$ with respect to system parameters.

1-With respect to all system parameters

From Figures 10 and 11 we can easily observe that the biggest impact almost happened at the same time for λ_2, μ_1 and μ_2 but it's happen at shorter time for λ_1. Moreover, we find λ_1 is the most

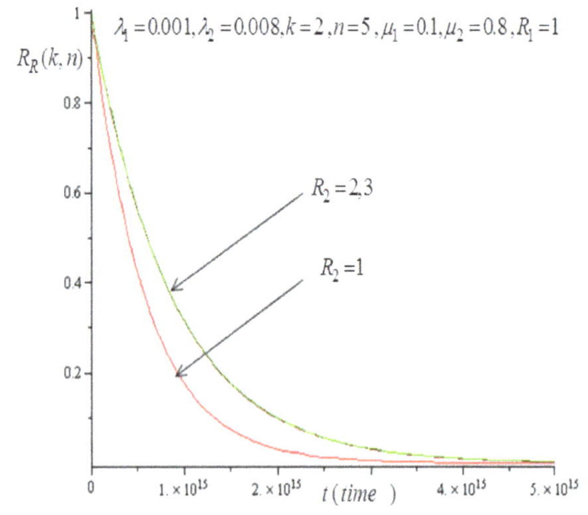

Figure 9: Repairable system reliability for different numbers of repairmen in second service line.

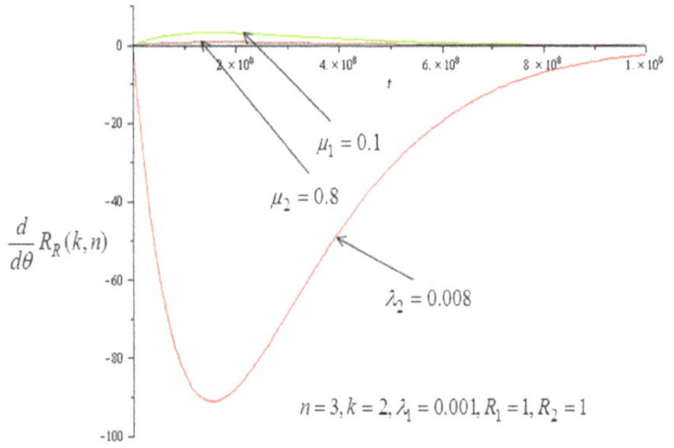

Figure 10: Sensitivity of system reliability with respect to λ_2, μ_1 and μ_2.

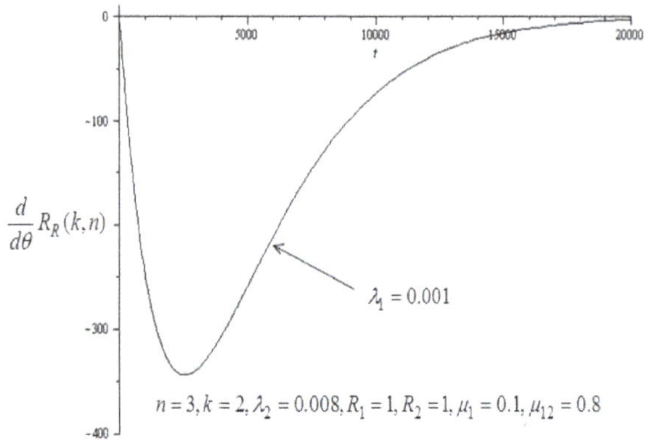

Figure 11: Sensitivity of system reliability with respect to λ_1.

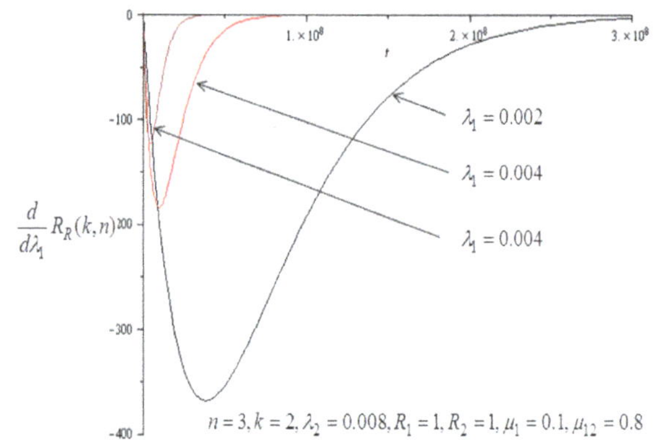

Figure 12: Sensitivity of system reliability with respect to λ_1 at various values of λ_1.

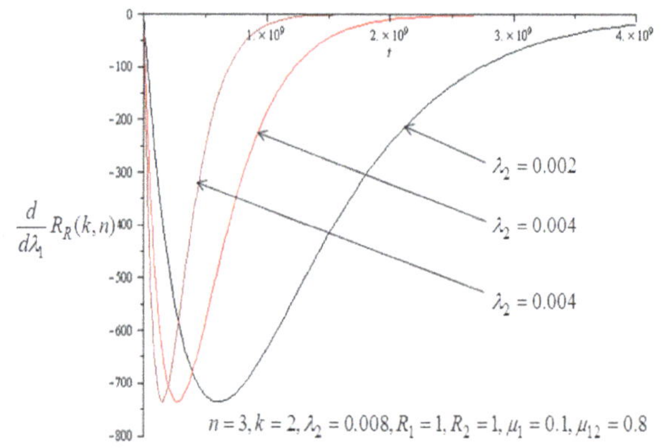

Figure 13: Sensitivity of system reliability with respect to λ_1 at various values of λ_2.

prominent parameter whileλ_2, μ_1 and μ_2 are the second, the third and the fourth respectively in magnitude.

2-With respect to λ_1 at various values of λ_1

Figure 12 shows that when λ_1 decreases its impact on reliability $R_Y(t)$ happened at longer interval time, and the biggest impact almost happened at longer time.

3-With respect to λ_1 at various values of λ_2

From Figure 13 we can also observe that as λ_2 decreases the impact of λ_1 on reliability $R_Y(t)$ happened at longer interval time, and the biggest impact almost happened at longer time, we can observe that the biggest impact of λ_1 on reliability $R_Y(t)$ is not affected by the value of λ_2 but it's happen at longer time as λ_2 decreases.

Conclusions

In this paper, we studied reliability and mean time to system failure for k-out-of-n models with degradation facility. Mathematical model were constructed for these models. Results indicate that the reliability of k-out-of-n nonrepairable system with degradation increases as k or n increases. In repairable system we observe that the order of magnitude of the effect is ($\lambda_1 > \lambda_1 > \mu_1 > \mu_2$).

References

1. Galikowsky C, Sivazlian BD, Chaovalitwongse P (1996) Optimal redundancies

for reliability and availability of series systems. Microelectronics and Reliability 36: 1537-1546.

2. Wang KH, Sivazlian BD (1989) Reliability of a system with warm standbys and repairmen. Microelectronics and Reliability 29: 849-860.
3. ke JB, Chen JW, Wang KH (2011) Reliability Measures of a Repairable System with Standby Switching Failures and Reboot Delay. Quality Technology & Quantitative Management 8: 15-26.
4. Yuge T, Maruyama M, Yanagi S (2016) Reliability of a k-out-of-n system with common-cause failures using multivariate exponential distribution. Procedia Computer Science 96: 968-976.
5. Zhang T, Xie M, Horigome M (2006) Availability and reliability of k-out-of-(M+N): G warm standby systems. Reliability Engineering & System Safety 91: 381-387.

Characterizations of Certain Doubly Truncated Distribution Based on Order Statistics

Shawky AI[1]* and Badr MM[2]

[1]Department of Statistics, Faculty of Science, King Abdulaziz University, Saudi Arabia
[2]Department of Statistics, Faculty of Science for Girls, King Abdulaziz University, Saudi Arabia

Abstract

In this paper, we characterize doubly truncated classes of absolutely continuous distributions by considering the conditional expectation of functions of order statistics. Specific distributions considered as a particular case of the general class of distributions are Weibull, Pareto, Power function, Rayleigh and Inverse Weibull.

Keywords: Double truncated; Order statistics; Conditional expectation; Weibull, Pareto; Rayleigh; Inverse Weibull distributions

AMS 2000 Subject Classification: 62E10; 62G30

Introduction

The order statistics arise naturally in many real life applications and it is considered as an increasingly important subject. Articles relating to this area have appeared in numerous different publications. Many authors have studied order statistics; for example, David [1], Balakrishnan and Cohen [2], Arnold et al. [3], David [4], David and Nagaraja [5] and Mahmoud et al. [6,7]. Several authors discussed conditional expectations, for example, Balakrishnan and Sultan [8], Mohie El-Din et al. [9], Abu-Youssef [10], Abd- El-Mougod [11], Shawky and Abu-Zinadah [12], Shawky and Bakoban [13] and Pushkarna et al. [14].

Let $X_{1:n} \le X_{2:n} \le \ldots \le X_{n:n}$ be the first n order statistics based on distribution with probability density function (pdf) f(x) and cumulative distribution function (cdf) F(x). Then the pdf of the rth order statistics, $X_{r:n}$, $1 \le r \le n$, is given by (see David (1981))

$$f_r(x) = C_r\left[F(x)\right]^{r-1}\left[1-F(x)\right]^{n-r} f(x), \qquad (1.1)$$

where $C_r = \dfrac{n!}{(r-1)!(n-r)!}$, $-\infty < x < \infty$,

and the joint pdf of two order statistics $X_{r:n}$ and $X_{s:n}$, $1 \le r \le s \le n$ is given by

$$f_{r,s}(x,y) = C_{r,s}\left[F(x)\right]^{r-1}\left[F(y)-F(x)\right]^{s-r-1}\left[1-F(y)\right]^{n-s} f(x)f(y), -\infty < x < y < \infty, \qquad (1.2)$$

where $C_{r,s} = \dfrac{n!}{(r-1)!(s-r-1)!(n-s)!}$.

The doubly truncated case of a distribution is the most general case since it includes the right truncated, left truncated and non-truncated distributions as special cases, Joshi [15], Balakrishnan and Joshi [16], Khan and Ali [17] and Ahmad [18], among others, investigated doubly truncated distributions.

Suppose that the random variable X has a cdf F(x) and pdf f(x), where $\alpha \le x$ β. Let, for given ε and

$$\int_\alpha^\varepsilon f(x)dx = F(\varepsilon) = P \text{ and } \int_\alpha^\gamma f(x)dx = F(\gamma) = Q.$$

Then the doubly truncated pdf of X, say g(x), and cdf, say G(x), are given respectively by

$$g(x) = \frac{f(x)}{l_1}, \alpha \le \varepsilon < x < \gamma \le \beta, \qquad (1.3)$$

$$G(x) = \frac{F(x)-P}{l_1}, \varepsilon < x < \gamma, \qquad (1.4)$$

where

$l_1 = Q - P$, $G(\varepsilon)=0$ and $G(\gamma)=1$.

The conditional density function of $X_{s:n=}y$, given that $X_{r:n=}x$ is given [3] by

$$f_{s|r}(y|X_{r:n}=x) = \frac{(n-r)!}{(s-r-1)!(n-s)![1-G(x)]^{n-r}}.$$
$$[G(y)-G(x)]^{s-r-1}[1-G(y)]^{n-s}g(y),$$
$$\varepsilon < x < y < \gamma. \qquad (1.5)$$

Also, the conditional density function of $X_{r:n=}x$, given that $X_{s:n=}y$ is *given by*

$$f_{r|s}(x|X_{s:n}=y) = \frac{(s-1)!}{(r-1)!(s-r-1)![G(y)]^{s-1}}.$$
$$[G(x)]^{r-1}[G(y)-G(x)]^{s-r-1}g(x) \qquad (1.6)$$

Let

$$\mu_{s|r} = E[\varphi(X_{s:n})|X_{r:n}=x] \text{ and } \mu_{r|s} = E[\varphi(X_{r:n})|X_{s:n}=y],$$

where $\varphi(.)$ is a monotonic, continuous and differentiable function on the interval (α,β). For abbreviation, we will denote

$$\mu_{s|r} = E_{s|r}[\varphi(Y)|X_{r\ n}=x] \text{ and } \mu_{r|s} = E_{r|s}[\varphi(X)|X_{s:n}=y]. \qquad (1.7)$$

Main Results

In this section, we characterize three general classes of distributions,

$F(x) = 1-[b-ae^{-\varnothing(x)}]^c, \alpha < x < \beta$, i.e.,

***Corresponding author:** Shawky AI, Department of Statistics, Faculty of Science, King Abdulaziz University, P.O. Box 80203, Jeddah 21589, Saudi Arabia
E-mail: aishawky@yahoo.com

$$G(x)=\frac{1}{l}\{\left[b-ae^{-\varnothing(x)}\right]^{c}-v\},\varepsilon<x<\gamma, \quad (2.1)$$

where

$v=1-P$, $l=P-Q$, $P=F(\varepsilon)$, $Q=F(\gamma)$, $G(\varepsilon)=0$, $G(\gamma)=1$,

$F(\alpha)=0$, $F(\beta)=1$, and $a \neq 0$, c 0, b are finite constants.

$F(x)=1-[a-b\,\varnothing(x)]^{c}$, $\alpha<x<\beta$, i.e.,

$$G(x)=\frac{1}{l}\{[a-b\varnothing(x)]^{c}-v\},\varepsilon<x<\gamma, \quad (2.2)$$

where

$v=1-P, l=P-Q, P=F(\varepsilon), Q=F(\gamma), G(\varepsilon)=0, G(\gamma)=1,$

$F(\alpha)=0$, $F(\beta)=1$, and $b \neq 0$, $c \neq 0$, a are finite constants.

$F(x)=1-[b-ae^{-C\varnothing(x)}],\alpha<x<\beta$, i.e.,

$$G(x)=\frac{1}{l}\{[b-ae^{-C\varnothing(x)}]-v\},\varepsilon<x<\gamma, \quad (2.3)$$

where

$v=1-P$, $l=P-Q$, $P=F(\varepsilon)$, $Q=F(\gamma)$, $G(\varepsilon)=0$, $G(\gamma)=1$,

$F(\alpha)=0$, $F(\beta)=1$, and $C \neq 0$, $a>0$, $b>0$ are finite constants.

Note: If we put $l=-1$, $v=1$, thus G(x) reduces to complete cdf of x, i.e. F(x), $\alpha<x<\beta$.

Let X be an absolutely continuous random variables with pdf g(x), cdf G(x) and $\varnothing(x)$ is a monotonic, continuous and differentiable function on (ε,γ).

Theorems 1-4 given below characterize the general class given by (2.1), Theorems 5-8 characterize the general class given by (2.2), while Theorems 9-12 characterize the general class given by (2.3).

Theorem 1

Referring to (1.6), (1.7) and (2.1), then

$$\mu_{r|s}=\mu_{r+1|s}+\frac{1}{acr}\left\{aE_{r|s}\left[V(X)\middle|X_{s:n}=y\right]-bE_{r|s}\left[V(X)e^{\varphi(X)}\middle|X_{r:s}=y\right]\right\}, \quad (2.4)$$

where

$$V(x)=\frac{lG(x)}{lG(x)+v}. \quad (2.5)$$

Proof

It is clear from (1.6) and (1.7) that

$$\mu_{r|s}=\frac{(s-1)!}{(r-1)!(s-r-1)![G(y)]^{s-1}}\int_{\varepsilon}^{y}\varphi(x)\left[G(x)\right]^{r-1}\left[G(y)-G(x)\right]^{s-r-1}g(x)dx. \quad (2.6)$$

Integrating (2.6) by parts, we get

$$\mu_{r|s}=\mu_{r+1|s}-\frac{(s-1)!}{r!(s-r-1)![G(y)]^{s-1}}\int_{\varepsilon}^{y}\varphi(x)\left[G(x)\right]^{r}\left[G(y)-G(x)\right]^{s-r-1}dx. \quad (2.7)$$

Differentiating (2.1) with respect to x, we have

$$\varphi(x)=\frac{be^{\varphi(x)}-a}{ac}\frac{lg(x)}{lG(x)+v}. \quad (2.8)$$

From (2.7) and (2.8), we obtain

$$\mu_{r|s}=\mu_{r+1|s}-\frac{(s-1)!}{(ac)r!(s-r-1)!\left[G(y)\right]^{s-1}}\int_{\varepsilon}^{y}(be^{\varphi(x)}-a)V(x)\left[G(x)\right]^{r-1}\left[G(y)-G(x)\right]^{s-r-1}g(x)dx$$
$$=\mu_{r+1|s}-\frac{1}{acr}\int_{\varepsilon}^{y}\left[be^{\varphi(x)}-a\right]V(x)g_{r|s}(x|y)dx. \quad (2.9)$$

Simplifying (2.9), we get (2.4). Thus, the theorem is proved.

Theorem 2

Referring to (1.6), (1.7), then (2.1) if and only if

$$\mu_{r|r+1}=\varphi(y)+\frac{1}{ac}\sum_{i=0}^{\infty}(-1)^{i+1}\left(\frac{l}{v}\right)^{i+1}\frac{[G(y)]^{i+1}}{r+i+1}\{bE_{r+i+1|r+i+2}\left[e^{\varphi(X_{r+i+1:n})}\middle|X_{r+i+2:n}=y\right]-a\}. \quad (2.10)$$

Proof

It is clear that

$$\mu_{r|r+1}=\frac{r}{[G(y)]^{r}}\int_{\varepsilon}^{y}\varphi(x)\left[G(x)\right]^{r-1}g(x)dx.$$

Integrating by parts, we get

$$\mu_{r|r+1}=\varphi(y)-\frac{1}{[G(y)]^{r}}\int_{\varepsilon}^{y}\varphi(x)[G(x)]^{r}dx \quad (2.11)$$

Compensation for (2.8) in (2.11), we have

$$\mu_{r|r+1}=\varphi(y)-\frac{l}{ac[G(y)]^{r}}\{\int_{\varepsilon}^{y}\left(be^{\varphi(x)}-a\right)\frac{g(x)}{lG(x)+v}\left[G(x)\right]^{r}dx. \quad (2.12)$$

Expand $\frac{1}{lG(x)+v}$ and compensation for (2.12), after some simplification, we get (2.10). Thus (2.1) implies (2.10). Now from (1.6) and (2.10), we obtain

$$\frac{r}{[G(y)]^{r}}\int_{\varepsilon}^{y}\varphi(x)\left[G(x)\right]^{r-1}g(x)dx=$$
$$\varphi(y)+\frac{1}{ac}\sum_{i=0}^{\infty}(-1)^{i+1}\left(\frac{l}{v}\right)^{i+1}\frac{[G(y)]^{i+1}}{r+i+1}\{b\int_{\varepsilon}^{y}e^{\varphi(x)}\left[G(x)\right]^{r+i}g(x)dx-a\}. \quad (2.13)$$

Taking the derivative, we get

$$\varphi(y)=\frac{(\frac{l}{v})g(y)}{1+\frac{l}{v}G(y)}\left[\frac{b}{ac}e^{\varphi(y)}-\frac{1}{c}\right],$$

which gives

$$\frac{ac\varphi(y)e^{-\varphi(y)}}{b-ae^{-\varphi(y)}}=\frac{lg(y)}{v+lG(y)}. \quad (2.14)$$

Integrate (2.14), hence G(y) has the form (2.1), and so (2.10) implies (2.1).

Special case:

Return to the (2.10), if we put $l=-1$, $v=1$we get

$$\mu_{r|r+1}=\varphi(y)+\frac{1}{ac}\sum_{i=0}^{\infty}\frac{[F(y)]^{i+1}}{i+r+1}\left\{bE_{r+i+1|r+i+2}\left[e^{\varphi(X_{r+i+1:n})}\middle|X_{r+i+2:n}=y\right]-a\right\},$$

$$\alpha<x<y<\beta \quad (2.15)$$

the relation (2.15) is before doubly truncated case.

Theorem 3

Referring to (1.5), (1.7) and (2.1), then

$$\mu_{s|r}=\mu_{s-1|r}+\frac{l}{ac(n-s+1)}$$
$$\left\{aE_{s|r}\left[N(X_{s:n})\middle|X_{r:n}=x\right]-bE_{s|r}\left[e^{\varphi(X_{s:n})}N(X_{s:n})\middle|X_{r:n}=x\right]\right\}, \quad (2.16)$$

where

$$N(y)=\frac{[1-G(y)]}{[lG(y)+v]}\cdot \tag{2.17}$$

It is clear from (1.5) and (1.7) that

$$\mu_{s|r}=\frac{(n-r)!}{(s-r-1)!(n-s)![1-G(x)]^{n-r}}\int_x^{\gamma}\varphi(y)\left[G(y)-G(x)\right]^{s-r-1}\left[1-G(y)\right]^{n-s}g(y)dy\,.$$

Integrating by parts, we get

$$\mu_{s|r}=\mu_{s-1|r}-\frac{(n-r)!}{(s-r-1)!(n-s+1)![1-G(x)]^{n-r}}\int_x^{\gamma}. \tag{2.18}$$

$$\varphi(y)[G(y)-G(x)]^{s-r-1}[1-G(y)]^{n-s+1}dy$$

Substituting (2.7) in (2.18), we get

$$\mu_{s|r}=\mu_{s-1|r}-\frac{(n-r)!}{(s-r-1)!(n-s+1)![1-G(x)]^{n-r}}\int_x^{\gamma}\frac{e^{\varphi(y)}}{ac}\left[b-ae^{-\varphi(y)}\right]N(y)[G(y)-G(x)]^{s-r-1}$$

$$\mu_{s|r}=\mu_{s-1|r}-\frac{(n-r)!}{(s-r-1)!(n-s+1)![1-G(x)]^{n-r}}\int_x^{\gamma}\frac{e^{\varphi(y)}}{ac}\left[b-ae^{-\varphi(y)}\right]N(y)[G(y)$$
$$-G(x)]^{s-r-1}\times[1-G(y)]^{n-s}dy \tag{2.19}$$
$$=\mu_{s-1|r}+\frac{l}{ac(n-s+1)}\{\int_x^{\gamma}\left[a-be^{\varphi(y)}\right]N(y)g_{s|r}(y|x)dy$$

After some simplification, we get (2.16).

Theorem 4

Referring to (1.5), (1.7), then (2.1) if and only if

$$\mu_{r+1|r}=\varphi(x)+\frac{lb}{ac(n-r)}E_{r+1|r}\left[N(y)e^{\varphi(y)}\middle|X_r=x\right]-\frac{l}{c(n-r)}E_{r+1|r}\left[N(y)\middle|X_{r:n}=x\right], \tag{2.20}$$

where $N(y)$ is defined in (2.17).

Proof

It is clear that

$$\mu_{r+1|r}=\frac{(n-r)}{[1-G(x)]^{n-r}}\int_x^{\gamma}\varphi(y)\left[1-G(y)\right]^{n-r-1}g(y)dy\,\cdot \tag{2.21}$$

Integrating by parts, we get

$$\mu_{r+1|r}=\varphi(x)+\frac{1}{[1-G(x)]^{n-r}}\int_x^{\gamma}\varphi(y)[1-G(y)]^{n-r}dy\,\cdot \tag{2.22}$$

Compensation for (2.8) in (2.22), we have

$$\mu_{r+1|r}=\varphi(x)+\frac{l}{ac[1-G(x)]^{n-r}}\int_x^{\gamma}\frac{be^{\varphi(y)}-a}{\left[lG(y)+v\right]}\cdot\left[1-G(y)\right]^{n-r}g(y)dy\,\cdot \tag{2.23}$$

Simplifying (2.23), we obtain (2.20). Thus (2.1) implies (2.20), i.e. the necessary condition is proved. To prove the sufficient condition, from (2.20) and (1.7), we have

$$\frac{(n-r)}{[1-G(x)]^{n-r}}\int_x^{\gamma}\varphi(y)\left[1-G(y)\right]^{n-r-1}g(y)dy=\varphi(x)+\frac{lb}{ac[1-G(x)]^{n-r}}\int_x^{\gamma}e^{\varphi(y)}$$
$$\times\left[1-G(y)\right]^{n-r-1}N(y)g(y)dy-\frac{l}{c[1-G(x)]^{n-r}}\int_x^{\gamma}N(y)\left[1-G(y)\right]^{n-r-1}g(y)dy \tag{2.24}$$

Taking the derivative of (2.24) with respect to x, we get (2.8), and integrate it we have (2.1), thus (2.20) implies (2.1). Then, the Theorem is proved.

Special case

Return to (2.17), if we put $l=-1$, $v=1$ we get

$$\mu_{r+1|r}=\varphi(x)-\frac{b}{ac(n-r)}E_{r+1|r}\left[e^{\varphi(y)}\middle|X_r=x\right]+\frac{1}{c(n-r)},\quad \alpha<x<y<\beta\,,$$

it is before doubly truncated case (Table 1).

Theorem 5

Referring to (1.6), (1.7) and (2.2), then

$$\mu_{r|s}=\mu_{r+1|s}-\frac{1}{bcr}\left\{bE_{r|s}\left[\varphi(X_{r:n})V(X_{r:n})\middle|X_{s:n}=y\right]-aE_{r|s}\left[V(X_{r:n})\middle|X_{s:n}=y\right]\right\}, \tag{2.25}$$

where $V(x)$ is defined in (2.5).

Proof

As before in Theorem (1), differentiate (2.2) with respect to x, we have

$$\varphi(x)=\frac{b\varphi(x)-a}{bc}\frac{lg(x)}{lG(x)+v}\cdot \tag{2.26}$$

Compensation for (2.26) in (2.7), we get

$$\mu_{r|s}=\mu_{r+1|s}-\frac{(s-1)!}{r!(s-r-1)![G(y)]^{s-1}}\int_{\varepsilon}^{y}\left\{\frac{b\varphi(x)-a}{bc}\right\}$$
$$V(x)[G(x)]^{r-1}\left[G(y)-G(x)\right]^{s-r-1}g(x)dx \tag{2.27}$$
$$=\mu_{r+1|s}-\frac{1}{bcr}\int_{\varepsilon}^{y}\left(b\varphi(x)-a\right)V(x)g_{r|s}(x|y)^{r-1}dx$$

Simplifying (2.27), we obtain (2.25). Thus, the Theorem is proved.

Name	[$lG(x)+v$]	$\varphi(x)$	(l,v)	(a,b,c)
Weibull	$e^{-\theta x^p};\alpha\le\varepsilon<x<\gamma\le\beta,\varepsilon=0,\gamma\to\infty$	x^p θx^p	$\left(e^{-\theta\varepsilon^p}-e^{-\theta\gamma^p},e^{-\theta\varepsilon^p}\right)$	(−1,0,θ) (−1,0,1)
Pareto	$\theta^p x^{-p}$; $\alpha\le\varepsilon<x<\gamma\le\beta$, $\varepsilon=0$, $\gamma\to\infty$	$\ln(x)$ $\ln[x^{-p}]$	$(\theta^p(\gamma^{-p}-\varepsilon^{-p}),\ \theta^p\varepsilon^{-p})$	$(-\theta,0,p)$ $(-\theta^p,0,-1)$
Power function	$1-\theta^{-p}x^p,\alpha\le\varepsilon<x<\gamma\le\beta$, $\varepsilon=0$, $\gamma\to\infty$	$\ln\left[\frac{x}{\theta}\right]^{-p}$ $\ln[x^p]$	$(\theta^{-p}(\varepsilon^p-\gamma^p),\ 1-\theta^{-p}\varepsilon^p)$	(1,1,1) $(\theta^{-p},1,1)$
Rayleigh	$e^{(-\theta x^2)};a\le\varepsilon<x<\gamma\le\beta$	x^2	$\left(e^{-\theta\gamma^2}+e^{-\theta\varepsilon^2},e^{-\theta\gamma^2}\right)$	$(-1,0,\theta)$
Inverse Weibull	$e^{(-\theta x^{-p})};a\le\varepsilon<x<\gamma\le\beta$	θx^{-p}	$\left(e^{-\theta\gamma^{-p}}+e^{-\theta\varepsilon^{-p}},e^{-\theta\varepsilon^{-p}}\right)$	(−1,0,1)

Table 1: Example of $G(x)=\frac{1}{l}\{[b-ae^{-\varphi(x)}]^c-v\}$ distributions.

Theorem 6

Referring to (1.6), (1.7), then (2.2) if and only if

$$\mu_{r|r+1} = \varphi(y) - \frac{1}{bcr}\left\{bE_{r|r+1}\left[\varphi(X_{r:n})V(X_{r:n})\middle|X_{r+1:n}=y\right] - aE_{r|r+1}\left[V(X_{r:n})\middle|X_{r+1:n}=y\right]\right\}, \quad (2.28)$$

Where $V(x)$ is defined in (2.5).

Proof

As before in Theorem (2), from (2.26) and (2.11), we have

$$\mu_{r|r+1} = \varphi(y) - \frac{1}{[G(y)]^r}\int_\varepsilon^y \frac{b\varphi(x)-a}{bc}\frac{lG(x)}{lG(x)+v}[G(x)]^{r-1}g(x)dx \quad . \quad (2.29)$$
$$= \varphi(y) - \frac{1}{bcr}\int_\varepsilon^y (b\varphi(x)-a)V(x)g_{r|r+1}(x|y)dx$$

Therefore, we get (2.28), then (2.2) implies (2.28). To prove the sufficient condition, from (2.28) and (1.7), we obtain

$$\frac{r}{[G(y)]^r}\int_\varepsilon^y \varphi(x)[G(x)]^{r-1}g(x)dx = \varphi(y) - \quad . \quad (2.30)$$
$$\frac{1}{bcr}\int_\varepsilon^y (b\varphi(x)-a)\frac{lG(x)}{lG(x)+v}[G(x)]^{r-1}g(x)dx$$

Taking the derivative, we get (2.26) and we obtain, after integration, (2.2). Thus (2.28) implies (2.2).

Special case

Return to (2.28), then put l=−1, v=1, we get

$$\mu_{r|r+1} = \varphi(y) + \frac{1}{bcr}\{aE_{r|r+1}\left[V(X_{r:n})\middle|X_{r+1}=y\right] - bE_{r|r+1}\left[\varphi(X_{r:n})V(X_{r:n})\middle|X_{r+1}=y\right],$$
$$\alpha < x < y < \beta,$$

it is before doubly truncated case.

Theorem 7

Referring to (1.5), (1.7) and (2.2), then

$$\mu_{s|r} = \mu_{s-1|r} - \frac{l}{bc(n-s+1)} \quad , \quad (2.31)$$
$$\left\{bE_{s|r}\left[\varphi(X_{s:n}x)N(X_{s:n})\middle|X_{r:n}=x\right] - aE_{s|r}\left[N(X_{s:n})\middle|X_{r:n}=x\right]\right\}$$

where $N(y)$ is defined in (2.17).

Proof

As before in Theorem (3), compensation for (2.26) in (2.18), we have

$$\mu_{s|r} = \mu_{s-1|r} - \frac{(n-r)!}{bc(s-r-1)!(n-s+1)![1-G(x)]^{n-r}}\int_x^\gamma (b\varphi(y)-a)N(y)[G(y)-G(x)]^{s-r-1} \quad (2.32)$$
$$\times[1-G(y)]^{n-s}g(y)dy = \mu_{s-1|r} - \frac{l}{bc(n-s+1)}\{\frac{1}{c}\int_x^\gamma (b\varphi(y)-a)N(y)g_{s|r}(y|x)dy$$

After simplification, we get (2.31). Then (2.2) implies (2.31).

Theorem 8

Referring to (1.5), (1.7), then (2.2) if and only if

$$\mu_{r+1|r} = \varphi(x) + \frac{l}{bc(n-r)} \quad , \quad (2.33)$$
$$\left\{bE_{r+1|r}\left[\varphi(X_{r+1:n})N(X_{r+1:n})\middle|X_{r:n}=x\right] - aE_{r+1|r}\left[N(X_{r+1:n})\middle|X_{r:n}=x\right]\right\}$$

where $N(y)$ is defined in (2.17).

Proof

As before in Theorem (4), compensation for (2.26) in (2.23), we have

$$\mu_{r+1|r} = \varphi(x) + \frac{l}{bc\,[1-G(x)]^{n-r}} \quad (2.34)$$
$$\int_x^\gamma [b\varphi(y)-a]N(y)[1-G(y)]^{n-r-1}g(y)dy$$

Then, we obtain (2.35). Thus (2.2) implies (2.33). Now from (1.7) and (2.33) we get

$$\frac{(n-r)}{[1-G(x)]^{n-r}}\int_x^\gamma \varphi(y)[1-G(y)]^{n-r-1}g(y)dy = \varphi(x) + \frac{l}{c[1-G(x)]^{n-r}}\int_x^\gamma \varphi(y)N(y) \quad (2.35)$$
$$\times[1-G(y)]^{n-r-1}g(y)dy - \frac{al}{bc[1-G(x)]^{n-r}}\int_x^\gamma N(y)[1-G(y)]^{n-r-1}g(y)dy$$

Taking the derivative with respect to x, we obtain (2.26), and integrate it we have (2.2), thus (2.33) implies (2.2).

Hence, the Theorem is proved. Special caseReturn to (2.33), then put l=−1, v=1, we get

$$\mu_{r+1|r} = \varphi(x) - \frac{1}{bc(n-r)}$$
$$\left\{bE_{r+1|r}\left[\varphi(Y)\middle|x\right]-a\right\},\ \alpha < x < y < \beta$$

it is before doubly truncated case (Table 2).

Theorem 9

Referring to (1.6), (1.7) and (2.3), then

$$\mu_{r|s} = \mu_{r+1|s} - \frac{l}{acr}\left\{(b-v)E_{r|s}\left[e^{c\varphi(X_{r:n})}\middle|X_{r:s}=y\right]-a\right\} \quad (2.36)$$

Name	[lG(x) + v]	ϕ(x)	(l,v)	(a,b,c)
Weibull	$\left(e^{-\theta\varepsilon^p} - e^{-\theta\gamma^p}, e^{-\theta\varepsilon^p}\right)$	e^{-x^p} $e^{-\theta x^p}$	$e^{-\theta x^p}$	(-1,0,θ)
Power function	$1-\theta^{-p}x^p$, $\alpha \le \varepsilon < x < \gamma \le \beta$, $\varepsilon = 0$, $\gamma \to \infty$	$\left(\frac{x}{\theta}\right)^p$ x^p	$(\theta^{-p}(\varepsilon^p + \gamma^p), 1-\theta^{-p}\varepsilon^p)$	(1,1,1) $(\theta^{-p}$,1,1)
Rayleigh	$e^{-\theta x^2};\alpha \le \varepsilon < x < \gamma \le \beta,$ $\varepsilon = 0, \gamma \to \infty$	e^{-x^2}	$\left(e^{-\theta\gamma^2} + e^{-\theta\varepsilon^2}, e^{-\theta\gamma^2}\right)$	(−1,0,θ)
Inverse Weibull	$e^{-\theta x^{-p}};\alpha \le \varepsilon < x < \gamma \le \beta,$ $\varepsilon = 0, \gamma \to \infty$	$e^{-\theta x^{-p}}$	$\left(e^{-\theta\gamma^{-p}} + e^{-\theta\varepsilon^{-p}}, e^{-\theta\varepsilon^{-p}}\right)$	(0,−1,1)

Table 2: Example of $G(x) = \frac{1}{l}\{[b - a\varphi(x)]^c - v\}$ distributions.

Name	**$[lG(x) + v]$**	**$\phi(x)$**	**(l,v)**	**(a,b,c)**
Power distribution	$1-\theta^{-p} x^p, \alpha \le \varepsilon < x < \gamma \le \beta, \varepsilon = 0, \gamma = \theta$	$Ln[1-\theta^{-p}x^p]$	$(\theta^{-p}(\varepsilon^p + \gamma^p), 1-\theta^{-p}\varepsilon^p)$	(1,0,–1)
Weibull	$e^{-\theta xp}, \alpha \le \varepsilon < x < \gamma \le \beta, \varepsilon = 0, \gamma \to \infty$	θx^p	$(e^{-\theta\gamma^p} + e^{-\theta\varepsilon^p}, e^{-\theta\varepsilon^p})$	(–1,0,1)
Burr	$(1+\theta x^p)^{-\gamma}, \alpha \le \varepsilon < x < \gamma \le \beta, \varepsilon = 0, \gamma \to \infty$	$\ln[1 + \theta x^p]$	$(\theta\gamma^{-p}) + \theta^{\varepsilon p} + 1), 1+\theta^{\varepsilon p})$	(–1,0,γ)
Inverse Weibull	$e^{-\theta x^{-p}}, \alpha \le \varepsilon < x < \gamma \le \beta, \varepsilon = 0, \gamma \to \infty$	$e^{-\theta x^{-p}}$	$(e^{-\theta\gamma^p} + e^{-\theta\varepsilon^p}, e^{-\theta\varepsilon^{-p}})$	(–1,0,1)

Table 3: Example of $G(x) = \frac{1}{l}\{b - ae^{-c\varphi(x)} - v\}$ distributions.

Proof

As given in Theorems (1) and (5), , differentiate (2.3) with respect to x, we have

$$\varphi(x) = \frac{l}{ac} e^{c\,\varphi(x)} g(x) \cdot \quad (2.37)$$

Compensation for (2.37) in (2.7), we get

$$\mu_{r|s} = \mu_{r+1|s} - \frac{1}{acr} \int_{\varepsilon}^{y} \left[(b-v) e^{c\varphi(x)} - a \right] g_{r|s}(x|y) dx \, , \quad (2.38)$$

which gives (2.36). Thus, (2.3) implies (2.36).

Theorem 10

Referring to (1.6) and (1.7), then (2.3) if and only if

$$\mu_{r|r+1} = \varphi(y) - \frac{1}{acr} \left\{ (b-v) E_{r|r+1} \left[e^{c\varphi(X_{r:n})} \middle| X_{r+1:n} = y \right] - a \right\}, \quad (2.39)$$

Proof

As given previously of Theorems (2) and (6), substituting from (2.26) in (2.11), we have

$$\mu_{r|r+1} = \varphi(y) - \frac{1}{acr} \int_{\varepsilon}^{y} e^{c\varphi(x)} \left[b - v - ae^{-c\varphi(x)} \right] g_{r|r+1}(x|y) dx \quad (2.40)$$

After some simplification, we get (2.36). Then (2.3) implies (2.36). Now from (1.7) and (2.36), we obtain

$$r \int_{\varepsilon}^{y} \varphi(x) \left[G(x) \right]^{r-1} g(x) dx = [G(y)]^r$$

$$\left[\varphi(y) - \frac{(b-v)}{acr} \int_{\varepsilon}^{y} \frac{r \left[G(x) \right]^{r-1} e^{c\varphi(x)} g(x)}{\left[G(y) \right]^r} dx + \frac{1}{cr} \right]$$

Taking the derivative with respect to y, we get

$$\frac{l\, g(y)}{l\, \mathrm{G}(y)} = \frac{ace^{-c\,\varphi(y)} \varphi(y)}{b - v - ae^{-c\,\varphi(y)}} \, . \quad (2.41)$$

Integrate (2.41), we obtain (2.3).

Special case

Return to (2.36), then put l=–1, v=1, we get

$$\mu_{r|r+1} = \varphi(y) - \frac{1}{acr} \left\{ (b-1) E_{r|r+1} \left[e^{c\varphi(X_{r:n})} \middle| X_{r+1:n} = y \right] - a \right\}, \alpha < x < y < \beta$$

it is before doubly truncated case.

Theorem 11

Referring to (1.5), (1.7) and (2.3), then

$$\mu_{s|r} = \mu_{s-1|r} - \frac{l}{ac(n-s+1)} \{ (l-b+v) E_{s|r} \left[e^{c\varphi(x_{s:n})} \middle| X_{r:n} = x \right] + a \, . \quad (2.42)$$

Proof

As previously in Theorems (3) and (7), from (2.37) in (2.18), we have

$$\mu_{s|r} = \mu_{s-1|r} - \frac{1}{ac(n-s+1)} \int_{x}^{\gamma} \left[(l-b+v) e^{c\varphi(y)} + a \right] g_{s|r}(y|x) dy \, ,$$

which gives (2.42).

Theorem 12

Referring to (1.5) and (1.7), then (2.3) if and only if

$$\mu_{r+1|r} = \varphi(x) + \frac{(l-b+v)}{ac(n-r)} E_{r+1|r} \left[e^{c\varphi(X_{r+1:n})} \middle| X_{r:n} = x \right] + \frac{1}{c(n-r)} . (2.43)$$

Proof

Similarly as given in Theorems (4) and (8), we easily prove it.

Special case

Return to (2.43), then put l=–1, v=1, we get

$$\mu_{r+1|r} = \varphi(x) - \frac{b}{ac(n-r)} E_{r+1|r} \left[e^{c\varphi(X_{r+1:n})} \middle| X_{r:n} = x \right] + \frac{1}{c(n-r)}, \ \alpha < x < y < \beta \, .$$

The relation before doubly truncated case (Table 3).

Conclusion

It was obtained recurrence relations based on order statistics without truncated and doubly truncated, and have been getting function of various distributions new by using certain parameters.

Acknowledgement

This paper was founded by the Deanship of Scientific Research (DSR), King Abdulaziz University,Jeddah, under grant No. (65-363-D1431).The authors, therefore, acknowledge with thanks DSR technical and financial support.

References

1. David HA (1981) Order Statistics. (2ndedn) John Wiley, New York.
2. Balakrishnan N, Cohen AC (1991) Order Statistics and Inference: Estimation Methods. Academic Press, San Diego.
3. Arnold BC, Balakrishnan N, Nagaraja HN (1992) A first Course in order Statistics. John Wiley, New York.
4. David HA (1995) On recurrence relations for order statistics. Statistics and Probability Letters 24: 133-138.
5. David HA, Nagaraja HN (2003) Order Statistics. (3rdedn) John Wiley, New York.
6. Mahmoud MAW, Sultan KS, Amer SM (2003) Order statistics from inverse Weibull distribution and associated inference. Comput Statist Data Analysis 42: 49-163.
7. Mahmoud MAW, Sultan KS, Amer SM (2003) Order statistics from inverse Weibull distribution and characterizations. Metron LXI 3: 389-401.
8. Balakrishnan N, Sultan KS (1998) Recurrence relations and identities for moments of order statistics. In: Balakrishnan N and Rao CR (eds.) Handbook of statistics16: 149-228.
9. El-Din M, Mahmoud MM, Abu-Youssef MAW, Sultan SE (1997) Order statistics from the doubly truncated linear-exponential distribution and its characterizations. Commun Statist- Simul Comput 26: 281-290.
10. Abu-Youssef SE (2003) On characterization of certain distribution of record values. Applied Mathematics and Computation 145: 445-450.

11. Abd-El-Mougod GA (2005) On Record Values and Its Applications. Sohag University, Egypt.

12. Shawky AI, Abu-Zinadah HH (2008) General recurrence relations and characterizations of certain distributions based on order statistics. Journal of Statistical Theory and Applications 7: 93-117.

13. Shawky AI, Bakoban RA (2009) Conditional expectation of certain distributions of record values. Int Journal of Math Analysis 3: 829-838.

14. Pushkarna N, Saran J, Tiwari R (2012) Recurrence relations for higher moments of order statistics from doubly truncated exponential distribution. International Mathematical Forum, 7(4), 193-201.

15. Joshi PC (1979) A note on the moments of order statistics from doubly truncated exponential distribution. Annals of the Institute of Statistical Mathematics 31: 321-324.

16. Balakrishnan N, Joshi PC (1984) Product moments of order statistics from the doubly truncated exponential distribution. Naval Research Logistics Quarterly: 31: 27-31.

17. Khan AH, Ali MM (1987) Characterization of probability distributions through higher order gap. Communications in Statistics-Theory and Methods 16: 1281-1287.

18. Ahmad AA (2001) Moments of order statistics from doubly truncated continuous distribution. Statistics 35: 479-494.

New Technique for Solving the Advection-diffusion Equation in Three Dimensions using Laplace and Fourier Transforms

Essa KSM[1]*, Marrouf AA[1], El-Otaify MS[1], Mohamed AS[2] and Ismail G[2]

[1]*Mathematics and Theoretical Physics, NRC, Atomic Energy Authority, Cairo, Egypt*

[2]*Department of Mathematics, Faculty of Science, Zagazig University, Egypt*

Abstract

A steady-state three-dimensional mathematical model for the dispersion of pollutants from a continuously emitting ground point source in moderated winds is formulated by considering the eddy diffusivity as a power law profile of vertical height. The advection along the mean wind and the diffusion in crosswind and vertical directions was accounted. The closed form analytical solution of the proposed problem has obtained using the methods of Laplace and Fourier transforms. The analytical model is compared with data collected from nine experiments conducted at Inshas, Cairo (Egypt). The model shows a best agreement between observed and calculated concentration.

Keywords: Advection-diffusion equation; Laplace transform; Fourier transform; Bessel function

Introduction

Environmental problems caused by the huge development and the big progress in industrial, which cause's a lot of pollutions. The transport of these pollutants can be adequately described by the advection–diffusion equation. In the last few years, there has been increased research interest in searching for analytical solutions for the advection–diffusion equation (ADE). Therefore, it is possible to construct a theoretical model for the dispersion from a continuous point source from an Eulerian perspective, given adequate boundary and initial conditions and the knowledge of the mean wind velocity field and of the concentration turbulent fluxes. The exact solution of the linear advection–dispersion (or diffusion) transport equation for both transient and steady-state regimes has been obtained [1].

The two- and three-dimensional advection–diffusion equation with spatially variable velocity and diffusion coefficients has been provided analytically [2]. A mathematical treatment has been proposed for the ground level concentration of pollutant from the continuously emitted point source [3]. Essa and El-Otaify studied a mathematical model for hermitized atmospheric dispersion (self adjoined by itself) in low winds with eddy diffusivities as linear functions of the downwind distance [4]. More recently the generalized analytical model describing the crosswind-integrated concentrations is presented [5]. Also, an analytical scheme is described to solve the resulting two dimensional steady-state advection–diffusion equation for horizontal wind speed as a generalized function of vertical height above the ground and eddy diffusivity as a function of both downwind distance from the source and vertical height.

On the other hand, the literature presents several methods to analytically solve the partial differential equations governing transport phenomena [6-10]. For example, the method of separation-of-variables is one of the oldest and most widely used techniques. Similarly, the classical Green's function method can be applied to problems with source terms and inhomogeneous boundary conditions on finite, Semi-infinite, and infinite regions [10,11]. Integral transform techniques, such as the Laplace and Fourier transform methods, employ a mathematical operator that produces a new function by integrating the product of an existing function and a kernel function between suitable limits.

In this study, we obtained a mathematical model for dispersion of air pollutants in moderated winds by taking into account the diffusion in vertical height direction and advection along the mean wind. The eddy diffusivity is assumed to be power law in the vertical length. We provided analytical solutions to the advection–diffusion equation for three-dimensional with the physically relevant boundary conditions. The moderate data collected during the convective conditions. From nine experiments conducted at Inshas site, Cairo-Egypt [4], which used to investigate the analytical solution.

Mathematical Treatment

The dispersion of pollutants in the atmosphere is governed by the basic atmospheric diffusion equation. Under the assumption of incompressible flow, atmospheric diffusion equation based on the Gradient transport theory can be written in the rectangular coordinate system as:

$$\frac{\partial C}{\partial t}+u\frac{\partial C}{\partial x}+v\frac{\partial C}{\partial y}+w\frac{\partial C}{\partial z}=\frac{\partial}{\partial x}\left(K_x\frac{\partial C}{\partial x}\right)+\frac{\partial}{\partial y}\left(K_y\frac{\partial C}{\partial y}\right)+\frac{\partial}{\partial z}\left(K_z\frac{\partial C}{\partial z}\right)+S+R \quad (1)$$

where C(x, y, z) is the mean concentration of a pollutant (Bq/m^3), (μg/m^3) and (ppm); in which t is the time, S and R are the source and removal terms, respectively; (u, v, w) and (k_x, k_y, k_z) are the components of wind and diffusivity vectors in x, y and z directions, respectively, in an Eulerian frame of reference.

The following assumptions are made in order to simplify equation (1):

1) Steady -state conditions are considered, i.e., $\partial C/\partial t = 0$

2) We are going to study Eq. (1) in case, when the components of

***Corresponding author:** Essa KSM, Mathematics and Theoretical Physics, NRC, Atomic Energy Authority, Cairo, Egypt
E-mail: mohamedksm56@yahoo.com

wind (v, w) tends to zero

3) Source and removal (physical/chemical) pollutants are ignored so that S=0 and R=0

4) Under the moderate to strong winds, the transport due advection dominates over that due to longitudinal diffusion: $\left|u\frac{\partial C}{\partial x}\right| >> \left|\frac{\partial}{\partial x}\left(k_x\frac{\partial C}{\partial x}\right)\right|$

With the above assumptions, equation (1) reduces to:

$$u\frac{\partial \tilde{C}(x,y,z)}{\partial x} = \frac{\partial}{\partial y}\left(K_y\frac{\partial \tilde{C}(x,y,z)}{\partial y}\right) + \frac{\partial}{\partial z}\left(K_z\frac{\partial \tilde{C}(x,y,z)}{\partial z}\right) \tag{2}$$

Under the following boundary conditions:

$$\tilde{C}(x,y,z) = 0 \qquad \forall \qquad x,y,z \Rightarrow \infty \tag{3}$$

$$\tilde{C}(x,y,z) = \frac{Q}{u}\delta_x\delta_y \quad at \quad z = h \tag{4}$$

$$\tilde{C}(x,y,z) = \frac{Q}{u}\delta_y\delta_z \quad at \quad x = 0 \tag{5}$$

$$\frac{\partial \tilde{C}(x,y,z)}{\partial y} = 0 \quad at \quad y = \pm\infty \tag{6}$$

where δ (…) is Dirac's delta function, and h is the mixing height.

The wind speed u and eddy diffusivity k_y, k_z is expressed as a functions of power law of z as:

$$u = z^m;\ z \neq 0;\quad u = u_0;\ z = 0;$$
$$k_y = \alpha z^m;\quad \alpha \neq 0; \tag{7}$$
$$k_z = \beta z^n;\quad n < 1;\quad \beta \neq 0;$$

Where α, β, m, n are turbulence parameters and depend on atmospheric stability.

The Analytical Solution

Eq. (2) can solve analytically as follows:

Transform the variable x to s by applying the Laplace transform on Eq. (2) to become

$$-Q\delta(y)\delta(z-z_0) + sz^m\tilde{C}(s,y,z) = \alpha z^m\frac{\partial^2\tilde{C}(s,y,z)}{\partial y^2} + \beta z^n\frac{\partial^2\tilde{C}(s,y,z)}{\partial z^2} + \beta nz^{n-1}\frac{\partial\tilde{C}(s,y,z)}{\partial z} \tag{8}$$

Again transform the variable y to λ by applying the Fourier transform on Eq. (8) to become

$$-Q\delta(z-z_0) + sz^m\tilde{C}_*(s,\lambda,z) = -\alpha\lambda^2 z^m\tilde{C}_*(s,\lambda,z) + \beta z^n\frac{\partial\,\tilde{C}_*(s,\lambda,z)}{\partial z} + \beta nz^{n-1}\frac{\partial\tilde{C}_*(s,\lambda,z)}{\partial z} \tag{9}$$

Eq. (9) simplified to the form

$$\beta z^n\frac{\partial^2\tilde{C}_*(s,\lambda,z)}{\partial z^2} + \beta nz^{n-1}\frac{\partial\tilde{C}_*(s,\lambda,z)}{\partial z} - (\alpha\lambda^2 + s)z^m\tilde{C}_*(s,\lambda,z) = -Q\delta(z-z_0) \tag{10}$$

Now we will solve the homogeneous equation of Eq. (10) which takes the form

$$z^2\frac{\partial^2\tilde{C}_*(s,\lambda,z)}{\partial z^2} + nz\frac{\partial\tilde{C}_*(s,y,z)}{\partial z} - \frac{(\alpha\lambda^2 + s)}{\beta}z^{m-n+2}\tilde{C}_*(s,\lambda,z) = 0 \tag{11}$$

Transform $z\,to\,z_* : z_* = z^{\frac{m-n+2}{2}}$ then Eq. (11) becomes

$$z_*^2\frac{\partial^2\tilde{C}_*}{\partial z_*^2} + \frac{m+n}{m-n+2}z_*\frac{\partial\tilde{C}_*}{\partial z_*} - \frac{4(\alpha\lambda^2 + s)}{\beta(m-n+2)^2}z_*^2\tilde{C}_* = 0 \tag{12}$$

Again transform $\tilde{C}_*\,to\,\tilde{C}_{**} : \tilde{C}_* = z_*^{\frac{1-n}{m-n=2}}\tilde{C}_{**}$ then Eq. (12) become

$$z_*^2\frac{\partial^2\tilde{C}_{**}}{\partial z_*^2} + z_*\frac{\partial\tilde{C}_{**}}{\partial z_*} - ((\eta z_*)^2 + \nu^2)z_*^2\tilde{C}_{**} = 0 \tag{13}$$

where $\nu = \frac{1-n}{m-n=2};\ and\,\eta = \sqrt{\frac{4(s+\alpha\lambda^2)}{\beta(m-n+2)^2}}$

But Eq. (13) is modified Bessel equation which has solution [12].

$$\tilde{C}_{**} = AI_\nu(\eta z_*) + BK_\nu(\eta z_*) \tag{14}$$

$$\tilde{C}_* = z_*^\nu\tilde{C}_{**} = z_*^\nu[AI_\nu(\eta z_*) + BK_\nu(\eta z_*)] \tag{15}$$

where A and B are constant

Now the general solution of the non-homogeneous Eq. (10) takes the form

$$\tilde{C}_* = z_*^\nu[A_*I_\nu(\eta z_*) + B_*K_\nu(\eta z_*)] \tag{16}$$

where A_* and B_* are constants

Apply the boundary condition Eq. (3) on Eq. (16) which become

$$\tilde{C}_*(s,\lambda,z) = B_*z_*^\nu K_\nu(\eta z_*) \tag{17}$$

Apply the boundary condition Eq. (5) on Eq. (16) which gives

$$B_* = \frac{Q}{u}\frac{\Gamma(1-\nu)\sin(\nu\pi)}{\frac{\pi}{2}(\frac{\eta}{2})^{-\nu}} \tag{18}$$

Substitute B_* Eq. (16) in Eq. (17) which gives:

$$\tilde{C}_*(s,\lambda,z) = \frac{2Q}{\pi u_0}(\frac{\eta}{2})^\nu\Gamma(1-\nu)\sin(\nu\pi)z_*^\nu K_\nu(\eta z_*) \tag{19}$$

Apply inverse Fourier and inverse Laplace respectively on Eq. (19) we get:

$$C(x,y,z) = \frac{2\sqrt{2}Q\Gamma(1-\nu)\sin(\nu\pi)}{u_0\sqrt{\alpha\pi}}(2x)^{\nu-\frac{1}{2}}\left(\frac{z^{\frac{m-n+2}{2}}}{(m-n+2)\sqrt{\beta}}\right)^\nu\left(\frac{4z^{m-n+2}}{(m-n+2)^2\beta} + \frac{y^2}{\alpha}\right)^{-\nu-\frac{1}{2}}e^{\left[-\left(\frac{z^{m-n+2}}{(m-n+2)^2 x\beta} + \frac{y^2}{4\alpha x}\right)\right]} \tag{20}$$

Source Data

The diffusion data for the estimating were gathered during ^{135}I isotope tracer nine experiments in moderate wind with unstable conditions at Inshas, Cairo. During each run, the tracer was released from source has height 43 m for twenty four hours working, where the air samples were collected during half hour at a height 0.7 m. We collected air samples from 92 m to 184 m around the source in AEA, Egypt. The study area is flat, dominated by sandy soil with poor vegetation cover. The air samples collected were analyzed in Radiation Protection Department, NRC, AEA, Cairo, Egypt using a high volume air sampler with 220 V/50 Hz bias [13]. Meteorological data have been provided by the measurements done at 10 m and 60 m.

For the concentration computations, we require the knowledge of wind speed, wind direction, source strength, the dispersion parameters, mixing height and the vertical scale velocity. Wind speeds are greater than 3 m/s most of the time even at 10 m level. Further the variation wind direction with time is also visible. Thus in the present study, we have adopted dispersion parameters for urban terrain which are based on power law functions. The analytical expressions depend upon downwind distance, vertical distance and atmospheric stability. The atmospheric stability has been calculated from Monin-Obukhov length scale (1/L) [14] based on friction velocity, temperature, and surface heat flux.

Results and Discussion

The concentration is computed using data collected at vertical distance of a 30 m multi-level micrometeorological tower. In all a test runs were conducted for the purpose of computation. The concentration at a receptor can be computed in the following way:

Applying formula Eq. (21) which contains eddy diffusivities as function with power law at y = 0.0 for half hourly averaging.

As an illustration, results computed from these approaches are shown in Table 1, for nine typical tests conducted at Ins has site,

Test	Downwind distance (m)	Observed conc. (Bq/m³)	Predicted conc. (Bq/m³)
1	100	0.025	0.047
2	98	0.037	0.076
3	115	0.091	0.127
4	135	0.197	0.225
5	99	0.272	0.282
6	184	0.188	0.196
7	165	0.447	0.503
8	134	0.123	0.162
9	96	0.032	0.039

Table 1: Observed and predicted concentrations for run 9 experiments.

Cairo-Egypt [4]. This table shows that the observed and predicted concentrations for ^{135}I using Eq. (20) with power law of eddy diffusivities and the wind speed are very near to each other of ^{135}I.

Figure 1 shows the variation of predicted and observed concentration of ^{135}I with the downwind distance. One gets very good agreement between observed and predicted concentration.

Figure 2 shows that the predicted concentrations which are estimated from Eq. (20) are a factor of two with the observed concentration.

Statistical Methods

Now, the statistical method is presented and comparison among analytical, statically and observed results will be offered [13]. The following standard statistical performance measures that characterize the agreement between prediction ($C_p = C_{pred}$) and observations ($C_o = C_{obs}$):

1. Normalized mean square error (NMSE), It is an estimator of the overall deviations between predicted and observed concentrations. Smaller values of NMSE indicate a better model performance. It is defined as:

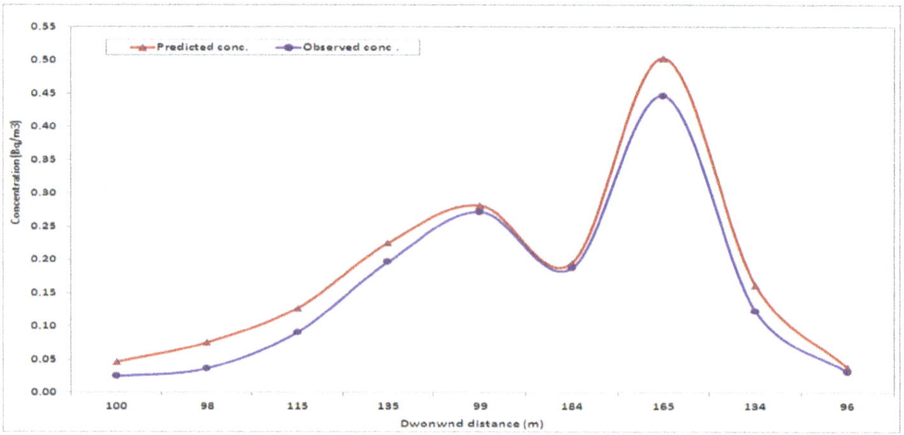

Figure 1: Maximum computed concentrations compared with observed maximum value for each test run.

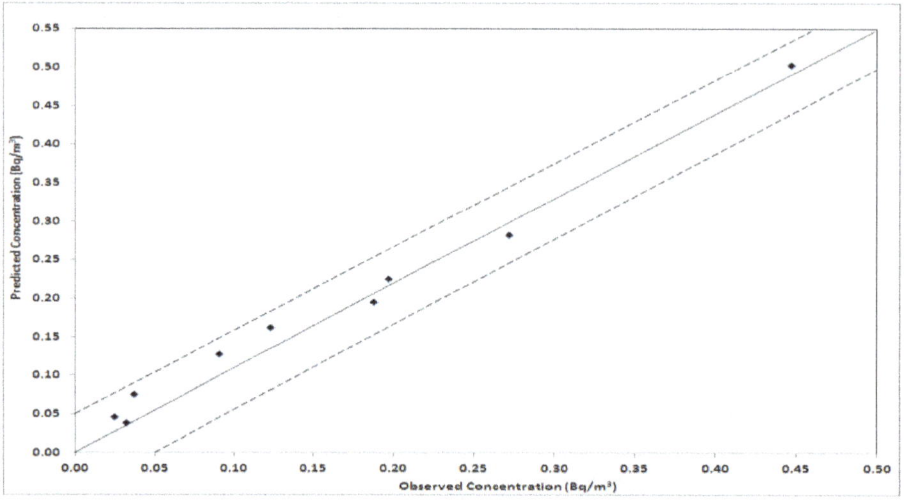

Figure 2: Diagram of predicted model for Eq. (20) with corresponding observation. Solid lines indicate one to one and dashed lines a factor of two.

Statistical functions	^{135}I			
	NMSE	FB	COR	FAC2
Predicated Concentrations model	0.03	-0.16	0.99	1.36

Table 2: Comparison between averages predicted isotopes for ^{135}I and observed concentrations.

$$NMSE = \frac{\overline{(C_o - C_p)^2}}{\overline{C}_o \overline{C}_p}$$

2. Fractional bias (FB): It provides information on the tendency of the model to overestimate or underestimate the observed concentrations. The values of FB lie between -2 and +2 andit has a value of zero for an ideal model. It is expressed as:

$$FB = \frac{(\overline{C}_o - \overline{C}_p)}{0.5(\overline{C}_o + \overline{C}_p)}$$

3. Correlation coefficient (R): It describes the degree of association between predicted and observed concentrations and is given by:

$$R = \frac{\overline{(C_o - \overline{C}_o)(C_p - \overline{C}_p)}}{\sigma_o \sigma_p}$$

4. Fraction within a factor of two (FAC2) is defined as:

FAC2 = fraction of the data for which

$0.5 \leq (C_p/C_o) \leq 2$

Where σ_p and σ_o are the standard deviations of C_p and C_o respectively. Here the over bars indicate the average over all measurements (Nm). A perfect model would have the following idealized performance: NMSE = FB = 0 and COR = FAC2 = 1.0

From the statistical method of Table 2, we find that the predicted concentrations for ^{135}I lie inside factor of 2 with observed data. Regarding to NMSE, FB and COR the predicted concentrations for ^{135}I are better with observed data.

Conclusion

In this paper, a steady-state three-dimensional mathematical model for the dispersion of pollutants from a continuously emitting ground point source in moderated winds is formulated. Besides advection along the mean wind, the model takes into account the diffusion in crosswind and vertical directions. The eddy diffusivity and the wind speed are assumed to be power law in the vertical height z.

The closed form analytical solution of the proposed problem has obtained using the methods of Laplace and Fourier transforms.

In general, the present model is compared with data collected from nine experiments conducted at Inshas, Cairo (Egypt). One gets the predicted concentrations are in a best agreement with the corresponding observation. Moreover, the Statistical results here are in agreement with the analytical results.

Acknowledgements

This work has been completed by the support of Egyptian Atomic Energy Authority, and the authors thank for this support. The First author's thank is extended to all members of Mathematics and theoretical Physics for providing the experimental data of ^{135}I.

References

1. Guerrero JSP, Pimentel LCG, Skaggs TH, van Genuchten MT (2009) Analytical solution of the advection-diffusion transport equation using a change-of-variable and integral transform technique. International Journal of Heat and Mass Transfer 52: 3297-3304.
2. Zoppou C, Knight JH (1999) Analytical solution of a spatially variable coefficient advection diffusion equation in up to three dimensions. Applied Mathematical Modeling 23: 667-685.
3. Embaby M, Mayhoub AB, Essa KSM, Etman S (2002) Maximum ground level concentration of air pollutant. Atmosfera 15: 185-191.
4. Essa KSM, El-Otaify MS (2007) Mathematical model for hermitized atmospheric dispersion in low winds with eddy diffusivities linear functions downwind distance. Meteorology and Atmospheric Physics 96: 265-275.
5. Kumar P, Sharan M (2014) An analytical model for dispersion of pollutants from a continuous source in the atmospheric boundary layer. Proc R Soc A pp: 1-24.
6. Courant D, Hilbert D (1953) Methods of Mathematical Physics. Wiley Interscience Publications, New York, USA.
7. Morse PM, Feshbach H (1953) Methods of Theoretical Physics. McGraw-Hill, New York, USA.
8. Carslaw HS, Jaeger JC (1959) Conduction of Heat in Solids (2nd edn.). Oxford University press, Oxford.
9. Sneddon IN (1972) The Use of Integral Transforms. McGraw-Hill.
10. Ozisik MN (1980) Heat Conduction. Wiley, New York.
11. Leij FJ, van Genuchten MT (2000) Analytical modeling of non-aqueous phase liquid dissolution with Green's functions. Transport in Porous Media 38: 141-166.
12. Irving J, Mullineux (1959) Mathematics in Physics and Engineering. Pure and applied physics.
13. Essa KSM, Mubarak F, Khadra SA (2005) Comparison of Some Sigma Schemes for Estimation of Air Pollutant Dispersion in Moderate and Low Winds. Atmospheric Science Letter 6: 90-96.
14. Donald G (1972) Relation among Stability Parameters in the Surface Layer. Boundary Layer Meteorology 3: 47-58.

Laying Chicken Algorithm: A New Meta-Heuristic Approach to Solve Continuous Programming Problems

Eghbal Hosseini*

Philosophy Doctor of Operational Research and Optimization, Department of Mathematics, UAE

Abstract

A concept for the optimization of continuous programming problems using an approach based on behavior of laying chicken, to produce chicken, is introduced. Laying chicken algorithm (LCA) is used for solving different kinds of linear and non-linear programming problems. The presented approach gives efficient and feasible solutions in an appropriate time which has been evaluated by solving test problems. The comparison between LCA and both of meta-heuristic and classic approaches are proposed.

Keywords: Laying chicken algorithm; Meta-heuristic approaches; Optimization problems

Introduction

In the recent decades, optimization and computer scientists have been designing several algorithms based on behavior of animals and insects because the natural systems are very efficient. Swarm Intelligence (SI) was introduced in 1989 as a novel approach in the global optimization [1]. Ant Colony Optimization (ACO) was proposed to solve discrete problems as a new meta-heuristic optimizer [2]. Particle Swarm Optimization (PSO) introduced for solving continuous programming problems [3]. These algorithms have been found acceptable solutions to optimization problems. Therefore, the meta-heuristic algorithms, such as Artificial Bee Colony Algorithm [4], krill herd algorithm [5], Bat Algorithm (BA) [6], social spider optimization [7], Chicken Swarm Optimization (CSO) [8], firefly algorithm [9] have attracted by many researchers.

This paper suggests a novel optimizer for optimization of linear and non-linear programming problems in continuous state. The approach has been discovered in simulating of a natural bi-inspiring model. The paper introduces the laying chicken algorithm concept, strongly discusses the steps of its extension from bi-inspiring simulation model to optimizer. Finally, by proposed numerical results of linear and non-linear test problems, it is easy to see that the simulated model has been succeeded.

Laying chicken algorithm is related to two principle concepts. In general, LCA ties to artificial agent or artificial life obviously, and to, laying fishes, laying turtles, laying snakes and laying chickens in particular (not SI behavior). It comes from both evolutionary programming and genetic algorithms. In this paper, relationships between LCA and above concepts are obvious.

Proposed laying chicken algorithm by the author includes an easy natural theory and concept, and performance of steps can be displayed in some lines of MATLAB code. It needs just an array, to store feasible solutions, and initial mathematical factors. So it has an acceptable computational complexity in both of memory and time. Initial testes have realized the enforcement to be feasible and effective using different classes of problems. In the rest of paper, performances of steps and their MATLAB code will be presented. Finally use of approach to solve several kinds of problems, such as constraint and unconstraint programming problems in different states, is discussed.

Simulation of Laying Chicken Behavior

The hens and their eggs are a great source of food as one of the most extensive tame animals [8]. This paper focuses on behavior of laying hen and answer of this question: "how does she convert the egg to the chicken?" In this paper, same as eggs to the chicken, the feasible solutions have been changed to the optimal solution. In fact, each egg displays a feasible solution in continuous programming problem and a chicken describes optimal solution in the problem.

Farmers use a false egg sometimes to encourage hens to stay in the nest. Because hens often prefer in the same location and not empty nest to lay, in fact they try to do that in the nest that already contain eggs. This is a great idea to create an initial feasible solution and to generate first population near that.

Pheromone of ants in ant colony, individual members or global best in particle swarm optimization, crossover or combination of genes in genetic, are the fundamental concepts of some of the meta-heuristic algorithms. Here hens try to warm their eggs; this concept is base to development of laying chicken algorithm. Same as temperature of eggs objective function of solutions will be improved. Rotation of eggs is the next concept which will be simulated by little change of solutions.

The Laying Chicken Algorithm Concept

The laying chicken algorithm optimizer may the best proposed using describe its conception development. As mentioned, LCA comes from laying hen as an original naturel event, so in this section the main concept of LCA and its relationship with the bio-inspiring event is discussed.

The initial solution

The simulated approach already was written based on two main

***Corresponding author:** Hosseini E, Philosophy Doctor of Operational Research and Optimization, Department of Mathematics, UAE
E-mail: eghbal_math@yahoo.com

concepts: initial solution and population. Same as the first egg in the nest, the initial feasible solution was necessary. So it has been created randomly. If it is not feasible, a loop in the MATLAB code is repeated to create a feasible one. Initial solution for some optimization problems is created in MATLAB are as follows:

The initial population

In the first iteration, initial population of solutions has been created near the initial feasible solution as possible. In fact, the next factor of the simulation defines "the initial neighborhood," an n-dimensional neighborhood of R^n, this is defined as follows:

$$\|X-Y\| \le k \quad (1)$$

or

$$\sqrt{(x_1-y_1)^2+(x_2-y_2)^2+\ldots+(x_n-y_n)^2} \le k$$

Which, X is initial solution, Y is an n-dimensional vectors and k is a positive constant. Here the initial population of eggs has been created randomly in the possible nearest neighborhood of the initial solution.

Each member of initial population has to be in this neighborhood of the initial solution. We try to generate solutions very near the initial solution. This is because hens usually like to stay in their nest with their eggs. In fact, they prefer to convert their eggs to chicken than other animal eggs. Figure 1 shows 500 eggs (feasible solutions) near to the initial solution for a given problem with k=1 in R^2 (Figure 1).

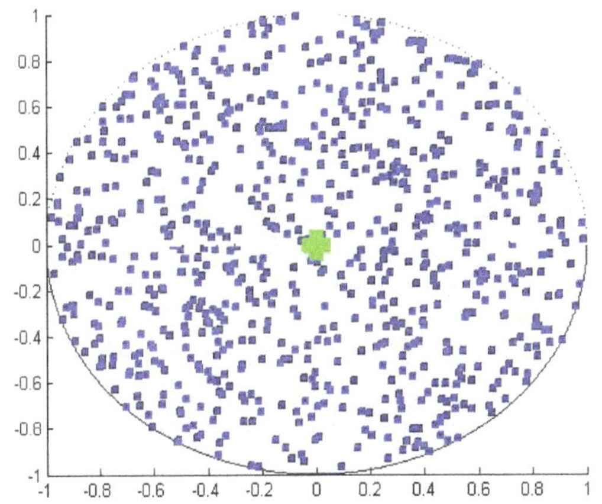

Figure 1: 500 eggs (blue points) have been created near initial solution (green point).

The algorithm will be more efficient when k be very small. This is because it does not miss many solutions near initial solution small k.

Improving of population

Each solution in population, which its objective function is not better than objective function of initial solution, should be changed in direction initial solution while it will be better than initial solution. In fact, value of particles have been changed in direction vector which connects its and the initial solution. These solutions have been modified as follows:

$$x_{j+1}=x_j+\alpha d_{j0} \quad (2)$$

Which, d_{j0} is the vector from xj to x_0 and $f(x_j)< f(x_0)$, $0 \le \alpha \le 2k$ in maximization problems.

All states of α have been described in Table 1 and according to that, the feasible interval for α as follows:

$$0 \le \alpha \le 2k \quad (3)$$

It is easy to show that $\alpha \to 0$ does not change solutions very well, so interval $0<\alpha<k/4$ has been removed and the following is the best:

$$k/4<\alpha \le 2k \quad (4)$$

But according to the gradient theorem in Figure 2a objective function of blue points are not better than initial solution (large red point) and small red points are better than it, in a given problem that its optimal solution is in right hand side of the initial solution. So interval of α has been modified as follows:

$$k \le \alpha \le 2k \quad (5)$$

This is because the author wants to move all blue solutions in Figure 2a in direction initial solution such that they will be better than it. Green points in Figure 2b are these solutions after their movement. By this stage all solution in population will be better than initial solution Figure 2c. The best solution in this iteration will be initial solution in the next iteration. So in the next population and after this step, all solution will be better than the best solution of current population. This is the main idea of the algorithm which every population is better than previous population. Pseudo code of this stage has been shown in Figure 3.

Changing the solutions

The last trait of the simulated method has been inspired from rotation of the eggs by the hen. She rotates the eggs three or four times every day. In this stage except the best solution, all member

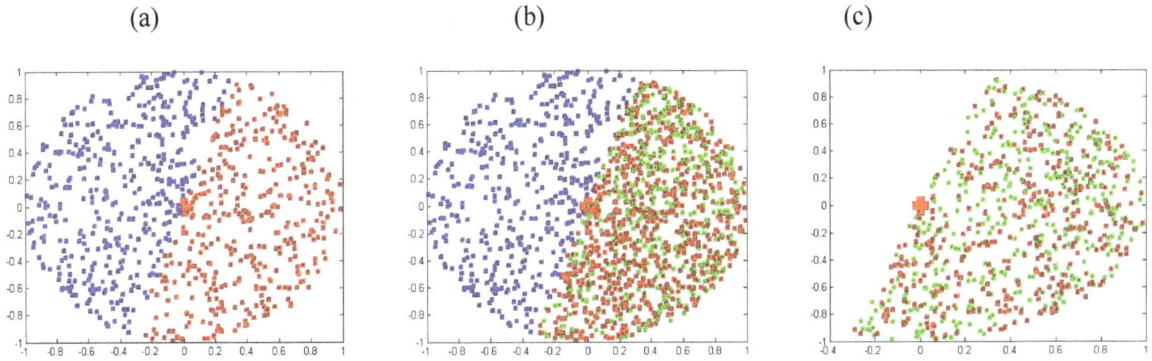

Figure 2: Process of changing solutions which are not better than initial solution in direction it.

```
begin
    Objective function f(x), x = (x1, ..., xd)T
    Generate initial population of eggs xi (i = 1, 2, ..., n) near x0
    Temperature Ei at xi is determined by f(xi)
        for j = 1 : n (all n eggs)
            while (Ej < E0), warm egg j;
            end
            Evaluate new solutions and update temperature
        end
end
```

Figure 3: Pseudo code of improving of solution.

```
begin
        for j=1 : n (all n eggs)
         if (f(xj) < objective function of the best solution), change xj a
            little;
         end
            Evaluate new solutions and update temperature(objective
            function)
        end
end
```

Figure 4: MATLAB code of changing the solutions.

of population have been little changed as follows. ε is a given small positive number.

Some of solutions have been selected randomly and changed as follows:

$(xi+1,yi+1)=(x_i \mp \varepsilon, yi \mp \varepsilon)$

Each solution j which $x_j < x_{best}$ has been changed as follows:

$(x_j+1, y_j+1)=(x_j+\varepsilon,y_j)$

Each solution k which $x_k > x_{best}$ has been changed as follows:

$(x_{k+1}, y_{k+1})=(x_k-\varepsilon,y_k)$

At each iteration, the best solution has been saved and other solutions selected near that in the next population. There are two states for current stage: If this stage creates the better solution from the best one (best in this iteration), it will substitute the best and in the next iteration solutions should be selected near that. Otherwise the best solution will not change. In fact by this stage, the best solution will be better or not changed. Figure 4 shows code of this step.

This stage is useful because it causes to generate more random solutions except the best solution. In fact, the algorithm has more choices to select the best solution by more random solutions.

Steps of the algorithm

The main steps of the algorithm in R^2 as follows:

- The initial feasible solution (x_0,y_0), is created. Number of iteration, N, and an arbitrary small positive number, ε_1 are given.
- Initial population near (x_0,y_0), is generated.
- Each solution in step 2, which its objective function is not better than (x_0,y_0), should be changed in direction (x_0,y_0) and found the best solution (x_{best},y_{best})
- All solutions, except the best one, have been very little changed.
- Objective function of solutions and the best solution is updated. Let $(x_0,y_0)=(x_{best},y_{best})$, go back to step 2.
- If $|f(x_{ibest})-f(x_{(i+1)best})|<\varepsilon 1$ or the number of iteration is more than M the algorithm will be finished, x_{ibest}, $x_{(i+1)best}$ are the best solutions in two consecutive generations. Figure 5 shows the process of the algorithm to gain optimal solution from a given feasible solution. Explanation of Figure 5 as follows: initial solution is generated in eqn (1), Red point is the optimal solution and the green point is an arbitrary feasible solution. Eqn (2) shows first population near initial solution with k=1, red points are better than the initial solution and blue points are not. Blue points move in eqn (3) and convert to green points which are better than initial solution. The algorithm continues with eqn (4). All solutions except the best solution have been little changed according eqn (5). Next population will be created near the best solution.

The process of the algorithm in R^3 has been shown in Figure 6.

Convergence

Convergent parameter set includes initial solution, small positive number ε and constants k and α. BBA is run several times to determine convergence rate and convergent parameter set of the algorithm. The convergence rate is top if various results are gained by more

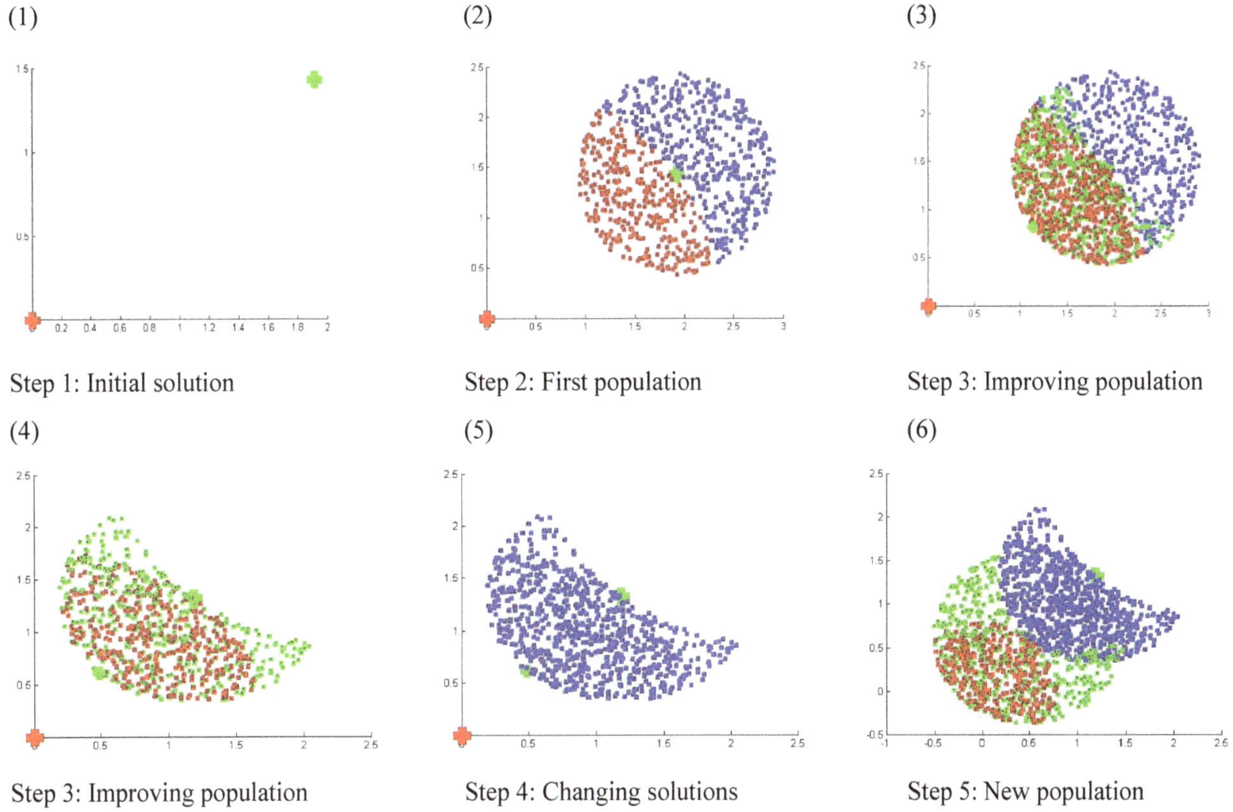

Figure 5: Steps of the algorithm to obtain optimal solution R^2.

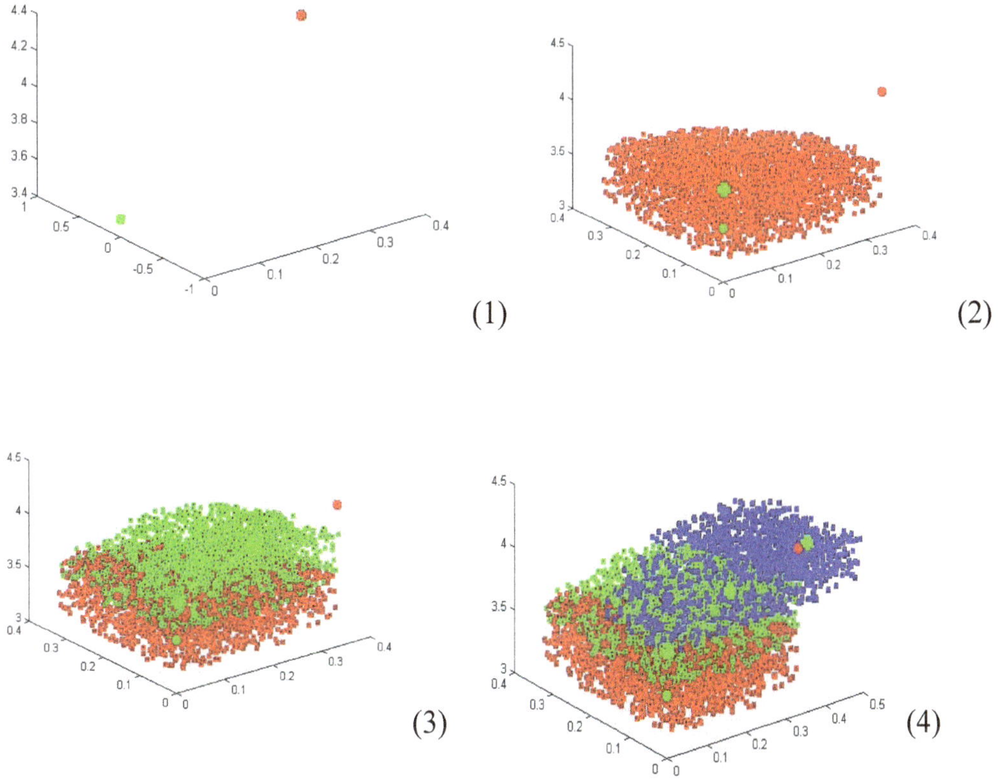

Figure 6: Steps of the algorithm to obtain optimal solution in R^3.

performances. Large number of eggs and slow convergent parameter set must be used in this state. The convergence rate is low if after large number of iterations same result is gained. Small number of eggs and quick convergent parameter set should be used here. Finally, if suitable results are gained, convergence rate is well and the common parameter set should be used. According to the computational results, convergence rate of the algorithm is completely high rather than other proposed meta-heuristic approaches.

Computational Results

Example 1 [9]

Consider the following problem:

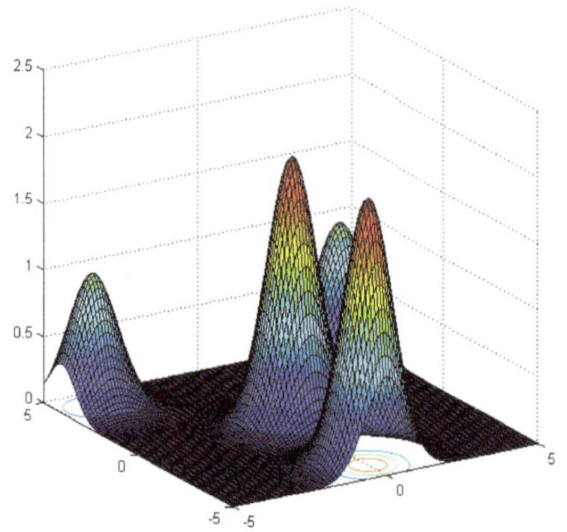

Figure 7: Behavior of the function: Example 1.

$$\min \ \exp(-(x-4)^2-(y-4)^2)+\exp(-(x+4)^2-(y-4)^2)+2\exp(-x^2-y^2) + 2\exp(-x^2--(y+4)^2)$$

Figure 7 shows behavior of objective function in Example 1. To solve the problem, all efficient factors to obtain optimal solution are: number of eggs, stochastic constant (k), small positive number ε to change solutions, and initial feasible solution. According to the Table 2 the proposed meta-heuristic approach has presented a solution with less time and number of eggs than firefly algorithm. Behavior of agents to obtain optimal solution has been shown in Figure 8.

Example 2 [10]

Consider the following linear programming problem:

$$\min -3x_1+x_2$$

$$x_1+2x_2 \leq 4$$

$$-x_1+x_2 \geq 0$$

Comparison LCA and exact methods has been proposed in Table 3. Figure 9 shows to move generations to optimal solution in feasible region.

Example 3 [11]

Consider the following non-linear programming problem:

$$\min -(x_1-4)^2-(x_2-4)^2$$

$$x_1+3 \leq 0$$

$$-x_1+x_2-2 \leq 0$$

$$x_1+x_2-4 \leq 0$$

$$x_1,x_2-2 \geq 0$$

Comparison LCA and other methods by example 3 have been proposed in Table 4. Behavior of generations has been shown in Figure 10.

States of α	$x_{j+1}=x_j+\alpha d_{j0}$	Explanation	Logical decision	P. Infeasible solutions
$\alpha >> 2k$	$\alpha\to\infty \Rightarrow x_{j+1}\to\infty$	x_{j+1} will be infeasible.	This state should not be selected.	100%
$\alpha << 2k$	$\alpha\to-\infty \Rightarrow x_{j+1}\to-\infty$	x_{j+1} will be infeasible.	This state should not be selected.	100%
$\alpha=0$	$x_{j+1}=x_j$	x_{j+1} will not be changed.	α should not be near zero.	–
$\alpha=k$	$x_{j+1}=x_0$	x_0 is already in population.	This state should not be selected.	–
$\alpha=2k$	$x_{j+1}=x_j+kd_{j0}$	x_{j+1} will be feasible.	This state can be selected.	0%
$\alpha<k$	$x_{j+1}=x_j+2kd_{j0}$	x_{j+1} till be feasible.	This state can be selected.	0%

Table 1: States of α description.

Algorithms	N.Eggs/Firefly	N. Iterations	Optimal Solution	F Max	K	ε	x_0
LCA	24	2	(–0.03,–0.02)	1.99	1	0.01	(0.80,0.90)
LCA	20	3	(–0.10,–0.01)	1.95	1	0.01	(1.81,1.90)
FA[9]	25	20	(0,0)	2	–	–	–

Table 2: Comparison of LCA and firefly algorithm: Example 1.

Algorithms	N. Eggs	N. Iterations	Optimal Solution	F Max	K	ε	x_0
LCA	100	4	(3.85,0.59)	–11.50	1	0.01	(0.80,0.90)
Exact methods[10]	–	–	(4,0)	–12	–	–	–

Table 3: Comparison of LCA and exact methods: Example 2.

Algorithms	N. Eggs	N. Iterations	Optimal Solution	F Max	k	ε	x_0
LCA	100	2	(0.04,0.02)	–31.47	1	0.01	(0.81,0.90)
LCA	40	3	(0.04,0.11)	–30.73	1	0.01	(0.81,0.90)
LCA	100	4	(0.00,0.17)	–30.6	1	0.01	(3,1)
Classic methods [11]			(0,0)	–32	–		

Table 4: Comparison of LCA and other methods: Example 3.

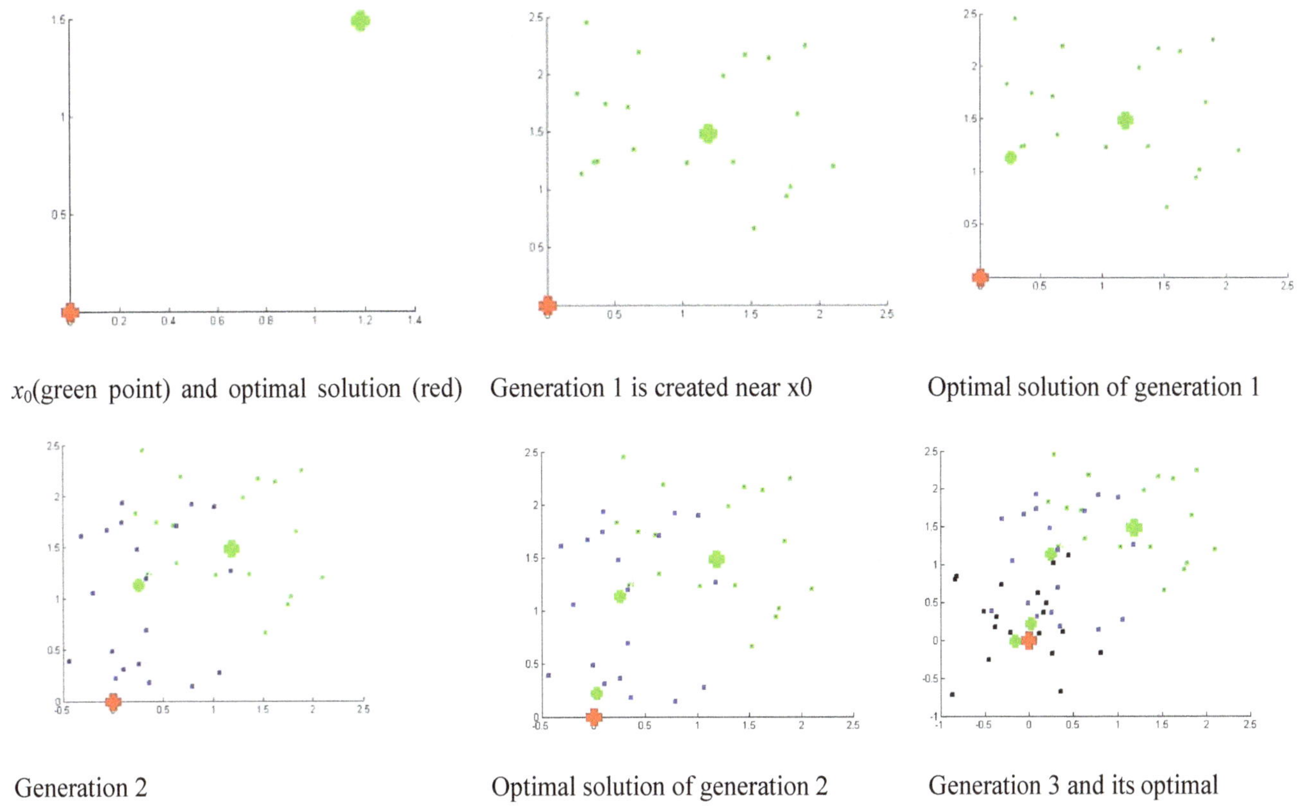
x_0(green point) and optimal solution (red)
Generation 1 is created near x0
Optimal solution of generation 1
Generation 2
Optimal solution of generation 2
Generation 3 and its optimal

Figure 8: Generations have been moved to find optimal solution: Example 1.

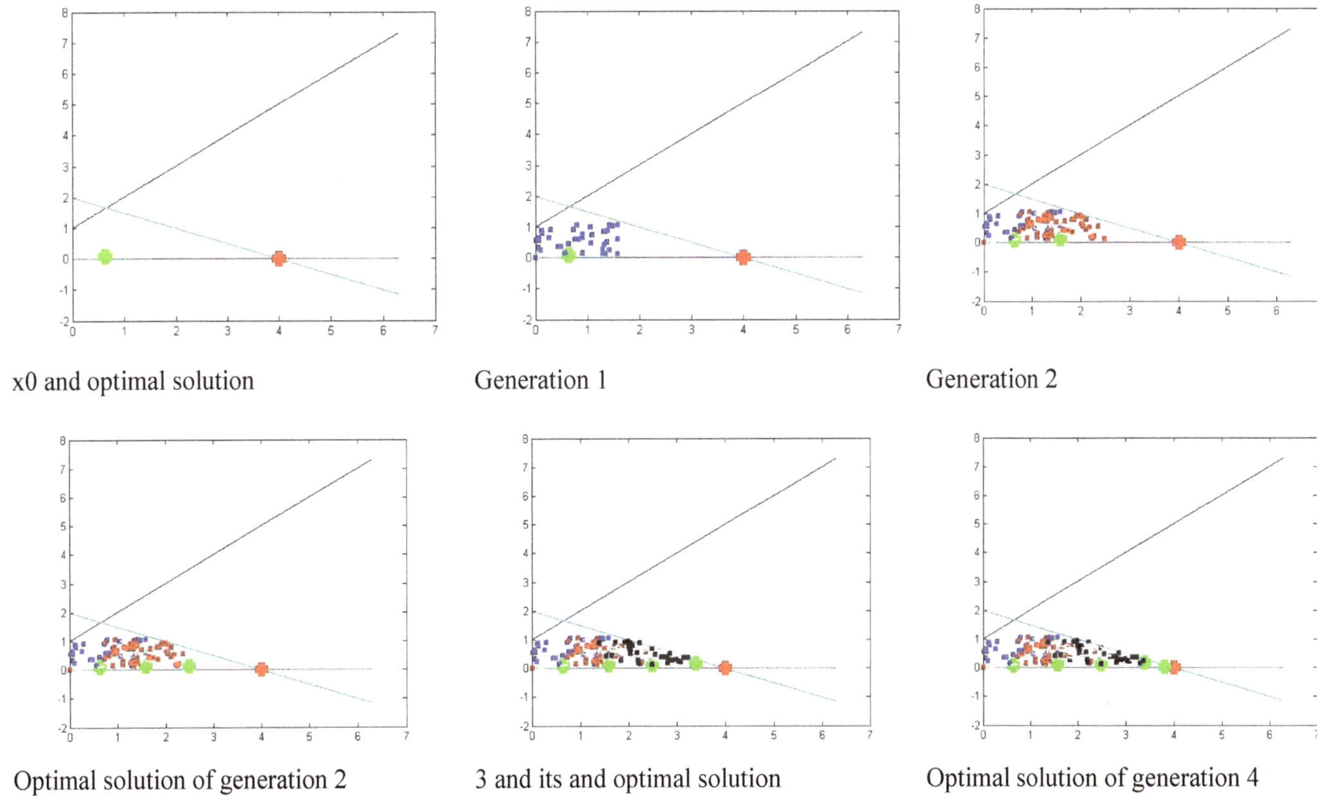
x0 and optimal solution
Generation 1
Generation 2
Optimal solution of generation 2
3 and its and optimal solution
Optimal solution of generation 4

Figure 9: Generations have been created in feasible region and moved to optimal solution: Example 2.

The proposed algorithm is efficient for problems by more than two variables according to the following example.

Example 4 [10]

Consider the following linear programming problem:

Min $x_1+x_2-4x_3$

$x_1+x_2+2x_3 \le 9$

$x_1+x_2-x_3 \le 2$

$-x_1+x_{2+}x_3 \le 4$

$x_1,x_2,x_3 \ge 0$

Comparison LCA and exact methods has been proposed in Table 5. Behavior of generations to find optimal solution has been shown in Figure 11.

Conclusion

Laying chicken algorithm is an easy meta-heuristic approach which optimizes different kinds of functions and optimization programming problems. Also, it seems efficient according to the examples. LCA was proposed as a natural event algorithm, not based on swarm intelligence unlike most of pervious meta-heuristic approaches. It ties to behavior of hen in process of produce chickens from eggs. In fact, LCA relates to both of biological and evolution computation because of its evolution and stochastic process.

LCA was successful because it does not miss the great solutions near initial solution particularly when k is small. The number of generations would be less according to a suitable feasible solution such as x_0 in fact consuming time to find optimal solution is much better than other meta-heuristic approaches.

Finally, there are many different NP-Hard problems which can be solved by meta-heuristic approaches especially using laying chicken algorithm. The simple MATLAB code of the LCA can be interested in the future researches especially for problems in large size. However, the proposed solution by LCA is near to optimal solution, but it is an approximate approach and the better algorithms can be proposed in the future researches.

Algorithms	N. Eggs	N. Iterations	Optimal Solution	F Max	K	ε
LCA	27	7	(0.31,0.00,4.29)	–16.87	1	0.01
Exact methods[10]	–	–	(0.33,0,4.33)	-17	–	–

Table 5: Comparison of LCA and other methods: Example 4.

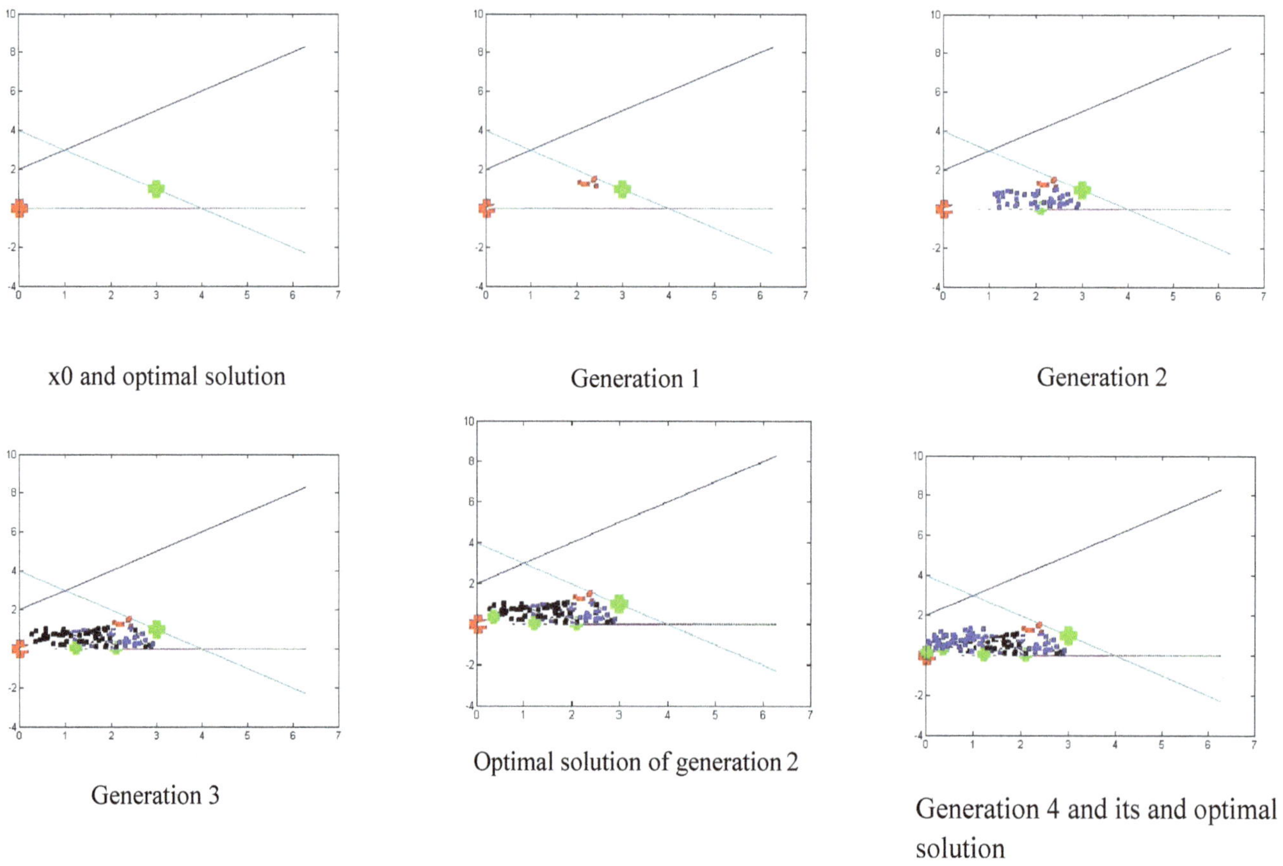

Figure 10: Generations moves in feasible region to find optimal solution: Example 3.

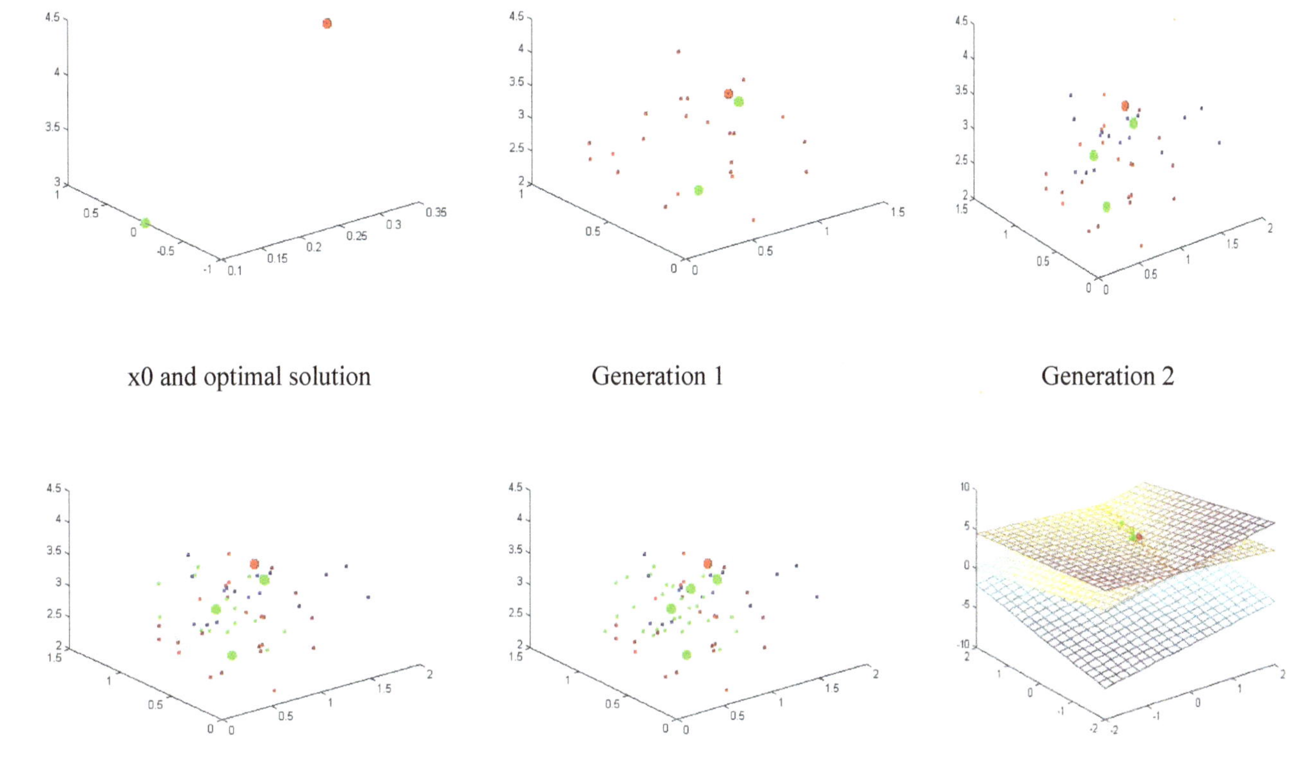

x0 and optimal solution | Generation 1 | Generation 2

Generation 3 | Generation 3 and its and optimal solution | Constraints and optimal solution

Figure 11: Generations moves in the feasible region to ind optimal solution: Example 4.

References

1. Beni G, Wang J (1989) Swarm intelligence in cellular robotic systems. In NATO Advanced Workshop on Robots and Biological Systems, Il Ciocco, Tuscany, Italy, pp: 703-712.
2. Colorni A, Dorigo M, Maniezzo V, Trubian M (1994) Ant system for job-shop scheduling. Belgian Journal of Operations Research. Statistics and Computer Science 34: 39-53.
3. Shi Y (2001) Particle swarm optimization: developments, applications and resources. In evolutionary computation 2001. Proceedings of the 2001 Congress 1: 81-86.
4. Karaboga D, Basturk B (2007) A powerful and efficient algorithm for numerical function optimization: artificial bee colony (ABC) algorithm. Journal of Global Optimization 39: 459-471.
5. Gandomi AH, Alavi AH (2012) Krill herd: a new bio-inspired optimization algorithm. Communications in Nonlinear Science and Numerical Simulation 17: 4831-4845.
6. Yang XS, He X (2013) Bat algorithm: literature review and applications. International Journal of Bio-Inspired Computation 5: 141-149.
7. Cuevas E, Cienfuegos M, Zaldívar D, Pérez-Cisneros M (2013) A swarm optimization algorithm inspired in the behavior of the social-spider. Expert Systems with Applications 40: 6374-6384.
8. Meng X, Liu Y, Gao X, Zhang H (2014) A new bio-inspired algorithm: chicken swarm optimization. In International Conference in Swarm Intelligence. Springer International Publishing, pp: 86-94.
9. Yang XS (2010) Nature-inspired metaheuristic algorithms. Luniver press.
10. Bazzara M (2010) linear programming and Network Flows, Wiley. Inc, New York.
11. Bazzara M (2007) Non-linear programming Theory and Algorithms. Wiley Inc. New York.

Application of Septic B-Spline Collocation Method for Solving the Coupled-BBM System

Raslan KR[1], EL-Danaf TS[2] and Ali KK[1*]

[1]*Mathematics Department, Faculty of Science, Al-Azhar University, Nasr-City, Cairo, Egypt*
[2]*Mathematics Department, Faculty of Science, Menoufia University, Shebein El-Koom, Egypt*

Abstract

In the present paper, a numerical method is proposed for the numerical solution of a coupled-BBM system with appropriate initial and boundary conditions by using collocation method with septic B-spline on the uniform mesh points. The method is shown to be unconditionally stable using von-Neumann technique. To test accuracy the error norms L_2, L_∞ are computed. Furthermore, interaction of two and three solitary waves are used to discuss the effect of the behavior of the solitary waves after the interaction. These results show that the technique introduced here is easy to apply. We make linearization for the nonlinear term.

Keywords: Collocation method; Septic B-Splines method; Coupled-BBM system

Introduction

In this paper, we consider the Coupled-BBM system, which belongs to the class of Boussinesq systems, modeling two-way propagation of long waves of small amplitude on the surface of water in a channel. The system is a good candidate for modeling long waves of small to moderate amplitude. The Coupled-BBM system is given by Bona and Chen [1],

$$v_t + u_x + (vu)_x - \frac{1}{6} v_{xxt} = 0, \tag{1}$$

$$u_t + v_x + u u_x - \frac{1}{6} u_{xxt} = 0, \tag{2}$$

Where subscripts x and t denote differentiation x distance and t time, is considered, $v(x,t)$ is a dimensionless deviation of the water surface from its undisturbed position and $u(x,t)$ is the dimensionless horizontal velocity above the bottom of the channel.

Boundary conditions

$$\begin{aligned} &u(a,t) = \alpha_1, \quad u(b,t) = \alpha_2, \\ &v(a,t) = \beta_1, \quad v(b,t) = \beta_2, \quad 0 \le t \le T. \\ &u_x(a,t) = 0, \quad u_x(b,t) = 0, \\ &v_x(a,t) = 0, \quad v_x(b,t) = 0, \quad 0 \le t \le T. \end{aligned} \tag{3}$$

And initial conditions.

$$\begin{aligned} &u(x,0) = f(x) \\ &v(x,0) = g(x) \quad a \le x \le b. \end{aligned} \tag{4}$$

One of the advantages that equation (1) has over alternative Boussinesq-type systems is the easiness with which it may be integrated numerically [2]. Furthermore, it was proved in [2,3] that the initial value problem either for $x \in R$ or with boundary conditions ($x \in [a,b]$) for (1) is well posed in certain natural function classes. The initial-boundary value problem of the form (1) posed on a bounded smooth plane domain with homogenous Dirichlet or Neumann or reflective (mixed) boundary conditions which is locally well-posed [4]. The existence and uniqueness of the system have been proved in Bona et al. [3]. They investigated the solution of the system as integral equation, while Chen [5] in his article established the existence of solitary waves for several Boussinesq types, including the Coupled-BBM system. Various numerical techniques including the finite element method have been used for the solution of Bona-Smith system of Boussinesq type in Antonopoulos et al. [6]. SS Behzadi and A Yildirim, using Quintic B-Spline Collocation Method for Solving the Coupled-BBM System [7]. ES Al-Rawi and MAM Sallal, using finite element method to fiend the Numerical solution of Coupled-BBM system [8]. Chen fined the exact traveling-wave solutions to bidirectional wave equations [9]. The numerical solutions of coupled nonlinear systems are very important in applied science, for example, the hirota-satsuma coupled KDV equation which admits soliton solution and it has many applications in communication and optical fibers; this system has been discussed numerically by Raslan et al. finite element methods [10]. Also, the Hirota equation has been solving by Raslan et al. using finite element methods [11]. A finite element algorithm based on the collocation method with trial functions taken as septic *B*-spline functions over the elements will be constructed. The septic *B*-spline basis together with finite element methods are shown to provide very accurate solutions in solving some partial differential equations and have been used before by several authors. In this article we are going to derive a numerical solution of the coupled BBM-system. The brief outline of this paper is as follows. In Section 2, septic B-spline collocation scheme is explained. In Sections 3 and 4, the method is described and applied to the coupled BBM-system. In Section 5, stability of the method is discussed. In Section 6, numerical examples are included to establish the applicability and accuracy of the proposed method computationally. Conclusion is given in Section 7 that briefly summarizes the numerical outcomes.

***Corresponding author:** Ali KK, Mathematics Department, Faculty of Science, Al-Azhar University, Nasr-City, Cairo, PN Box 11884, Egypt
E-mail: khalidkaram2012@yahoo.com

Septic B-spline Functions

To construct numerical solution, consider nodal points (x_j, t_n) defined in the region $[a,b]\times[0,T]$ where

$$a = x_0 < x_1 < \ldots < x_N = b, \qquad h = x_{j+1} - x_j = \frac{b-a}{N}, \qquad j = 0,1,\ldots,N.$$
$$0 = t_0 < t_1 < \ldots < t_n < \ldots < T, \qquad t_{j+1} - t_j = \Delta t, \quad t_n = n\Delta t, \qquad n = 0,1,\ldots .$$

The septic B-spline basis functions at knots are given by:

$$B_j(x) = \frac{1}{h^7}\begin{cases} a_1 = (x - x_{j-4})^7 & ,x_{j-4} \le x \le x_{j-3} \\ a_2 = a_1 - 8(x - x_{j-3})^7 & ,x_{j-3} \le x \le x_{j-2} \\ a_3 = a_2 + 28(x - x_{j-2})^7 & ,x_{j-2} \le x \le x_{j-1} \\ a_4 = a_3 - 56(x - x_{j-1})^7 & ,x_{j-1} \le x \le x_j \\ b_4 = b_3 - 56(x_{j+1} - x)^7 & ,x_j \le x \le x_{j+1} \\ b_3 = b_2 + 28(x_{j+2} - x)^7 & ,x_{j+1} \le x \le x_{j+2} \\ b_2 = b_1 - 8(x_{j+3} - x)^7 & ,x_{j+2} \le x \le x_{j+3} \\ b_1 = (x_{j+3} - x)^7 & ,x_{j+3} \le x \le x_{j+4} \\ 0 & ,otherwise \end{cases} \tag{5}$$

Using septic B-spline basis function (5) the values of $B_j(x)$ and its derivatives at the knots points can be calculated, which is tabulated in Table 1.

Solution of Coupled-BBM System

To apply the proposed method, we rewrite (1) and (2) as

$$\frac{\partial v(x,t)}{\partial t} + \frac{\partial u(x,t)}{\partial x} + \left(u(x,t)\frac{\partial v(x,t)}{\partial x} + v(x,t)\frac{\partial u(x,t)}{\partial x}\right) - \frac{1}{6}\left[\frac{\partial v^3(x,t)}{\partial x^2 \partial t}\right] = 0,$$
$$\frac{\partial u(x,t)}{\partial t} + \frac{\partial v(x,t)}{\partial x} + \left(u(x,t)\frac{\partial u(x,t)}{\partial x}\right) - \frac{1}{6}\left[\frac{\partial u^3(x,t)}{\partial x^2 \partial t}\right] = 0,$$

we take the approximations $u(x,t) = U_j^n$ and $v(x,t) = V_j^n$, then from famous Cranck-Nicolson scheme and forward finite difference approximation for the derivative t, [12]. We get

$$\frac{V_j^{n+1} - V_j^n}{k} + \frac{U_{xj}^{n+1} + U_{xj}^{n}}{2} + \left[\frac{(UV_x)_j^{n+1} + (UV_x)_j^n}{2} + \frac{(VU_x)_j^{n+1} + (VU_x)_j^n}{2}\right] - \frac{1}{6}\left[\frac{(V_{xx})_j^{n+1} + (V_{xx})_j^n}{k}\right] = 0, \tag{6}$$

$$\frac{U_j^{n+1} - U_j^n}{k} + \frac{V_{xj}^{n+1} + V_{xj}^{n}}{2} + \left[\frac{(U\ \ _x)_j^{n+1} + (U\ \ _x)_j^n}{2}\right] - \frac{1}{6}\left[\frac{(U_x\ )_j^{n+1} + (U_x\ )_j^n}{k}\right] = 0, \tag{7}$$

Where $k=\Delta t$ is the time step (Table 1).

In the Crank-Nicolson scheme, the time stepping process is half explicit and half implicit. So the method is better than simple finite difference method.

The nonlinear terms in Eqs. (6) and (7) is linearized using the form given by Rubin and Graves [13] as: we take linearization of the nonlinear term as follows

$$\begin{aligned} (UV_x)_j^{n+1} &= U_j^n V_{xj}^{n+1} + U_j^{n+1} V_{xj}^n - U_j^n V_{xj}^n, \\ (VU_x)_j^{n+1} &= V_j^n U_{xj}^{n+1} + V_j^{n+1} U_{xj}^n - V_j^n U_{xj}^n, \\ (UU_x)_j^{n+1} &= U_j^n U_{xj}^{n+1} + U_j^{n+1} U_{xj}^n - U_j^n U_{xj}^n \end{aligned} \tag{8}$$

Expressing $U(x,t)$ and $V(x,t)$ by using septic B-spline functions $B_j(x)$ and the time dependent parameters $c_j(t)$ and $\delta_j(t)$, for $U(x,t)$ and $V(x,t)$ respectively, the approximate solution can be written as:

$$U_N(x,t) = \sum_{j=-3}^{N+3} c_j(t)\, B_j(x), \quad V_N(x,t) = \sum_{j=-3}^{N+3} \delta_j(t)\, B_j(x), \tag{9}$$

Using approximate function (9) and septic B-spline functions (5), the approximate values $U(x)$, $V(x)$ and their derivatives up to second order are determined in terms of the time parameters $c_j(t)$ and $\delta_j(t)$, respectively, as

$$\begin{aligned} U_j &= U(x_j) = c_{j-3} + 120c_{j-2} + 1191c_{j-1} + 2416c_j + 1191c_{j+1} + 120c_{j+2} + c_{j+3}, \\ U'_j &= U'(x_j) = \frac{7}{h}(-c_{j-3} - 56c_{j-2} - 245c_{j-1} + 245c_{j+1} + 56c_{j+2} + c_{j+3}), \\ U''_j &= U''(x_j) = \frac{42}{h^2}(c_{j-3} + 24c_{j-2} + 15c_{j-1} - 80c_j + 15c_{j+1} + 24c_{j+2} + c_{j+3}), \\ V_j &= V(x_j) = \delta_{j-3} + 120\delta_{j-2} + 1191\delta_{j-1} + 2416\delta_j + 1191\delta_{j+1} + 120\delta_{j+2} + \delta_{j+3}, \\ V'_j &= V'(x_j) = \frac{7}{h}(-\delta_{j-3} - 56\delta_{j-2} - 245\delta_{j-1} + 245\delta_{j+1} + 56\delta_{j+2} + \delta_{j+3}), \\ V''_j &= V''(x_j) = \frac{42}{h^2}(\delta_{j-3} + 24\delta_{j-2} + 15\delta_{j-1} - 80\delta_j + 15\delta_{j+1} + 24\delta_{j+2} + \delta_{j+3}). \end{aligned} \tag{10}$$

On substituting the approximate solution for U,V and its derivatives from Eq. (10) at the knots in Eqs. (6) and (7) yields the following difference equation with the variables $c_j(t)$ and $\delta_j(t)$.

$$\begin{aligned} &A_1\delta_{j-3}^{n+1} + A_2\delta_{j-2}^{n+1} + A_3\delta_{j-1}^{n+1} + A_4\delta_j^{n+1} + A_5\delta_{j+1}^{n+1} + A_6\delta_{j+2}^{n+1} + A_7\delta_{j+3}^{n+1} + A_8c_{j-3}^{n+1} + \\ &A_9c_{j-2}^{n+1} + A_{10}c_{j-1}^{n+1} + A_{11}c_j^{n+1} + A_{12}c_{j+1}^{n+1} + A_{13}c_{j+2}^{n+1} + A_{14}c_{j+3}^{n+1} = A_{15}\delta_{j-3}^n + \\ &A_{16}\delta_{j-2}^n + A_{17}\delta_{j-1}^n + A_{18}\delta_j^n + A_{17}\delta_{j+1}^n + A_{16}\delta_{j+2}^n + A_{15}\delta_{j+3}^n + A_{19}c_{j-3}^n + \\ &A_{20}c_{j-2}^n + A_{21}c_{j-1}^n - A_{21}c_{j+1}^n - A_{20}c_{j+2}^n - A_{19}c_{j+3}^n, \end{aligned} \tag{11}$$

$$\begin{aligned} &B_1c_{j-3}^{n+1} + B_2c_{j-2}^{n+1} + B_3c_{j-1}^{n+1} + B_4c_j^{n+1} + B_5c_{j+1}^{n+1} + B_6c_{j+2}^{n+1} + B_7c_{j+3}^{n+1} - B_8\delta_{j-3}^{n+1} - \\ &B_9\delta_{j-2}^{n+1} - B_{10}\delta_{j-1}^{n+1} + B_{10}\delta_{j+1}^{n+1} + B_9\delta_{j+2}^{n+1} + B_8\delta_{j+3}^{n+1} = B_{11}c_{j-3}^n + B_{12}c_{j-2}^n + \\ &B_{13}c_{j-1}^n + B_{14}c_j^n + B_{13}c_{j+1}^n + B_{12}c_{j+2}^n + B_{11}c_{j+3}^n + B_8\delta_{j-3}^n + B_9\delta_{j-2}^n + B_{10}\delta_{j-1}^n - \\ &B_{10}\delta_{j+1}^n - B_9\delta_{j+2}^n - B_8\delta_{j+3}^n \end{aligned} \tag{12}$$

where

$$A_1 = 1 - \frac{7\Delta t}{2h}z_1 + \frac{7\Delta t}{2h}z_2 - \frac{7}{h^2},\ A_2 = 120 - 56\frac{7\Delta t}{2h}z_1 + 120\frac{7\Delta t}{2h}z_2 - 24\frac{7}{h^2},$$
$$A_3 = 1191 - 245\frac{7\Delta t}{2h}z_1 + 1191\frac{7\Delta t}{2h}z_2 - 15\frac{7}{h^2},\ A_4 = 2416 + 2416\frac{7\Delta t}{2h}z_2 + 80\frac{7}{h^2},$$
$$A_5 = 1191 + 245\frac{7\Delta t}{2h}z_1 + 1191\frac{7\Delta t}{2h}z_2 - 15\frac{7}{h^2},\ A_6 = 120 + 56\frac{7\Delta t}{2h}z_1 + 120\frac{7\Delta t}{2h}z_2 - 24\frac{7}{h^2},$$
$$A_7 = 1 + \frac{7\Delta t}{2h}z_1 + \frac{7\Delta t}{2h}z_2 - \frac{7}{h^2},\ A_8 = -\frac{7\Delta t}{2h} + \frac{7\Delta t}{2h}z_4 - \frac{7\Delta t}{2h}z_3,$$
$$A_9 = -56\frac{7\Delta t}{2h} + 120\frac{7\Delta t}{2h}z_4 - 56\frac{7\Delta t}{2h}z_3,$$
$$A_{10} = -245\frac{7\Delta t}{2h} + 1191\frac{7\Delta t}{2h}z_4 - 245\frac{7\Delta t}{2h}z_3,\ A_{11} = 2416\frac{7\Delta t}{2h}z_4,$$
$$A_{12} = 245\frac{7\Delta t}{2h} + 1191\frac{7\Delta t}{2h}z_4 + 245\frac{7\Delta t}{2h}z_3,\ A_{13} = 56\frac{7\Delta t}{2h} + 120\frac{7\Delta t}{2h}z_4 + 56\frac{7\Delta t}{2h}z_3,$$
$$A_{14} = \frac{7\Delta t}{2h} + \frac{7\Delta t}{2h}z_4 + \frac{7\Delta t}{2h}z_3,\ A_{15} = 1 - \frac{7}{h^2},\ A_{16} = 120 - 24\frac{7}{h^2},$$
$$A_{17} = 1191 - 15\frac{7}{h^2},\ A_{18} = 2416 + 80\frac{7}{h^2},\ A_{19} = \frac{7\Delta t}{2h},\ A_{20} = 56\frac{7\Delta t}{2h},$$
$$A_{21} = 245\frac{7\Delta t}{2h},$$

$$B_1 = 1 - \frac{7\Delta t}{2h}z_1 + \frac{7\Delta t}{2h}z_2 - \frac{7}{h^2},\ B_2 = 120 - 56\frac{7\Delta t}{2h}z_1 + 120\frac{7\Delta t}{2h}z_2 - 24\frac{7}{h^2},$$
$$B_3 = 1191 - 245\frac{7\Delta t}{2h}z_1 + 1191\frac{7\Delta t}{2h}z_2 - 15\frac{7}{h^2},\ B_4 = 2416 + 2416\frac{7\Delta t}{2h}z_2 + 80\frac{7}{h^2},$$
$$B_5 = 1191 + 245\frac{7\Delta t}{2h}z_1 + 1191\frac{7\Delta t}{2h}z_2 - 15\frac{7}{h^2},\ B_6 = 120 + 56\frac{7\Delta t}{2h}z_1 + 120\frac{7\Delta t}{2h}z_2 - 24\frac{7}{h^2},$$
$$B_7 = 1 + \frac{7\Delta t}{2h}z_1 + \frac{7\Delta t}{2h}z_2 - \frac{7}{h^2},\ B_8 = \frac{7\Delta t}{2h},\ B_9 = 56\frac{7\Delta t}{2h},\ B_{10} = 245\frac{7\Delta t}{2h},$$
$$B_{11} = 1 - \frac{7}{h^2},\ B_{12} = 120 - 24\frac{7}{h^2},\ B_{13} = 1191 - 15\frac{7}{h^2},\ B_{14} = 2416 + 80\frac{7}{h^2},$$
$$z_1 = c_{j-3} + 120c_{j-2} + 1191c_{j-1} + 2416c_j + 1191c_{j+1} + 120c_{j+2} + c_{j+3},$$
$$z_2 = -c_{j-3} - 56c_{j-2} - 245c_{j-1} + 245c_{j+1} + 56c_{j+2} + c_{j+3},$$
$$z_3 = \delta_{j-3} + 120\delta_{j-2} + 1191\delta_{j-1} + 2416\delta_j + 1191\delta_{j+1} + 120\delta_{j+2} + \delta_{j+3},$$
$$z_4 = -\delta_{j-3} - 56\delta_{j-2} - 245\delta_{j-1} + 245\delta_{j+1} + 56\delta_{j+2} + \delta_{j+3},$$

The system thus obtained on simplifying Eqs. (11) and (12) consists of $(2N + 2)$ linear equations in the $(2N + 14)$ unknowns $(c_{-3}, c_{-2}, c_{-1}, c_0, \ldots \ldots, c_N, c_{N+1}, c_{N+2}, c_{N+3})^T$, $(\delta_{-3}, \delta_{-2}, \delta_{-1}, \delta_0, \ldots\ldots, \delta_N, \delta_{N+1}, \delta_{N+2}, \delta_{N+3})^T$.

To obtain a unique solution to the resulting system six additional constraints are required. These are obtained by imposing boundary conditions. Eliminating $c_{-3}, c_{-2}, c_{-1}, c_0, \ldots\ldots, c_N, c_{N+1}, c_{N+2}, c_{N+3}$ and $\delta_{-3}, \delta_{-2}, \delta_{-1}, \delta_0, \ldots\ldots, \delta_N, \delta_{N+1}, \delta_{N+2}, \delta_{N+3}$ the system get reduced to a matrix system of dimension $(2N + 2) \times (2N + 2)$ which is the septa-diagonal system that can be solved by any algorithm.

Initial Values

To find the initial parameters c_j^0 and δ_j^0, the initial conditions and the derivatives at the boundaries are used in the following way

$$(U')(x_0,0)=\frac{7}{h}(-c_{-3}-56c_{-2}-243c_{-1}+243c_1+56c_2+c_3)=f'(x_0),$$

$$(U'')(x_0,0)=\frac{42}{h^2}(c_{-3}+24c_{-2}+15c_{-1}-80c_0+15c_1+24c_2+c_3)=f''(x_0),$$

$$(U''')(x_0,0)=\frac{210}{h^3}(-c_{-3}-8c_{-2}+19c_{-1}-19c_1+8c_2+c_3)=f'''(x_0),$$

$$(U)(x_j,0)=c_{j-3}+120c_{j-2}+1191c_{j-1}+2416c_j+1191c_{j+1}+120c_{j+2}+c_{j+3}=f(x_j),$$

$$(U')(x_N,0)=\frac{7}{h}(-c_{N-3}-56c_{N-2}-245c_{N-1}+245c_{N+1}+56c_{N+2}+c_{N+3})=f'(x_N),$$

$$(U'')(x_N,0)=\frac{42}{h^2}(c_{N-3}+24c_{N-2}+15c_{N-1}-80c_N+15c_{N+1}+24c_{N+2}+c_{N+3})=f''(x_N),$$

$$(U''')(x_N,0)=\frac{210}{h^3}(-c_{N-3}-8c_{N-2}+19c_{N-1}-19c_{N+1}+8c_{N+2}+c_{N+3})=f'''(x_N),$$

$$(V')(x_0,0)=\frac{7}{h}(-\delta_{-3}-56\delta_{-2}-245\delta_{-1}+245\delta_1+56\delta_2+\delta_3)=g'(x_0),$$

$$(V'')(x_0,0)=\frac{42}{h^2}(\delta_{-3}+24\delta_{-2}+15\delta_{-1}-80\delta_0+15\delta_1+24\delta_2+\delta_3)=g''(x_0),$$

$$(V''')(x_0,0)=\frac{210}{h^3}(-\delta_{-3}-8\delta_{-2}+19\delta_{-1}-19\delta_1+8\delta_2+\delta_3)=g'''(x_0),$$

$$(V)(x_j,0)=\delta_{j-3}+120\delta_{j-2}+1191\delta_{j-1}+2416\delta_j+1191\delta_{j+1}+120\delta_{j+2}+\delta_{j+3}=g(x_j),$$

$$(V')(x_N,0)=\frac{7}{h}(-\delta_{N-3}-56\delta_{N-2}-245\delta_{N-1}+245\delta_{N+1}+56\delta_{N+2}+\delta_{N+3})=g'(x_N),$$

$$(V'')(x_N,0)=\frac{42}{h^2}(\delta_{N-3}+24\delta_{N-2}+15\delta_{N-1}-80\delta_N+15\delta_{N+1}+24\delta_{N+2}+\delta_{N+3})=g''(x_N),$$

$$(V''')(x_N,0)=\frac{210}{h^3}(-\delta_{N-3}-8\delta_{N-2}+19\delta_{N-1}-19\delta_{N+1}+8\delta_{N+2}+\delta_{N+3})=g'''(x_N).$$

Which forms a linear block septa-diagonal system for unknown initial conditions c_j^0 and δ_j^0, of order $(2N + 2)$ after eliminating the functions values of c and δ. This system can be solved by any algorithm. Once the initial vectors of parameters have been calculated, the numerical solution of coupled BBM system U and V can be determined from the time evaluation of the vectors c_j^n and δ_j^n, by using the recurrence relations

$$U(x_j,t_n)=c_{j-3}+120c_{j-2}+1191c_{j-1}+2416c_j+1191c_{j+1}+120c_{j+2}+c_{j+3},$$

$$V(x_j,t_n)=\delta_{j-3}+120\delta_{j-2}+1191\delta_{j-1}+2416\delta_j+1191\delta_{j+1}+120\delta_{j+2}+\delta_{j+3}.$$

Stability Analysis of the Method

The stability analysis of nonlinear partial differential equations is not easy task to undertake. Most researchers copy with the problem by linearizing the partial differential equation. Our stability analysis will be based on the Von-Neumann concept in which the growth factor of a typical Fourier mode defined as

$$c_j^n = A\zeta^n \exp(ij\varphi), \qquad \delta_j^n = B\zeta^n \exp(ij\varphi), \tag{13}$$

$$g=\frac{\zeta^{n+1}}{\zeta^n},$$

Where A and B are the harmonics amplitude, $\varphi=kh$, k is the mode number, $i=\sqrt{-1}$ and g is the amplification factor of the schemes. We will be applied the stability of the septic schemes by assuming the nonlinear term as a constants λ_1, λ_2. This is equivalent to assuming that all the c_j^n and δ_j^n as a local constants λ_1, λ_2 respectively. At $x=x_j$ systems (11) and (12) can be written as

$$\begin{aligned}&a_1\delta_{j-3}^{n+1}+a_2\delta_{j-2}^{n+1}+a_3\delta_{j-1}^{n+1}+a_4\delta_j^{n+1}+a_5\delta_{j+1}^{n+1}+a_6\delta_{j+2}^{n+1}+a_7\delta_{j+3}^{n+1}-a_8c_{j-3}^{n+1}-\\&a_9c_{j-2}^{n+1}-a_{10}c_{j-1}^{n+1}+a_{10}c_{j+1}^{n+1}+a_9c_{j+2}^{n+1}+a_8c_{j+3}^{n+1}=a_7\delta_{j-3}^n+\\&a_6\delta_{j-2}^n+a_5\delta_{j-1}^n+a_4\delta_j^n+a_3\delta_{j+1}^n+a_2\delta_{j+2}^n+a_1\delta_{j+3}^n+a_8c_{j-3}^n+\\&a_9c_{j-2}^n+a_{10}c_{j-1}^n-a_{10}c_{j+1}^n-a_9c_{j+2}^n-a_8c_{j+3}^n,\end{aligned} \tag{14}$$

where

$$a_1=1-\frac{7\Delta t}{2h}\lambda_1-\frac{7}{h^2},\ a_2=120-56\frac{7\Delta t}{2h}\lambda_1-24\frac{7}{h^2},$$

$$a_3=1191-245\frac{7\Delta t}{2h}\lambda_1-15\frac{7}{h^2},\ a_4=2461+80\frac{7}{h^2},$$

$$a_5=1191+245\frac{7\Delta t}{2h}\lambda_1-15\frac{7}{h^2},\ a_6=120+56\frac{7\Delta t}{2h}\lambda_1-24\frac{7}{h^2},$$

$$a_7=1+\frac{7\Delta t}{2h}\lambda_1-\frac{7}{h^2},\ a_8=\frac{7\Delta t}{2h}(1+\lambda_2),$$

$$a_9=56\frac{7\Delta t}{2h}(1+\lambda_2),\ a_{10}=245\frac{7\Delta t}{2h}(1+\lambda_2),$$

$$\begin{aligned}&b_1c_{j-3}^{n+1}+b_2c_{j-2}^{n+1}+b_3c_{j-1}^{n+1}+b_4c_j^{n+1}+b_5c_{j+1}^{n+1}+b_6c_{j+2}^{n+1}+b_7c_{j+3}^{n+1}-b_8\delta_{j-3}^{n+1}-\\&b_9\delta_{j-2}^{n+1}-b_{10}\delta_{j-1}^{n+1}+b_{10}\delta_{j+1}^{n+1}+b_9\delta_{j+2}^{n+1}+b_8\delta_{j+3}^{n+1}=b_7c_{j-3}^n+b_6c_{j-2}^n+\\&b_5c_{j-1}^n+b_4c_j^n+b_3c_{j+1}^n+b_2c_{j+2}^n+b_1c_{j+3}^n+b_8\delta_{j-3}^n+b_9\delta_{j-2}^n+b_{10}\delta_{j-1}^n-\\&b_{10}\delta_{j+1}^n-b_9\delta_{j+2}^n-b_8\delta_{j+3}^n\end{aligned} \tag{15}$$

where

$$b_1=1-\frac{7\Delta t}{2h}\lambda_1-\frac{7}{h^2},\ b_2=120-56\frac{7\Delta t}{2h}\lambda_1-24\frac{7}{h^2},$$

$$b_3=1191-245\frac{7\Delta t}{2h}\lambda_1-15\frac{7}{h^2},\ b_4=2461+80\frac{7}{h^2},$$

$$b_5=1191+245\frac{7\Delta t}{2h}\lambda_1-15\frac{7}{h^2},\ b_6=120+56\frac{7\Delta t}{2h}\lambda_1-24\frac{7}{h^2},$$

$$b_7=1+\frac{7\Delta t}{2h}\lambda_1-\frac{7}{h^2},\ b_8=\frac{7\Delta t}{2h},$$

$$b_9=56\frac{7\Delta t}{2h},\ b_{10}=245\frac{7\Delta t}{2h},$$

Substituting (13) into the difference (14), we get

$$\zeta^{n+1}\left[\begin{array}{l}B\left[\begin{array}{l}2\left(1-\frac{7}{h^2}\right)\cos 3\varphi+2\left(120-24\frac{7}{h^2}\right)\cos 2\varphi+2\left(11191-15\frac{7}{h^2}\right)\cos\varphi+\\\left(2416+80\frac{7}{h^2}\right)\end{array}\right]+\\ i\left[\begin{array}{l}B\left[\left(\frac{7\Delta t}{h}\lambda_1\right)\sin 3\phi+\left(56\frac{7\Delta t}{h}\lambda_1\right)\sin 2\phi+\left(245\frac{7\Delta t}{h}\lambda_1\right)\sin\phi\right]\\A\left[\left(\frac{7\Delta t}{h}(1+\lambda_2)\right)\sin 3\phi+\left(56\frac{7\Delta t}{h}(1+\lambda_2)\right)\sin 2\phi+\left(245\frac{7\Delta t}{h}(1+\lambda_2)\right)\sin\phi\right]\end{array}\right]\end{array}\right]=$$

$$\zeta^{n}\left[\begin{array}{l}B\left[\begin{array}{l}2\left(1-\frac{7}{h^2}\right)\cos 3\varphi+2\left(120-24\frac{7}{h^2}\right)\cos 2\varphi+2\left(11191-15\frac{7}{h^2}\right)\cos\varphi+\\\left(2416+80\frac{7}{h^2}\right)\end{array}\right]-\\ i\left[\begin{array}{l}B\left[\left(\frac{7\Delta t}{h}\lambda_1\right)\sin 3\phi+\left(56\frac{7\Delta t}{h}\lambda_1\right)\sin 2\phi+\left(245\frac{7\Delta t}{h}\lambda_1\right)\sin\phi\right]\\A\left[\left(\frac{7\Delta t}{h}(1+\lambda_2)\right)\sin 3\phi+\left(56\frac{7\Delta t}{h}(1+\lambda_2)\right)\sin 2\phi+\left(245\frac{7\Delta t}{h}(1+\lambda_2)\right)\sin\phi\right]\end{array}\right]\end{array}\right],$$

we get

$$g=\frac{X-iY}{X+iY}, \tag{16}$$

where

$$X=B\left[\begin{array}{l}2\left(1-\frac{7}{h^2}\right)\cos 3\varphi+2\left(120-24\frac{7}{h^2}\right)\cos 2\varphi+2\left(11191-15\frac{7}{h^2}\right)\cos\varphi+\\\left(2416+80\frac{7}{h^2}\right)\end{array}\right]$$

and

$$Y=\left[\begin{array}{l}B\left[\left(\frac{7\Delta t}{h}\lambda_1\right)\sin 3\phi+\left(56\frac{7\Delta t}{h}\lambda_1\right)\sin 2\phi+\left(245\frac{7\Delta t}{h}\lambda_1\right)\sin\phi\right]\\A\left[\left(\frac{7\Delta t}{h}(1+\lambda_2)\right)\sin 3\phi+\left(56\frac{7\Delta t}{h}(1+\lambda_2)\right)\sin 2\phi+\left(245\frac{7\Delta t}{h}(1+\lambda_2)\right)\sin\phi\right]\end{array}\right].$$

Similar substituting (13) into the difference (15), we get

$$\zeta^{n+1}\begin{bmatrix} A\begin{bmatrix}2\left(1-\frac{7}{h^2}\right)\cos3\varphi+2\left(120-24\frac{7}{h^2}\right)\cos2\varphi+2\left(11191-15\frac{7}{h^2}\right)\cos\varphi+\\ \left(2416+80\frac{7}{h^2}\right)\end{bmatrix}+ \\ i\begin{bmatrix} A\left[\left(\frac{7\Delta t}{h}\lambda_1\right)\sin3\phi+\left(56\frac{7\Delta t}{h}\lambda_1\right)\sin2\phi+\left(245\frac{7\Delta t}{h}\lambda_1\right)\sin\phi\right]\\ B\left[\left(\frac{7\Delta t}{h}\right)\sin3\phi+\left(56\frac{7\Delta t}{h}\right)\sin2\phi+\left(245\frac{7\Delta t}{h}\right)\sin\phi\right]\end{bmatrix}\end{bmatrix} = \zeta^{n}\begin{bmatrix} A\begin{bmatrix}2\left(1-\frac{7}{h^2}\right)\cos3\varphi+2\left(120-24\frac{7}{h^2}\right)\cos2\varphi+2\left(11191-15\frac{7}{h^2}\right)\cos\varphi+\\ \left(2416+80\frac{7}{h^2}\right)\end{bmatrix}- \\ i\begin{bmatrix} A\left[\left(\frac{7\Delta t}{h}\lambda_1\right)\sin3\phi+\left(56\frac{7\Delta t}{h}\lambda_1\right)\sin2\phi+\left(245\frac{7\Delta t}{h}\lambda_1\right)\sin\phi\right]\\ B\left[\left(\frac{7\Delta t}{h}\right)\sin3\phi+\left(56\frac{7\Delta t}{h}\right)\sin2\phi+\left(245\frac{7\Delta t}{h}\right)\sin\phi\right]\end{bmatrix}\end{bmatrix},$$

we get

$$g=\frac{X_1-iY_1}{X_1+iY_1}, \tag{17}$$

where

$$X_1=B\begin{bmatrix}2\left(1-\frac{7}{h^2}\right)\cos3\varphi+2\left(120-24\frac{7}{h^2}\right)\cos2\varphi+2\left(11191-15\frac{7}{h^2}\right)\cos\varphi+\\ \left(2416+80\frac{7}{h^2}\right)\end{bmatrix}$$

and

$$Y_1=\begin{bmatrix} B\left[\left(\frac{7\Delta t}{h}\lambda_1\right)\sin3\phi+\left(56\frac{7\Delta t}{h}\lambda_1\right)\sin2\phi+\left(245\frac{7\Delta t}{h}\lambda_1\right)\sin\phi\right]\\ A\left[\left(\frac{7\Delta t}{h}\right)\sin3\phi+\left(56\frac{7\Delta t}{h}\right)\sin2\phi+\left(245\frac{7\Delta t}{h}\right)\sin\phi\right]\end{bmatrix}.$$

From (16) and (17) we get $|g|\leq 1$, hence the schemes are unconditionally stable. It means that there is no restriction on the grid size, i.e. on h and Δt, but we should choose them in such a way that the accuracy of the scheme is not degraded.

Numerical Tests and Results of Coupled-BBM system

In this section, we present some numerical examples to test validity of our scheme for solving coupled-BBM system.

The norms L_2-norm and L_∞-norm are used to compare the numerical solution with the analytical solution [14].

$$L_2=\left\|u^E-u^N\right\|=\sqrt{h\sum_{i=0}^{N}(u_j^E-u_j^N)^2}, \qquad L_\infty=\max_j\left|u_j^E-u_j^N\right|,\ j=0,1,\cdots,N. \tag{18}$$

Where u^E is the exact solution u and u^N is the approximation solution U_N.

Now we can study our scheme from this problem.

Single soliton

Consider the coupled-BBM system (1) and (2) with the following initial and boundary conditions:

$$u(x,0)=f(x),$$
$$v(x,0)=g(x), \qquad a\leq x\leq b.$$

And

$$u(a,t)=0,\quad u(b,t)=0,$$
$$v(a,t)=0,\quad v(b,t)=0,$$
$$u_x(a,t)=0,\quad u_x(b,t)=0,$$
$$v_x(a,t)=0,\quad v_x(b,t)=0,\quad 0\leq t\leq T.$$

The exact solution is

$$u(x,t)=\left(1-\frac{g}{6}\right)c+\frac{c\,g}{2}\operatorname{sech}^2\left(\frac{\sqrt{g}}{2}(x+x_0-ct)\right),\quad v(x,t)=-1,$$

Now, for comparison, we consider a test problem where, $g=6, c=\frac{1}{3}, x_0=0, k=0.001$ and $-20\leq x$ 40. The Errors, at time 5 are satisfactorily small L_2-error=7.11457×10^{-6} and L_∞-error=9.35827×10^{-6} for approximation solution of $u(x,t)$ and L_2-error and L_∞-error approach to zero for approximation solution of $v(x,t)$ at h=0.2. The Errors, at time 5 are satisfactorily small L_2-error=1.4783910^{-7} and L_∞-error=1.47839×10^{-7} for approximation solution of $u(x,t)$ and L_2-error and L_∞-error approach to zero for approximation solution of $v(x,t)$ at h=0.1. Our results are recorded in Table 2. The motion of solitary wave using our scheme is plotted at times t=0,10,20 in Figure 1. These results illustrate that the scheme has a highest accuracy (Table 2 and Figure 1).

$-20\leq x\leq 40$ at t=5.

In Table 3 we show that our results are better than the results in [7] (Table 3).

Interaction of two solitary waves

The interaction of two solitary waves having different amplitudes and traveling in the same direction is illustrated. We consider Coupled-BBM system with initial conditions given by the linear sum of two well separated solitary waves of various amplitudes.

x	x_{j-4}	x_{j-3}	x_{j-2}	x_{j-1}	x_j	x_{j+1}	x_{j+2}	x_{j+3}	x_{j+4}
B_j	0	1	120	1191	2416	1191	120	1	0
B'_j	0	$\frac{-7}{h}$	$\frac{-392}{h}$	$\frac{-1715}{h}$	0	$\frac{392}{h}$	$\frac{392}{h}$	$\frac{7}{h}$	0
B''_j	0	$\frac{42}{h^2}$	$\frac{1008}{h^2}$	$\frac{630}{h^2}$	$\frac{-3360}{h^2}$	$\frac{630}{h^2}$	$\frac{1008}{h^2}$	$\frac{42}{h^2}$	0

Table 1: The values of septic B-spline and its first and second derivatives at the knots points.

h	T	u(x,t)		v(x,t)	
		L_2- norm	L_∞- norm	L_2- norm	L_∞- norm
h=0.2	0.0	0.00000000	0.00000000	0.00000000	0.00000000
	1.0	4.97858E-6	7.92026E-6	0.00000000	0.00000000
	2.0	6.53347E-6	9.35827E-6	0.00000000	0.00000000
	3.0	6.59617E-6	6.24577E-6	0.00000000	0.00000000
	4.0	6.84312E-6	7.04553E-6	0.00000000	0.00000000
	5.0	7.11457E-6	7.56269E-6	0.00000000	0.00000000
h=0.1	0.0	0.00000000	0.00000000	0.00000000	0.00000000
	1.0	4.69497E-8	5.92814E-8	0.00000000	0.00000000
	2.0	7.70501E-7	1.11922E-8	0.00000000	0.00000000
	3.0	9.93037E-7	1.12545E-8	0.00000000	0.00000000
	4.0	1.23042E-7	1.24181E-7	0.00000000	0.00000000
	5.0	1.47839E-7	1.45966E-7	0.00000000	0.00000000

Table 2: L_2- norm and L_∞- norm for t=5.0, $g=6, c=\frac{1}{3}, x_0=0, k=0.001$ and –20 ≤ x ≤ 40.

$$u(x,0)=\sum_{j=1}^{2}\left(1-\frac{g_j}{6}\right)c_j+\frac{c_j g_j}{2}\operatorname{sec}h^2\left(\frac{\sqrt{g_j}}{2}(x+x_j)\right),\ v(x,0)=-1, \qquad (19)$$

Where j=1,2,g_j,x_j and c_j are arbitrary constants. In our computational work. Now, we choose $g_1=6, g_2=6, c_1=1, c_2=\frac{1}{3}$, $x_1=0, x_2=-10, h=0.1, k=0.01$ with interval [-20, 40]. In Figure 2, the interactions of these solitary waves are plotted at different time levels (Figure 2).

Interaction of three solitary waves

The interaction of three solitary waves having different amplitudes and traveling in the same direction is illustrated. We consider Coupled-BBM system with initial conditions given by the linear sum of three well separated solitary waves of various amplitudes

$$u(x,0)=\sum_{j=1}^{3}\left(1-\frac{g_j}{6}\right)c_j+\frac{c_j g_j}{2}\operatorname{sec}h^2\left(\frac{\sqrt{g_j}}{2}(x+x_j)\right),\ v(x,0)=-1, \qquad (20)$$

Where j=1, 2, 3, g_j, x_j and c_j are arbitrary constants. In our computational work. Now, we choose $g_1=6,\ g_2=6, g_3=6, c_1=1,\ c_2=\frac{2}{3}, c_3=\frac{1}{3}$, $x_1=0, x_2=-5, x_3=-10, h=0.1, k=0.01$ with interval [-20, 40]. In Figure 3, the interactions of these solitary waves are plotted at different time levels (Figure 3).

Conclusions

In this paper a numerical treatment for the nonlinear Coupled-BBM system is proposed using a collection method with the septic B-splines. The stability analysis of the method is shown to be unconditionally stable. We make linearization for the nonlinear term. We tested our schemes through a single solitary wave in which the analytic solution

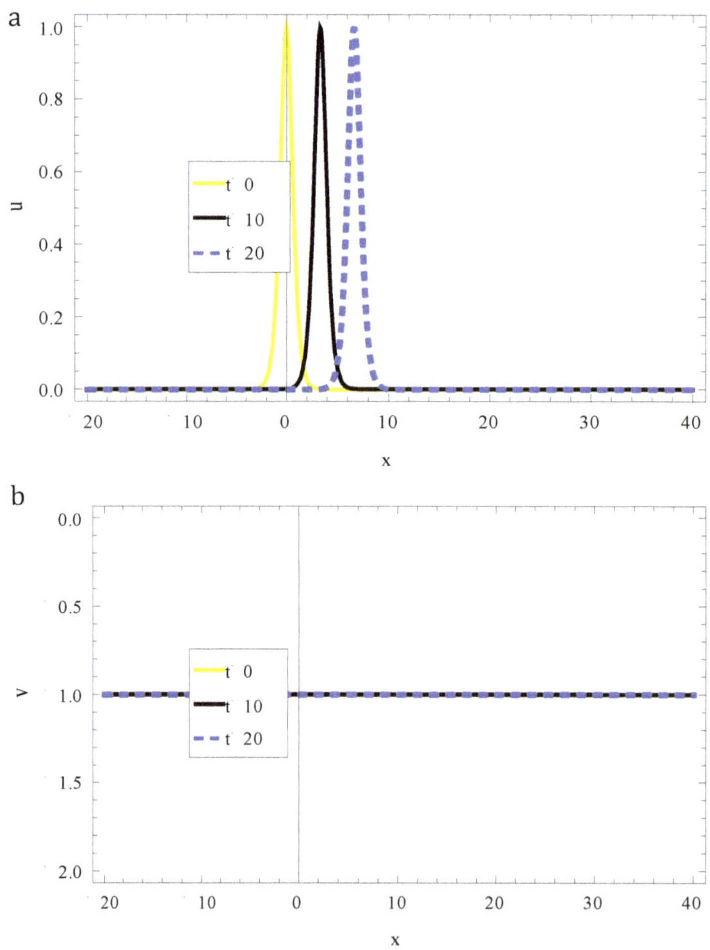

Figure 1: Single solitary wave with $g=6, c=\frac{1}{3}, x_0=0, k=0.001$ and $-20 \le x$ 40 at times t=0, 10, 20 respectively.

Schemes at t=5	$u(x,t)$ at h=0.2		$u(x,t)$ at h=0.1	
	L_2- norm	L_∞- norm	L_2- norm	L_∞- norm
our scheme	7.11457E-6	7.56269E-6	1.47839E-7	1.45966E-7
(Shadan [7])	8.85111E-4	8.97762E-4	7.99452E-4	8.77474E-4

Table 3: Comparison of numerical results of the problem with the results obtained from [7] for the variable u with, $g=6, c=\frac{1}{3}, x_0=0, k=0.001$.

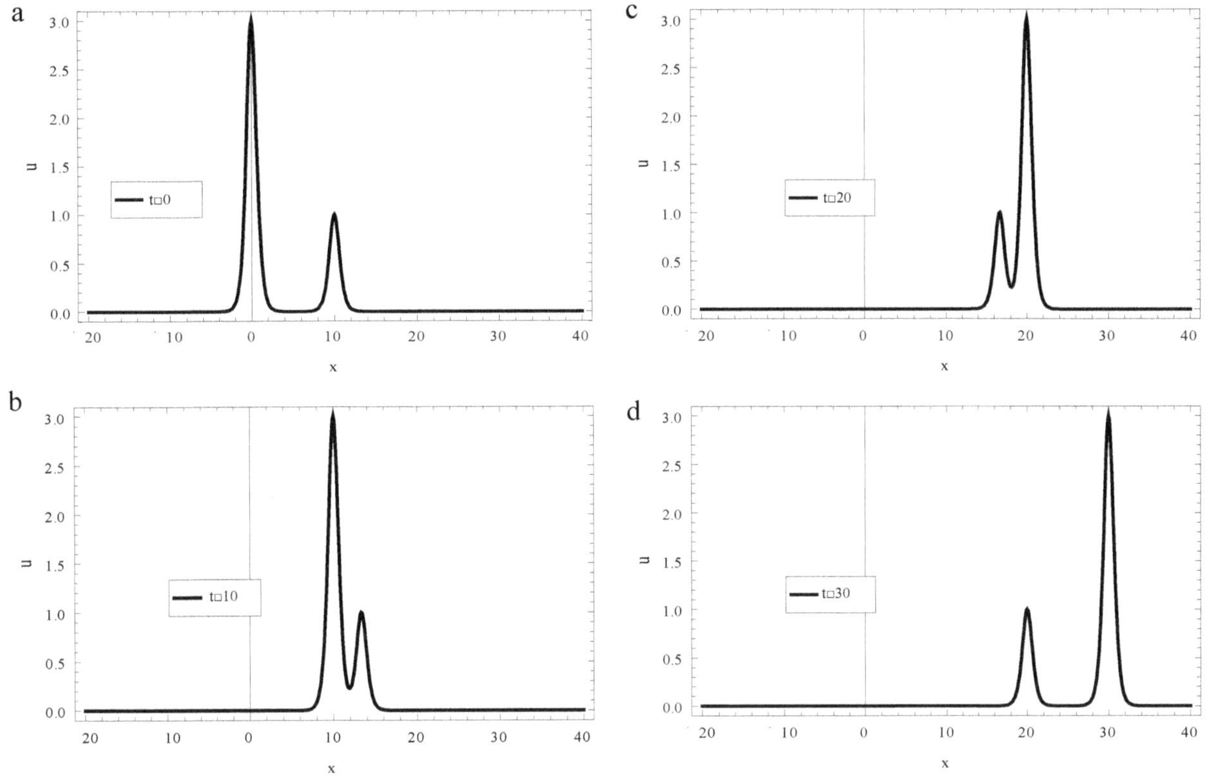

Figure 2: interaction two solitary waves with $g_1 = 6,\ g_2 = 6, c_1 = 1,\ c_2 = \frac{1}{3},$ $h = 0.1,$ $k = 0.01,$ $-20 \le x \le 40$ for value *u* at times t=0, 10, 20, 30 respectively.

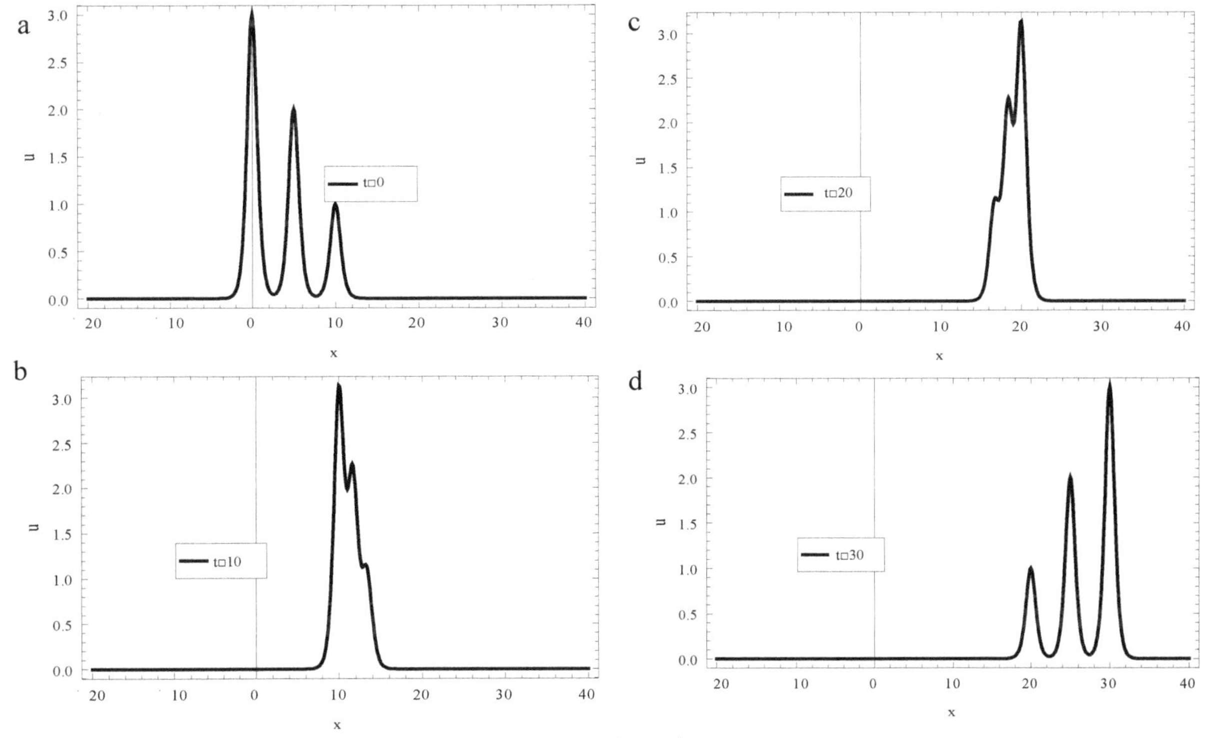

Figure 3: Interaction three solitary waves with $g_1 = 6,\ g_2 = 6, g_3 = 6, c_1 = 1, c_2 = \frac{2}{3},\ c_3 = \frac{1}{3}, x_1 = 0,\ x_2 = -5, x_3 = -10, h = 0.1,\ k = 0.01,\ -20 \le x \le 40$ for values u at times t=0, 10, 20, 30 respectively.

is known, then extend it to study the interaction of solitons where no analytic solution is known during the interaction. The accuracy of our scheme was shown by calculating error norms L_2 and L_∞.

References

1. Bona JL, Chen M (1998) A Boussinesq system for two way propagation of Nonlinear dispersive waves. Physica D 116: 191-224.
2. Bona JL, Chen M, Saut JC (2002) Boussinesq equations and other Systems for mall - amplitude long waves in nonlinear dispersive media. Derivation and linear theory. Journal of Nonlinear Science 12: 283-318.
3. Bona JL, Chen M, Saut JC (2004) Boussinesq equations and other Systems for small amplitude long waves in nonlinear dispersive media II. The Nonlinear theory, Nonlinearity 17: 925-952.
4. Dougalis VA, Mitsotakis DE, Saut JC (2009) On initial boundary value problem for Boussinesq system of BBM - BBM type in plane domain. Discrete and Continuous Dynamical Systems 23: 1191-1204.
5. Chen M (2000) Solitary Wave and multi pulsed traveling wave solutions of Boussinesq systems. Applicable Analysis 75: 213-240.
6. Antoropoulos DC, Dougalis VA, Mitsotakis DE (2010) Numerical Solution of Boussinesq systems of Bona-Smith family. Applied Numerical Mathematics 30: 314-336.
7. Behzadi SS, Yildirim A (2013) Application of Quintic B-Spline Collocation Method for Solving the Coupled-BBM System. Middle-East Journal of Scientific Research 15: 1478-1486.
8. Al-Rawi ES, Sallal MAM (2012) Numerical solution of BBM-system using finite element method. International journal of recent scientific research 3: 1026-1029.
9. Chen M (1998) Exact Traveling-Wave Solutions to Bidirectional Wave Equations. International Journal of Theoretical Physics 37: 5.
10. Raslan KR, El-Danaf TS, Ali KA (2016) Collocation method with quintic b-spline method for solving hirota-satsuma coupled KDV equation. International Journal of Applied Mathematical Research 5: 123-131.
11. Raslan KR, El-Danaf TS, Ali KA (2016) Collocation Method with Quintic *B*-Spline Method for Solving the Hirota equation. Journal of Abstract and Computational Mathematics 1: 1-12.
12. EL-Danaf TS, Raslan KR, Ali KK (2016) Collocation method with cubic B-Splines for solving the GRLW equation. Int J of Num Meth and Appl 15: 39-59.
13. Rubin SG, Graves RA (1975) Cubic spline approximation for problems in fluid mechanics. Nasa TRR-436. Washington, DC.
14. EL-Danaf TS, Raslan KR, Ali KK (2014) New Numerical treatment for the Generalized Regularized Long Wave Equation based on finite difference scheme. Int J of S Comp and Eng 4: 16-24.

Solving Nonlinear Integral Equations by using Adomian Decomposition Method

Mohedul Hasan Md* and Abdul Matin Md

Department of Mathematics, Dhaka University, Dhaka, Bangladesh

Abstract

In this paper, we propose a numerical method to solve the nonlinear integral equation of the second kind. We intend to approximate the solution of this equation by Adomian decomposition method using He's polynomials. Several examples are given at the end of this paper with the exact solution is known. Also the error is estimated.

Keywords: Nonlinear Fredholm integral equation; Adomian decomposition method; Adomian polynomials; Approximate solutions; OriginPro 8 and MATHEMATICA v9 softwares

Introduction

Several scientific and engineering applications are usually described by integral equations. Integral equations arise in the potential theory more than any other field. Integral equations arise also in diffraction problems, conformal mapping, water waves, scattering in quantum mechanics, and population growth model. The electrostatic, electromagnetic scattering problems and propagation of acoustical and elastically waves are scientific fields where integral equations appear [1]. The Fredholm integral equation is of widespread use in many realms of engineering and applied mathematics [2].

Consider the general form non-linear Fredholm integral equation of the second kind

$$y(x)=f(x)+\lambda\int_a^b K(x,t)G(y(t))dt,\ a\le x\le b$$

where $y(x)$ is the unknown solution, a and b are real constants. The kernel $K(x,t)$ and $f(x)$ are known smooth functions on R^2 and R, respectively. The parameter λ is a real (or complex) known as the eigenvalue when λ is a real parameter, and G is a nonlinear function of y.

Adomian Decomposition Method

Consider the following non-linear Fredholm integral equation of the second kind of the form

$$y(x)=f(x)+\lambda\int_a^b K(x,t)G(y(t))dt,\quad a\le x\le b \tag{1}$$

We assume $G(y(t))$ is a nonlinear function of $y(x)$. That means that the nonlinear Fredholm integral equation (1) contains the nonlinear function presented by $G(y(t))$. Assume that the solution of equation (1) can be written in the form

$$y=\sum_{i=0}^{\infty}p^i y_i(x)=y_0+p^1y_1+p^2y_2+p^3y_3+\dots \tag{2}$$

The comparisons of like powers of p give solutions of various orders and the best approximation is

$$y=\lim_{p\to 1}\sum_{i=0}^{\infty}p^i y_i(x)=y_0+y_1+y_2+y_3+\dots$$

The nonlinear term $G(y(t))$ can be expressed in Adomian polynomials [3-5] as

$$G(y)=\sum_{k=0}^{\infty}p^k H_k(y_0,y_1,y_2,\dots,y_k) \tag{3}$$

$$=H_0(y_0)+p^1H_1(y_0,y_1)+\dots+p^kH_k(y_0,y_1,\dots\dots,y_k)$$

where H_k's are the so called Adomian polynomials which can be calculated by using the formula

$$H_k(y_0,y_1,\dots\dots,y_k)=\frac{1}{k!}\frac{d^k}{dp^k}\left[G\left(\sum_{i=0}^{k}p^i y_i\right)\right],k=0,1,2\dots\dots \tag{4}$$

Using (2), (3) and (4) into (1), we have

$$\sum_{i=0}^{\infty}p^i y_i(x)=f(x)+\lambda\int_a^b K(x,t)\sum_{j=0}^{\infty}(p^jH_j)dt \tag{5}$$

Equating the term with identical power of p in equation (5),

p^0: $y_0(x)=f(x)$

$$p^1: y_1(x)=\lambda\int_a^b K(x,t)H_0(t)dt$$

and so on.

and in general form we have

$$\begin{cases} y_0(x)=f(x) \\ y_{k+1}(x)=\lambda\int_a^b K(x,t)H_k(t)dt, k=0,1,2,\dots \end{cases} \tag{6}$$

Using the recursive scheme (6), the n-term approximation series solution can be obtained as follows:

$$\varphi_n(x)=\sum_{j=0}^{n}y_j(x) \tag{7}$$

Numerical Implementations

In this section, we will apply the Adomian decomposition method

***Corresponding author:** Mohedul Hasan Md, Department of Mathematics, Dhaka University, Dhaka, Bangladesh
E-mail: mohedul.math@gmail.com; drmatin1_du@yahoo.com

to compute a numerical solution for non-linear integral equation of the Fredholm type. Then we will compare between the results which we obtain by the numerical solution technique and the results of the exact solution. To illustrate this, we consider the following example:

Example 1

Consider the following nonlinear Fredholm integral equation of the second kind

$$y(x)=\frac{7}{8}x+\frac{1}{2}\int_0^1 xty^2(t)\mathrm{d}t \tag{8}$$

where the exact solution of the equation is $y(x)=x$. In the following, we will compute Adomian polynomials for the nonlinear terms $y^2(t)$ that arises in nonlinear integral equation.

For $k=0$, equation (4) becomes

$$H_0=\frac{1}{0!}\frac{d^0}{dp^0}\left[G\left(\sum_{i=0}^{\infty}p^i y_i\right)\right]_{p=0}$$

$$=\left[G\left(p^0y_0+p^1y_1+p^2y_2+\ldots\right)\right]_{p=0}$$

The Adomian polynomials for $G(y)=y^2$ are given by

$$=(y_0+py_1+p^2y_2+\ldots)^2\,|_{(p=0)}$$

$$\therefore H_0=y_0^{\ 2}$$

By using the MATHEMATICA software, the next few terms, we have

$$H_1=2y_0y_1$$
$$H_2=2y_0y_2+y_1^{\ 2}$$
$$H_3=2(y_0y_3+2y_1y_1)$$
$$H_4=2(y_1y_3+2y_0y_4)+y_2^{\ 2}$$
$$H_5=2(y_2y_3+2y_1y_4+y_0y_5)$$
$$H_6=2(y_2y_4+y_1y_5+y_0y_6)+y_3^{\ 2}$$
$$H_7=2(y_3y_4+y_2y_5+y_1y_6+y_0y_7)$$
$$H_8=2(y_3y_5+y_2y_6+y_1y_7+y_0y_8)+y_4^{\ 2}$$
$$H_9=2(y_4y_5+y_3y_6+y_2y_7+y_1y_8)$$

and so on.

Applying the technique as stated above in equation (6), we have

$$p^0:y_0(x)=\frac{7}{8}x$$

$$p^1:y_1(x)=\frac{1}{2}\int_0^1 xt\,H_0(t)dt=\frac{x}{2}\int_0^1 t\,y_0^{\ 2}(t)dt=\frac{x}{2}\int_0^1 t\left(\frac{7}{8}t\right)^2 dt=\frac{49}{512}x=\frac{7^2}{8^3}x$$

$$p^2:y_2(x)=\frac{1}{2}\int_0^1 xt\,H_1(t)dt=\frac{x}{2}\int_0^1 2ty_0(t)y_1(t)dt=\frac{x}{2}\int_0^1 2t\left(\frac{7t}{8}\right)\left(\frac{49t}{512}\right)dt=\frac{7^3}{4\times 8^4}x$$

In a similar manner, we stop the iteration at the tenth step. Therefore we can write

$$y(x)=\left(\frac{7}{8}+\frac{7^2}{8^3}+\frac{7^3}{4\times 8^4}+\frac{5\times 7^4}{8^7}+\frac{7^6}{4\times 8^8}+\frac{3.7^7}{4\times 8^{10}}+\frac{1811\times 7^6}{2\times 8^{14}}\right.$$
$$\left.+\frac{5\times 283\times 7^8}{4\times 8^{15}}+\frac{5\times 3673\times 7^9}{2\times 8^{18}}+\frac{5\times 798101\times 7^{10}}{2\times 8^{22}}x\right)\approx 0.999947x$$

The table under shows the approximate solutions obtained by applying the Adomian Decomposition method giving to the value of x, which is in the interval [0-1] (Table 1 and Figure 1).

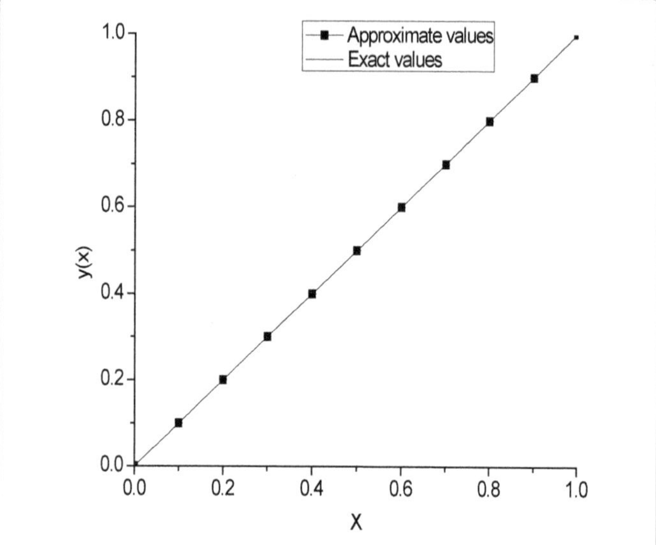

Figure 1: Numerical and exact solutions to the integral equation (8).

Nodes (*x*)	Exact solutions	Approximate solutions	Absolute error
0	0	0	0
0.10	0.100000	0.0999947	0.0000053
0.20	0.200000	0.1999890	0.0000110
0.30	0.300000	0.2999840	0.0000160
0.40	0.400000	0.3999790	0.0000210
0.50	0.500000	0.4999740	0.0000260
0.60	0.600000	0.5999680	0.0000320
0.70	0.700000	0.6999630	0.0000370
0.80	0.800000	0.7999580	0.0000420
0.90	0.900000	0.8999520	0.0000480
1.0	1.000000	0.9999470	0.0000530

Table 1: Numerical and exact solutions to the integral equation (8).

Example 2

Consider the following nonlinear Fredholm integral equation of the second kind

$$y(x)=3+0.6625x+\frac{x}{20}\int_0^1 ty^2(t)dt \tag{9}$$

Applying above procedure, we have

$$p^0:y_0(x)=3+0.6625x$$

$$p^1:y_1(x)=\frac{x}{20}\int_0^1 ty_0^{\ 2}(t)dt=0.296736x$$

In a similar manner, we stop the iteration at the tenth step. Therefore we can write

$$y(x)=3+(0.6625+0.296736+0.0345883+0.0356889+0.00441658+0.000794517+$$
$$0.000156235+0.0000438455+0.0000069959+0.000002199)\approx 3+1.03493x$$

The exact solution of the equation is $3+x$. The table below shows the approximate solutions obtained by applying the Adomian decomposition method according to the value of x, which is confined between zero and one. We compared these results with the results which were obtained by the exact solution (Table 2 and Figure 2).

Example 3

Consider the following nonlinear Fredholm integral equation

$$y(x)=\sin(\pi x)+\frac{1}{5}\int_0^1\cos(\pi x)\sin(\pi t)(y(t))^3\,dt,\quad x\in[0,1] \tag{10}$$

Nodes (*x*)	Exact solutions	Approximate solutions	Absolute error
0	3	3	0
0.10	3.100000	3.10349	0.00349
0.20	3.200000	3.20699	0.00699
0.30	3.300000	3.31048	0.01048
0.40	3.400000	3.41397	0.01397
0.50	3.500000	3.51747	0.01747
0.60	3.600000	3.62096	0.02096
0.70	3.700000	3.72445	0.02445
0.80	3.800000	3.82794	0.02794
0.90	3.900000	3.93144	0.03144
1.0	4.000000	4.03493	0.03493

Table 2: Numerical and exact solutions to the integral equation (9).

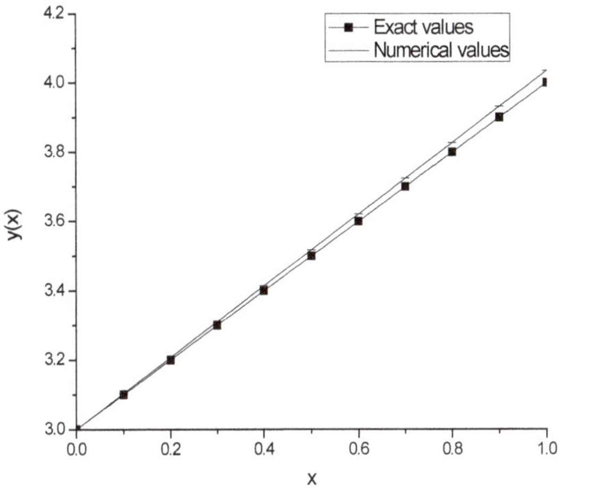

Figure 2: Numerical and exact solutions to the integral equation (9).

Nodes	Exact values	Approximate values	Absolute error
0.00	0.07542668890493687	0.07542668888687896	1.8057902×10^{-11}
0.05	0.23093252624133365	0.23093252622349808	1.7835566×10^{-11}
0.10	0.38075203836055493	0.3807520383433809	1.7174039×10^{-11}
0.15	0.5211961716517719	0.5211961716356822	1.6089685×10^{-11}
0.20	0.6488067254459994	0.6488067254313902	1.4609202×10^{-11}
0.25	0.7604415043936764	0.7604415043809076	1.2768786×10^{-11}
0.30	0.8533516897425216	0.8533516897319074	1.0614176×10^{-11}
0.35	0.9252495243780194	0.9252495243698213	8.198108×10^{-12}
0.40	0.9743646449962113	0.9743646449906311	5.580202×10^{-12}
0.45	0.9994876743237375	0.9994876743209126	2.824962×10^{-12}
0.50	1	1	0
0.55	0.9758890068665379	0.9758890068693629	2.824962×10^{-12}
0.60	0.9277483875940958	0.927748387599676	5.580202×10^{-12}
0.65	0.8567635239987162	0.8567635240069144	8.198108×10^{-12}
0.70	0.7646822990073733	0.7646822990179875	1.0614176×10^{-11}
0.75	0.6537720579794186	0.6537720579921874	1.2768786×10^{-11}
0.80	0.526763779138947	0.5267637791535562	1.4609202×10^{-11}
0.85	0.38678482782732176	0.38678482784341145	1.6089685×10^{-11}
0.90	0.23728195038933994	0.237281950406514	1.7174067×10^{-11}
0.95	0.08193640383912822	0.08193640385696378	1.7835566×10^{-11}
1.00	-0.07542668890493687	-0.07542668888687896	1.8057902×10^{-11}

Table 3: Numerical and exact solutions to the integral equation (10).

The exact solution of the equation (10) is $y(x)=\sin(\pi x)+\frac{1}{3}\left(20-\sqrt{391}\right)\cos(\pi x)$. In the following, we will calculate Adomian polynomials for the nonlinear terms $y^3(t)$ that arises in nonlinear integral equation.

For k=0, equation (4) becomes

$$H_0=\frac{1}{0!}\frac{d^0}{dp^0}\left[G\left(\sum_{i=0}^{\infty}p^i y_i\right)\right]_{p=0}$$

$$=\left[G\left(\sum_{i=0}^{\infty}p^i y_i\right)\right]_{p=0}$$

$$=\left[G\left(p^0y_0+p^1y_1+p^2y_2+\ldots\right)\right]_{p=0}$$

The Adomian polynomials for $G(y)=y^3$ are given by

$$=\left(y_0+py_1+p^2y_2+\ldots\right)^3\Big|_{p=0}$$

$$\therefore\ H_0=y_0^{\ 3}$$

By using the MATHEMATICA v9 software, the next few terms, we have

$$H_1=3y_0^{\ 2}y_1$$
$$H_2=3(y_0y_1^{\ 2}+y_0^{\ 2}y_2)$$
$$H_3=y_1^{\ 3}+6y_0y_1y_2+3y_0^{\ 2}y_3)H_4=3(y_1^{\ 2}y_2+y_0y_2^{\ 2}+y_0^{\ 2}y_4)+6y_0y_1y_3$$
$$H_5=3\left(y_1y_2^{\ 2}+y_1^{\ 2}y_3+y_0^{\ 2}y_5\right)+6(y_0y_2y_3+y_0y_1y_4)$$
$$H_6=y_2^{\ 3}+6\left(y_1y_2y_3+y_0y_2y_4+y_0y_1y_5\right)+3\left(y_0y_3^{\ 2}+y_1^{\ 2}y_4+y_0^{\ 2}y_6\right)$$
$$H_7=3\left(y_2^{\ 2}y_3+y_1y_3^{\ 2}+y_1^{\ 2}y_5+y_0^{\ 2}y_7\right)+6(y_1y_2y_4+y_0y_3y_4+y_0y_2y_5+y_0y_1y_6)$$
$$H_8=3\left(y_2y_3^{\ 2}+y_4y_2^{\ 2}+y_4^{\ 2}y_0+y_1^{\ 2}y_6+y_0^{\ 2}y_8\right)+6(y_1y_3y_4+y_1y_2y_5+y_0y_3y_5\ \ +y_0y_2y_6+y_0y_1y_7$$

and so on.

Applying the procedure as stated above in equation (6), we have

$$p^0:y_0(x)=\sin(\pi x)$$
$$p^1:y_1(x)=\frac{\cos(\pi x)}{5}\int_0^1\sin(\pi t)H_0(t)dt=\frac{\cos(\pi x)}{5}\int_0^1\sin(\pi t)y_0^{\ 3}(t)dt=\frac{3}{40}\cos(\pi x)$$

In a similar manner, we stop the iteration at the ninth step. Therefore we can write

$$y(x)=\sin(\pi x)+\left(\frac{3}{40}+0+\frac{27}{64000}+0+\frac{243}{51200000}+0+\frac{2187}{32768000000}+0+\frac{137781}{131072000000000}\right)\cos(\pi x)$$

$$=\sin(\pi x)+\frac{9886326965781}{131072000000000}\cos(\pi x)$$

The table under shows the approximate solutions obtained by applying the Adomian decomposition method according to the value of x, which is in the interval [0-1] (Table 3 and Figure 3).

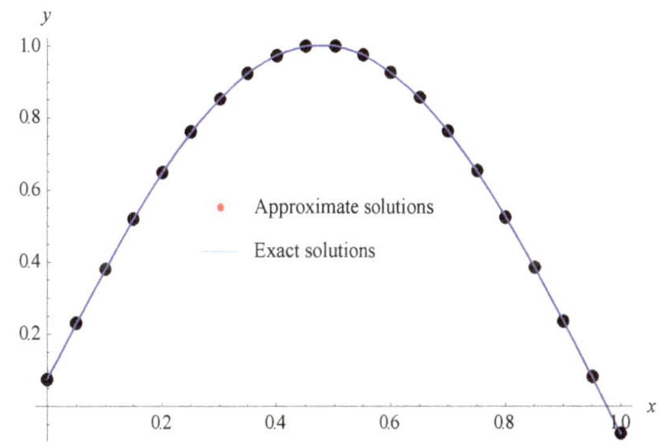

Figure 3: Numerical and exact solutions to the integral equation (10).

Conclusion

This paper presents a technique to find the result of a nonlinear Fredholm integral equation by Adomian decomposition method (ADM). The estimated solutions obtained by the ADM are compared with exact solutions. It can be concluded that the ADM is effective and accuracy of the numerical results demonstrations that the proposed method is well suited for the solution of such kind problems.

References

1. Wazwaz AM (2015) A First Course in Integral Equations. World Scientific.
2. Jerri AA (1985) Introduction to Integral Equations with Applications. Marcel Dekker Inc, New York.
3. Abbasbandy S, Shivanian E (2011) A new Analytical Technique to Solve Fredholm's Integral Equations. Numerical Algorithms 56: 27-43.
4. Ghorbani A (2009) Beyond Adomian polynomials: He polynomials. Chaos, Solitons and Fractals 39: 1486-1492.
5. Ghorbani A, Saberi-Nadjafi J (2007) He's homotopy perturbation method for calculating adomian polynomials. International Journal of Nonlinear Sciences and Numerical Simulation 8: 229-232.

Stability of Fourier Solutions of Nonlinear Stochastic Heat Equations in 1D

Hazaimeh MH*

Department of Mathematics, Zayed University, Dubai, UAE

Abstract

The main focus of this article is studying the stability of solutions of nonlinear stochastic heat equation and give conclusions in two cases: stability in probability and almost sure exponential stability. The main tool is the study of related Lyapunov-type functionals. The analysis is carried out by a natural *N*-dimensional truncation in isometric Hilbert spaces and uniform estimation of moments with respect to *N*.

Nonlinear stochastic heat equation, additive space-time noise, Lyapunov functional, Fourier solution, finite-dimensional approximations, moments, stability.

Keywords: Nonlinear stochastic heat equation; Additive space-time noise; Lyapunov functional; Fourier solution; Finite-dimensional approximations; Moments; Stability

Introduction

In this article we study the stability of solutions of semi-linear stochastic heat equations

$$u_t = \sigma^2 \Delta u + A(u) + B(u)\frac{dW}{dt}$$

with cubic nonlinearities $A(u)$ in one dimensions in terms of all systems parameters, i.e., with non-global Lipschitz continuous nonlinearities. Our study focusses on stability of analytic solution $u=u(x,t)$ under the geometric condition

$$\sigma^2 \frac{\pi^2}{l^2} - a_1 = \frac{\sigma^2 \pi^2 l^2 - a_1 l^2}{l^2} := \gamma > 0,$$

where $0 \leq x \leq 1$ such that D=[0,l].

Many authors have treated stochastic heat equations (e.g. [1,2]), semi-linear stochastic heat equations (e.g. [1-3]) or nonlinear stochastic evolution equations (e.g. [4,5]). Also, some authors study the stability of stochastic heat equations like Fournier and Printems [6] study the stability of the mild solution. Walsh reats the stochastic heat equations in one dimension. Chow [1] studies that the null solution of the stochastic heat equation is stable in probability by using the definition. Recall that:

$$\mathbb{L}^2(\mathbb{D}) = \{f : \mathbb{D} \to \mathbb{R} | \int_{\mathbb{D}} |f(x)|^2 \, d\mu(x) < \infty\},$$

Where μ is the Lebesgue measure in one dimensions. The paper is organized as follows. Section 2 states that the strong Fourier solution of equation (1) is proved. We write the solution using the finite-dimensional truncated system verifies properties of finite-dimensional Lyapunov functional. Section 3 discusses the stability of the strong solution of equation (1) is stable in probability and almost sure exponential stability. Eventually, Section 4 summarizes the most important conclusions on the well-posedness and behaviour of the original infinite-dimensional system (1).

Truncated Fourier Series Solution and Finite-dimensional Lyapunov Functional

Consider the stochastic nonlinear heat equation with additive noise

$$u_t = \sigma^2 u_{xx} + (a_1 - a_2 \| u \|^2_{\mathbb{L}^2(\mathbb{D})})u + b\frac{dW}{dt} \tag{1}$$

with the initial condition $u(x,0)=f(x)$ with $f \in L^2(D)$ (initial position) and $W(x,t) = \sum_{n=1}^{\infty} \alpha_n W_n(t) e_n(x)$ and $e_n = \sqrt{\frac{2}{l}} sin(\frac{n\pi x}{l})$ driven by *i.i.d.* standard Wiener processes W_n with $E[W_n(t)]^2=0$, $E[W_n(t)]^2=t$. The solution of equation (1) in terms of Fourier series is proved by Schurz [3] and given by

$$u(x,t) = \sum_{n=1}^{\infty} c_n(t) e_n(x). \tag{2}$$

Theorem 1

Assume that $\sum_{n,m=1}^{\infty} \alpha_n^2 < \infty$, $\forall u \in \mathbb{L}^2(\mathbb{D}) \cap C^1(\mathbb{D} \times \mathbb{R}^1_+)$ *with* $u_x \in L^2(D)$ *and* $W(x,t) = \sum_{n,m=1}^{\infty} \alpha_n W_n(t) e_n(x)$, *then for all* $t \geq 0$, $x \in D=(0,l_x)$, *the Fourier-series solutions (2) have Fourier coefficients* c_n *satisfying (a.s.)*

$$\frac{d}{dt} c_n(t) = [-\sigma^2 \frac{\pi^2 n^2}{l^2} + a_1 - a_2 \sum_{k=1}^{\infty} (c_k)^2] c_n + b\alpha_n \frac{dW_n}{dt}. \tag{3}$$

Proof. See Schurz [3].

We need to truncate the infinite series (2) for practical computations. So, we have to consider finite-dimensional truncations of the form

$$u_N(x,t) = \sum_{n=1}^{N} c_n e_n \tag{4}$$

with Fourier coefficients c_n satisfying the naturally truncated system of stochastic differential equations (SDEs).

$$\frac{d}{dt} c_n(t) = [-\sigma^2 \frac{\pi^2 n^2}{l^2} + a_1 - a_2 \sum_{k=1}^{\infty} (c_k)^2] c_n + b\alpha_n \frac{dW_n}{dt}. \tag{5}$$

where $\lambda_n = (\frac{n\pi}{l})^2$.

Assume that $\sigma^2 \frac{\pi^2}{l^2} > a_1$. Define the Lyapunov functional V_N as follows

***Corresponding author:** Hazaimeh MH, Department of Mathematics, University College, Zayed University, P.O.Box 19282, Dubai, UAE
E-mail: haziem67@gmail.com

$$V_N(c)=V_N((c_n)_{n=1,\dots,N}):=\sum_{n=1}^{N}(\sigma^2\lambda_n-a_1)(c_n)^2+\frac{a_2}{2}\left(\sum_{n=1}^{N}(c_n)^2\right)^2 \quad (6)$$

for $N \in \mathbb{N}$.

This functional is a modification of a functional appeared in Schurz [7]. It is clear that this function is of Lyapunov-type because it is nonnegative and smooth as long as $a_2 \geq 0$, radially unbounded if additionally $\sigma^2\pi^2>a_1l^2$. Equipped with Euclidean norm

$$|c|_{l_N^2}=\sqrt{\sum_{n=1}^{N}c_n^2}$$

Lemma 2

Consider the Lyapunov functional defined in equation (6), and let

$$\sigma^2\frac{\pi^2}{l^2}-a_1=\frac{\sigma^2\pi^2-a_1l^2}{l^2}=:\gamma>0.$$

Then $\forall u\in L^2(D)$:

$$V_N(u)\geq\gamma\|u\|^2_{\mathbb{L}^2(\mathbb{D})} \quad (7)$$

Proof. See [7].

Lemma 3

Assume that $a_2 \geq 0$. Then, $\forall N \in \mathbb{N}$, the functional V_N is

(a) *nonnegative and positive semi-definite* if $\sigma^2\pi^2>a_1l^2$ or $a_2 \geq 0$.

(b) *positive-definite* if $\sigma^2\pi^2>a_1l^2$,

and

(c) satisfies the condition of radial unboundedness

$$\lim_{|c|_{l_N^2}\to+\infty}V_N(c)=+\infty,\text{ if }[\sigma^2\pi^2-a_1l^2]_++a_2>0.$$

Proof. See [7].

Stability of Fourier Solutions

Recall equation (5) governed by

$$\frac{d}{dt}c_n(t)=[-\sigma^2\frac{\pi^2n^2}{l^2}+a_1-a_2\sum_{k=1}^{\infty}(c_k)^2]c_n+b\alpha_n\frac{dW_n}{dt} \quad (8)$$

$$=\left[-\sigma^2\lambda_n+a_1-a_2\|u_N\|^2\right]c_n+b\alpha_n\frac{dW_n}{dt}. \quad (9)$$

To simplify, let

$$f(u_N)=-\sigma^2\lambda_n+a_1-a_2\|u_N\|^2$$

and

$$g(u_N)=b\alpha_n$$

Definition: *The trivial solution of system (8) (in terms of norm $\|u\|_{\mathbb{L}^2(\mathbb{D})}$) is said to be stochastically stable or stable in probability, if for $0<\varepsilon<1$ and $r>0$, $\exists$ a $\delta=\delta(\varepsilon,r)$ such that, $\forall t\geq\delta$, we have*

$$\mathbb{P}\{\|u(t)\|_{\mathbb{L}^2(\mathbb{D})}<r\}\geq1-\varepsilon. \quad (10)$$

whenever $\delta > 0$.

Lemma 4

If $\exists$ a positive-definite function $V\in C^{2,1}(R^d\times[0.\ \infty),R_+)$ such that $LV(x,t)\leq 0$ and $\forall(x,t)\in R^d\times[0.\ \infty)$, then the trivial solution of the equation.

$$dX(t)=f(x(t),t)dt+g(x(t),t)dw(t) \quad (11)$$

is *stochastically stable.*

Proof. See Arnold [8].

Theorem 5

Let $V(u(t))=\sigma^2\|\nabla u\|^2_{\mathbb{L}^2(\mathbb{D})}-a_1\|u\|^2_{\mathbb{L}^2(\mathbb{D})}+\frac{a_2}{2}\|u\|^4_{\mathbb{L}^2(\mathbb{D})}$.

If $(1-a_2\sum_{n=1}^{N}c_n^2)\sum_{n=1}^{N}c_n>0$, then the trivial solution of equation (8) is stochastically stable i.e., stable in probability.

Proof. From Lemma 3, we know that $V_N(u(t))$ is positive-definite if $\forall n\mathbb{N},\sigma^2\lambda_n-a_1>0$. Define the linear operator L as in Schurz [3]

$$\mathcal{L}=\sum_{n=1}^{N}\left[-\sigma^2\frac{\pi^2n^2}{l^2}+a_1-a_2\sum_{k=1}^{N}(c_k)^2\right]c_n\frac{\partial}{\partial c_n}+\frac{b^2}{2}\sum_{n=1}^{N}\alpha_n^2\frac{\partial^2}{\partial c_n^2}.$$

The first and second partial derivative of $V_N(t)$ with respect to c_n are

$$\frac{\partial V_N}{\partial c_n}=2\sum_{n=1}^{N}\left[(\sigma^2\lambda_n-a_1)+a_2\left(\sum_{n=1}^{N}c_n^2\right)\right]c_n$$

and

$$\frac{\partial^2V_N}{\partial c_n^2}=2\sum_{n=1}^{N}(\sigma^2\lambda_n-a_1)c_n+4a_2\left(\sum_{n=1}^{N}c_n\right)^2+2a_2\sum_{n=1}^{N}c_n^2.$$

Then

$$\mathcal{L}V_N(c_n(t))=-2\left(\sum_{n=1}^{N}(\sigma^2\lambda_n-a_1)c_n\right)^2-2a_2\sum_{n=1}^{N}(\sigma^2\lambda_n-a_1)\left(\sum_{n=1}^{N}c_n^2\right)^2$$

$$-2a_2\sum_{n=1}^{N}c_n^2\left(1-a_2\sum_{n=1}^{N}c_n^2\right)\sum_{n=1}^{N}\left[(\sigma^2\lambda_n-a_1)\right]c_n.$$

But by our assumption that

$$(1-a_2\sum_{n=1}^{N}c_n^2)\sum_{n=1}^{N}c_n>0,$$

Then thus

$$LV_N(c_n(t))\leq 0.$$

So by Lemma 4, applied to truncation of (8), the trivial solution of system (8) is *stochastically stable.*

Corollary 6

Let $p\geq2$ and let V be as above. Imposing the same assumptions as in Theorem 5 with $N\to+\infty$, then we have $\forall 0\leq t\leq T$,

$$\mathbb{E}\|u(t)\|^p_{\mathbb{L}^2(\mathbb{D})}\leq\frac{1}{\min(1,\gamma)}\mathbb{E}V^{\frac{p}{2}}(u(0)).$$

Proof. We know, from the definition of $V(u)$, and Lemma 2 that $\|u\|^2_{\mathbb{L}^2(\mathbb{D})}\leq\frac{V(u)}{\gamma}$. it is easy to show that

$$\mathbb{E}\|u(t)\|^p_{\mathbb{L}^2(\mathbb{D})}\leq\frac{1}{\min(1,\gamma)}\mathbb{E}V^{\frac{p}{2}}(u(0)).$$

Corollary 7

$\forall p\geq 2$ and $\forall 0\leq t\leq T$, with $\sigma^2\lambda_1-a_1>0$, we have $\forall 0\leq t\leq T$.

1) If $a_2 \geq 0$, then

$$\mathbb{E}\|u(t)\|^p_{\mathbb{L}^2(\mathbb{D})}\leq\frac{\mathbb{E}V^{\frac{p}{2}}(u(0))}{[\sigma^2\lambda_1-a_1]^{\frac{p}{2}}}.$$

2) If $a_2 > 0$, then

$$\mathbb{E}\|u(t)\|_{\mathbb{L}\,(\mathbb{D})} \le (-)^{\frac{p}{4}}\mathbb{E}V^{\frac{p}{4}}(u(0)).$$

Proof. 1) Note that we have $(\sigma^2\lambda_1 - a_1)\|u(.,t)\|^2_{\mathbb{L}^2(\mathbb{D})} \le V_N(u(t))$. Since λ_n is increasing in n,

$$\left[\sigma^2\lambda_1 - a_1\right]\|u(t)\|^2_{\mathbb{L}^2(\mathbb{D})} \le V_N(u(t)).$$

So,

$$\|u(t)\|^2_{\mathbb{L}^2(\mathbb{D})} \le \frac{V_N(u(t))}{\sigma^2\lambda_1 - a_1}$$

Pull over expectation, then

$$\mathbb{E}\|u(t)\|^2_{\mathbb{L}^2(\mathbb{D})} \le \frac{\mathbb{E}V(u(t))}{\sigma^2\lambda_1 - a_1}.$$

By using Corollary 6, we have

$$\mathbb{E}\|u(t)\|^2_{\mathbb{L}^2(\mathbb{D})} \le \frac{\mathbb{E}V(u(0))}{\sigma^2\lambda_1 - a_1}.$$

2) From the definition of $V(u(t))$, it is clear that $\frac{a_2}{2}\|u(t)\|^4_{\mathbb{L}^2(\mathbb{D})} \le V(u(t))$, so

$$(\frac{a_2}{2})\|u(t)\|^4_{\mathbb{L}^2(\mathbb{D})} \le V_N(u(0)).$$

Now, take the expectation to both sides, and we get

$$(\frac{a_2}{2})\mathbb{E}\|u(t)\|^4_{\mathbb{L}^2(\mathbb{D})} \le \mathbb{E}V_N(u(0)),$$

i.e., $\forall 0 \le t \le T$,

$$\mathbb{E}\|u(t)\|^2_{\mathbb{L}^2(\mathbb{D})} \le (\frac{2}{a_2})^{\frac{1}{2}}\mathbb{E}V^{\frac{1}{2}}(u(0)).$$

Remark: The corollary 7 means that $\forall t \ge 0$:

$$\mathbb{E}\|u(t)\|^2_{\mathbb{L}^2(\mathbb{D})} \le \min\{\frac{\mathbb{E}V_N(u(0))}{[\sigma^2\lambda_1 - a_1]}, (\frac{2}{a_2})^{\frac{1}{2}}\mathbb{E}V^{\frac{1}{2}}(u(0))\}.$$

Definition: *The trivial solution of system (8) is said to be a.s. exponentially stable if*

$$\theta(u_N) := \limsup_{t\to\infty}\frac{1}{t}\log\|u(t)\|_{\mathbb{L}^2(\mathbb{D})} < 0 \quad (a.s.) \tag{12}$$

$\forall u(0) \in D$. The quantity of the left hand side of (12) is called *the sample top Lyapunov exponent of u.*

Lemma 8

Let v(t) be a nonnegative integrable function such that [9]

$$v(t) \le C + A\int_0^t v(s)\,ds, \quad 0 \le t \le T \tag{13}$$

for some constants C, A. Then $C \ge 0$ and

$$v(t) \le C\exp(At),\ 0 \le t \le T \tag{14}$$

Theorem 9

Let $V(u(t))$ as in Theorem 5. If $(1 - a_2\sum_{n=1}^N c_n^2)\sum_{n=1}^N c_n > 0$, then the norm of the trivial solution of N-dimensional system (8) is a.s. exponentially stable with sample top Lyapunov exponent

$$\theta(u_N) \le 0.$$

Proof. Return to the analysis of finite N-dimensional equation (5). Recall that

$$V(u_N(t)) = \sigma^2\|\nabla u_N\|^2_{\mathbb{L}^2(\mathbb{D})} - a_1\|u_N\|^2_{\mathbb{L}^2(\mathbb{D})} + \frac{a_2}{2}\|u_N\|^4_{\mathbb{L}^2(\mathbb{D})}$$

$$= \sum_{n=1}^N\left[\sigma^2\lambda_n - a_1\right](c_n(t))^2 + \frac{a_2}{2}(\sum_{n=1}^N(c_n(t))^2)^2$$

where $V(u_N) = V_N(c)$ and from Theorem 5 we know that

$$\mathcal{L}V_N(c_n(t)) = -2\left(\sum_{n=1}^N(\sigma^2\lambda_n - a_1)c_n\right)^2 - 2a_2\sum_{n=1}^N(\sigma^2\lambda_n - a_1)\left(\sum_{n=1}^N c_n^2\right)^2$$

$$-2a_2\sum_{n=1}^N c_n^2\left(1 - a_2\sum_{n=1}^N c_n^2\right)\sum_{n=1}^N\left[(\sigma^2\lambda_n - a_1)\right]c_n.$$

But by our assumption that

$$(1 - a_2\sum_{n=1}^N c_n^2)\sum_{n=1}^N c_n > 0,$$

so

$$\mathcal{L}V_N(c_n(t)) \le -k,$$

where $k \ge 0$.

Using Dynkin's formula, we find that [10-16]

$$\mathbb{E}V_N(c(t)) = \mathbb{E}V_N(c(0)) + \mathbb{E}\int_0^t \mathcal{L}V_N(c_n(s))\,ds$$

$$\le \mathbb{E}V_N(c,v)(0) - kt$$

so

$$\mathbb{E}\|u_N(t)\|^2_{\mathbb{L}^2(\mathbb{D})} \le \mathbb{E}V_N(c(t)) \le \mathbb{E}V_N(c(0)) - kt$$

using extended Gronwall lemma, Lemma 8, gives us

$$\mathbb{E}\|u_N(t)\|^2_{\mathbb{L}^2(\mathbb{D})} \le \mathbb{E}V_N(c(0))e^{-kt}$$

hence

$$\log\mathbb{E}\|u_N(t)\|^2_{\mathbb{L}^2(\mathbb{D})} \le \log\mathbb{E}V_N(c(0)) - kt$$

thus

$$\limsup_{t\to\infty}\frac{\log\mathbb{E}\|u_N(t)\|^2_{\mathbb{L}^2(\mathbb{D})}}{t} \le \limsup_{t\to\infty}\frac{\log\mathbb{E}V_N(c(0))}{t} - k$$

$$\le b_2^2\|\alpha\|^2_{\mathbb{L}^2_{4N\times N}} - 2\kappa \le -k. \tag{15}$$

If $(1 - a_2\sum_{n=1}^N c_n^2)\sum_{n=1}^N c_n > 0$ then the left side of identity (12) is negative and the trivial solution of the velocity v of N-dimensional system (8) is *a.s.* exponential stable.

Finally, we observe that all the previous estimates are uniformly bounded as $N \to \infty$. Hence, we arrive at

$$\limsup_{t\to\infty}\frac{\log\mathbb{E}\|u(t)\|^2_{\mathbb{L}^2(\mathbb{D})}}{t} < 0. \tag{16}$$

Corollary 10

Let $V(u(t))$ as in Theorem 5. If $\left(1 - a_2\sum_{n=1}^N c_n^2\right)$, then the norm of the v-component of the trivial solution of infinite-dimensional system (1) is a.s. exponentially stable with sample top Lyapunov exponent

$$\theta(v_N) \le -k < 0.$$

Proof. Return to the proof of previous Theorem 9 and take the limit N to $+\infty$ after the estimation process (16) in the sample Lyapunov exponent $\theta(v_N)$.

Conclusion

By analyzing appropriate N-dimensional truncations of the original semi-linear heat equations (1), we can verify the asymptotic stability of random Fourier series solutions with strongly unique, Markovian, continuous time Fourier coefficients under the presence of cubic nonlinearities. For this purpose, we introduced and studied an appropriate Lyapunov. The analysis is basicly relying on the fact that all estimations of moments of Lyapunov functional are made independent of dimensions N of their finite-dimensional truncations. Thus, the techniques of our proof are finite-dimensional in character, however the conclusions can be drawn to the original infinite-dimensional semi-linear equation.

References

1. Chow PL (2007) Stochastic Partial Differential Equations. Chapman and Hall/CRC, Boca Raton.
2. Da Prato G, Zabzcyk J (1992) Stochastic Equations in Infinite Dimensions, Cambridge University Press, Cambridge, London.
3. Schurz H (2010) Nonlinear stochastic heat equations with cubic nonlinearity and additive Q-reglar noise in R1. Electronic Journal of Differential Equations Conf 19: 221-233.
4. Grecksch W, Tudor C (1995) Stochastic Evolutions: A Hilbert Space Approach. Akademie-Verlag, Berlin.
5. Schurz H (2007) Existence and uniqueness of solutions of semilinear stochastic infinite-dimensional differential systems with H-Regular noise. J Math Anal Appl 332: 334-345.
6. Fournier N, Printems J (2011) Stability of the Stochastic Heat Equation in L1([0,1]). Elect Comm in Probab 16: 337-352.
7. Schurz H, Hazaimeh HM (2014) Existence, Uniqueness, and stability of stochastic wave equation with cubic nonlinearities in two dimensions. Journal of Mathematical Analysis and Applications 148: 775-795.
8. Arnold L (1974) Stochastic Differential Equations: Theory and Applicants. John Wiley and Sons, Inc., New York.
9. Ksendal BØ (2003) Stochastic Differential Equations: An Introduction With applications, Springer-Verlag, Berlin.
10. Dalang R, Frangos NE (1998) The stochastic wave equation in two spatial dimensions. Ann Probab 26: 187-212.
11. Friedman A (1975) Stochastic Differential Equations and Applications. Dover Publications Inc. Mineola, New York.
12. Gard TC (1988) Introduction to Stochastic Differential Equations. Marcel Dekker, Basel.
13. Karatzas I, SchreveSE (1991) Brownian Motion and Stochastic Calculus. Springer-Verlag, New York.
14. Millet A, Sanz-Sole M (1999) A stochastic wave equation in two space dimension: Smoothness of the law. Ann Probab 27: 803-844.
15. Shiryaev (1996) Probability (2ndedn) Springer-Verlag, Berlin.
16. Walsh JB (1986) An introduction to stochastic partial differential equations. École d'été de probabilités de Saint-Flour, XIV-1984, Springer-Verlag, Berlin pp: 265-439.

Coupled Fixed Points of α - Ψ -Contractive Type Multi Functions

Mohammadi B* and Alizade E

Department of Mathematics, Marand Branch, Islamic Azad University, Tehran, Iran

Abstract

Recently, Samet, Vetro and Vetro introduced α-ψ-contractive mappings and gave some results on fixed points of the mappings . In fact, their technique generalized some ordered fixed point results . Also they have proved some results on coupled fixed points of α-ψ-contractive mappings. In 1974 Ciric introduced quasicontractive mappings and obtained an important generalization of Banach's contraction principle. Recently Mohammadi, Rezapour and Shahzad have proved some fixed point results on α-ψ-contractive and α-ψ-quasicontractive multifunction's. By using the main idea of, we give some new results for coupled fixed points of α-ψ-contractive multifunction.

Keywords: Coupled fixed point; α-ψ-Contractive; Multifunction

Introduction

Denote by Ψ the family of non-decreasing functions $\psi: [0, +\infty) \to [0, +\infty)$ such that $\sum_{n=1}^{+\infty} \psi^n(t) < +\infty$ for all $t > 0$. It is well known that $\psi(t) < t$ for all $t > 0$.

Definition 1.1: Let (X, d) be a metric space and $\alpha: X \times X \to [0, \infty)$ be a map. We say that X satisfies condition (C_α), if for any sequence $\{x_n\}$ in X, that $x_n \to x$ and $\alpha(x_n, x_{n+1}) \geq 1$ for all n, then $\alpha(x_n, x) \geq 1$ for all n. Also, we say that the selfmap T on X is α-admissible whenever $\alpha(x, y) \geq 1$ implies $\alpha(T_x, T_y) \geq 1$.

Definition 1.2: Let X is an arbitrary space and $\alpha: X^2 \times X^2 \to [0, +\infty)$ a map. A mapping $F: X^2 \to X$ is said to be α-admissible whenever $\alpha((x, y), (u, v)) \geq 1$ implies

$$\alpha((F(x, y), F(y, x)), (F(u, v), F(v, u))) \geq 1.$$

Definition 1.3: Let (X, d) be a metric space and $\alpha: X^2 \times X^2 \to [0, \infty)$. We say that X satisfies condition (C_α^*) if for any two sequences $\{x_n\}$ and $\{y_n\}$ in X, that $x_n \to x$, $y_n \to y$ and

$$\alpha((x_n, y_n), (x_{n+1}, y_{n+1})) \geq 1,\ \alpha((y_{n+1}, x_{n+1}), (y_n, x_n)) \geq 1$$

For all n, then we have

$$\alpha((x_n, y_n), (x, y)) \geq 1,\ \alpha((y, x), (y_n, x_n)) \geq 1$$

for all n.

Definition 1.4: Let X is an arbitrary space and $F: X^2 \to X$ is a mapping. We say that $(x^*, y^*) \in X^2$ is a coupled fixed point of F, if we have $F(x^*, y^*)=x^*$ and $F(y^*, x^*)=y^*$.

In 2011, Samet, Vetro and Vetro have proved the following theorem [1].

Theorem 1.1: Let (X, d) be complete a metric space, $\alpha: X^2 \times X^2 \to [0, \infty)$ a function, $\psi \in \Psi$ and $F: X^2 \to X$ an α-admissible mapping such that

$$\alpha((x, y), (u, v))d(F(x, y), F(u, v)) \leq \frac{1}{2}\psi(d(x, u) + d(y, v)).$$

for all $(x, y), (u, v) \in X^2$. Assume that the following assertions hold. (i) There exists $(x_0, y_0) \in X^2$ such that

$$\alpha((x_0, y_0), (F(x_0, y_0), F(y_0, x_0))) \geq 1,$$

$$\alpha((F(y_0, x_0), F(x_0, y_0)), (y_0, x_0)) \geq 1.$$

(ii) Either F is continuous or X satisfies condition (C_α^*). Then F has a coupled fixed point in X^2.

Let (X, d) be a metric space. Define the metric δ on X^2 by $\delta((x, y), (u, v))=d(x, u) + d(y, v)$, for all $(x, y), (u, v) \in X^2$. Also if $F: X^2 \to X$ then put [2].

$$m((x, y), (u, v))=\max\{\delta((x, y), (u, v)),\ \delta((x, y), (F(x, y), F(y, x))),\ \delta((u, v), (F(u, v), F(v, u)))$$

$$\frac{1}{2}[\delta((x,y),(F(u,v),F(v,u)))+\delta((u,v),(F(x,y),F(y,x)))]\},$$

for all $(x, y), (u, v) \in X^2$. It is easy to see that $m((x, y), (u, v))=m((v, u), (y, x))$. Recently Rezapour and H. Asl have extended theorem 1.1 to quasi-contractions as follow [3].

Theorem 1.2: Let (X, d) be a complete metric space, $\alpha: X^2 \times X^2 \to [0, +\infty)$ a function, $\psi \in \Psi$ and $F: X^2 \to X$ an α-admissible mapping such that [4]

$$\alpha((x, y), (u, v))d(F(x, y), F(u, v)) \leq \frac{1}{2}\psi(m((x, y), (u, v))).$$

for all $(x, y), (u, v) \in X^2$. Assume that the following assertions hold. (i) There exists $(x_0, y_0) \in X^2$ such that

$$\alpha((x_0, y_0), (F(x_0, y_0), F(y_0, x_0))) \geq 1,$$

$$\alpha((F(y_0, x_0), F(x_0, y_0)), (y_0, x_0)) \geq 1.$$

(ii) Either F is continuous or ψ is right upper semi continuous and X satisfies condition (C_α^*).

Then F has a coupled fixed point in X^2.

Definition 1.5: Let (X, d) be a metric space and $F: X^2 \to X$. We say that F is orbitally continuous if for any two sequences $\{x_n\}$ and $\{y_n\}$ in X, that $x_n \to x$, $y_n \to y$ and $x_{n+1}=F(x_n, y_n)$, $y_{n+1}=F(y_n, x_n)$, then $F(x_n, y_n) \to F(x, y)$.

Obviously theorem 1.2 is true if F be orbitally continuous instead of continuity. Now we give the following example to show that theorem 1.2 is a real generalization of theorem 1.1, which is there exist mappings

***Corresponding author:** Mohammadi B, Department of Mathematics, Marand Branch, Islamic Azad University, Tehran, Iran
E-mail: bmohammadi@marandiau.ac.ir

that we can use theorem 1.2, but we can't use theorem 1.1 for them [5].

Example 1.6:

Let $M1=\{\frac{m}{n}: m=0, 1, 3, 9, \dots$ and $n=3k+1$ $(k \geq 0)\}$, $M2=\{\frac{m}{n}: m=1, 3, 9, 27, \dots$ and $n=3k+2$ $(k \geq 0)\}$. Set $M=M1 \cup M2$, $d(x, y)=|x-y|$ and define $F: M^2 \to M$ by

$$F(x,y)=\begin{cases} 3x/13 & x,y\in M_1, x\geq y \\ x/8 & x,y\in M_2, x\geq y \\ 0 & otherwise \end{cases} \quad (1)$$

and $\alpha: M^2 \times M^2 \to [0, +\infty)$ by

$$\alpha((x,y),(u,v))=\begin{cases} 1 & (x,y),(u,v)\in M_1^2\cup M_2^2, x\geq y, u\geq v \\ 0 & otherwise. \end{cases} \quad (2)$$

If $(x, y) \in M_1^2$, $x \geq y$ and $(u, v) \in M_2^2$, $u \geq v$, then

$$d(F(x,y),F(u,v))=\left|\frac{3x}{13}-\frac{u}{8}\right|=\frac{3}{13}\left|x-\frac{13u}{24}\right|.$$

Now we consider two cases. If $x>\frac{13}{24}u$, then we have

$$\frac{3}{13}\left|x-\frac{13u}{24}\right|=\frac{3}{13}(x-\frac{13u}{24})\leq\frac{3}{13}(x-\frac{u}{8})=\frac{1}{2}(\frac{12}{13})\frac{d(x,F(u,v))}{8}.$$

$$\leq\frac{1}{2}(\frac{12}{13})\frac{\delta(x,y,F(u,v),F(u,v)))}{2}\leq\frac{1}{2}(\frac{12}{13})m(x,y),(u,v)).$$

Hence

$$d(F(x,y),F(u,v))\leq\frac{1}{2}(\frac{12}{13})m(x,y),(u,v)).$$

If $x<\frac{13}{24}u$, then

$$\frac{3}{13}\left|x-\frac{13u}{24}\right|=\frac{3}{13}(\frac{13u}{24}-x)\leq\frac{3}{13}(u-x)$$

$$\leq\frac{1}{2}(\frac{6}{13})\delta(x,y),(u,v))\leq\frac{1}{2}(\frac{12}{13})m(x,y),(u,v)).$$

Hence

$$d(F(x,y),F(u,v))\leq\frac{1}{2}(\frac{12}{13})m(x,y),(u,v)).$$

If $(x, y),(u, v)\in M_1^2$, $x \geq y$, $u \geq v$, then

$$d(F(x,y),F(u,v))=\left|\frac{3x}{13}-\frac{3u}{13}\right|=\frac{3}{13}d(x,u)\leq\frac{1}{2}(\frac{12}{13})m(x,y),(u,v)).$$

If $(x, y),(u, v)\in M_2^2$, then

$$d(F(x,y),F(u,v))=\left|\frac{x}{8}-\frac{u}{8}\right|=\frac{1}{8}d(x,u)\leq\frac{1}{2}(\frac{12}{13})m(x,y),(u,v)).$$

Now put $\psi(t)=\frac{12t}{13}$, for all $t \geq 0$, and then we see that

$$\alpha((x,y),(u,v))d(F(x,y),F(u,v))\leq\frac{1}{2}\psi(m((x,y),(u,v))),$$

for all $(x, y), (u, v) \in M^2$. Also

$\alpha((1, 1), (F(1, 1), F(1, 1)))=\alpha((1, 1), (3/13, 3/13))=1,$

$\alpha((F(1, 1), F(1, 1)), (1, 1))=\alpha((3/13, 3/13), (1, 1))=1.$

To show that F is α-admissible, assume $\alpha((x, y), (u, v)) \geq 1$. Then $(x, y),(u, v)\in M_1^2\cup M_2^2$, $x \geq y$, $u \geq v$. Hence either $x>y$ and $F(y, x)=0$ or $x=y$ and $F(x, y)$ $F(y, x)$. However $F(x, y) \geq F(y, x)$. Similarly $F(u, v) \geq F(v, u)$. On the other hand $F(x, y), F(y, x), F(u, v), F(v, u) \in M_1$. Hence

$\alpha((F(x, y), F(y, x)), (F(u, v), F(v, u))) \geq 1.$

It is easy to check that F is orbitally continuous. Now by theorem 1.2 we can say that F has a coupled fixed point in X^2. In fact (0, 0) is a coupled fixed point of F [6].

Now, we show that we cannot apply theorem 1.1 in this example. To see this put [7]

$x=y=9$, $u=\frac{243}{26}$, $v=\frac{243}{29}$. Then

$$\alpha((x,y),(u,v))d(F(x,y),F(u,v))=\left|\frac{27}{13}-\frac{243}{26\times 8}\right|=\frac{189}{208}=0.9.$$

On the other hand

$$\frac{1}{2}\psi(\delta((x,y),(u,v)))\leq\frac{1}{2}\delta((x,y),(u,v))=\frac{1}{2}(|x-u|+|y-v|)$$

$$\frac{1}{2}\left|9-\frac{243}{26}\right|+\left|9-\frac{243}{26}\right|)=\frac{729}{1508}=0.48.$$

Hence

$$\alpha((x,y),(u,v))d(F(x,y),F(u,v))>\frac{1}{2}\psi(\delta((x,y),(u,v))).$$

Let (X, d) be a metric space and $C B(X)$ is the set of all nonempty closed bounded subsets of X, $\alpha: X \times X \to [0, \infty)$ a mapping and $T: X \to C B(X)$ a multifunction. We say that T is α-admissible whenever for each $x \in X$ and $y \in Tx$ with $\alpha(x, y) \geq 1$ we have $\alpha(y, z) \geq 1$ for all $z \in Ty$ ([4]). Recall that T is continuous whenever $H(Tx_n, Tx) \to 0$ for all sequence $\{x_n\}$ in X with $x_n \to x$, where H is the Hausdorff metric on $CB(X)$ defined by $H(A, B)=max\{sup_{x\in A} d(x, B), sup_{y\in B} d(y, A)\}$, for all $A, B \in C B(X)$. Also we say that T is orbitally continuous whenever $H(Tx_n, T_x) \to 0$ for all sequence $\{x_n\}$ in X with $x_{n+1} \in Tx_n$ for all n and $x_n \to x$. Recently Mohammadi, Rezapour and Shahzad have proved the following lemma [8].

Lemma 1.3: ([4] Let (X, d) be a complete metric space, $\alpha: X \times X \to [0, \infty)$ a function, $\psi \in \Psi$ a strictly increasing map and $T: X \to C B(X)$ an α-admissible multifunction such that $\alpha(x, y)H(Tx, Ty) \leq \psi(d(x, y))$ for all $x, y \in X$ and there exist $x_0 \in X$ and $x_1 \in Tx_0$ with $\alpha(x_0, x_1) \geq 1$. If T is continuous or X satisfies the condition (C_α), then T has a fixed point. Note that if T be orbitally continuous instead of continuity, then the lemma1.3 is also true [2].

Main Results

Now, we are ready to state and prove our main results.

Definition 2.1: Let X is an arbitrary space and $\alpha: X^2 \times X^2 \to [0, +\infty)$ a map. A multifunction $F: X^2 \to C B(X)$ is said to be α-admissible whenever if $(x, y) \in X^2, (u, v) \in F(x, y) \times F(y, x)$ and $\alpha((x, y), (u, v)) \geq 1$, $\alpha((v, u), (y, x)) \geq 1$, then $\alpha((u, v), (w, z)) \geq 1$, $\alpha((z, w), (v, u)) \geq 1$, for all $(w, z) \in F(u, v) \times F(v, u)$.

Definition 2.2: Let (X, d) be a metric space and $F: X^2 \to CB(X)$ is a multifunction. We say that $(x^*, y^*) \in X^2$ is a coupled fixed point of F if we have $x^* \in F(x^*, y^{**})$ and $y^* \in F(y^*, x^*)$.

Theorem 2.1: Let (X, d) be a complete metric space, $\alpha: X^2 \times X^2 \to [0,$

$+\infty)$ a function, $\psi \in \Psi$ a strictly increasing map and F: $X_2 \to CB(X)$ an α-admissible multifunction such that [7]

$$\alpha((x,y),(u,v))H(F(x,y),F(u,v)) \leq \frac{1}{2}\psi(d(x,u)+d(y,v)).$$

for all $(x, y), (u, v) \in X^2$. Assume that the following assertions hold.

(i) There exists $(x_0, y_0) \in X_2$ and $(x_1, y_1) \in F(x_0, y_0) \times F(y_0, x_0)$ such that

$$\alpha((x_0, y_0),(x_1, y_1)) \geq 1, \quad \alpha((y_1, x_1),(y_0, x_0)) \geq 1.$$

(ii) Either F is continuous or X satisfies condition (C^*). Then F has a coupled fixed point in X^2.

Proof Define $\beta: X_2 \times X_2 \to [0, +\infty)$ by

$$\beta((\xi_1, \xi_2), (\eta_1, \eta_2)) = \min\{\alpha((\xi_1, \xi_2), (\eta 1, \eta 2)), \alpha((\eta_1, \eta_2), (\xi_1, \xi_2))\} \tag{3}$$

for all

$\xi=(\xi_1, \xi_2) \in X^2$, $\eta=(\eta_1, \eta_2) \in X^2$.

Also suppose $T: X^2 \to C\,B(X^2)$ is defined by $T(x, y)=F(x, y) \times F(y, x)$. Obviously the metric space (X^2, δ) is complete. Now since F is α-admissible, it is easy to see that T is β-admissible. Also by assumption

$$\alpha((x,y),(u,v))H(F(x,y),F(u,v)) \leq \frac{1}{2}\psi(d(x,u)+d(y,v)),$$

$$\alpha((v,u),(y,x))H(F(v,u),F(y,x)) \leq \frac{1}{2}\psi(d(v,y)+d(u,x)).$$

for all $(x, y), (u, v) \in X^2$. By adding the above two relations we obtain.

$$\beta((x, y), (u, v))[H(F(x, y), F(u, v)) + H(F(v, u), F(y, x))] \leq \psi(\delta((x, y), (u, v))). \tag{4}$$

Assume that H_δ is the Hausdorff metric on (X^2, δ). We should show that

$$H_\delta(T(x, y), T(u, v)) \leq H(F(x, y), F(u, v)) + H(F(v, u), F(y, x)). \tag{5}$$

For this we have

$$H_\delta(T(x, y), T(u, v)) = H_\delta(F(x, y) \times F(y, x), F(u, v) \times F(v, u))$$

$$= \max\{\sup_{(\xi_1,\xi_2)\in F(x,y)\times F(y,x)} \delta((\xi_1,\xi_2), F(u,v)\times F(v,u)),$$

$$\sup_{(\eta_1,\eta_2)\in F(u,v)\times F(v,u)} \delta((\eta_1,\eta_2), F(x,y)\times F(y,x))\}.$$

Let $(\xi_1, \xi_2) \in F(x, y) \times F(y, x)$. Then

$$\delta((\xi_1,\xi_2), F(u,v)\times F(v,u)) = \inf_{(\eta_1,\eta_2)\in F(u,v)\times F(v,u)} \delta((\xi_1,\xi_2),(\eta_1,\eta_2)),$$

$$= \inf_{(\eta_1,\eta_2)\in F(u,v)\times F(v,u)} [d(\xi_1,\eta_1)+d(\xi_2,\eta_2)].$$

$$= \inf_{\eta 1\in F(u,v)} d(\xi_1,\eta_1) + \inf_{\eta_2\in F(v,u)} d(\xi_2,\eta_2)$$

$$= d(\xi_1, F(u,v)) + d(\xi_2, F(v,u)) \leq H(F(x,y),F(u,v)) + H(F(y,x),F(v,u)).$$

Similarly for $(\eta_1, \eta_2) \in F(u, v) \times F(v, u)$ we have

$$\delta((\eta_1, \eta_2), F(x, y) \times F(y, x)) \leq H(F(x, y), F(u, v)) + H(F(y, x), F(v, u)).$$

Hence (5) holds. By (4) and (5) we have

$$\beta((x, y), (u, v))H_\delta(T(x, y), T(u, v)) \leq \psi(\delta((x, y), (u, v))). \tag{6}$$

Hence for any $\xi=(\xi_1, \xi_2) \in X^2$, $\eta=(\eta_1, \eta_2) \in X^2$ we have

$$\beta(\xi, \eta)H_\delta(T\xi, T\eta) \leq \psi(\delta(\xi, \eta)).$$

So $T: X^2 \to C\,B(X^2)$ is a β-ψ-contractive multifunction. Put $P_0=(x_0, y_0)$, $P_1=(x_1, y_1)$. Then $P1 \in T\,P_0$, $\beta(P_0, P_1) \geq 1$. Now let F be continuous. We show that T is continuous. To see this let $\{(x_n, y_n)\}$ be a sequence in X^2 such that $\delta((x_n, y_n), (x, y)) \to 0$. Then $d(x_n, x)+d(y_n, y) \to 0$ and hence $d(x_n, x) \to 0$, $d(y_n, y) \to 0$. Now since F is continuous, hence

$$H(F(x_n, y_n), F(x, y)) \to 0, \quad H(F(y_n, x_n), F(y, x)) \to 0.$$

Therefore

$$H_\delta(T(x_n, y_n), T(x, y)) \leq H(F(x_n, y_n), F(x, y)) + H(F(y_n, x_n), F(y, x)) \to 0.$$

This shows that $T: X^2 \to C(X^2)$ is continuous.

Also if X has (C^*_α) condition, it is easy to see that X^2 has condition C_β. Now all of the conditions of lemma 1.3 hold. Hence by the lemma there exists $(x^*, y^*) \in X^2$ such that $(x^*, y^*) \in T(x^*, y^*)=F(x^*, y^*) \times F(y^*, x^*)$, hence $x^* \in F(x^*, y^*)$, $y^* \in F(y^*, x^*)$. That is (x^*, y^*) is a coupled fixed point of F in X^2. Now by using the following simple definitions we want to prove another version of theorem 2.1.

Definition 2.3: Let (X, d) is a metric space and $\alpha: X^2 \times X^2 \to [0, +\infty)$ is a map. A multifunction $F: X^2 \to C\,B(X)$ is said to be modified α-admissible whenever if $(x, y) \in X^2, (u, v) \in F(x, y) \times F(y, x)$ and $\alpha((x, y), (u, v)) \geq 1$, then $\alpha((u, v), (w, z)) \geq 1$, for all $(w, z) \in F(u, v) \times F(v, u)$.

Definition 2.4: Let (X, d) be a metric space and $\alpha: X\,2 \times X\,2 \to [0, \infty)$. We say that X satisfies condition (B^*_α) if for any two sequences $\{x_n\}$ and $\{y_n\}$ in X, that $x_n \to x$, $y_n \to y$ and $\alpha((x_n, y_n), (x_{n+1}, y_{n+1})) \geq 1$ for all n, then we have $\alpha((x_n, y_n), (x, y)) \geq 1$ for all n.

Theorem 2.2: Let (X, d) be a complete metric space, $\alpha: X^2 \times X^2 \to [0, +\infty)$ a function, $\psi \in \Psi$ a strictly increasing map and $F: X^2 \to C\,B(X)$ a modified α- admissible multifunction such that

$$\alpha((x,y),(u,v))H(F(x,y),F(u,v)) \leq \frac{1}{2}\psi(d(x,u)+d(y,v)).$$

For all $(x, y), (u, v) \in X^2$. Assume that the following assertions hold.

(i) There exist $(x_0, y_0) \in X^2$ and $(x_1, y_1) \in F(x_0, y_0) \times F(y_0, x_0)$ such that

$$A((x_0, y_0), (x1, y1)) \geq 1, \quad \alpha((y_0, x_0), (y_1, x_1)) \geq 1.$$

(ii) Either F is continuous or X satisfies condition (B^*_α).

Then F has a coupled fixed point in X^2.

Proof Define $\beta: X^2 \times X^2 \to [0, +\infty)$ by

$$\beta((\xi_1, \xi_2), (\eta_1, \eta_2)) = \min\{\alpha((\xi_1, \xi_2), (\eta_1, \eta_2)), \alpha((\xi_2, \xi_1), (\eta_2, \eta_1))\} \tag{7}$$

for all

$$\xi = (\xi_1, \xi_2) \in X^2, \eta = (\eta_1, \eta_2) \in X^2$$

The remain of proof is completely similar to proof of theorem 2.1

Example 2.5: Let X be the space of real numbers with the usual metric $d(x, y)=|x - y|$ and define $F: X^2 \to C\,B(X)$ by

$$F(x,y) = \begin{cases} \left[0, \frac{x-y}{3}\right] & x \geq y \\ \{0\} & x < y \end{cases} \tag{8}$$

Also define $\alpha: X^2 \times X^2 \to [0, +\infty)$ by

$$\alpha((x,y),(u,v)) = \begin{cases} 1 & x\geq y, u\geq v \\ 0 & otherwise \end{cases} \quad (9)$$

If $(x, y), (u, v) \in X^2$ and $x \geq y$, $u \geq v$, then

$$H(F(x,y),F(u,v)) = H([0,\frac{x-y}{3}],[0,\frac{u-v}{3}])$$

$$=\frac{1}{3}|x-y-(u-v)| \leq \frac{1}{2}(\frac{2}{3})(d(x,u)+d(y,v)).$$

Now put $\psi(t)=\frac{2t}{3}$. We see that

$$\alpha((x,y),(u,v))H(F(x,y),F(u,v)) \leq \frac{1}{2}\psi(d(x,u)+d(y,v)).$$

Also for $(x_0, y_0)=(1, 1)$, $(x_1, y_1)=(0, 0)$ we have

$F(x_0, y_0)=F(y_0, x_0)=\{0\}$, $(x_1, y_1)=(0, 0) \in \{(0, 0)\}=F(x_0, y_0) \times F(y_0, x_0)$,

$\alpha((x_0, y_0), (x_1, y_1))=\alpha((1, 1), (0, 0))=1$,

$\alpha((y_0, x_0), (y_1, x_1))=\alpha((1, 1), (0, 0))=1$.

Now let $(x, y) \in X^2$, $(u, v) \in F(x, y) \times F(y, x)$ and $\alpha((x, y), (u, v)) \geq 1$. Then $x \geq y$, $u \geq v$. Hence $F(v, u)=\{0\}$. Now if $(w, z) \in F(u, v)\times F(v, u)$, then $w \geq 0=z$. Therefore $\alpha((u, v), (w, z)) \geq 1$. This shows that F is modified α-admissible. It is easy to see that F is continuous. Hence by theorem 2.2, F has a coupled fixed point in X^2. For example $(0,0)$ is a coupled fixed point of F.

Example 2.6: Let X be the space of real numbers with the usual metric $d(x, y)=|x - y|$ and define $F: X^2 \to C B(X)$ by $F(x, y)=[0, 2|x - y|]$ for all $x, y \in X$. Also define $\alpha: X^2 \times X^2 \to [0, \infty)$ by

$$\alpha((x,y),(u,v)) = \begin{cases} \frac{1}{8} & (x,y,u,v)\neq(0,0,0,0) \\ 1 & x=y=u=v=0 \end{cases}$$

Now put $\psi(t)=\frac{t}{2}$. We see that

$$\alpha((x,y),(u,v))H(F(x,y),F(u,v)) \leq \frac{1}{2}\psi(d(x,u)+d(y,v)).$$

Also for $(x_0, y_0)=(x_1, y_1)=(0, 0)$ we have

$F(x_0, y_0)=F(y_0, x_0)=\{0\}$, $(x_1, y_1)=(0, 0) \in \{(0, 0)\}=F(x_0, y_0) \times F(y_0, x_0)$, $\alpha((x_0, y_0), (x_1, y_1))=\alpha((0, 0), (0, 0))=1$,

$\alpha((y_0, x_0), (y_1, x_1))=\alpha((0, 0), (0, 0))=1$.

It is easy to see that F is modified α-admissible. Obviously F is continuous. Hence by theorem 2.2, F has a coupled fixed point in X^2. For example (0,0) is a coupled fixed point of F. Note that coupled fixed point of F is not unique for example (2, 1) is a coupled fixed point of F too. In fact for any $x, y \geq 0$ that $x \geq 2y$ or $y \geq 2x$, (x, y) is a coupled fixed point of F.

Corollary 2.3: Let (X, d) be a complete metric space and $\preceq$ be an order on X^2. Suppose $\psi \in \Psi$ a strictly increasing map and $F: X^2 \to C B(X)$ a multifunction such that

$$H(F(x,y),F(u,v)) \leq \frac{1}{2}\psi(d(x,u)+d(y,v)),$$

for all comparable elements (x, y), $(u, v) \in X^2$. Assume that the following assertions hold.

(i) If $(x, y) \in X^2$, $(u, v) \in F(x, y) \times F(y, x)$ and (x, y), (u, v) are comparable, then (u, v), (w, z) are comparable, for all $(w, z) \in F(u, v) \times F(v, u)$.

(ii) There exist $(x_0, y_0) \in X^2$ and $(x_1, y_1) \in F(x_0, y_0)\times F(y_0, x_0)$ such that (x_0, y_0), $(x1, y_1)$ are comparable and (y_0, x_0), (y_1, x_1) are comparable.

(iii) Either F is continuous or for any two sequences $\{x_n\}$ and $\{y_n\}$ in X, that $x_n \to x$, $y_n \to y$ and (x_n, y_n), (x_{n+1}, y_{n+1}) are comparable for all n, then (x_n, y_n), (x, y) are comparable for all n.

Then F has a coupled fixed point in X^2.

Proof Define $\alpha: X^2 \times X^2 \to [0, +\infty)$ by $\alpha((x, y), (u, v))=1$ if (x, y), (u, v) are com- parable and $\alpha((x,y), (u, v))=0$ otherwise and apply theorem 2.2

Corollary 2.4: Let (X, d) be a complete metric space and $\preceq$ be an order on X^2. Fix $(x^*, y^*) \in X^2$. Suppose $\psi \in \Psi$ a strictly increasing map and $F: X^2 \to C B(X)$ a multifunction such that $H(F(x,y),F(u,v)) \leq \frac{1}{2}\psi(d(x,u)+d(y,v))$, *for all comparable elements* (x, y), $(u, v) \in X^2$ *with* $(\bar{x}, y)$. Assume that the following assertions hold.

(i) If $(x, y) \in X^2$, $(u, v) \in F(x, y) \times F(y, x)$ and (x, y), (u, v) are comparable with (x^*, y^*), then (u, v), (w, z) are comparable with (x^*, y^*), for all $(w, z) \in F(u, v) \times F(v, u)$.

(ii) There exist $(x_0, y_0) \in X^2$ and $(x_1, y_1) \in F(x_0, y_0)\times F(y_0, x_0)$ such that (x_0, y_0), (x_1, y_1) are comparable with (x^*, y^*) and (y_0, x_0), (y_1, x_1) are comparable with (x^*, y^*).

(iii) Either F is continuous or for any two sequences $\{x_n\}$ and $\{y_n\}$in X, that $x_n \to x$, $y_n \to y$ and (x_n, y_n), (x_{n+1}, y_{n+1}) are comparable with (x^*, y^*) for all n, then (x_n, y_n), (x, y) are comparable with (x^*, y^*) for all n.Then F has a coupled fixed point in X^2.

Proof Define $\alpha: X^2 \times X^2 \to [0, +\infty)$ by $\alpha((x, y), (u, v))=1$ *if* (x, y), (u, v) are comparable with (x^*,y^*) and $\alpha((x, y), (u, v))=0$ otherwise and apply theorem 2.2.

Corollary 2.5: Let $(X, \preceq, d)$ be a partial ordered complete metric space. Suppose $\psi \in \Psi$ a strictly increasing map and $F: X^2 \to C B(X)$ a multifunction such that

$$H(F(x,y),F(u,v)) \leq \frac{1}{2}\psi(d(x,u)+d(y,v)),$$

for all elements (x, y), (u, v) $\in X^2$ that $x \geq u$ *or* $y \geq v$.

Assume that the following assertions hold.

(i) If $(x, y) \in X^2$, $(u, v) \in F(x, y) \times F(y, x)$ and $x \geq u$ *or* $y \geq v$, then $u \geq w$ *or* $v \geq z$, for all $(w, z) \in F(u, v) \times F(v, u)$.

(ii) There exist $(x_0, y_0) \in X^2$ and $(x_1, y_1) \in F(x_0, y_0) \times F(y_0, x_0)$ such that $x_0 \geq x_1$ *or* $y_0 \geq y_1$.

(iii) Either F is continuous or for any two sequences $\{x_n\}$ and $\{y_n\}$ in X, that $x_n \to x$, $y_n \to y$ and $x_n \geq x_{n+1}$ *or* $y_n \geq y_{n+1}$ for all n, then $x_n \geq x$ *or* $y_n \geq y$ for all n.

Then F has a coupled fixed point in X^2.

Proof Define $\alpha: X^2 \times X^2 \to [0, +\infty)$ by $\alpha((x, y), (u, v))=1$ *if* $x \geq u$ *or* $y \geq v$ are comparable and $\alpha((x, y), (u, v))=0$ otherwise and apply theorem 2.2.

References

1. Alikhani H, Rezapour SH, Shahzad N (2013) Fixed points of a new type contractive mappings and multi functions. To appear in Filomat 27: 1315-1319.
2. Ciric LB (1974) A generalization of Banachs contraction principle. Proc Amer Math Soc 45: 267-273.
3. Jachymski J (2011) Equivalent conditions for generalized contractions on (ordered) metric spaces. Nonlinear Analysis 74: 768-774.

4. Mohammadi B, Rezapour SH, Shahzad N (2013) Some results on fixed points of α-ψ-Ciric generalized multi functions. Fixed Point Theory Appl 2013:24.

5. Mohammadi B, Dinu S, Rezapour SH (2013) Fixed points of Suzuki type quasi-contractions. U.P.B Sci Bull Series A 75: 3-12.

6. Rezapour SH, Asl HJ (2013) A simple method for obtaining coupled fixed points of α-ψ-contractive type mappings. Internat J Analysis.

7. Salimi P, Latif A, Hussain N (2013) Modified α-ψ-contractive mappings with applications. Fixed Point Theory Appl 2013:151.

8. Samet B, Vetro C, Vetro P (2012) Fixed point theorems for α-ψ-contractive type mappings. Nonlinear Analysis 75: 2154-2165.

Extending a Chebyshev Subspace to a Weak Chebyshev Subspace of Higher Dimension and Related Results

Mansour Alyazidi-Asiry*

Department of Mathematics, College of Sciences, King Saud University, Riyadh, Saudi Arabia

Abstract

Let G={g1,…,gn} be an n-dimensional Chebyshev sub-space of C[a, b] such that $1 \notin G$ and $U=(u_0, u_1,\ldots,u_n)$ be an (n+1)-dimensional subspace of C[a, b] where $u_0=1$, $u_i=g_i$, i=1….. n. Under certain restriction on *G*, we proved that *U* is a Chebyshev subspace if and only if it is a Weak Chebyshev subspace. In addition, some other related results are established.

Keywords: Chebyshev system; Weak Chebyshev system

Introduction

The finite set of functions $\{g_1,\ldots,g_n\}$ and C[a, b] is called a Chebyshev system on [a, b] if it is linearly independent and $D\begin{pmatrix} g_1,\ldots, & g_n \\ x_1,\ldots, & x_n \end{pmatrix} = Det\left[g_i(x_j)\right] \geq 0, i,j=1,\ldots,n$ for all $\{x_j\}_{j=1}^n$ such that $a \leq x_1 < x_2 < \ldots < x_n \leq b$, and the n-dimensional subspace $G=[g_1,\ldots,g_n]$ of C[a, b] will be called a Chebyshev subspace [1-4]. Using the continuity of the determinant, it can be shown that the sign of the determinant is constant [5], so we will assume that the sign of the determinant is always positive through this paper (replace g_1 by $-g_1$ if necessary). And the finite set of functions $\{g_1,\ldots,g_n\}$ and C[a, b] is called a Weak Chebyshev system on [a, b] if it is linearly independent and $D\begin{pmatrix} g_1,\ldots, & g_n \\ x_1,\ldots, & x_n \end{pmatrix} = Det\left[g_i(x_j)\right] \geq 0, i,j=1,\ldots,n$ for all $\{x_j\}_{j=1}^n$ such that $a \leq x_1 < x_2 < \ldots < x_n \leq b$ and the n-dimensional subspace $G=\{g_1,\ldots,g_n\}$ of C[a, b] will be called a weak Chebyshev subspace, C[a, b] is the space of all real-valued continuous functions. Extending an n-dimensional Chebyshev subspace which does not contain a constant function to an (n+1)-dimensional Chebyshev subspace containing a constant function was investigated [4]. In what follows is the statement of the problem considered in this paper: Let $G=[g_1,\ldots,g_n]$ be a Chebyshev subspace of C[a, b] such that $1 \notin G$ and $U=\{u_0, u_1,\ldots,u_n\}$ be an (n+1)-dimensional subspace of C[a, b] where $u_0=1$, $u_i=g_i$, i=1,….,n [6-8]. Our main purpose is to prove that, under certain restriction on G, U is a Chebyshev subspace of C[a, b] if and only if it is a Weak Chebyshev subspace of C[a, b]. An example illustrating that the preceding assertion is not true in general is presented and some related results are given at the end of the last section.

Preliminary

We start this section by the following well known theorem [3,5].

Theorem

For an n-dimensional subspace G of C[a,b], the following statements are equivalent.

(i) G is a Chebyshev subspace.

(ii) Every nontrivial function $g \in G$ has at most n-1 distinct zeros in [a, b].

(iii) For all points $a=t_0 \leq t_1 < \ldots < t_{n-1} \leq t_n=b$, there exists a function $g \in G$ such that

$g(t)=0$, $t \in \{t_1,\ldots,t_{n-1}\}$

$g(t) \neq 0$, $t \notin \{t_1,\ldots,t_{n-1}\}$

$(-1)^i g(t)>0$, $t \in (t_{i-1}, t_i)$, i=1,….,n

We need the following definitions:

Definition 1: Let *U* be a subspace of C[a, b], $x \in [a, b]$ and $f \in U$ such that f(x)=0. We call x an essential zero of *f* with respect to *U*, if and only if there is a $g \in U$ with $g(x) \neq 0$.

If no confusion arises, the term "with respect to *U* "will be omitted.

Definition 2: Let $f \in C[a, b]$ and $a \leq t_1 < \ldots < t_n \leq b$ be zeros of *f*. we say that these zeros are separated if and only if there are $s_1,\ldots,s_{n-1}$ in [a, b] with

$t_i < s_i < t_{i+1}$

Such that

$f(s_i) \neq 0$, i=1,……n-1.

The following theorem is a version of theorem 1 of Stockenberg [6].

Theorem

Let **G** be an n-dimensional Weak Chebyshev subspace of C[a, b]. Then the following statements hold.

1. If there is a $g \in G$ with n separated, essential zeros $a \leq t_1 < \ldots < t_n \leq b$, then g(t)=0 for all t with $t \leq t_1 \geq t_n$.

2. No $g \in G$ has more than n separated, essential zeros.

The Main Result

We start this section with the following lemma.

Lemma

Let $G=\{g_1,\ldots,g_n\}$ be an n –dimensional Chebyshev subspace of

***Corresponding author:** Mansour Alyazidi-Asiry, Department of Mathematics, College of Sciences, King Saud University, P.O.Box 2455,Riyadh 11451,, Saudi Arabia, E-mail: yazidi@ksu.edu.sa

C[a, b] such that $1 \notin$ and $U=\{u_0,u_1,\ldots,u_n\}$ be an (n+1)-dimensional subspace of C[a,b] where $u_0=1$, $u_i=g_i$, I=1,…..,n. If there are two nontrivial functions h, k∈ U and a set of n points $\{x_j\}_{j=1}^n$ with

$a \le x_1 < x_2 < \ldots < x_n \le b$

such that

$h(x_i)=k(x_i)=0, \quad i=1,\ldots, n,$

then there is a nonzero constant λ such that h(x) – λk(x) for every x∈ [a, b].

Proof: Write $h = a_0 + \sum_{i=1}^{n} aigi$, if $a_0=0$ then h ∈ G and from theorem (1) h(x)=0 for every x ∈[a, b], so $a_0 \neq 0$ then h $(x_i)=0$,i=1,….,n, where

$$\bar{h} = \frac{I}{a_0} h = 1 + \sum_{i=1}^{n} \frac{a_i}{a_0} gi,$$

Similarly, if $\bar{k} = b_o + \sum_{b_0}^{n} b_i g_i$, then $b_0 \neq 0$ and $\bar{k}$ $(x_i)=0$, I=1, …,n, where

$$\bar{k} = \frac{1}{b_o} k = 1 + \sum_{b_0}^{n} \frac{b_i}{b_0} g_i$$

Now let $f = \bar{h} - \bar{k}$, then f is an element of the n-dimensional Chebyshev subspace G with $f(x_i)=0$, i=1,….,n, so $f_{\equiv 0}$ and $\bar{h} = \bar{k}$, taking $\lambda = \frac{a_0}{b_0}$, we have λ ≠ 0 and h(x)=λ k(x) for every x ∈ [a, b].

Assumption A: We say that the subspace G of C[a, b] satisfies assumption A if for each f ∈ G such that f(x)=f(y) for some x, y ∈[a, b] with x< y there is appoint z, x<z<y such v that f (z) ≠ f (x).

Lemma

Let $G=(g_1,\ldots, g_n)$ be an n-dimensional Chebyshev subspace of C[a, b] such that 1∈G and $U=(u_0,u_1,\ldots,u_n)$ be an (**n**+1)-dimensional subspace of C[a, b] where $u_0=1$, $u_i=g_i$, i=1,…, n. If **G** satisfies Assumption A, then the zeros of each nontrivial function h ∈ U are separated and essential.

Proof: Let h be a nontrivial element of U such that h(x)=h(y)=0 for some x, y with a ≤ x < y ≤ b. If h ∈ G, then n ≤ 3, for otherwise h ≡ 0, and since G is an n-dimensional Chebyshev subspace of C[a, b], there is a point z ∈ (x, y) such that h(z) ≠ 0. If h∉ G, then h=α+g, where α ≠ 0 and g∈G, hence g(x)=g(y)=-α, but G satisfies Assumption A, so there is a point z∈ (x, y) such that g(z) ≠-α that is h (z)≠ 0, this shows that the zeros of h are separated. For the second part of the assertion of the lemma, it is clear that each zero of any nontrivial element of u is an essential zero, that is because 1∈ U.

Remark 1: Note that if 1 ∉ U and 0 ∉ f ∈U, f(x)=0, then since G is a Chebyshev space, x is an essential zero for f. Indeed, there is an element g ∈G such that g(x) ≠ 0.

Theorem

Let $G=(g_1,\ldots g_n)$ be an n-dimensional Chebyshev subspace of C[a, b] such that 1 ∈ G and $U=(u_{0,}\, u_{1,\ldots,}\, u_{n)}$ be an (n+1)-dimensional subspace of C[a, b] where $u_0=1$, $u_i=g_i$, i=1,…., n. If G satisfies Assumption A, then U is a Chebyshev subspace of C[a, b] if and only if it is a Weak Chebyshev subspace of C[a, b].

Proof: One direction is trivial.

For the other direction, suppose $U=(u_0,\, u_1,\, \ldots\, u_n)$ is an (n+1) – dimensional Weak Chesbyshev subspace of C [a, b] where $u_{0=}1$, $u_i=g_i$, I=1, n and $G=(g_1,\ldots g_n)$ is an n-dimensional Chesbyshev subspace of C[a, b] satisfying Assumption A. Let ū be a nontrivial element of U such that we

$\bar{u}(x_i) = 0, \ldots i = 1, \ldots d,$

$a \le x_1 <. x_{n+1} \le b,$

If d>n+1, then by lemma (2) together with theorem (2) we must have $\bar{u} \equiv 0$, so d ≥ n+1, if d≤ n, then is nothing to prove, so to this end, we will assume that d=n+1 and

$\bar{u}(x_i) = 0$, i=1,….n+1,

$a \le x_1 < \ldots < x_{n+1} \le b,$

again from lemma (2) and theorem (2) we must have $a=x_1$ or $x_{n+1=}b$ and ū(x) ≠ 0 for all $x \in [a,b] \setminus \{x_j\}_{j=1}^n$ Writing $\hat{u} = \alpha_0 + \sum_{i=1}^{n} \bar{\alpha}_i g_i$, then $\alpha_0 \neq 0$ that is because G is an n-dimensional Chebyshev subspace of C[a, b].

Taking $u = \frac{1}{\alpha_0} \bar{u}$, then

$$u = 1 + \sum_{i=1}^{n} \alpha_i g_i, \text{where } \alpha_i = \frac{\bar{\alpha}}{\alpha_0} i = 1, \ldots n$$

$u(x_i)=0$, I=1,…, n+1

and

$u(x) \neq 0$ for all $x \in [a,b] \setminus \{x_j\}_{j=1}^n$

The rest of the proof is divided into several cases.

Case A: $a=x_1$ and $x_{1+1=}b$.

Since G is an n-dimensional Chebyshev subspace of C [a, b], then for any point q ∈ (x_n, b) there is a function $g = \sum_{i=1}^{n} \beta_i g_i$. ∈ G such that

$g(y_i)=1$, I=1, ….., n, where $y_i=x_{i,\, i=1,\ldots\ldots n-1}$ and $y_{n=}q$.

Taking

$$v = 1 - g = 1 - \sum_{i=1}^{n} \beta_i g_i,$$

Then v is a nontrivial element of U with

$v(y_i)=0$, i=1,…..n,

$a=y_1 < \ldots < y_n < b,$

and if there is a point $t \in [a,b] \setminus \{y_i\}_{i=1}^n$ such that v(t)=0, then by theorem (2) we must have t=b, hence u and v are two dimensional elements of U such that

$u(x_{n+1})=v(x_{n+1})=0,$

$u(x_i)=v(x_i)=0$, i=1, ….,n-1

And by lemma (1) there is a non-zero constant λ such that u=λ v, this implies that

$u(t_i)=0$, I=1, ….n+2

Where $t_i=x_i$, i=1,….n, $t_{n+1}=y_n$

And $t_{n+2=}x_{n+1=}b$

This means that u has at least n+2 separated zeros in [a, b] which implies that u=v ≡ 0 contradicting the fact u and v are nontrivial elements of U, hence $v(x)^{1}0$, $x \in [a,b] \setminus \{y_i\}_{i=1}^n$. It is clear that

$u(x) \neq 0, x \in (x_n, b), u(x_n)=u(b)=0$

and

$v(y_n)=0, y_n \in (x_n, b), v(t) \neq 0$ for all $t \in [x_n, b] \setminus \{y_n\}$,

and if $x \in [x_n, y_n], y \in (y_n, b)$ then sign $v(x)$=-sign $v(y)$, subsequently,

We treat four different subcases.

Case A1: $u(x) < 0$ for all $x \in (x_n, b)$ and $v(x) > 0$ for all $x \in [x_n, y_n]$,

then $v(x) < 0$ for all $x \in (y_n, b]$, taking $w=u-v$, we have

$w(x_n)=-v(x_n) < 0$ and $w(y_n)=u(y_n) > 0$

by the continuity of w, there is a point $s \in (x_n, y_n)$ such that $w(s)=0$,

hence we have:

$w(z_i)=0$, $i=1,\ldots,n$, where $z_i=x_i$, $I=1,\ldots,n-1$ and $z_n=s$.

But w belongs to the n-dimensional Chebyshev subspace G of C [a,b].

Hence $w \equiv 0$ and it follows that u=v and

$u(t_i)=0$, $I=1, \ldots n+2$

Where

$t_i=x_i$, $I=1, \ldots n$,

$t_{n+1}=y_n$ and $t_{n+2}=x_{n+1}=b$

So u must be identically zero.

Case A2: $u(x) > 0$ for all $x \in (x_n, b)$ and $v(x) < b0$ for all $x \in [x_n, y_n]$, then $v(x) < 0$ for all $x \in [x_n, y_n]$,

then $v(x) > 0$ for all $x \in (y_n, b]$, again taking $w=u-v$, we have

$w(y_n)=u(y_n) > 0$ $w(b)=-v(b) < 0$,

and there is a point $s \in (y_n, b]$, such that $w(s)=0$, so w has at least n distinct zeros in [a,b]. A similar argument as in case A1 shows that u must be identically zero.

Case A3: $u(x) < 0$ for all $x \in (x_n, b)$ and $v(x) < 0$ for all $x \in [x_n, y_n)$,

then $v(x) > 0$ for all $x \in (y_n, b]$, taking $w=u-v$, we have

$w(y_n)=-v(x_n) > 0$ and $w(y_n)=u(y_n) < 0$,

and continuing exactly as in case A1, we conclude that u must be identically zero.

Case A4: $u(x) < 0$ for all $x \in (x_n, b)$ and $v(x) > 0$ for all $x \in [x_n, y_n)$,

then $v(x) < 0$ for all $x \in (y_n, b]$, taking $w=u-v$, we have

$w(y_n)=u(y_n) < 0$ and $w(b)=-v(b) > 0$ and an argument similar to that of case A2 shows be identically zero.

Case B: $a < x_1$ and $x_{n+1}=b$

As in case A, for any $q \in (x_n, b)$ there is a function $g=\sum_{i=1}^{n} \beta_i g_i. \in G$

Such that $g(y_i)=1$, $I=1,\ldots,n$, where

$y_i=x_i$, $i=1,\ldots n-1$ and $y_n=q$

Taking $v=1-g=1-\sum_{i=1}^{n} \beta_i g_i$,

Then v is a nontrivial element of U with

$v(y_i)=0$, $I=1, \ldots, n$,

$a < y_1 < \ldots\ldots < y_n < b$

and if there is a point $t \in [a,b]\{y_i\}_{i=1}^{n}$ such that $v(t)=0$,then by theorem (2) we must have t=b or t=a.

If t=b, then u and v are two nontrivial elements of U such that

$U(x_{n+1})=v(x_{n+1})=0$,

$u(x_i)=v(x_i)=0$, $I=1, \ldots n-1$

and by lemma 1 there is a nonzero constant λ such that $u=\lambda v$, this implies that

$u(t_i)=0$, $i=1,\ldots n+2$

where $t_i=x_i$, $i=1, \ldots n$,

$t_{n+1}=y_n$

and $t_{n+2}=x_{n+1}=b$

so u has at least n+2 separated zeros in[a, b] which implies that $u=v \equiv 0$ and this is a contradiction.

so $t \neq b$ and the situation becomes exactly as in case A, proceedings as in case A we conclude that u must be identically zero.

Case C: $a=x_1$ and $x_{n+1} < b$

The proof of this case requires that $n \geq 2$ and the proof for n=1 will be given in remark (2).

Now, for any point $p \in (a,x_1)$ there is a function $g=\sum_{i=1}^{n} \beta_i g_i \in G$ such that

$G(y_i)=1$, $i=1,\ldots, n$,

Where $y_1=p$ and $y_{i-1}=x_i$, $i=3,\ldots n+1$.

Taking

$$v=1-g=1-\sum_{i=1}^{n} \beta_i g_i,$$

Then v is a nontrivial element of U with

$v(y_i)=0$, $i=1,\ldots,n$,

$a < y_1 < \ldots < y_n < b$,

and if there is a point $t \in [a,b]/\{y_i\}_{i=1}^{n}$ such that $v(t)=0$, then by therome (2)we must have t=a or t=b.

If t=a, then u and v are two nontrivial elements of U such that

$U(x_1)=v(x_1)=0$, $u(x_i)=0$, $i=3,\ldots,n+1$.

A similar argument to that of the other cases leads to a contradiction.

So $t \neq a$ and on the interval $[a, x_2]$ we have

$u(a)=u(x_2)=0$, $u(x) \neq 0$, $x \in (a,x_2)$

and

$v(y_1)=0$, $y_1 \in (a,x_2)$,$v(t) \neq 0$ for every $t \in [a,x_2] \setminus \{y_1\}$.

If $x \in [a,y_1)$, $y \in (y_1,x_2]$ then sign $v(x)$=-sign $v(y)$, and as in the other cases we are presented with four different subcases. In each case, a similar argument to that of the cases in A can be used to show that the function u-v in G has at least n zeros which leads to the conclusion that u must be identically zero. Hence U is a Chebyshev subspace of C [a,b].

Remark 2: The following is the proof for theorem (3) when n=1 which is somehow more direct:

Suppose g is a non constant continuous function on [a,b] such that G=[g] is a Chebyshev subspace of C[a,b] of dimension 1 satisfying Assumption A and U=[1,g] is a subspace of C[a,b] of dimension 2. If *U* is not a Chebyshev subspace, then there is a nontrivial element u=α +βg of *U* such that $u(x_1)=u(x_2)=0$ where $a \le x_1 < x_2 \le b$, clearly α ≠ 0 and β ≠ 0, so $g(x_1) = g(x_2) = c = -\frac{\alpha}{\beta} \neq 0$.

By lemma (2) there is a point y_1, $x_1 < y_1 < x_2$ such that $g(y_1)=d \neq c$. Taking $x_1=z_1$, $y_1=z_2$ and $x_2=z_3$, then

$$a \le z_1 < z_2 < z_3 \le b$$

$$D\begin{pmatrix} 1 & g \\ z_1 & z_2 \end{pmatrix} = \text{Det}\begin{pmatrix} 1 & 1 \\ c & d \end{pmatrix} = d - c \neq 0$$

And

$$D\begin{pmatrix} 1 & g \\ z_2 & z_3 \end{pmatrix} = c - d \neq 0,$$

Hence

$$sign\, D\begin{pmatrix} 1 & g \\ z_1 & z_2 \end{pmatrix} = -sign\, D\begin{pmatrix} 1 & g \\ z_2 & z_3 \end{pmatrix}$$

This shows that *U* is not a weak Chebyshev subspace and the theorem is proved.

The followings example illustrates that theorem (3) is not true in general is proved.

Example 1

Let

$$g(x) = \begin{pmatrix} 1 & 0 \le x \le 1 \\ x & 1 < x \le 2 \end{pmatrix}$$

G={g} is a Chebyshev Subspace of C[0,2] of dimension 1, if *U*=(1,g) and $\le x_1 < x_2 \le 2$, then

$$D\begin{pmatrix} 1 & g \\ x_1 & x_2 \end{pmatrix} = 0 \text{ if } x_2 \in [0,1]$$

And

$$D\begin{pmatrix} 1 & g \\ x_1 & x_2 \end{pmatrix} > 0 \text{ if } x_2 \in (1,2]$$

That is *U* is a 2-dimensional weak Chebyshev Subspace of C[0,2] but not a Chebyshev Subspace.

If H is n-dimensional subspace of C[a, b], then it is possible that H is a Chebyshev subspace on one of the intervals (a, b] or [a, b) but not on the closed interval [a,b] as illustrated in the following example.

Example 2

Let H=(sin x, cos x), it can be easily checked that H is a Chebyshev subspace of dimension 2 on each of the intervals (0,π] of dimension 2.

In next result we give a necessary and sufficient condition for an n-dimensional Chebyshev H on (a,b] or [a,b) to be a chebyshev subspace on the closed interval [a,b].

Theorem

Let H be an n-dimensional subspace of C [a, b] such that H is a Chebyshev subspace on (a,b] or on [a,b), then H is a Chebyshev subspace on [a, b] if and only if each function h_i, I=1,...n can have at most n-1 distinct zeros on [a,b] whenever $H=[h_1,...h_n]$.

Proof: If $(h_1,...h_n)$ is a basis of H such that for some $s \in \{1,...n\}$, h_s has at least n zeros on [a, b], then clearly H is not a Chebyshev Subspace on [a,b]. For the other direction, suppose H is Chebyshev subspace on.

I=(a, b] but not on [a, b], there is a non-trivial element u ∈ *U* such that

$u(z_1) \neq 0$, $u(z_i)=0$, i=2, ...n.

Since H is a Chebyshev subspace on (a, b], there is a subset $E=\{h_1,...., h_n\}$ and H such that

$$h_i(z_j) = \delta_{ij} = \begin{cases} 1 & if\ i=j \\ 0 & \text{otherwise} \end{cases}$$

The elements of E are linearly independent and $H=\{h_1,....h_2\}$, write

$$u = \sum_{i=1}^{n} a_i h_i, a_i \in \mathbb{R}\, i = 1,...n,$$

Then

$0 \neq u(z_1)=a_1h_1(z_1)=a_1,$

$0=u(z_i)=a_ih_i(z_i)=a_i$, I=2, ...n,

Hence $h_1 = \frac{1}{a_1 u}$ which has n zeros on [a, b] and this is a contradiction.

Using a similar argument when *I*=[a, b) leads to a contradiction and the theorem is proved.

References

1. Cheney EW (1966) Introduction to Approximation Theory. McGraw-Hill.
2. Haverkamp R, Zielke R (1980) On the Embedding Problem for Cebysev Systems. Journal of Approximation Theory. 30: 155-156.
3. Karlin S, StuddenW (1966) Tchebyche Systems: with Applications in Analysis and Statistics. Interscience.
4. Mansour Alyazidi-Asiry (2016) Adjoining a Constant Function to N-Dimensional Chebyshev Space. Journal of Function Spaces, p: 3.
5. Nurnberger G. (1989) Approximation by Spline Functions. Springer-Verlag Berlin Heidelberg.
6. Stockenberg S (1977) On the Number of Zeros of Functions in a Weak Tchebyshev-Space. Mathematische Zeitschrift, 156: 49-57.
7. Zalik RA (1975) Existence of Tchebyche Extension. Journal of Mathematical Analysis and Application,51: 68-75.
8. Zielke R (1979) Discontinuous Cebysev Systems" Lecture Notes in Mathematics No 707. Springer-Verlag Berlin Heidelberg.

Seeing as Understanding: The Importance of Visual Mathematics for our Brain and Learning

Jo Boaler*, Lang Chen, Cathy Williams and Montserrat Cordero

Stanford Graduate School of Education & Stanford Cognitive & Neuroscience Lab, Stanford University, USA

Abstract

A few weeks ago the silence of my Stanford office was interrupted by a phone call. A mother called to tell me that her 5-year old daughter had come home from school crying because her teacher had not allowed her to count on her fingers. A few weeks afterwards, when I told my undergraduate mathematics class that visual mathematics was really important, one of them asked: but it is only for low levels of math, isn't it?

Keywords: Visual mathematics; Brain and learning

Introduction

The teacher and student referenced above are reflecting what is a common belief in education - that visual mathematics is for lower level work, and for struggling or younger students, and that students should only work visually as a prelude to more advanced or abstract mathematics. As Thomas West, author, states, there is a centuries-old belief that words and mathematical symbols are "for serious professionals - where - as pictures and diagrams" are "for the lay public and children". This idea is an example of a damaging myth in education, and this paper will present compelling brain evidence to help dispel the myth. We will also provide examples of ways that visual mathematics may be integrated into curriculum materials and teaching ideas across grades K-16. The provision of ways to see, understand and extend mathematical ideas has been under developed or missed in most curriculum and standards in the US, that continue to present mathematics as an almost entirely numerical and abstract subject. Yet when students learn through visual approaches, mathematics changes for them, and they are given access to deep and new understandings. The brain evidence we will share, helps us understand the impact of visualizing and seeing, to all levels of mathematics, and suggests an urgent need for change in the ways mathematics is offered to learners.

Good mathematics teachers typically use visuals, manipulative and motion to enhance students' understanding of mathematical concepts, and the US national organizations for mathematics, such as the National Council for the Teaching of Mathematics (NCTM) and the Mathematical Association of America (MAA) have long advocated for the use of multiple representations in students' learning of mathematics. But for millions of students in US mathematics classes, mathematics is presented as an almost entirely numeric and symbolic subject, with a multitude of missed opportunities to develop visual understandings. Students who display a preference for visual thinking are often labeled as having special educational needs in schools, and many young children hide their counting on fingers, as they have been led to believe that finger counting is babyish or just wrong. This short paper, collaboration between a neuroscientist and mathematics educators, shares stunning new evidence from the science of the brain, showing the necessity and importance of visual thinking - and, interestingly, finger representations - to all levels of mathematics.

What Does the Brain Science Say?

In recent years, scientists have developed a more nuanced understanding of the ways our brains work when we study and learn mathematics. Our brains are made up of 'distributed networks', and when we handle knowledge, different areas of the brain light up and communicate with each other. When we work on mathematics, in particular, brain activity is spread out across a widely-distributed network, which include two visual pathways: the ventral and dorsal visual pathways. Neuroimaging has shown that even when people work on a number calculation, such as 12×25, with symbolic digits (12 and 25) our mathematical thinking is grounded in visual processing (Figure 1).

A widely distributed brain network underpins the mental processing of mathematics knowledge [1]. The area of the brain shown in green, which is part of the dorsal visual pathway, has reliably been shown to be involved when both children and adults work on mathematics tasks. This area of the brain particularly comes into play when students consider visual or spatial representations of quantity, such as a number line [2]. A number line representation of number quantity has been shown in cognitive studies to be particularly important for

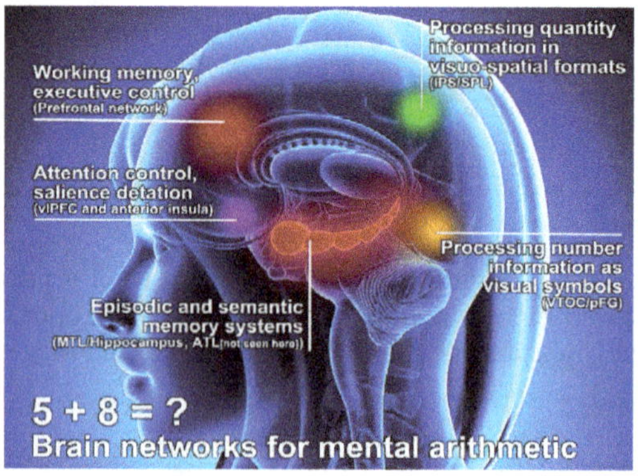

Figure 1: Distributed networks.

***Corresponding author:** Jo Boaler, Professor of Mathematics Education, Stanford Cognitive and Systems Neuroscience Lab, Stanford University, USA
E-mail: joboaler@stanford.edu

the development of numerical knowledge and a precursor of children's academic success [2-5].

Researchers even found that after four 15-minute sessions of playing a game with a number line, differences in knowledge between students from low-income backgrounds and those from middle-income backgrounds were eliminated [6].

The researchers in the study highlighted the importance of students learning numerical knowledge through linear representations and visuals. This is just one of many studies that show that visual mathematics problems help students and raise achievement. The brain research sheds light on this, as it is showing that the dorsal visual pathway is the core brain region for representing the knowledge of quantity.

One yet-to-be published study from our colleagues at Stanford, with children between the ages of 8 and 14, showed that as children get older they develop part of the ventral visual pathway, shown in orange in Figure 1, and the brain becomes more sensitive and specialized in representing visual number forms. The study also showed an important and increased interaction between the two visual pathways. This indicates that as children learn and develop, the brain becomes more interactive, connecting the visual processing of symbolic number forms, such as the number 10, with visuo-spatial knowledge of quantity, such as an array of dots or another visual representation [7]. Different areas of the brain are involved when we think mathematically, including the frontal networks shown in red and purple, the medial temporal lobe and, importantly, the hippocampus - the horseshoe shaped area in red. The important point that we want to stress in this paper is that the neurobiological basis of mathematics cognition involves complicated and dynamic communication between the brain systems for memory, control and detection and the visual processing regions of the brain.

A compelling and rather surprising example of the visual nature of mathematical activity in the brain comes from a new study on the ways that the brain uses representations of fingers, well beyond the time and age that people use their fingers to count. The different studies on the brain's use of finger representations give fascinating insights into human learning and clear implications for mathematics classrooms.

Mathematical Understanding and Fingers

Berteletti and Booth [8] studied one specific region of our brain that is dedicated to the perception and representation of fingers, known as the somatosensory finger area of the brain. Remarkably brain researchers know that we "see" a representation of our fingers in our brains, even when we do not use fingers in a calculation. Berteletti and Booth found that when 8-13 year olds were given complex subtraction problems, the somatosensory finger area lit up, even though the students did not use their fingers.

We "see" a representation of fingers in our brains when we calculate.

The researchers also found that this finger representation area was involved to a greater extent with more complex problems that involved higher numbers and more manipulation. Penner-Wilger [9] found that even university students' somatosensory knowledge of fingers predicts their calculation scores. She also found that finger perceptions in Grade 1 predict performance on number comparison and estimation in Grade 2 (2009). Researchers assess whether children have a good awareness of their fingers by touching the finger of a student - without the student seeing which finger is touched - and asking them which finger is being touched to perceive and represent their own fingers, they develop better representations of their fingers, which leads to higher mathematics achievement [10,11]. Researchers found that when 6 year old's improved the quality of their finger representation they improved in arithmetic knowledge, particularly subitizing1, counting and number ordering. Remarkably the 6 year old's finger representation was a better predictor of future mathematics success than their scores on tests of cognitive processing. Subitizing is the process of estimating small quantities such as 1, 2 or 3 without counting.

One of the recommendations of the neuroscientists conducting these important studies is that schools focus on finger discrimination. The researchers not only point out the importance of number counting on fingers, for brain development and future mathematics success, they advocate that schools help students' discriminate between their fingers. This seems particularly significant to us given that schools pay no attention to finger discrimination now and no published curriculum that we know of encourages this kind of mathematical work. Instead, many teachers have been led to believe that finger use is babyish or to be moved on from, as quickly as possible. Kumon, an after school program used by thousands of parents in 49 countries, tells parents that finger counting is a "no no" and parents who see children counting on their fingers should report them to the instructor [12].

There is debate among neuroscientists about the precise mechanisms by which finger knowledge improves mathematics achievement, but clear agreement on one thing, development of finger representations is critically important. Brian Butterworth, leading brain researcher in this area, states that if students are not learning about numbers through thinking about their fingers, numbers "will never have a normal representation in the brain" [13]. Despite the clear evidence on the importance of finger use, dangerous instructions to ban finger use are communicated to teachers and parents. Telling students not to use their fingers to count or represent quantities is akin to halting their mathematical development. Fingers are probably our most useful visual aid, critical to mathematical understanding, and brain development, that endures well into adulthood. The need for and importance of finger perception could even be a part of the reason that pianists, and other musicians, often display higher mathematical understanding (see http://www.livescience.com/51370-does-music-give-you-math-skills.html). The neuroscientist's recommend that fingers be regarded as the link between numbers and their symbolic representation, and an external support for learning arithmetic problems.

No US curriculum materials that we know of include activities for helping students develop finger discrimination, so we have developed a range of activities for use in classroom and homes that can be accessed in the appendix below and at https://www.youcubed.org/category/visual-math/, to help this development in children and prompt further ideas and work in this area. Importantly teachers should celebrate and encourage finger use among younger learners and enable learners of any age to strengthen this brain capacity through finger counting and use. This does not mean that learners should keep counting on fingers as they move through school, it means that anyone who needs to advance their perception and knowledge of their fingers and count on their fingers should do so, at any age, as it is critical for their brain development. It is important to remove the stigma from counting on fingers and to see this activity as inherently important and valuable.

Embodied Cognition

The evidence that is accumulating, showing the importance of visual pathways and the connections between different pathways in the brain, resonates with an area of research that is known as 'embodied cognition'. Most people think of the mind and the body as completely separate entities with the mind holding knowledge and abstractions, and the body passively taking ideas from the mind to the physical world by, for example, saying ideas out loud or writing them down. But embodied cognition researchers point out that many of our mathematical concepts are held in visual and sensory motor memories.

Embodied cognition researchers note the ways we posture, gaze, gesture, point, and use tools when expressing mathematical ideas as evidence of our holding mathematical ideas in the motor and perceptual areas of the brain [14] - which is now supported by brain evidence. The researchers point out that when we explain ideas, even when we don't have the words we need, we tend to draw shapes in the air. We might also use space around us to "spread out" our ideas. For example, deciding that one side of a table represents an idea, and pointing back to it when we want to refer to that idea even though there is nothing actually there, just our previous motions designating the space [15]. The researchers in this field don't assume the separation of mind and body, and have concluded that the body is an intrinsic part of cognition, that the parts of our brain that control perception and movement of our bodies, are also involved in knowledge representation [16]. It is fairly well known that knowledge of dance, or sport is held in sensory motor areas of our brain (Figure 2).

Some embodied cognition researchers have concluded that as we use gestures when we think mathematically, teachers should use gestures to ground mathematical thinking, alongside their verbal explanations [15] but researchers do not advocate 'giving' gesture schemes to students in mathematics classrooms and prefer that students be given opportunities to develop their own. We gesture because we see, experience and remember mathematics physically and visually, and greater emphasis on visual and physical mathematics will help students understand mathematics. Giving students someone else's gestures seems counterproductive in this regard. Instead we should give students more experiences of visual and even physical mathematics, as we expand upon below.

Implications for Classrooms and Homes

The new knowledge that we have, showing the visual processing of mathematical ideas, may explain the many research studies indicating that teachers who emphasize visual mathematics and who use well-chosen manipulative encourage higher achievement for students, not only in elementary school but middle school, high school and college [17]. Consistent with this, if we ask the best teachers about the

Figure 2: A teacher drawing a circle in the air when describing circumference to students.

importance of visual representations they will usually share the rich knowledge they hold, of the deep understanding that is enabled - both from teachers introducing mathematical ideas visually, and students using visuals to think and make sense of mathematics. Entire volume from the Mathematical Association of America (MAA) have been devoted to the encouragement of visual mathematics in college [18]. When our team at youcubed (a Stanford center dedicated to giving research based mathematics resources to teachers and parents) created a free set of visual mathematics lessons for grades 3-9 last summer, they were downloaded one quarter of a million times by teachers and used in every state across the US. Eighty-eight per cent of teachers said they would like more of the activities, and 83% of students reported that the visual activities enhanced their learning of mathematics.

Despite the prevalence of the idea that drawing, visualizing or working with models is low level or for young children, some of the most interesting and high level mathematics is predominantly visual. Maryam Mirzakhani made headlines across the world recently when she became the first woman to win the coveted Field's medal, the greatest prize in mathematics. Her work is almost entirely visual. Other mathematicians described her theories as "beautiful", "stunning" and connecting previously unconnected theories in mathematics. Children may go through hundreds of hours of calculating only ever seeing numbers and symbols but mathematicians rarely if ever, solve a problem without visual representations. As West reflects: "It's masochism for a mathematician to do without pictures" [4].

Yet another reason that visual mathematics should be used in schools to a greater extent is the nature of the knowledge needed for today's high-tech world. Years ago workplace knowledge was based on words and numbers, but the new knowledge of the world is based largely on images, that are 'rich in content and information' [19]. Most companies now have large amounts of data, known as "big data" and the largest growing job of the future is the task of making sense of the data, including seeing data patterns, visually. Computer scientists and mathematicians at Stanford and elsewhere now see patterns in data that could never have been picked up by statistical techniques. When we trialed our visual mathematics activities that we shared on youcubed.org in a local middle school, a parent stopped me and asked what we had done in class that week. She told me that her daughter had always said she disliked and couldn't do math, but after working on our visual tasks she came home saying she had changed her mind and she could see a future in math. Why? The math was open, creative and visual [20]. Such activities not only offer deep engagement, new understandings, and visual brain activity, but they show students that mathematics can be an open and beautiful subject, rather than a fixed, closed and impenetrable subject. Visual mathematics is not important only for some students - struggling or so called "visual" thinkers, nor is it only a prelude for abstract mathematics - visual mathematics is important for everyone, at all levels of mathematics.

It is hardly surprising that students feel that mathematics is inaccessible and uninteresting when they are plunged into a world of abstraction and numbers. This is particularly ironic when a different teaching approach of visual, creative mathematics - is available to all teachers and learners (https://www.youcubed. org/week-of-inspirational-math/). Mathematics classes in the US do not reflect the knowledge we have of the importance of visual mathematics approaches, as most curriculum standards and published textbooks do not invite visual thinking. Many textbooks provide pictures but these are often irrelevant to the mathematical ideas presented. The Common Core pays more attention to visual work in the K-8 standards, than many previous sets of standards, but the high school content of the Common Core leads teachers to remain committed to numerical and abstract thinking. Where the Common Core does encourage visual work, it is usually encouraged as a prelude to the development of abstract ideas, rather than a tool for seeing and extending mathematical ideas and strengthening important brain networks.

A few years ago Howard Gardner proposed a theory of multiple intelligences, suggesting that people have different approaches to learning, such as a visual, kinesthetic or logical approach. This idea helpfully expanded people's thinking about intelligence and competence, but was often used in unfortunate ways in schools, leading to the labeling of students as particular type of learners who were then taught in different ways. But people who are not strong visual thinkers probably need visual thinking more than anyone. Everyone uses visual pathways when we work on mathematics and we all need to develop the visual areas of our brains.

The problem of mathematics in schools is it has been presented, for decades, as a subject of numbers and symbols, ignoring the potential of visual mathematics for transforming students' mathematical experiences and developing important brain pathways.

The new brain research showing the importance of visual thinking should also prompt changes in the ways we view students in schools. Mathematics classrooms promote the students who memorize and calculate well, even when these students are weak at visualizing, modeling or thinking about concepts visually.

When the converse is true and students are weak memorizers or number users, but produce strong visual ideas and representations, they are often referred to special education classes. This could be the reason that some of our greatest scientists - Albert Einstein and Thomas Edison for example were written off by teachers and even labeled as "stupid." Einstein often shared that all of his thinking was visual and he struggled, later, to turn his visual ideas into words and symbols [19]. Visual mathematics is widely thought of as being appropriate for younger or struggling students and as a prelude to the "more important" abstract mathematics. It is true that abstract ideas can come from and be aided by visual mathematics, but visual ideas can also come from abstract mathematics and extend them to much higher levels. They can also inspire students and teachers, to see mathematics differently, to see the creativity and beauty in mathematics and to understand mathematical ideas.

Putting Research Ideas into Practice

Over recent years I have worked with colleagues, teaching math summer camps to 7th and 8th grade students. Last summer we ran a math camp at Stanford, in which students had 18 math lessons with myself, Cathy Williams, and other teachers. At the end of camp the students described their experiences as transforming their views of mathematics and, importantly, their own potential. When they were given a district test that they had taken at the end of the school year there was an average of 50% improvement in test scores across the 81 students. A video of the camp can be seen here (https://www.youcubed.org/youcubed-summer-camp-2015/). In our math camps we teach the students visual algebra through pattern study and generalization, exploring the worlds of linear and quadratic functions.

Algebra classes are often dedicated to students rearranging symbols, and students approach important mathematical concepts, such as functions, through numbers and symbols, without any visual understandings. In our teaching we approached algebra visually, as well as numerically and symbolically. In one activity, for example, we asked students to look briefly at a border around a square and work out how many squares were in the border, without counting them [21] (Figure 3).

The students thought about the number of squares in the border in many different ways, shown below, which they described at first numerically and then algebraically.

The students' different ways of seeing were a resource for engaging discussions between students and the development of different algebraic generalizations, which students learned were equivalent. When students see patterns growing in different ways [22], they are equally fascinated and engaged, and they learn, deeply, about functional growth, a major area of the US curriculum.

In one striking example of powerful mathematics learning we asked the students to consider distance-time graphs, which is an area that is notoriously challenging, even for college students [23]. We invited students to learn about distance, time, and velocity, through actually walking the line of a distance-time graph, using a motion sensor that tracked their movement. Further details of these activities are given here (https://www.youcubed.org/category/visual-math/) (Figure 4).

The students stunned district visitors when a girl, who was one of the lowest achievers in her grade, gave a perfect explanation of the graphing of velocity, rejecting a common misconception that is held by millions of students. When the students explained the concept they gestured, with their hands, to show the movement, again showing that their understanding of the concept was held in sensory-motor memories. The teaching of velocity through movement was clearly

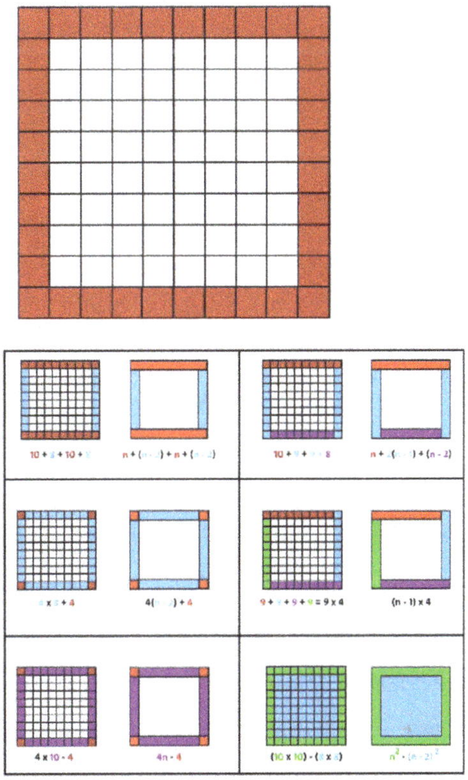

Figure 3: Students opinion about the number of squares in the border in many different ways.

Figure 4: Students to learn about distance, time, and velocity, through actually walking the line of a distance-time graph.

powerful for the students and motion is a helpful resource for teachers. But other visual teaching is just as important.

To engage students in productive visual thinking, they should be asked, at regular intervals, how they see mathematical ideas, and to draw what they see. Drawing mathematical ideas helps mathematics users of any level, including mathematicians, to formulate ideas and develop understandings. Students can be given activities with visual questions and they can be asked to provide visual solutions to questions (examples across K-16 grades can be seen in Visual Mathematics Activities and our Youcubed task page). New dynamic software and high quality apps and games (see https://www.youcubed.org/category/teaching-ideas/ math-apps/), are also powerful in developing students' visual brain pathways. Other suggestions teachers can give to students are to represent ideas in a multitude of ways, such as through pictures, models, graphs, even doodles or cartoons. More ideas for the visualization of mathematics are given in Boaler [20,22] and on youcubed.org.

At the end of camp one girl reflected that she had: "Never before seen a mathematical idea".

This seems to be a sad indictment of her US mathematics education. Another student reflected:

"It's like the way, the way our schools did it. It's like very black and white. And the way people do it here (in summer camp), it's like very colorful, very bright. You have very different varieties you're looking at. You can look at it one way, turn your head, and all of a sudden you see a whole different picture."

When mathematics classrooms focus on numbers, status differences between students often emerge, to the detriment of classroom culture and learning, with some students stating that work is "easy" or "hard" or announcing they have "finished" after racing through a worksheet. But when the same content is taught visually, it is our experience that the status differences that so often beleaguer mathematics classrooms, disappear. Thomas West also notes the equalizing effect of visual work, describing the time that various experts from academic disciplines came together to think visually, showing mutual respect towards each other and to different ideas, in ways that rarely happen when work is numerical [19]. It seems possible that visual mathematics may contribute to equity, in valuing students' thinking in different ways, as well as encouraging deep engagement, as we have found all students to be excited to see mathematical ideas, and from there they have developed higher levels of understanding and performance.

In our extensive work with school districts, teachers have also been inspired by visual and open mathematics. When we give teachers visual experiences of ideas that they have only previously encountered numerically and abstractly, such as multiplication facts or algebra, they gain insights into mathematical concepts and ideas they had never before experienced, and start to understand more deeply. They also feel empowered. Inviting people to think visually about mathematics is liberating for teachers and students alike. Mathematics is a multi-dimensional subject, and problems can be solved with numeric, abstract or visual mathematical pathways - we now know that our brain networks are correspondingly multi-dimensional and need to be developed and used. It is our belief that learners would develop stronger mathematical understanding if we helped them develop the visual networks in their brains, increasing their ability to work mathematically with a fully developed brain network.

Conclusion

Three recommendations for teaching and parenting,

The classroom and parental implications of the emerging science of the brain that shows that mathematical thinking centrally involves visual pathways - and finger representations - are important to consider. Here are 3 recommendations for educators and parents:

1. Encourage and celebrate students' visual approaches and replace the idea that strong mathematics learners are those who memorize and calculate well. Recent PISA evidence, from millions of students, tells us that the students who approach mathematics with a memorization approach are the lowest achieving students in the world [24]. We also need to end the myth that good mathematics performance is about calculating fast; a number of mathematicians are working to change this idea, explaining how they think slowly and deeply about mathematics [25]. Fast calculation is not what is needed in high-level mathematics work. Strong mathematics learners are those who think deeply, make connections and visualize. When I introduce math problems to my Stanford students I say - "I don't care about speed, in fact I am unimpressed by those who finish quickly, that shows you are not thinking deeply. Instead I would like to see interesting and creative representations of ideas". After a few lessons the students broaden their views of mathematics, and they start to blow me away with their creative thinking and insightful representations, and the new understandings they develop.

2. Focus on finger discrimination and encourage finger use. Successful mathematics users have well developed finger representations in their brains that they use into adulthood. Finger discrimination even predicts mathematics success. When we stop students using fingers we stop an important part of their mathematical development. Teachers who have stopped students using fingers are doing what they thought was best for children, as the idea that finger use is babyish, and needs to be discouraged, is widespread. But we now have the knowledge that should change this and encourage teachers to focus on finger discrimination and use in classrooms to a much greater extent.

3. Importantly, mathematics teaching and learning needs to become more visual - there is not a single idea or concept that cannot be illustrated or thought about visually. Elementary school, ironically more than higher-grade levels, is often obsessively numerical. Students are made to memorize math facts, and plough through worksheets of numbers, with few visual or creative representations of mathematics or invitations to work visually, often because of policy directives and faulty curriculum guides. By the time most students leave elementary school they have developed the idea that visuals and manipulative are babyish, fingers should never be used, and mathematical success is about memorizing numerical methods. As students move up through the grades they continue on overly numerical and symbolic pathways.

Algebra classes are often composed entirely of symbol manipulation and the idea that visuals or manipulative are a mere prelude to abstract mathematics becomes instantiated. What would mathematics look like if it were visual, instead of merely numerical? Our appendix and companion document, Visual Mathematics Activities, shows a range of examples, for different grade levels.

Some scholars note that it will be those who have developed visual thinking that will be 'at the top of the class' in our new high-tech workplace that increasingly draws upon information visualization technologies and techniques, in business, technology, art and science [19]. In our education system it is important not to prioritize any 'type of learner' over others or even to give the idea that it is productive to take one learning approach and focus upon it. The brain science supports this - work on mathematics draws from different areas of the brain and we want students to be strong with visuals, numbers, symbols and words. One of the aims of this paper is to point out that many schools are not encouraging this broad development in mathematics now and we urgently need to expand the ways we think about mathematics, and to teach it as the visual and multidimensional subject that it is.

Jo Boaler is professor of mathematics education at Stanford University, co-founder of youcubed, and author of the new book Mathematical Mindsets: Unleashing Students Potential through Creative Math, Inspiring Messages and Innovative Teaching (2016) Wiley.

Lang Chen is a postdoctoral scholar in the Department of Psychiatry and Behavioral Science at Stanford University. His research focuses on the development of knowledge representations, currently of math and language, in the brain.

Cathy Williams is co-founder and director of youcubed.

Monsterrat Cordero is an undergraduate student at Stanford University pursuing a BS in mathematics and interdisciplinary honors in education and a founding youcubed member.

References

1. Menon V (2014) Arithmetic in child and adult brain. In: Cohen KR, Dowker A (eds.) Handbook of mathmatical cognition. Oxford University Press, London.
2. Kucian K, Grond U, Rotzer S, Henzi B, Schönmann C, et al. (2011) Mental number line training in children with developmental dyscalculia. NeuroImage 57: 782-795.
3. Hubbard EM, Piazza M, Pinel P, Dehaene S (2005) Interactions between number and space in parietal cortex. Nature Reviews Neuroscience 6: 435-448.
4. Siegler RS, Booth JL (2004) Development of Numerical Estimation in Young Children. Child Development 75: 428-444.
5. Schneider M, Grabner RH, Paetsch J (2009) Mental number line, number line estimation, and mathematical achievement: Their interrelations in grades 5 and 6. Journal of Educational Psychology 101: 359.
6. Siegler RS, Ramani GB (2008) Playing linear numerical board games promotes low income children's numerical development. Developmental science 11: 655-661.
7. Battista C, Ngoon T, Chen T, Chen L, Kochalka J, et al. (2016) Longitudinal development and emergence of specialized functional brain systems supporting cognition in children.
8. Berteletti I, Booth JR (2015) Perceiving fingers in single-digit arithmetic problems. Frontiers in Psychology 6: 226.
9. Penner-Wilger M (2013) Symbolic and non-symbolic distance effects in number comparison and ordinarily tasks. Canadian Journal of Experimental Psychology/Revue Canadienne De Psychologie Expérimentale 67: 281-282.
10. Ladda AM, Pfannmoeller JP, Kalisch T, Roschka S, Platz T, et al. (2014) Effects of combining 2 weeks of passive sensory stimulation with active hand motor training in healthy adults. PloS one 9: e84402.
11. Gracia-Bafalluy M, Noël MP (2008) Does finger training increase young children's numerical performance? Cortex 44: 368-375.
12. Kumon Connections Newsletter (2014) News from your Center.
13. Brian B (1999) The mathematical brain. Macmillan, London.
14. Nemirovsky R, Rasmussen C, Sweeney G, Wawro M (2012) When the classroom floor becomes the complex plane: Addition and multiplication as ways of bodily navigation. Journal of the Learning Sciences 21: 287-323.
15. Alibali MW, Nathan MJ (2012) Embodiment in mathematics teaching and learning: Evidence from learners' and teachers' gestures. Journal of the learning sciences 21: 247-286.
16. Hall R, Nemirovsky R (2012) Introduction to the special issue: Modalities of body engagement in mathematical activity and learning. Journal of the Learning Sciences 21: 207-215.
17. Sowell EJ (1989) Effects of Manipulative Materials in Mathematics Instruction. Journal for Research in Mathematics Education 20: 498.
18. Zimmermann W, Cunningham S (1991) Visualization and the nature of mathematics. In Visualization in teaching and learning mathematics: A project. Washington DC: Mathematical Association of America.
19. West T (2014) Thinking like Einstein: Returning to our visual roots with the emerging revolution in computer information visualization. Prometheus Books, New York.
20. Boaler J (2016) The Many Colors of Math: Engaging Students through Collaboration and Agency. Journal of Mathematical Behavior.
21. Boaler J, Humphreys C (2005) Connecting Mathematical Ideas: Middle School Cases of Teaching and Learning. Ports-mouth: Heinemann.
22. Boaler J (2016) Mathematical Mindsets: Unleashing Students' Potential through Creative Math, Inspiring Messages and Innovative Teaching. Chappaqua, Jossey-Bass/Wiley, New York.
23. Clement J (1989) The Concept of Variation and Misconceptions in Cartesian Graphing. Focus on Learning Problems in Mathematics 11: 77-87.
24. Organisation for Economic Co-operation and Development (OECD) Learning by Heart May Not be For the Best.
25. Schwartz I (2001) A Mathematician Grappling with his Century Birkhäuser.

The Inverse Derivative - The New Algorithm of the Derivative

GuagSan Y*

Harbin Macro Dynamics Institute, P. R. China

Abstract

The Newton's second law and third law has proven to be wrong, his mistake is the derivative operation the limitation resulting. Therefore new derivative the computation method desiderate the creation, this is the Inverse Derivative algorithm. The Inverse Derivative is a kind of extension of the derivative computation, it is based on the derivative calculated generated, using most of the derivative formulas to converted adjusted. In the field of physics and mechanical engineering, Inverse Derivative may be a large number of employed, to replace original derivative computation method. Will produce the enormous and profound influence.

PACS: 02.10.a.v, 02.10.De, 02.70.2 c, 02.10.Ud, 03.65.Fd.

Keywords: Inverse derivative; Contrary derivative; Inverse differential; Contrary differential; Inverse integral

Introduction

This paper is in a basis of the derivative, bring forward a new calculation method-Inverse Derivative.

The Inverse Derivative Principle

For functions:

$$y=f(x) \quad (1)$$

In this case the derivative is:

$$f'(x_0)=y_{0'}=\frac{dy_0}{dx_0}=\lim_{\Delta x_0\to 0}\frac{\Delta y_0}{\Delta x_0}=\lim_{\Delta x_0\to 0}\frac{f(x+\Delta x_0)-f(x)}{\Delta x_0} \quad (2)$$

It shows that when the increment Δx_0 approaches zero, the limit of $\Delta y_0/\Delta x_0$ namely the y respect to x the derivatives. Here make incremental Δx_0 approaches zero, which is very important. It shows the derivative $f'(x_0)$ is a unit of differential dx_0 to calculated. For example, when the derivative is represented with speed, it is in units of time, That is in time of unit, by changing the distance traveled, to represent the rate of change of speed.

About Inverse Derivative, it is different from the derivative of it? First, look at the inverse function:

$$x=g(y) \quad (3)$$

Or:

$$x=f^{1}(y) \quad (4)$$

Well, that is the Inverse Derivative:

$$\left[f^{-1}(y_0)\right]^{+}=x_{0'}^{+}=\frac{qy_0}{qx_0}=\lim_{\Delta y_0\to 0}\frac{\Delta y_0}{\Delta x_0}=\lim_{\Delta y_0\to 0}\frac{\Delta y_0}{f^{-1}(y+\Delta y_0)-f^{-1}(y)} \quad (5)$$

So the Inverse Derivative is recorded as $[f^{1}(y_0)]^{+}$, $x_{0'}^{+}$, and qy_0/qx_0. Corresponds with the differential of derivative qy_0 and qx_0 also called Inverse Differential [1-5].

Clearly, in the Inverse Derivative operations, is when incremental Δy_0 approaches zero, for $\Delta y_0/\Delta x_0$ seek limit. Therefore, the Inverse Derivative $[f^{1}(y_0)]^{+}$ to indicate is with Inverse Differential qy_0 for the units to calculated. Here it with the derivative calculate, been have essentially different. For example, when by the Inverse Derivative represented speed, it is with move distance for unit. That is within the distance of units, through experienced time its different, and represent the speeds is different. This is with the derivative to record or described speed, is had significantly different. And it versus the thing the rule of physics, and even there will be different interpretations. So this is very important.

In the course of physics, many calculations should be use the Inverse Derivative to calculate, it will be correct. For example, concerning force the calculate, decided size of the force, moving objects the speed or acceleration, is by the object the displacement, to direct reflection the energy of the movement of objects. In this time denote the object move the velocity and acceleration, its at move the unit space by experienced the time, reflect then is the object locomotion energy to relatively to the density of the time. So the force to calculate, regardless is a velocity or acceleration, all should regard that object move space to be unit, been proceed the computation. Adopt the Inverse Derivative algorithm namely [6,7].

According to this principle, we discovered the Newton second law of the adoption derivative computation is wrong. And put forward to regard Inverse Derivative computation as fundamental, new that second laws of motion and new the third laws of motion.

Obviously, the principle of the Inverse Derivative be near to derivative very, been a derivative principle to expand. The Inverse Derivative quote the inverse function of the derivative, so it as if the reciprocal of the derivative operation. But increment Δy_0 and increment Δx_0 that the position, does not change however in Inverse Derivative. This is for the sake of the habit of the traditional physics in care concept, therefore just was changed by denominator the unit of the computation to numerator at this time, but denominator numerator of however the character did not change. For example while computing velocity or accelerations, is still the divide of distance of the move with undergo that the time.

***Corresponding author:** GuagSan Y, Harbin Macro Dynamics Institute, 150066, P. R. China, E-mail: 1951669731@qq.com

So the analytic formula of the Inverse Derivative is equals to will the analytic formula inside of the derivative, after exchange of the x of the variable y will the numerator denominator exchange. And will function symbol changes to inverse function symbol.

The Conversion of the Derivative Formula

Because the principle of the Inverse Derivative is that principle to base on the derivative, so usually be applicable to the derivative the circumstance, also will be applicable to the Inverse Derivative equally. For example concerning whether function can seek the derivative, the Inverse Derivative is opposite with derivative. When the image of function its shows whether can seek the derivative, well then at this image of function revolving 90°, its whether can seek the Inverse Derivative to is same with it.

It is obvious, the Inverse Derivative analytic formula just will the derivative analytic formula, the numerator denominator makes exchange. Therefore, from the analytic formula of the derivative, can will derivative and differential the formula, conversion through simple adjustment to Inverse Derivative and Inverse Differential the formula [8].

For example the function that the sum difference that the derivative formula:

$$\frac{d}{dx_0}(u \pm v) = \frac{du_0}{dx_0} \pm \frac{dv_0}{dx_0} \tag{6}$$

and

$$(u \pm v) = u' \pm v' \tag{7}$$

Can to convert the sum difference that the Inverse Derivative formula:

$$\frac{qy_0}{q(u \pm v)} = \frac{qy_0}{qu_0 \pm qv_0} \tag{8}$$

and

$$[u \pm v]^{+} = \lim_{\Delta y_0 \to 0} \frac{\Delta y_0}{[(u + \Delta u_0) - (u)] \pm [(v + \Delta v_0) - (v)]}$$

$$= \lim_{\Delta y_0 \to 0} \frac{\Delta y_0}{(\Delta u_0) \pm (\Delta v_0)} = \frac{qy_0}{qu_0 \pm qv_0} \tag{9}$$

And the product the derivative formula:

$$\frac{d}{dx_0} uv = u\frac{dv_0}{dx_0} + v\frac{du_0}{dx_0} \tag{10}$$

and

$$(uv) = uv' + vu' \tag{11}$$

Also can convert to the Inverse Derivative formula of the product:

$$\frac{qy_0}{q(uv)} = \frac{qy_0}{uqv_0 + vqu_0 + u_0 v_0} \tag{12}$$

and

$$[uv]^{+} = \lim_{\Delta y_0 \to 0} \frac{\Delta y_0}{(u + \Delta u_0)(v + \Delta v_0)} = \lim_{\Delta y_0 \to 0} \frac{\Delta y_0}{u\Delta v_0 + v\Delta u_0 + \Delta u_0 \Delta v_0}$$

$$= [u\Delta v_0 + v\Delta u_0 + \Delta u_0 \Delta v_0]^{+} = \frac{qy_0}{uqv_0 + vqu_0 + qu_0 qv_0} \tag{13}$$

And the derivative formula of the quotient:

$$\frac{d}{dx_0}\left(\frac{u}{v}\right) = \frac{v(du_0/dx_0) - u(dv_0/dx_0)}{v^2} \tag{14}$$

and

$$\left(\frac{u}{v}\right) = \frac{vu_{0'} - uv_{0'}}{v^2} \tag{15}$$

Conversion to the Inverse Derivative formula is then:

$$\frac{qy_0}{q\left(\frac{u}{v}\right)} = \lim_{\Delta y_0 \to 0} \frac{\Delta y_0}{\Delta x_0} = \lim_{\Delta y_0 \to 0} \frac{\Delta y_0 \cdot v(v + \Delta v_0)}{v\Delta u_0 - u\Delta v_0} = \frac{qy_0}{qx_0} = \frac{qy_0 \cdot v(v + qv_0)}{v(qu_0) - u(qv_0)} \tag{16}$$

and

$$\left[\frac{u}{v}\right]^{+} = \lim_{\Delta y_0 \to 0} \frac{\Delta y_0}{\left(\frac{(u + \Delta u_0)}{(v + \Delta v_0)} - \frac{u}{v}\right)} = \lim_{\Delta y_0 \to 0} \frac{(v + \Delta v_0)v\Delta y_0}{(u + \Delta u_0)v - u(v + \Delta v_0)}$$

$$= \lim_{\Delta y_0 \to 0} \frac{(v + \Delta v_0)v\Delta y_0}{[(u + \Delta u_0) - u]v - u[(v + \Delta v_0) - v]}$$

$$= \left[\frac{\Delta u_0 v - u\Delta v_0}{(v \ + v\Delta v\ )}\right] = \frac{qy_0 \cdot v(v + qv_0)}{v(qu_0) \quad u(qv_0)} \tag{17}$$

The derivative formula of the composite function:

$$\frac{dy_0}{dx_0} = \frac{dy_0}{du_0} \cdot \frac{du_0}{dx_0} \tag{18}$$

and

$$(f[g(x)]) = f'(u) \cdot g'(x) \Leftrightarrow \frac{dy_0}{dx_0} = \frac{dy_0}{du_0} \cdot \frac{du_0}{dx_0} \tag{19}$$

Can convert directly to the formula of the Inverse Derivative:

$$\lim_{\Delta y_0 \to 0} \frac{\Delta y_0}{\Delta x_0} = \lim_{\Delta y_0 \to 0} \frac{\Delta y_0}{\Delta u_0} \cdot \lim_{\Delta y_0 \to 0} \frac{\Delta u_0}{\Delta x_0} \Leftrightarrow \frac{qy_0}{qx_0} = \frac{qy_0}{qu_0} \cdot \frac{qu_0}{qx_0} \tag{20}$$

and

$$\left(f^{-1}\left[g^{-1}(y)\right]\right)^{+} = f'^{+}(y) \cdot g'^{+}(u) \Leftrightarrow \frac{qy_0}{qx_0} = \frac{qy_0}{qu_0} \cdot \frac{qu_0}{qx_0} \tag{21}$$

The formula of the derivative of higher order(second-order):

$$y'' = (y') = \frac{d}{dx}\left(\frac{dy}{dx}\right) = \frac{d^2 y}{dx^2} \tag{22}$$

Conversion the higher order(second-order) Inverse Derivative formula is then:

$$\left[f^{-1}(y_0)\right]^{+} = x_{0''}^{+} = \frac{1}{qx_0}\left(\frac{qy_0}{qx_0}\right) = \frac{qy_0}{q^2 x_0^2} \tag{23}$$

Above is that some in common use derivative formula, conversion to Inverse Derivative formula the method. In need can will more derivative principle, be to conversion and quote for that the Inverse Derivative.

The Inverse Derivative and the Partial Derivative the Conversion

The Inverse Derivative expressive, in reality can with the partial derivative to achieve. Therefore, the Inverse Derivative can convert to the partial derivative. Whereas, the partial derivative can also convert to the Inverse Derivative [9,10].

For example the function of second variables:

$$z = f(x,y) \tag{24}$$

To the y partial derivative is:

$$\frac{\partial z}{\partial y} = \lim_{\Delta y_0 \to 0} \frac{f(x, y + \Delta y_0) - f(x,y)}{\Delta y_0} = f_{y'}(x,y) = z_{y'} = \left.\frac{\partial z}{\partial y}\right|_{\substack{x = x_0 \\ y = y_0}} \tag{25}$$

Be to partial differential at this time:

$$d_y z = \frac{\partial z}{\partial y}dy = f_{y'}(x,y)\Delta y_0 = \lim_{\Delta y_0 \to 0} f(x, y+\Delta y_0) - f(x,y) \quad (26)$$

Then the Inverse Derivative is:

$$x_{0'}^{+} = \frac{qy_0}{qx_0} = \lim_{\Delta y_0 \to 0} \frac{\Delta y_0}{\Delta x_0} = \lim_{\Delta y_0 \to 0} \frac{\Delta y_0}{f^{-1}(y+\Delta y_0) - f^{-1}(y)} \quad (27)$$

Can treat Inverse Derivative as a function of second variables its obviously:

$$z = \frac{qy_0}{qx_0} = f(x_0, y_0) \quad (28)$$

Usher in to be partial differential will this function of second variables:

$$d_y z = \lim_{\Delta y_0 \to 0} f(x, y+\Delta y_0) - f(x,y) = \lim_{\Delta y_0 \to 0} f(x, \Delta y_0)$$

$$= \lim_{\Delta y_0 \to 0} f(x, \Delta y_0)\Big|_{x=\Delta x_0} \quad (29)$$

$$d_y z = \lim_{\Delta y_0 \to 0} f(\Delta x_0, \Delta y_0) = \lim_{\Delta y_0 \to 0} \frac{\Delta y_0}{\Delta x_0} = \frac{qy_0}{qx_0} \quad (30)$$

Then the Inverse Derivative be for a function of second variables, been it an partial differential. Therefore it can convert to partial differential of a partial derivative, some similar condition that the partial differential of partial derivative, also can convert to the Inverse Derivative.

The Applied of the Inverse Derivative - New Second Motion Laws and Third Motion Laws

The typical applied of the Inverse Derivative, it is in new second motion laws and third motion laws.

New the second laws of motion:

$$F = \lim_{\Delta l \to 0} m \cdot \frac{\Delta l}{\Delta^2 t^2} = m \cdot \frac{ql}{q^2 t^2} = ma^{+} \quad (31)$$

It too with mass and the product representation force of the acceleration, but its acceleration is second-order Inverse Derivative, record for a^{+} to express it is different with that the derivative. Because at this time second-order the Inverse Differential q^2t^2, just is among them the variable by the Inverse Derivative. So when the force changes, among them mass m with force is relation to geometric proportion, but its acceleration a^{+} then is relation to an inverse proportion square with force.

This will be caused a series of singular result, express for the same force, when the mass of the object is different, the mass of the force is with the value of the product of the acceleration is different, it will is. Theory deduce with experiment all denote, this is exactness is it. But Newton the second laws of motion and the third laws of motion, because is an adoption derivative operation, therefore in reality is wrong.

New the Inverse Derivative formula of the third laws of motion:

$$F_1 = m \cdot \frac{ql}{q^2 t^2} = ql \cdot m \frac{1}{q^2 t^2} \quad (32)$$

and

$$F_2 = \frac{m}{\beta} \cdot \frac{(\beta ql)\beta}{q^2 t^2} = \beta ql \cdot \frac{m}{\beta} \cdot \frac{\beta}{q^2 t^2} \quad (33)$$

It is explained two objects interactional, the small object in mass suffers the larger dynamic acting force to action. Then when two mass is different the object interactional, two objects produces the different momentum, therefore its momentum is not conservation. New the third laws of motion expresses, the momentum is not conservation, for this reason classical mechanics inside, the momentum conservation law also is wrong.

The above circumstance indicate, Inverse Derivative is in the physics and the mechanics to used, produce the strange result. There is evidence showing, Inverse Derivative is in electrodynamics or electromagnetics realms, also may have the very big use [11].

Inverse Integral

The integral of the Inverse Derivative computes, being called the Inverse Integral. It computes with common integral, should has no too big differentiate. It is will the common integral the compute that the differential symbol, changing the symbol of the inverse differential is then. The principle and method of its computation is similar is it. But although the superficial computing method is same, and because of the character of the Inverse Derivative, the result that computation come out will still have very big different.

For example compute the force to does work is it:

$$W_1 = \int F_1 qx = \int \left(m \cdot \frac{ql}{q^2 t^2} \right) qx = \frac{1}{2} mu^2 - \frac{1}{2} mu_0^2 \quad (34)$$

and

$$W_2 = \int F_1 qx = \int \left(\frac{m}{\beta} \cdot \frac{ql}{q^2 (t/\beta)(t/\beta)} \right) qx = \frac{m}{\beta} \cdot \beta^2 \cdot \left(\frac{1}{2} u^2 - \frac{1}{2} u_0^2 \right) \quad (35)$$

Express the same force action to the different object of two masses, the force to two the object those make the work is different. The object of small β times of the mass, by the force that big β times of make the work. This a kind circumstance gained the proof in of experiment of elasticity release of spring. This point also is very important. So operation in above integral and the experiment of the spring all expresses, the spring elasticity releases, to masses different the object, engender the different work. This a circumstance is shows, when the spring confront the different object in masses does the work, the energy is not conservation.

For this reason Inverse Derivative and the adduction of the Inverse Integral computation, cause some important physics law been by shaken. Among them of the meaning is enormous.

Conclusion

Therefore by complement and expand of the Inverse Derivative for derivative operation, is extremely important. In actuality adopt similar Inverse Derivative such method, to the thing come proceed measure or formulations, is also do not lack its example in fact. For example, the running that athletics compete the inside, is all in measure rule the distance, that consume the time. And is not in restrict time, run how much distances. For example in physics, usually the problem that discuss is, certain energy the action in stated the time. For example the force, electromagnetism force, electric quantity and so forth. This paper inside mentions, is the classical mechanics the second third laws of motion, and new the second third laws of motion that the force. Because of the adoption of the Inverse Derivative, the law of the some physics was wrong by the proof. That very much is may that the similar circumstance, will expand to include the other one physics course of electromagnetics and so on. And creation important influence.

Thereby discover and put forward this kind of new computing

method of the Inverse Derivative, the meaning is extremely important. To the development and progress of human science, inevitable creation profound and extensive influence.

Acknowledgement

The author thank his professors, for their support (Professor Shixu Guan, Chief editor Xinmin Zhu, Principal Lanxu Xu). The author also thank his the university for assistance (Department head Shuquan Wang, Department head Xinde Jiang, Associate Professor Risheng Piao, and many teachers).

References

1. GuagSan Yu (2014) The experiment of physics of mechanics.
2. GuagSan Yu (2014) The experiment of the Inertia-torque.
3. GuagSan Yu (2014) Analyze Mistake of the Newton Third Law.
4. GuagSan Yu (2014) The Newton third law is wrong.
5. GuagSan Yu (2015) New Newtonian Mechanics and New Laws of Motion.
6. Halliday D, Resnick R (1979) Physics foundation. Higher education publishing organization, Beijing.
7. Cheng S, Jiang Z (1961) Common physics. People's education publishing organization, Beijing.
8. Stenphen FH (2010) A mathematical bridge-An Intuitive Journey in Higher Mathematics. Shanghai Scientific Technological Education Publishing House, Shanghai.
9. Shere W, Love G (1974) Applied mathematics for engineering and science. Science publishing organization, Beijing.
10. Togqi University Mathematics Department (2007) Higher Mathematics. Higher Education Publishing Organization, Beijing.
11. YigChuan F (1958) Higher Mathematics Teaching Materials. Higher education publishing organization, Beijing.

Statistical Process Capability Design to Improve Process Stability of a Molding Machine

Joseph KA*

Federal University of Technology Akure, Nigeria

Abstract

Justification of production and manufacturing processes overtime, process capability in the concept of statistical control has been of great importance because it has promoted the production of products that satisfy the expectation of consumers. This paper aimed at promoting the adoption of quality process capability design in a bid to improve the process stability of a process and improve the process performance in the long run of production. In ensuring this, the technique of design of experiment is adopted using factorial design after which the capability analysis was carried out on the data on plastic containers produced by molding machines as deduced from the control X-bar and Range charts, it was established that the process was observed to be stable and in a state of statistical quality control and the plastic containers produced were observed to differ from one another as a result of variation on the part of the operators of the molding machines.

Keywords: Process capability design; Analysis of variance; Molding machine; Statistical quality control

Introduction

The taste of consumers for quality has evolved from time immemorial of a production processes. In satisfying quality, statistical methods have been adopted in promoting stability and capability of processes during this phase. Process capability analysis together with statistical process mechanism and design of experiments are statistical methods that have been used for decades with the goal to reduce the variability in industrial processes. The study of process capability in manufacturing industries has aided improvement in the production of products that are fit for purpose (quality). Moreover, use of statistical methods in industry is expanding by the introduction of quality management concepts specified as the Six Sigma program where statistical methods, including process capability analysis, are valuable parts. A process is a unique combination of tools, materials, methods and people occupied in producing a measurable output and involves a periodical of actions or operations influenced by several factors all contributing to its eventual outcome; an example is a manufacturing segment for automobile parts. All processes possess inherent statistical variableness which can be identified, evaluated and limited by statistical methods. Thus, in order to meet customer requirements, organizations must improve the quality by reducing variant in production processes because the lesser the variation in the process, the better the quality it provides. Process capability is the long-term performance level of the process after it has been brought under statistical control and it is determined by comparing the width of the process variation with the width of the specification limits and the inherent ability of a process to produce similar parts for a sustained period of time under a given set of conditions when operating in a state of statistical control. In specific, process capability deals with the uniformity of the process and it is often necessary to liken the process variation with the engineering or specification tolerances to determine the suitability of the process. The significance of process capability study is to isolate the inherent (random) variability which is always present from the assignable causes of variability which must be investigated. In a true process capability study, when there is exact observation of the process, inferences can be made about the stability of the process over time by directly controlling or monitoring data collection activity and understanding the time sequence of the data. Process capability is further defined as the spread within which most of the part values within a distribution will fall, generally described within plus or minus three standard deviation ($\pm 3\sigma$) and it involves the repeatability and consistence of a manufacturing process relative to customers' requirements. Process capability is a measurable attribute of a process to specification, which is commonly expressed as process capability index (C_{pk}) or as a process performance index (P_{pk}). This baseline definition enables us to compare process capability under actual manufacturing conditions with specification tolerances. Specifications or requirements are the numerical values within which the system is likely to operate, i.e. the minimum or maximum acceptable values. Process capability design is set up to see what the process is capable of doing under controlled conditions.

Capability analysis has been used in many facets of industrial processes. Capability study has been applied in the field of medicine, management for the evaluation of the best technique for management as well as to determine service operations for quality sustainability purposes. This capability study helps to create room for continuous improvement and development of products as well as improving the trust of customers in the company's products. Capability analysis is majorly summarized in indices; these indices can be monitored overtime to show how a system is changing. Capability indices that have been widely used in the manufacturing and industrial environment include C_p and C_{pk} explicitly and these indices provide a common metric to evaluate and predict the performance of a process.

In view of the deficiency in the study of process stability and capability in manufacturing industries within the country, this paper

***Corresponding author:** Joseph KA, Department of Statistics school of Sciences, The Federal University of Technology, Akureondo state, Nigeria
E-mail: jakupolusi@futa.edu.ng

focuses on how quality process capability designs promote process stability of the production of plastic containers in a bid to improve the application of capability study in the manufacturing industry. The work is aimed towards promotion of quality process capability design in the production of plastic containers that will improve process stability of the process.

Several authors have worked on process capability design in a bid to improve on the process stability of a production process.

Sullivan [1] employed the concept of process capability studies using the knowledge of capability index C_p in the automobile industry to enhance stability of automobiles produced to engineering specifications.

Taguchi [2] first introduced Taguchi capability value which was also a measure for estimating process capability with respect to target values. This method of capability analysis fits with his loss function approach. In cases where it is important to achieve a result when machining as close to the target value as possible, then the Taguchi capability index C_{pm} can be used. This concept of process capability design was used in machining systems.

Kane [3] further worked on process capability by introducing the capability index C_{pk}. This index was used to quantify process performance and takes into consideration cases where upper specification limits (USL) as well as lower specification limit (LSL) is relevant. He also investigated the test of hypothesis about process capability ratios and provided a table of sample sizes and critical values for C to aid in testing process capability. His work was also designed at obtaining the process capability estimate for a one sided specification. Measures of process capabilities were solely in terms of process variation and do not consider process location.

Refaie and Bata [4] proposed a procedure for assessing a measurement system and manufacturing process capabilities using Gage Repeatability (GR&R) designed experiments with four quality measures. The gage and part variance components are then estimated by conducting analysis of variance (ANOVA) on the GR&R measurement observations.

Van der Merwe and Chikobvu [5] considered process performance and developed a process potential index for average of observations from the new or unknown model. They derived theoretical and simulation results using Bayesian approach. They therefore removed the complexities of frequency distributions for measuring process performance by this approach.

Wang [6] developed a procedure for constructing multivariate process capability indices based on the principal component analysis (PCA) and Clements method for short-run production. PCA can identify correlations among multiple characteristics and determine independent components. This technique solves the problem of evaluating process quality performance with multiple quality characteristics in a short run production to help to meet the practical requirements of industry.

Chen, Huang and Huang [7] work on process capability as regards the construction of control chart of unilateral specification index C_{pl} and C_{pu} to monitor and evaluate the stability of process and process capability. This control chart contains upper control limit, lower control limit, upper warning limit, lower warning limit. This chart also supports a set of sensitizing rules for user to easily monitor the quality of process and the stability of the process.

Kaya and Kharaman [8,9] considered capability study involved using fuzzy set theory to add more information and flexibility to process capability analysis. The fuzzy formulation includes the development of C_p, C_{pk}, C_a, C_{pm}, C_{pmk} which are the most traditional used PCIs. The fuzzy PCA is developed when the specifications limits are represented by triangular or trapezoidal fuzzy numbers. The fuzzy process capability indices (FPCIs) are analyzed under the existence of correlation and hence, robust process capability indices are obtained. The fuzzy capability indices are improved for six-sigma approach.

Aslam et al. [10] worked on process stability and capability as regards to acceptance sampling. Their work was on the study of process capability as a justification of accepting or rejecting a lot or batch of a production process. In their work, three repetitive types of sampling plans using the generalized process capability index of multiple characteristics was adopted which included a repetitive sampling plan, a resubmitted sampling plan and a multiple dependent state repetitive sampling plan. The plan parameters of these sampling schemes are determined through the nonlinear optimization solution [10,11].

Research Methods

The basic goal of producers is to produce products with the aim of satisfying consumers' expectation in quality. For this research paper, it is aimed at promoting the adoption of quality process capability design in a bid to improve the process stability of a process and improve the process performance in the long run of production. In order to achieve this, technique of design of experiment is adopted using factorial design after which the capability analysis will be carried out on the data of plastic containers produced by molding machines and control X-bar and Range charts. Factorial design technique will be adopted for the experiment because we have two factors at various levels for the possible variability in the plastic containers produced. Hence, the process can be represented by an effect 2^k factorial model which is shown below:

$$y_{ijk}=\mu+\alpha_i+\beta_j+(\alpha\beta)_{ij}+\varepsilon_{ijk}$$

Where the model parameters $\alpha_i, \beta_j, (\alpha\beta)_{ij}$ and ε_{ijk} are all independent random variables that represent the effects of machines, effects of operators, effects of interaction of machines and operators, and random error respectively, μ is the overall mean and y_{ijk} is the response.

Sampling Technique and Tool for Analysis

A total number of 100 samples of secondary data on plastic containers produced by the molding machine are used for analysis, since the aim is to enhance production of plastic containers that are in-control (stable) and that are capable from the molding machine through quality process capability design. Therefore, the analysis of variance (ANOVA) and Factorial fit will be carried out using statistical package Minitab. This is done to identify the source of variability in the process and to test if factors (machines and operators) have individual or joint influence on the response of containers produced. After which variability that could result from various factors of production in the process will be controlled to ensure the improvement of production of plastic containers that satisfy quality.

Data Analysis

Figure 1 presents the control charts which establish the statistical stability of the process. It is deduced from the figure that the process is in a state of statistical control. It was also observed for Range chart that the process is in a state of statistical control because the mean of the individual subgroups are within the upper and lower control limits

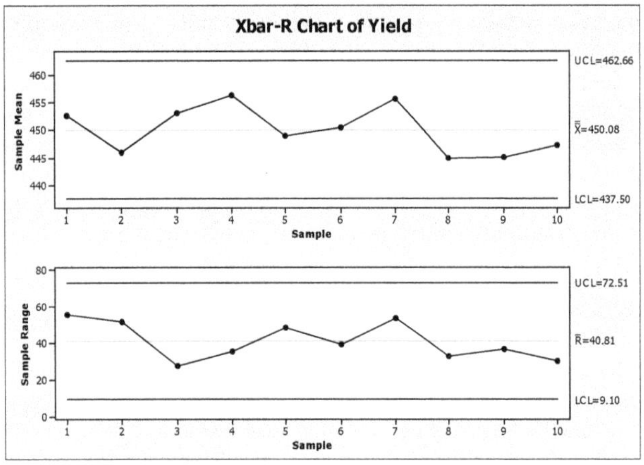

Figure 1: X-bar and R control charts for the plastic containers produced.

production (Table 1).

From Table 2, the p-value for the estimate of effects is greater than the significance level in the case of the machines as well as in the case of operators, inference can be drawn that there is a significant difference in the width dimension of the plastic containers produced as a result of the operation of the various machines as well as the various operators. It is also observed that there is a significant difference in the width of the plastic containers as a result of the joint operation of the machines and operators since the p-value is greater than the level of significance (Tables 1 and 2).

I proceeded to carry out the test for interaction between the two factors as was obtained that significant difference exist in the width of the containers due to joint operation of the machines and operators. From Table 2, conclusion can be drawn that there is no significant interaction between the machines and operators. Also, variability in the dimension of plastic containers produced result from the main effects of the operators and machines independently. This is established by the figure of main effects presented below:

It can be observed from Figure 2 that on the part of the machines, the mean of the respective subgroups are about the mean of the dimension of the plastic containers produced. Alternatively, on the part of the operators the mean of the respective subgroups is not about the mean of production. Therefore, it was deduced that the variability in the width of the plastic containers produced majorly results from the action of the various operators.

Table 3 as presented above shows the estimate of the coefficients of the parameters in the model of the factorial design.

Analysis of Variance (ANOVA)

Hypothesis to be tested

For machines

$H_{0:}\ \mu_1=\mu_2=\ldots=\mu_{10}$ (There is no significant effect of the machine in variability of containers)

$H_{1:}\ \mu_1 \neq \mu_2 \neq \ldots \neq \mu_{10}$ (There is a significant effect of the machine in variability of containers)

For operators;

$H_{0:}\ \mu_1=\mu_2=\ldots=\mu_{10}$ (There is no significant effect of the operator in variability of containers)

Estimated Effects and Coefficients for Yield (coded units)					
Term	Effect	Coef	SE Coef	T	P
Constant	-5.149	450.08	1.303	345.4	0.000
Machine	-6.305	-2.575	2.042	-1.26	0.21
Operator	-9.866	-3.153	2.042	-1.54	0.126
Machine*operator		-4.933	3.198	-1.54	0.126

Table 1: Factorial fit: yield versus machine, operator.

Analysis of variance for yield (coded units)						
Source	DF	Seq SS	Adj SS	Adj MS	F	P
Main Effects	2	675	675	337.5	1.99	0.143
2-Way interactions	1	403.9	403.9	403.9	2.38	0.126
Residual error	96	16300.5	16300.5	169.8		
Total	99	17379.4				

Table 2: Factorial fit: analysis of variance.

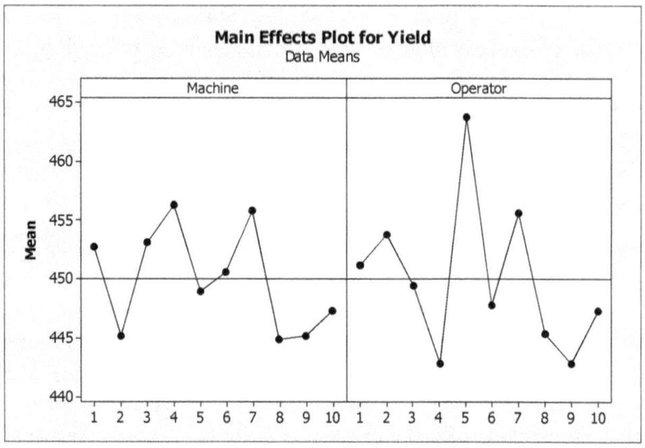

Figure 2: Main effects plot for yield.

$H_{1:}\ \mu_1 \neq \mu_2 \neq \ldots \neq \mu_{10}$ (There is a significant effect of the operator in variability of containers) (Table 4)

In the analysis of variance to determine the factor responsible for maximum variability in the width of the plastic containers, Table 5 shows that on the part of the operators since the p-value <0.05, the null hypothesis is rejected and was deduced that there is a significant effect of the operators in the variability of the plastic containers that is, the operators are responsible for the maximum variability in the plastic containers produced. While on the part of the machines, since the p-value >0.05 we accept the null hypothesis and conclude that there is no significant effect of the machines in the variability of the plastic containers that is, the machines are not responsible for variability in the plastic containers produced. Hence, variability in the dimension of the plastic containers majorly results from the operation of the operators.

Figure 3 presents the interaction plot of the various machines as well as various operators. On the part of the individual machines, it was observed that the variability in the width of the containers occurred majorly from machine 1 and machine 10 produce containers that are about the mean of production. On the other hand, in the case of the

Estimated Coefficients for Yield using data in uncoded units	
Term	**Coef**
Constant	449.711
Machine	0.767677
Operator	0.639192
Machine*Operator	-0.2436

Table 3: Factorial fit: estimated coefficients for yield.

Factor	Type	Levels	Values
Machine	random	10	1, 2, 3, 4, 5, 6, 7, 8, 9, 10
Operator	random	10	1, 2, 3, 4, 5, 6, 7, 8, 9, 10

Table 4: Hypothesis.

Analysis of Variance for Yield					
Source	**DF**	**SS**	**MS**	**F**	**P**
Machine	9	1709.7	190	1.24	0.283
Operator	9	3721.5	413.5	2.7	0.009
Error	81	12417.8	153.3		
Total	99	17849			

Table 5:ANOVA: yield versus machine, operator.

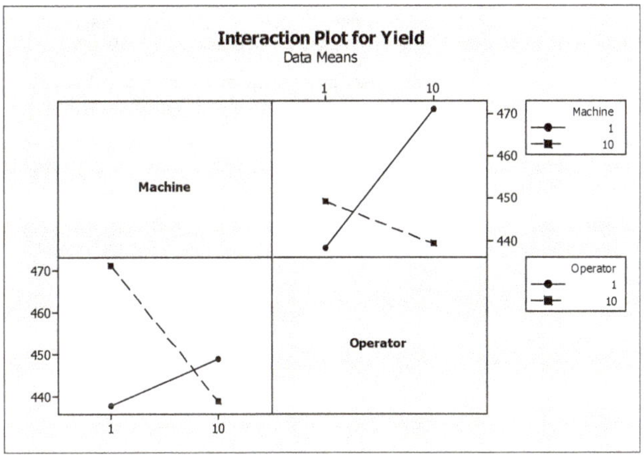

Figure 3: Interaction plot for yield.

operators it was observed that operator 10 is majorly responsible for the production of containers that vary from one another, while operator 1 produces containers that are about the mean of the production.

Process capability analysis

Process capability analysis is shown in Figure 4.

Computation for X-bar chart

Centre line (CL)=$\bar{\bar{X}}$

Upper control limit (UCL)=$\bar{\bar{X}} + A_2\bar{R}$

Lower control limit (LCL)=$\bar{\bar{X}} - A_2\bar{\bar{X}}$

Where: A_2=0.32 and $\bar{\bar{X}}$ and $\bar{R}$ are estimated as:

$$\bar{\bar{X}} = \frac{\sum_{i=1}^{10}\bar{X}i}{10} = \frac{4500.8}{10} = 450.08$$

$$\bar{R} = \frac{\sum_{i=1}^{10}\bar{R}i}{10} = \frac{408.1}{10} = 40.81$$

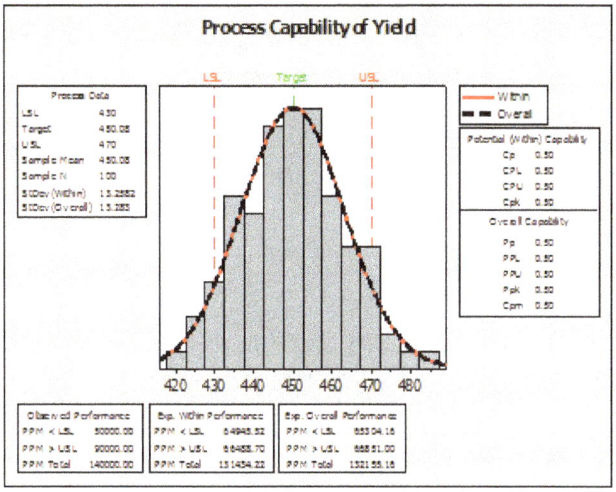

Figure 4: Process capability of yield.

Hence the control limits are estimated as:

CL=$\bar{\bar{X}}$=450.08

UCL=$\bar{\bar{X}} + A_2\bar{R}$ =450.08 + (0.32) 40.6=463

LCL=$\bar{\bar{X}} - A_2\bar{R}$ =450.08 – (0.32) 40.6=437

Computation for R chart

From the above table, $d_{4=}$1.777, $d_{3=}$0.223 and

Centre line=$\bar{R}_{=}$40.81. Hence, the control limits are estimated as:

Upper control limit (USL)=$d_4\bar{R}$ =1.777 * 40.81=72.51

Lower control limit (LCL)=$d_3\bar{R}_{=}$0.223 * 40.81=9.10

Analysis of the process capability using the company's specification limits:

USL=470

LSL=430

Therefore, the estimate of the process capability for the individual 10 subgroups is:

$$C_p = \frac{USL - LSL}{6};$$

where the values process capability indices for all the subgroups are estimated as: (Table 6).

It is observed that for all of the subgroups, the process was observed to be incapable in all the cases since all the C_p values are lesser than 1.

Furthermore, the overall process capability indices for the process are obtained by:

$$C_p = \frac{USL - LSL}{6} = \frac{470 - 430}{6\ (3032\ )} = 0.50$$

Also, the capability index C_{pk} is calculated as:

$$C_{pk} = \min\left[\frac{\bar{X} - LSL}{3\sigma}, \frac{USL - \bar{X}}{3\sigma}\right]$$

Where: $\bar{X}$ =450.08, LSL=430, USL=470 and σ=13.32

X_1	X_2	X_3	X_4	X_5	X_6	X_7	X_8	X_9	X_{10}
0.426	0.427	0.742	0.685	0.469	0.486	0.448	0.656	0.639	0.721

Table 6: Values for process capability.

$$C_{pk} = \min\left[\frac{450.08-430}{3(13.32)}, \frac{470-450.08}{3(13.32)}\right]$$

$C_{pk=}$min [0.50, 0.50]

C_{pk}=0.50

From the process capability plot for the plastic containers produced, it is observed that the process is not capable since the C_p=0.50<1 and the C_{pk}=0.50<1 and the process is also observed not to perform since process performance index P_p=0.50<1 even though the process is in a state of statistical control (Appendix 1).

Conclusion

As it was deduced from the control X-bar and Range charts in the analysis of this paper, it was established that the process was observed to be stable and in a state of statistical control. From the analysis carried out and the results obtained, the plastic containers produced were observed to differ from one another as a result of variation on the part of the operators of the molding machines. Therefore, conclusion can be drawn from the analysis that the process requires improvement to promote the process capability and process performance of the process in the long run of production in other to meet the required specification of customers as regards the plastic containers produced. Moreover, it was obtained that the specification limits of the company were off-centered since the capability index (C_p) was greater than the centering capability index (C_{pk}). Hence, improvement on the training of operators in respect of the use of the molding machines might be required since it is observed that the variability in the width of plastic containers resulted from the various operators. The variability on the part of the machines irrespective of operators is not significant since the plastic containers by the various machines are observed to be about the mean of production.

References

1. Sullivan IP (1984) Reducing variability-A new approach to quality. Quality Progress 17: 15-21.
2. Taguchi G (1986) Introduction to quality engineering: designing quality into products and processes p: 191.
3. Kane VE (1986) Process capability indices. Journal on Quality Tech 18: 41-52.
4. Refaie AA, Bata N (2010) Evaluating measurement and process capabilities by GR&R with four quality measures. International Journal of Measurement 43: 842-851.
5. Van der Merwe AJ, Chikobvu D (2010) A process capability index for averages of observations from new batches in the case of the balanced random effects model. Journal of Statistical Planning and Inference: 140: 20-29.
6. Wang CH (2005) Constructing multivariate process capability indices for short run production. International Journal of Advance Manufacturing Technology 26: 1306-1311.
7. Chen KS, Huang HL, Huang CT (2007) Control Charts for One-sided Capability Indices. Journal of Quality & Quantity 41: 413-427.
8. Kaya I, Kahraman C (2010) A new perspective on fuzzy process capability indices: Robustness. Expert Systems with Applications 37: 4593-4600.
9. Kahraman C, Kaya I (2008) Fuzzy process capability analyses: An application to teaching processes. Journal of Intelligent & Fuzzy Systems 19: 259-272.
10. Aslam M, Wu CW, Azam M, Jun CH (2012) Variable sampling inspection for resubmitted lots based on process capability index C_{pk} for normally distributed items. Applied Mathematical Modeling 3: 667-675.
11. Tatjana VS, Vidosav DM (2009) SPC and process capability analysis-case study. International Journal of Total Quality Management & Excellence 37: 1-2.

The Generalized Semi Normed Difference of χ^3 Sequence Spaces Defined by Orlicz Function

Mishra VN[1]*, Deepmala[2], Subramanian N[3] and Mishra LN[4]

[1]Applied Mathematics and Humanities Department, Sardar Vallabhbhai National Institute of Technology, 395 007, Surat, Gujarat, India
[2]SQC and OR Unit, Indian Statistical Institute, 203 B. T. Road, 700 108, Kolkata, West Bengal, India
[3]Department of Mathematics, SASTRA University, 613 401, Thanjavur, India
[4]Department of Mathematics, National Institute of Technology, 788 010, Silchar, District Cachar, Assam, India

Abstract

In this paper we introduced generalized semi normed difference of triple gai sequence spaces defined by an Orlicz function. We study their different properties and obtain some inclusion relations involving these semi normed difference triple gai sequence spaces.

2010 Mathematics Subject Classification: 40A05; 40C05; 40D05.

Keywords: Gai sequence; Analytic sequence; Triple sequence; Difference sequence

Introduction

Throughout the paper w,χ and Λ denote the classes of all, gai and analytic scalar valued single sequences, respectively. We write w^3 for the set of all complex triple sequences (x_{mnk}), where $m,n,k\in\mathbb{N}$, the set of positive integers. Then, w^3 is a linear space under the coordinatewise addition and scalar multiplication.

Some initial work on double series is found in Apostol [1] and double sequence spaces is found in Hardy [2], Subramanian et al. [3-9], and many others. Later on, some work on triple sequence spaces can also be found in Sahiner et al. [10] , Esi et al. [11-15], Subramanian et al. [16-19], Prakash et al. [20-24] and many others.

Let (x_{mnk}) be a triple sequence of real or complex numbers. Then the series $\sum_{m,n,k=1}^{\infty} x_{mnk}$ is called a triple series. The triple series $\sum_{m,n,k=1}^{\infty} x_{mnk}$ is said to be convergent if and only if the triple sequence (S_{mnk}) is convergent, where

$$S_{mnk} = \sum_{i,j,q=1}^{m,n,k} x_{ijq} \quad (m,n,k = 1,2,3,...).$$

A sequence $x=(x_{mnk})$ is said to be triple analytic if

$$sup_{m,n,k}\left|x_{mnk}\right|^{\frac{1}{m+n+k}} < \infty.$$

The vector space of all triple analytic sequences are usually denoted by Λ^3. A sequence $x=(x_{mnk})$ is called triple entire sequence if

$$\left|x_{mnk}\right|^{\frac{1}{m+n+k}} \to 0 \text{ as } m,n,k \to \infty.$$

The vector space of all triple entire sequences are usually denoted by Γ^3. The spaces Λ^3 and Γ^3 are metric spaces with the metric

$$d(x,y) = sup_{m,n,k}\left\{\left|x_{mnk} - y_{mnk}\right|^{\frac{1}{m+n+k}} : m,n,k : 1,2,3,...\right\}, \qquad (1)$$

for all $x=\{x_{mnk}\}$ and $y=\{y_{mnk}\}$ in Γ^3. Let φ be the set of finite sequences.

Consider a triple sequence $x=(x_{mnk})$. The $(m,n,k)^{th}$ section $x^{[m,n,k]}$ of the sequence is defined by $x^{[m,n,k]} = \sum_{i,j,q=0}^{m,n,k} x_{ijq}\Im_{ijq}$ for all $i,j,q\in\mathbb{N}$, where $\Im_{ijq}$ is a three dimensional matrix with 1 in the $(i,j,k)^{th}$ position and zero otherwise.

Let M and Φ be mutually complementary Orlicz functions. Then, we have:

(i) For all $u,y \geq 0$,

$$uy \leq M(u)+\Phi(y), \text{ (Young's inequality) [see [25]]} \qquad (2)$$

(ii) For all $u \geq 0$,

$$u\eta(u)=M(u)+\Phi(\eta(u)). \qquad (3)$$

(iii) For all $u \geq 0$, and $0<\lambda<1$,

$$M(\lambda u) \leq \lambda M(u) \qquad (4)$$

Lindenstrauss and Tzafriri [26] used the idea of Orlicz function to construct Orlicz sequence space

$$\ell_M = \left\{x \in w : \sum_{k=1}^{\infty} M\left(\frac{|x_k|}{\rho}\right) < \infty, \quad for \quad some \quad \rho > 0\right\}.$$

The space ℓ_M with the norm

$$\|x\| = inf\left\{\rho > 0 : \sum_{k=1}^{\infty} M\left(\frac{|x_k|}{\rho}\right) \leq 1\right\},$$

becomes a Banach space which is called an Orlicz sequence space. For $M(t)=t^p(1\leq p<\infty)$, the spaces ℓ_M coincide with the classical sequence space ℓ_p.

A sequence $f=(f_{mnk})$ of Orlicz functions is called a Musielak-Orlicz function. A sequence $g=(g_{mnk})$ defined by

$$g_{mnk}(v) = sup\{|v|u - (f_{mnk})(u) : u \geq 0\}, m,n,k = 1,2,\cdots$$

is called the complementary function of a Musielak-Orlicz function f. For a given Musielak-Orlicz function f, the Musielak-Orlicz sequence space t_f is defined as follows

$$t_f = \left\{x \in w^3 : M_f\left(\left|x_{mnk}\right|\right)^{1/m+n+k} \to 0, \quad as \quad m,n,k \to \infty\right\},$$

***Corresponding author:** Mishra VN, Applied Mathematics and Humanities Department, Sardar Vallabhbhai National Institute of Technology, Surat 395 007, Gujarat, India, E-mail: vishnunarayanmishra@gmail.com

where M_f is a convex modular defined by

$$M_f(x)=\sum_{m=1}^{\infty}\sum_{n=1}^{\infty}\sum_{k=1}^{\infty} f_{mnk}\left(\left|x_{mnk}\right|\right)^{1/m+n+k}, x=\left(x_{mnk}\right)\in t_f.$$

We consider t_f equipped with the Luxemburg metric

$$d(x,y)=\sum_{m=1}^{\infty}\sum_{n=1}^{\infty}\sum_{k=1}^{\infty} f_{mnk}\left(\frac{\left|x_{mnk}-y_{mnk}\right|^{1/m+n+k}}{mnk}\right)$$

is an extended real number.

If X is a sequence space, we give the following definitions:

(i) X= the continuous *dual of* X;

(ii) $X^{\alpha}=\left\{a=(a_{mnk}):\sum_{m,n,k=1}^{\infty}\left|a_{mnk}x_{mnk}\right|<\infty, \quad for \quad each \quad x\in X\right\};$

(iii) *üiŭüññijüiiñüijij*): $\sum_{m,n,k=1}^{\infty}$ $_{mnk}$ $_{mnk}$, $\in$ };

(iv) $X^{\gamma}=\left\{a=(a_{mnk}):sup_{mnk\geq 1}\left|\sum_{m,n,k=1}^{M,N,K}a_{mnk}x_{mnk}\right|<\infty, for \quad each \quad x\in X\right\};$

(v) $let \quad X \quad be \quad an \quad FK-space \supset \phi; \quad then \quad X^{f}=\left\{f(\Im_{mnk}):f\in X'\right\};$

(vi) $X^{\delta}=\left\{a=(a_{mnk}):sup_{mnk}\left|a_{mnk}x_{mnk}\right|^{1/m+n+k}<\infty, \quad for \quad each \quad x\in X\right\};$

X, X, X are called α-(or Kothe-Toeplitz) dual of X, β-(or generalized-Kothe-Toeplitz) dual of X, γ-dual of X, δ-dual of X respectively.

The notion of difference sequence spaces (for single sequences) was introduced by Kizmaz [27] as follows

$$Z(\Delta)=\left\{x=(x_k)\in w:(\Delta x_k)\in Z\right\}$$

for $Z=c,c_0$ and ℓ_{∞}, where $\Delta x_k=x_k-x_{k+1}$ for all $k\in\mathbb{N}$.

Later on the notion was further investigated by many others. We now introduce the following difference double sequence spaces defined by

$$Z(\Delta)=\left\{x=(x_{mn})\in w^2:(\Delta x_{mn})\in Z\right\}$$

where $Z=\Lambda^2,\chi^2$ and $\Delta x_{mn}=(x_{mn}-x_{mn+1})-(x_{m+1n}-x_{m+1n+1})=x_{mn}-x_{mn+1}-x_{m+1n}+x_{m+1n+1}$ for all $m,n\in\mathbb{N}$.

Let $w^3,\chi^3(\Delta_{mnk}),\Lambda^3(\Delta_{mnk})$ be denote the spaces of all, triple gai difference sequence space and triple analytic difference sequence space respectively and is defined as

$$\Delta_{mnk}=x_{mnk}-x_{m,n+1,k}-x_{m,n,k+1}+x_{m,n+1,k+1}-x_{m+1,n,k}+x_{m+1,n+1,k}+x_{m+1,n,k+1}-x_{m+1,n+1,k+1}$$ and $\Delta^0 x_{mnk}=\langle x_{mnk}\rangle$.

Definitions and Preliminaries

A sequence $x=(x_{mnk})$ is said to be triple analytic if $sup_{mnk}\left|x_{mnk}\right|^{\frac{1}{m+n+k}}<\infty$. The vector space of all triple analytic sequences is usually denoted by Λ^3. A sequence $x=(x_{mnk})$ is called triple entire sequence if $\left|x_{mnk}\right|^{\frac{1}{m+n+k}}\to 0$ as $m,n,k\to\infty$. The vector space of triple entire sequences is usually denoted by Γ^3. A sequence $x=(x_{mnk})$ is called triple gai sequence if $\left((m+n+k)!\left|x_{mnk}\right|\right)^{\frac{1}{m+n+k}}\to 0$ as $m,n,k\to\infty$. The vector space of triple gai sequences is usually denoted by χ^3. The space χ^3 is a metric space with the metric

$$d(x,y)=sup_{m,n,k}\left\{\left((m+n+k)!\left|x_{mnk}-y_{mnk}\right|\right)^{\frac{1}{m+n+k}}:m,n,k:1,2,3,...\right\} \quad (5)$$

for all $x=\{x_{mnk}\}$ and $y=\{y_{mnk}\}$ in χ^3.

Throughout the article $w^3,\chi^3(\Delta),\Lambda^3(\Delta)$ denote the spaces of all, triple gai difference sequence spaces and triple analytic difference sequence spaces respectively [28].

For a triple sequence $x\in w^3$, we define the sets

$$\chi^3(\Delta)=\left\{x\in w^3:\left((m+n+k)!\left|\Delta x_{mnk}\right|\right)^{1/m+n+k}\to 0, \quad as \quad m,n,k\to\infty\right\}$$

$$\Lambda^3(\Delta)=\left\{x\in w^3:sup_{m,n,k}\left|\Delta x_{mnk}\right|^{1/m+n+k}<\infty\right\}.$$

The space $\Lambda^3(\Delta)$ is a metric space with the metric

$$d(x,y)=sup_{m,n,k}\left\{\left|\Delta x_{mnk}-\Delta y_{mnk}\right|^{1/m+n}:m,n,k=1,2,\cdots\right\}$$

for all $x=(x_{mnk})$ and $y=(y_{mnk})$ in $\Lambda^3(\Delta)$.

The space $\chi^3(\Delta)$ is a metric space with the metric

$$d(x,y)=sup_{mnk}\left\{\left((m+n+k)!\left|\Delta x_{mnk}-\Delta y_{mnk}\right|\right)^{1/m+n+k}:m,n,k=1,2,\cdots\right\}$$

for all $x=(x_{mnk})$ and $y=(y_{mnk})$ in $\chi^3(\Delta)$.

Let $p=(p_{mnk})$ be a sequence of positive real numbers. We have the following well known inequality, which will be used throughout this paper:

$$\left|a_{mnk}+b_{mnk}\right|^{p_{mnk}}\leq D\left(\left|a_{mnk}\right|^{p_{mnk}}+\left|b_{mnk}\right|^{p_{mnk}}\right) \quad (6)$$

where a_{mnk} and b_{mnk} are complex numbers, $D=max\{1,2^{H-1}\}$ and $H=sup_{mnk}p_{mnk}<\infty$.

Spaces of strongly summable sequences were studied at the initial stage by Kuttner, Maddox and others. The class of sequences those are strongly Cesaro summable with respect to a modulus was introduced by Maddox as an extension of the definition of strongly Cesaro summable sequences. Jeff Connor further extended this definition to a definition of strongly A-summability with respect to a modulus when A is non-negative regular matrix.

Let $\eta=(\lambda_{abc})$ be a non-decreasing sequence of positive real numbers tending to infinity and $\lambda_{111}=1$ and $\lambda_{a+b+c+3}\leq\lambda_{a+b+c+3}+1$, for all $a,b,c\in\mathbb{N}$.

The generalized de la Vall`e e-Poussin means are defined by $t_{abc}(x)=\lambda_{abc}^{-1}\sum_{m,n,k\in I_{abc}}x_{mnk}$, where $I_{abc}=\left[abc-\lambda_{abc}+1,abc\right]$. A sequence $x=(x_{mnk})$ is said to (V,λ)-summable to a number L if $t_{abc}(x)\to L$, as $abc\to\infty$.

Throughout the article E will represent a semi normed space by a semi norm q. We define $w^3(E)$ to be the vector space of all E-valued sequences. Let f be an Orlicz function and $p=(p_{mnk})$ be any sequence of positive real numbers. Let $A=\left(a_{mn}^{jk}\right)$ be four dimensional infinite regular matrix of non-negative complex numbers such that $\sum_{m,n,k}a_{mn}^{jk}<\infty$.

We define the following sets of sequences in this article:

$$\left[V_{\lambda}^{E},A,\Delta_{\gamma}^{r},f,p\right]_{\chi^3}=\left\{x\in w^3(E):lim_{p,q,r\to\infty}\lambda_{pqr}^{-1}\sum_{mnk\in I_{pqr}}a_{mn}^{jk}\left[f\left(q\left((m+n+k)!\Delta_{\gamma}^{r}x_{mnk}\right)^{1/m+n+k}\right)\right]^{p_{mnk}}=0\right\}$$

uniformly in m,n,k.

$$\left[V_{\lambda}^{E},A,\Delta_{\gamma}^{r},f,p\right]_{\Gamma^3}=\left\{x\in w^3(E):lim_{p,q,r\to\infty}\lambda_{pqr}^{-1}\sum_{mnk\in I_{pqr}}a_{mn}^{jk}\left[f\left(q\left(\Delta_{\gamma}^{r}x_{mnk}\right)^{1/m+n+k}\right)\right]^{p_{mnk}}=0\right\}$$

uniformly in m,n,k

$$\left[V_{\lambda}^{E},A,\Delta_{\gamma}^{r},f,p\right]_{\Lambda^3}=\left\{x\in w^3(E):sup_{n,j,k}sup_{p,q,r}\lambda_{pqr}^{-1}\sum_{mnk\in I_{pqr}}a_{mn}^{jk}\left[f\left(q\left(\Delta_{\gamma}^{r}x_{mnk}\right)^{1/m+n+k}\right)\right]^{p_{mnk}}<\infty\right\}$$

For $\gamma=1$ these spaces are denoted by $\left[V_{\lambda}^{E},A,\Delta^{r},f,p\right]_{Z}$, for $Z=\chi^3,\Gamma^3$ and Λ^3 respectively. We define

$$\left[V_{\lambda}^{E},A,\Delta_{\gamma}^{r},f,p\right]_{\chi^3}=\left\{x\in w^3(E):lim_{p,q,r\to\infty}\lambda_{pqr}^{-1}\sum_{mnk\in I_{pqr}}\left[f\left(q\left((m+n+k)!\Delta_{\gamma}^{r}x_{mnk}\right)^{1/m+n+k}\right)\right]^{p_{mnk}}=0\right\}$$

Similarly $\left[V_{\lambda}^{E},\Delta_{\gamma}^{r},f,p\right]_{\Gamma^3}$ and $\left[V_{\lambda}^{E},\Delta_{\gamma}^{r},f,p\right]_{\Lambda^3}$ can be defined.

For $E=\mathbb{C}$, the set of complex numbers, $q(x)=|x|$; $f(x)=x^{1/m+n+k}$; $p_{mn}=1$, for all $m,n,k\in\mathbb{N}$, $r=0$, $\gamma=0$, the spaces $\left[V_{\lambda}^{E},\Delta_{\gamma}^{r},f,p\right]_{Z}$, for $Z=\chi^3,\Gamma^3$ and Λ^3

represent the spaces $[V,\lambda]_Z$, for $Z=\chi^3,\Gamma^3$ and Λ^3. These spaces are called as λ-strongly gai to zero, λ-strongly entire to zero and λ-strongly analytic by the de la Valle-Poussin method. In the special case, where $\lambda_{pqr}=pqr$, for all p,q,r=1,2,3.... the sets $[V,\lambda]_{\chi^3}$, $[V,\lambda]_{\Gamma^3}$ and $[V,\lambda]_{\Lambda^3}$ reduce to the sets $w^3_{\chi^3}$, $w^3_{\Gamma^3}$ and $w^3_{\Lambda^3}$.

In this chapter we introduced generalized semi normed difference of triple gai sequence spaces defined by an Orlicz function. We study their different properties and obtain some inclusion relations involving these semi normed difference triple gai sequence spaces.

Main Results

Theorem 1

Let the sequence $p=(p_{mnk})$ be analytic. Then the sequence space $\left[V_\lambda^E,A,\Delta_\gamma^r,f,p\right]_Z$, are linear spaces over the complex field $\mathbb{C}$, for $Z=\chi^3$ and Λ^3.

Proof: It is easy. Therefore the proof is omitted.

Theorem 2

Let f be an Orlicz function, then $\left[V_\lambda^E,A,\Delta_\gamma^r,f,p\right]_{\chi^3}\subset\left[V_\lambda^E,A,\Delta_\gamma^r,f,p\right]_{\Lambda^3}$

Proof: Let $x=(x_{mnk})\in\left[V_\lambda^E,A,\Delta_\gamma^r,f,p\right]_{\chi^3}$ will represent a semi normed space by a semi norm q. Here there exists a positive integer M_1 such that $q\le M_1$. Then we have

$$\lambda_{pqr}^{-1}\sum_{mnk\in I_{pqr}}a_{mn}^{jk}\left[f\left(q\left(\Delta_\gamma^r x_{mnk}\right)^{1/m+n+k}\right)\right]^{p_{mnk}}\le D\ \lambda_{pqr}^{-1}\sum_{mn\in I_{pqr}}a_{mn}^{jk}\left[f\left(q\left((m+n+k)!\Delta_\gamma^r x_{mnk}\right)^{1/m+n+k}\right)\right]^{p_{mnk}}+D\left(M_1,f(1)\right)^H\lambda_{pqr}^{-1}\sum_{m,n,k\in I_{pqr}}a_{mn}^{jk}$$

Thus $x\in\left[V_\lambda^E,A,\Delta_\gamma^r,f,p\right]_{\Lambda^3}$. Since $x\in\left[V_\lambda^E,A,\Delta_\gamma^r,f,p\right]_{\chi^3}$. This completes the proof.

Theorem 3

Let $p=(p_{mnk})\in\chi^3$, then $\left[V_\lambda^E,A,\Delta_\gamma^r,f,p\right]_{\chi^3}$ is a paranormed space with

$$g(x)=sup_{pqr}\left(\lambda_{pqr}^{-1}\sum_{mnk\in I_{pqr}}a_{mn}^{jk}\left[f\left(q\left((m+n+k)!\Delta_\gamma^r x_{mnk}\right)^{1/m+n+k}\right)\right]^{p_{mnk}}\right)^{1/H}$$

where $H=max(1,sup_{mnk}p_{mnk})$

Proof: From Theorem 3.2, for each $x\in\left[V_\lambda^E,A,\Delta_\gamma^r,f,p\right]_{\chi^3}$, $g(x)$ exists. Clearly $g(-x)=g(x)$. It is trivial that $\left((m+n+k)!\Delta_\gamma^r x_{mnk}\right)^{1/m+n+k}=\theta$ for $x=\bar{\theta}$. Hence, we get $g(\bar{\theta})=0$. By Minkowski inequality, we have $g(x+y)\le g(x)+g(y)$. Now we show that the scalar multiplication is continuous. Let α be any fixed complex number. By definition of f, we have $x\to\theta$ implies, $g(ax)\to 0$. Similarly we have for fixed X and $\alpha\to 0$ implies $g(\alpha x)\to 0$. Finally $x\to\theta$ and $\alpha\to 0$ implies $g(\alpha x)\to 0$. This completes the proof.

Theorem 4

If $r\ge 1$ then the inclusion $\left[V_\lambda^E,A,\Delta_\gamma^{r-1},f,p\right]_{\chi^3}\subset\left[V_\lambda^E,A,\Delta_\gamma^r,f,p\right]_{\chi^3}$ is strict. In general $\left[V_\lambda^E,A,\Delta_\gamma^j,f,p\right]_{\chi^3}\subset\left[V_\lambda^E,A,\Delta_\gamma^r,f,p\right]_{\chi^3}$ for $j=0,1,2,\dots r-1$ and the inclusions are strict.

Proof: The result follows from the following inequality

$$\lambda_{pqr}^{-1}\sum_{mnk\in I_{pqr}}a_{mn}^{jk}\left[f\left(q\left((m+n+k)!\Delta_\gamma^r x_{mnk}\right)^{1/m+n+k}\right)\right]^{p_{mnk}}\le$$
$$D\ \lambda_{pqr}^{-1}\sum_{mnk\in I_{pqr}}a_{mn}^{jk}\left[f\left(q\left((m+n+k)!x_{mnk}\right)^{1/m+n+k}\right)\right]^{p_{mnk}}+$$
$$D\ \lambda_{pqr}^{-1}\sum_{mnk\in I_{pqr}}a_{mn}^{jk}\left[f\left(q\left((m+n+k+1)!x_{mnk+1}\right)^{1/m+n+k+1}\right)\right]^{p_{mnk}}+$$
$$D\ \lambda_{pqr}^{-1}\sum_{mnk\in I_{pqr}}a_{mn}^{jk}\left[f\left(q\left((m+n+1+k)!x_{mn+1k}\right)^{1/m+n+1+k}\right)\right]^{p_{mnk}}$$
$$D\ \lambda_{pqr}^{-1}\sum_{mnk\in I_{pqr}}a_{mn}^{jk}\left[f\left(q\left((m+1+n+k)!x_{m+1n}\right)^{1/m+1+n+k}\right)\right]^{p_{mnk}}+$$
$$D\ \lambda_{pqr}^{-1}\sum_{mnk\in I_{pqr}}a_{mn}^{jk}\left[f\left(q\left((m+n+k+3)!x_{m+1n+1k+1}\right)^{1/m+n+k+3}\right)\right]^{p_{mnk}}$$

proceeding inductively, we have $\left[V_\lambda^E,A,\Delta_\gamma^j,f,p\right]_{\chi^3}\subset\left[V_\lambda^E,A,\Delta_\gamma^r,f,p\right]_{\chi^3}$ for $j=0,1,2,\dots r-1$. The inclusion is strict and it follows from the following example.

Example 1: Let $E=C$, $q(x)=|x|$; $\lambda_{pqr}=1$ for all $p,q,r\in\mathbb{N}$, $p_{mnk}=3$ for all $m,n,k\in\mathbb{N}$. Let $f(x)=x$, for all $x\in[0,\infty)$; $a_{mn}^{jk}=m^{-3}n^{-3}k^{-3}$ for all $m,n,k,j\in\mathbb{N}$; $\gamma=1$, $r\ge 1$. Then consider the sequence $x=(x_{mnk})$ defined by $x_{mnk}=\frac{1}{(m+n+k)!}(mnk)^{r(m+n+k)}$ for all $m,n,k\in\mathbb{N}$. Hence $(x_{mnk})\in\left[V_\lambda^C,A,\Delta^r,f,p\right]_{\chi^3}$ but $(x_{mnk})\notin\left[V_\lambda^C,A,\Delta^{r-1},f,p\right]_{\chi^3}$

Theorem 5

Let f be an Orlicz function, then

(a) Let $0\le p_{mnk}\le q_{mnk}$, for all $m,n,k\in\mathbb{N}$ and $\left(\frac{q_{mnk}}{p_{mnk}}\right)$ be analytic, then $\left[V_\lambda^E,A,\Delta_\gamma^r,f,q\right]_{\chi^3}\subset\left[V_\lambda^E,A,\Delta_\gamma^r,f,p\right]_{\chi^3}$

(b) If $0<inf_{mnk}p_{mnk}<p_{mnk}\le 1$ for all $m,n,k\in\mathbb{N}$ then $\left[V_\lambda^E,A,\Delta_\gamma^r,f,p\right]_{\chi^3}\subset\left[V_\lambda^E,A,\Delta_\gamma^r,f\right]_{\chi^3}$

(c) If $1\le p_{mnk}\le sup_{mnk}p_{mnk}<\infty$, then $\left[V_\lambda^E,A,\Delta_\gamma^r,f\right]_{\chi^3}\subset\left[V_\lambda^E,A,\Delta_\gamma^r,f,p\right]_{\chi^3}$

Theorem 6

Let f be an Orlicz function and s be a positive integer. Then, $\left[V_\lambda^E,A,\Delta_\gamma^r,f,q\right]_{\Lambda^3}\subset\left[V_\lambda^E,A,\Delta_\gamma^r,f,p\right]_{\Lambda^3}$

Proof: Let $\varepsilon>0$ be given and choose δ with $0<\delta<1$ such that $f(t)<\varepsilon$ for $0\le t\le\delta$. Write $y_{mnk}=f^{s-1}\left(q\left(\Delta_\gamma^r x_{mnk}\right)^{1/m+n+k}-M\right)$ and consider $\sum_{mnk\in I_r}a_{mn}^{jk}\left[f(y_{mnk})\right]^{p_{mnk}}=\sum_{mnk\in I_r,\ y_{mnk}\le\delta}a_{mn}^{jk}\left[f(y_{mnk})\right]^{p_{mnk}}+\sum_{mnk\in I_r}a_{mn}^{jk}\left[f(y_{mnk})\right]^{p_{mnk}}$.

Since f is continuous, we have

$$\sum_{mnk\in I_r,\ y_{mnk}\le\delta}a_{mn}^{jk}\left[f(y_{mnk})\right]^{p_{mnk}}\le\varepsilon^H\sum_{mnk\in I_r,\ y_{mnk}\le\delta}a_{mn}^{jk}\qquad(7)$$

and for $y_{mnk}>\delta$, we use the fact that, $y_{mnk}<\frac{y_{mnk}}{\delta}\le 1+\frac{y_{mnk}}{\delta}$ and so, by the definition of f, we have for $y_{mnk}>\delta$,

$$f(y_{mnk})<2f(1)\frac{y_{mnk}}{\delta}$$

Hence

$$\frac{1}{\lambda_{pqr}}\sum_{mnk\in I_r,\ y_{mnk}\le\delta}a_{mn}^{jk}\left[f(y_{mnk})\right]^{p_{mnk}}\le max\left(1,\left(2f(1)\delta^{-1}\right)^H\right)\frac{1}{\lambda_{pqr}}\sum_{mnk\in I_r,\ y_{mnk}\le\delta}a_{mn}^{jk}y_{mnk}^{p_{mnk}}\qquad(8)$$

From (7) and (8) we obtain $\left[V_\lambda^E,A,\Delta_\gamma^r,f,q\right]_{\Lambda^3}\subset\left[V_\lambda^E,A,\Delta_\gamma^r,f,p\right]_{\Lambda^3}$. This completes the proof.

Acknowledgement

The authors are extremely grateful to the anonymous learned referee(s) for their keen reading, valuable suggestion and constructive comments for the improvement of the manuscript. The authors are thankful to the editor(s) and reviewers of J Appl Comp Math.

Competing Interests

The authors declare that there is no conflict of interests regarding the publication of this research paper.

References

1. Apostol T (1978) Mathematical Analysis. Addison-Wesley, London.
2. Hardy GH (1917) On the convergence of certain multiple series. Proc Camb Phil Soc 19: 86-95.
3. Subramanian N, Misra UK (2011) Characterization of gai sequences via double Orlicz space. Southeast Asian Bulletin of Mathematics 35: 687-697.
4. Subramanian N, Priya C, Saivaraju N (2015) The $\int\chi^{2I}$ of real numbers over Musielak p-metric space. Southeast Asian Bulletin of Mathematics 39: 133-148.
5. Subramanian N, Anbalagan P, Thirunavukkarasu P (2013) The Ideal Convergence of Strongly of Γ^2 in *p*-Metric Spaces Defined by Modulus. Southeast Asian Bulletin of Mathematics 37: 919-930.
6. Subramanian N (2008) The Semi Normed space defined by Modulus function. Southeast Asian Bulletin of Mathematics 32: 1161-1166.
7. Deepmala, Subramanian N, Narayan Misra V (2016) Double almost (λ_m,μ_n) in ^{2}-Riesz space. Southeast Asian Bulletin of Mathematics in press.
8. Subramanian N, Tripathy BC, Murugesan C (2008) The double sequence space of Γ^2. Fasciculi Math 40: 91-103.
9. Subramanian N, Tripathy BC, Murugesan C (2009) The Cesaro of double entire sequences. International Mathematical Forum 4: 49-59.
10. Sahiner A, Gurdal M, Duden FK (20007) Triple sequences and their statistical convergence. Selcuk J Appl Math 8: 49-55.
11. Esi A (2014) On some triple almost lacunary sequence spaces defined by Orlicz functions. Research and Reviews:Discrete Mathematical Structures 1: 16-25.
12. Esi A, Necdet Catalbas M (2014) Almost convergence of triple sequences. Global Journal of Mathematical Analysis 2: 6-10.
13. Esi A, Savas E (2015) On lacunary statistically convergent triple sequences in probabilistic normed space. Appl Math and Inf Sci 9: 2529-2534.
14. Esi A (2013) Statistical convergence of triple sequences in topological groups. Annals of the University of Craiova, Mathematics and Computer Science Series 40: 29-33.
15. Savas E, Esi A (2012) Statistical convergence of triple sequences on probabilistic normed space. Annals of the University of Craiova, Mathematics and Computer Science Series 39: 226-236.
16. Subramanian N, Esi A (2015) The generalized triple difference of χ^3 sequence spaces. Global Journal of Mathematical Analysis 3: 54-60.
17. Subramanian N, Esi A (2016) The study on χ^3 sequence spaces. Songklanakarin Journal of Science and Technology.
18. Subramanian N, Esi A (2016) Characterization of Triple χ^3 sequence spaces. Mathematica Moravica.
19. Subramanian N, Esi A (2015) Some New Semi-Normed Triple Sequence Spaces Defined By A Sequence Of Moduli. Journal of Analysis and Number Theory 3: 79-88.
20. Shri Prakash TVG, Chandramouleeswaran M, Subramanian N (2015) The Triple Almost Lacunary Γ^3 sequence spaces defined by a modulus function. International Journal of Applied Engineering Research 10: 94-99.
21. Shri Prakash TVG, Chandramouleeswaran M, Subramanian N (2015) The triple entire sequence defined by Musielak Orlicz functions over *p*-metric space. Asian Journal of Mathematics and Computer Research,International Press 5: 196-203.
22. Shri Prakash TVG, Chandramouleeswaran M, Subramanian N (2016) The Random of Lacunary statistical on Γ^3 over metric spaces defined by Musielak Orlicz functions. Modern Applied Science 10: 171-183.
23. Shri Prakash TVG, Chandramouleeswaran M, Subramanian N (2015) The Triple Γ^3 of tensor products in Orlicz sequence spaces. Mathematical Sciences International Research Journal 4: 162-166.
24. Shri Prakash TVG, Chandramouleeswaran M, Subramanian N (2016) The strongly generalized triple difference Γ^3 sequence spaces defined by a modulus. Mathematica Moravica.
25. Kamthan PK, Gupta M (1981) Sequence spaces and series. Lecture notes, Pure and Applied Mathematics.
26. Lindenstrauss J, Tzafriri L (1971) On Orlicz sequence spaces. Israel J Math 10: 379-390.
27. Kizmaz H (1981) On certain sequence spaces. Canadian Mathematical Bulletin 24: 169-176.
28. Nakano H (1953) Concave modulars. Journal of the Mathematical society of Japan 5: 29-49.

Symmetric 2-Step 4-Point Hybrid Method for the Solution of General Third Order Differential Equations

Kayode SJ and Obarhua FO*

Department of Mathematical Sciences, Federal University of Technology, Nigeria

Abstract

This research considers a symmetric hybrid continuous linear multistep method for the solution of general third order ordinary differential equations. The method is generated by interpolation and collocation approach using a combination of power series and exponential function as basis function. The approximate basis function is interpolated at both grid and off-grid points but the collocation of the differential function is only at the grid points. The derived method was found to be symmetric, consistent, zero stable and of order six with low error constant. Accuracy of the method was confirmed by implementing the method on linear and non-linear test problems. The results show better performance over known existing methods solved with the same third order problems. AMS 2010 Subject Classification: 65D05; 65L05; 65L06.

Keywords: Symmetric; Hybrid method; Power series and exponential function; Continuous predictor-corrector method

Introduction

In this paper, we considered the solution of initial value problems for general third order ordinary differential equations of the form

$$y'''=f(t, y, y', y''), y(t_0)=y_0, y'(t_0)=y_1, y''(t_0)=y_2. \quad (1)$$

Where $t, y, f \in R$.

The numerical and theoretical studies of eqn. (1) have appeared in literature severally. The direct approach for solving this type of ordinary differential equations have been studied and appeared in different literatures [1-7]. This direct approach has demonstrated advantages over the popular approach (reduction to system of first order approach) in terms of speed and accuracy [8,9]. Many authors have focused on direct solution of general second order ivps of odes of the form

$$y''=f(t, y, y') \quad (2)$$

Majid et al. [10] proposed two point four step direct implicit block method for the solution of second order system of ordinary differential equations (ODEs), using variable step size. The method estimated the solutions of initial value problems at two points simultaneously by using four backward steps but with lower order of accuracies. Akinfenwa [11] presented ninth order hybrid block integrator for solving second order ordinary differential equations. In the paper, the proposed block integrator discretizes the problem using the main and the additional methods to generate system of equations. The resulting system was solved simultaneously in a block-by-block fashion but the order of accuracies is low compare to the order of the method. The authors came up with direct implementation of predictor-corrector methods [3,4,7]. The authors emphasized the need to develop the same order of accuracy of the main predictors and that of the correctors to ensure good accuracy of the method. The order of accuracies in these works improved significantly compare to the existing methods with lower order of their main predictors.

Attempts have also been made by these scholars [12-14,6,7,15]. Olabode [13] proposed a 5-step block scheme for the solution of special type of eqn. (1). The order of accuracy in Olabode et al. [13] improves more than that of Olabode et al [13]. Awoyemi et al. [6], developed a four-point implicit method for the numerical integration of third order ODEs using power series polynomial function [16]. Kuboye and Omar [7] proposed numerical solution of third order ordinary differential equations using a seven-step block method to improve on Awoyemi et al. [6] and Olabode [13] which are of lower order of accuracy. Furthermore, a symmetric hybrid linear multistep method of order six having two off-step points for the solution of eqn. (1) directly was presented by Obarhua and Kayode [15].

To improve on the study of Obarhua and Kayode [15] a symmetric of two-step four-point hybrid method for the solution of third order initial value problems of ordinary differential equations directly is therefore proposed using the combination of power series and exponential function as the approximate basis function [17].

Derivation of the Method

This research work considers the derivation of 2-step 4-point hybrid method for the solution of general third order initial value problems of ordinary differential equations. The approach is to solve eqn. (1) directly without reducing it to a system of first order differential equation. A combination of power series and exponential function is used as the basis function for eqn. (1). The approximate solution eqn. (1) and the resulting differential systems are respectively given as

$$y(t)=\sum_{j=0}^{r+s-1}\lambda_j t^j+\lambda_{r+s}\sum_{j=0}^{r+s}\frac{\alpha^j t^j}{j!}. \quad (3)$$

Where r and s are the number of interpolation and collocation points respectively.

The third derivative of eqn. (3) as compared with eqn. (1) gives

$$f(t,\ y,\ y'\ y'')=\sum_{j=3}^{r+s-1}\left(j(j-1)(j-2)\lambda_j t^{j-3}\right)+\lambda_{r+s}\sum_{j=3}^{r+s-1}\frac{\alpha^j t^{j-3}}{(j-3)!}. \quad (4)$$

***Corresponding author:** Obarhua FO, Department of Mathematical Sciences, Federal University of Technology, P. M. B. 704, Akure, Nigeria
E-mail: obarhuafo@futa.edu.ng

Collocating eqn. (4) at only the grid points, t_{n+j}, j=0(1)2, and interpolating (3) at both grid and off-grid points, $t_{n+j}\, j = 0\left(\frac{1}{3}\right)2$, leads to the following system of equations.

$$At=b \tag{5}$$

where $t=[\lambda_0 ... \lambda_8]^T$; $b=[y_n ... f_{n+2}]^T$

$$A = \begin{bmatrix} 1 & t_n & t_n^2 & \cdots & t_n^k & \delta_1 \\ 1 & t_{n+\frac{1}{3}} & t^2_{n+\frac{1}{3}} & \cdots & t^k_{n+\frac{1}{3}} & \delta_2 \\ 1 & t_{n+\frac{2}{3}} & t^2_{n+\frac{2}{3}} & \cdots & t^k_{n+\frac{2}{3}} & \delta_3 \\ 1 & t_{n+1} & t^2_{n+1} & \cdots & t^k_{n+1} & \delta_4 \\ 1 & t_{n+\frac{4}{3}} & t^2_{n+\frac{4}{3}} & \cdots & t^k_{n+\frac{4}{3}} & \delta_5 \\ 1 & t_{n+\frac{5}{3}} & t^2_{n+\frac{5}{3}} & \cdots & t^k_{n+\frac{5}{3}} & \delta_6 \\ 0 \; 0 \; 0 \; 6 & 24t_n & \cdots & \psi t_n^4 & & \lambda_1 \\ 0 \; 0 \; 0 \; 6 & 24t_{n+1} & \cdots & \psi t^4_{n+1} & & \lambda_2 \\ 0 \; 0 \; 0 \; 6 & 24t_{n+2} & \cdots & \psi t^4_{n+2} & & \lambda_3 \end{bmatrix} \quad \text{Where,}$$

$$\delta_1 = [1 + \alpha t_n + \frac{\alpha^2 t_n^2}{2!} + \ldots + \frac{\alpha^8 t_n^8}{8!}]$$

$$\delta_2 = [1 + \alpha t_{n+\frac{1}{3}} + \frac{\alpha^2 t^2_{n+\frac{1}{3}}}{2!} + \ldots + \frac{\alpha^8 t^8_{n+\frac{1}{3}}}{8!}]$$

$$\delta_3 = [1 + \alpha t_{n+\frac{2}{3}} + \frac{\alpha^2 t^2_{n+\frac{2}{3}}}{2!} + \ldots + \frac{\alpha^8 t^8_{n+\frac{2}{3}}}{8!}]$$

$$\delta_4 = [1 + \alpha t_{n+1} + \frac{\alpha^2 t^2_{n+1}}{2!} + \ldots + \frac{\alpha^8 t^8_{n+1}}{8!}]$$

$$\delta_5 = [1 + \alpha t_{n+\frac{4}{3}} + \frac{\alpha^2 t^2_{n+\frac{4}{3}}}{2!} + \ldots + \frac{\alpha^8 t^8_{n+\frac{4}{3}}}{8!}]$$

$$\delta_6 = [1 + \alpha t_{n+\frac{5}{3}} + \frac{\alpha^2 t^2_{n+\frac{5}{3}}}{2!} + \ldots + \frac{\alpha^8 t^8_{n+\frac{5}{3}}}{8!}]$$

$$\lambda_1 = [\alpha^3 + \alpha^4 t_n + \frac{\alpha^5 t_n^2}{2!} + \ldots + \frac{\alpha^8 t_n^5}{5!}]$$

$$\lambda_2 = [\alpha^3 + \alpha^4 t_{n+1} + \frac{\alpha^5 t^2_{n+1}}{2!} + \ldots + \frac{\alpha^8 t^5_{n+1}}{5!}]$$

$$\lambda_3 = [\alpha^3 + \alpha^4 t_{n+2} + \frac{\alpha^5 t^2_{n+2}}{2!} + \ldots + \frac{\alpha^8 t^5_{n+2}}{5!}]$$

$\psi = j\,(j-1)\,(j-2)$ as j=5,6.

Solving eqn. (5) for λ_j's and substituting back into eqn. (3), with some manipulation yields, a linear multistep method with continuous coefficients in the form:

$$y(t) = \sum_{j=0}^{k-1} \alpha_j(t) y_{n+j} + \{\tau_1(t) y_{n+r} + \tau_2(t) y_{n+s} + \tau_3(t) y_{n+u} + \tau_4(t) y_{n+v}\} + h^3 \sum_{j=0}^{k} \beta_j(t) f_{n+j}. \tag{6}$$

Taking k=2, the coefficients $\alpha_j(t)$ and $\beta_j(t)$ are expressed as function of $v = \frac{t - t_{n+1}}{h}$ as follows:

$$\alpha_0(v) = \frac{913}{5530}v^2 - \frac{1215}{632}v^4 + \frac{12879}{3160}v^6 - \frac{729}{553}v^8$$

$$\alpha_1(v) = 1 - \frac{28213}{2212}v^2 + \frac{11907}{316}v^4 - \frac{11421}{316}v^6 + \frac{22577}{2212}v^8$$

$$\tau_1(v) = \frac{61}{1092}v - \frac{239437}{172536}v^2 + \frac{124497}{8216}v^4 - \frac{243}{52}v^5 - \frac{204039}{8216}v^6 + \frac{243}{182}v^7 + \frac{447849}{57512}v^8$$

$$\tau_2(v) = -\frac{440}{273}v + \frac{172033}{21567}v^2 - \frac{149769}{4108}v^4 + \frac{243}{26}v^5 + \frac{197721}{4108}v^6 - \frac{243}{91}v^7 - \frac{209709}{14378}v^8$$

$$\tau_3(v) = \frac{440}{273}v + \frac{271939}{43134}v^2 - \frac{139563}{8216}v^4 - \frac{243}{26}v^5 + \frac{56295}{8216}v^6 + \frac{243}{91}v^7 - \frac{17739}{14378}v^8$$

$$\tau_4(v) = -\frac{61}{1092}v - \frac{262273}{862680}v^2 + \frac{20817}{8216}v^4 + \frac{243}{52}v^5 + \frac{78813}{81080}v^6 - \frac{243}{182}v^7 - \frac{49815}{57512}v^8$$

$$\beta_0(v) = \frac{4}{331695}v + \frac{41359}{104815620}v^2 - \frac{77}{16432}v^4 - \frac{1}{780}v^5 + \frac{1327}{123240}v^6 + \frac{1}{364}v^7 - \frac{575}{115024}v^8$$

$$\beta_1(v) = -\frac{4778}{331695}v - \frac{423632}{26203905}v^2 + \frac{1}{6}v^3 + \frac{580}{3081}v^4 - \frac{67}{195}v^5 - \frac{6148}{15405}v^6 + \frac{17}{182}v^7 + \frac{928}{7189}v^8 \tag{7}$$

$$\beta_2(v) = \frac{4}{331695}v + \frac{9769}{104815620}v^2 - \frac{46}{49296}v^4 - \frac{1}{780}v^5 + \frac{157}{123240}v^6 + \frac{1}{364}v^7 + \frac{127}{115024}v^8$$

Evaluating eqn. (7) at v=1 gives the discrete method:

$$y_{n+2} = \frac{256}{39}y_{n+\frac{5}{3}} - \frac{395}{39}y_{n+\frac{4}{3}} + \frac{395}{39}y_{n+\frac{2}{3}} - \frac{256}{39}y_{n+\frac{1}{3}} + y_n + \frac{h^3}{9477}(28f_{n+2} - 1856f_{n+1} + 28f_n). \tag{8}$$

The first and second derivatives of eqn. (8) are:

$$hy'_{n+2} = \frac{100048}{35945}y_{n+\frac{5}{3}} - \frac{1452853}{28756}y_{n+\frac{4}{3}} - \frac{5480}{553}y_{n+1} + \frac{493522}{7189}y_{n+\frac{2}{3}} - \frac{307952}{7189}y_{n+\frac{1}{3}} + \frac{72421}{11060}y_n + \frac{h^3}{8734635}(222904f_{n+2} - 10653608f_{n+1} + 170254f_n). \tag{9}$$

$$h^2y''_{n+2} = \frac{8220428}{107835}y_{n+\frac{5}{3}} - \frac{11174753}{86268}y_{n+\frac{4}{3}} - \frac{47276}{553}y_{n+1} + \frac{12093337}{43134}y_{n+\frac{2}{3}} - \frac{3607720}{21567}y_{n+\frac{1}{3}} + \frac{284317}{11060}y_n + \frac{h^3}{52407810}(9331249f_{n+2} - 234051128f_{n+1} + 4055719f_n). \tag{10}$$

Applying the truncation error formula in Awoyemi et al. [6], associated with eqn. (6) by the difference operator eqn. (9) to determine the order and error constant of the methods:

$$L[y(t):h] = \sum_{j=0}^{k} \begin{bmatrix} \alpha_j y(t_n + jh) + \{\tau_1 y(t_n + jhr) + \tau_2 y(t_n + jhs) + \tau_3 y(t_n + jhu) + \tau_4 y(t_n + jhv)\} \\ -h^3 \beta_j y'''(t_n + jh) \end{bmatrix} \tag{11}$$

Where $y(x)$ is assumed to be continuously differentiable of high order. Therefore, expanding eqn. (11) in Taylor's series and comparing the coefficient of h to give the expression

$$L[y(x):h] = c_0 y(x) + c_1 hy'(x) + c_2 h^2 y'(x) + \ldots + c_p h^p y^{(p)}(x) + \ldots + c_{p+3} h^{p+3} y^{(p+3)}(x) \tag{12}$$

The linear operator L and the associated methods are said to be of order P if $c_0=c_1=c_2=\ldots=c_p=\ldots=c_{p+2}=0$, $c_{p+3}\neq 0$ c_{p+3}. is equal to the error constant. For the purpose of this work, expanding methods (8), (9) and (10) in Taylor's series and comparing the coefficient of h gives both methods of order p=6 and error constant, $c_{p+3}=3.633772 \times 10^{-6}$, $c_{p+3}=1.1804754 \times 10^{-3}$ and $c_{p+3}=1.029201 \times 10^{-4}$ respectively.

Equations (8-10) are of order six, symmetric, consistent, low error constants and capable of handling oscillatory problems.

Implementation of the Method

To implement the implicit linear 2-step 4-point discrete scheme eqn. (8) and its first and second derivatives eqns.(9) and (10), respectively, the following symmetric explicit schemes and their derivatives are also developed by the same procedure for the evaluation of y_{n+2}, y'_{n+2} and y''_{n+2}.

$$y_{n+2}=\frac{10392}{2629}y_{n+\frac{5}{3}}-\frac{15753}{2629}y_{n+\frac{4}{3}}+\frac{11312}{2629}y_{n+1}-\frac{3903}{2629}y_{n+\frac{1}{3}}-\frac{115}{2629}y_n+\frac{h^3}{354915}\left(6636f_{n+\frac{5}{3}}-4780f_{n+1}-56f_n\right). \quad (13)$$

$p=6$ $c_{p+3}=3.107634\times10^{-7}$.

$$hy'_{n+2}=\frac{46152}{8365}y_{n+\frac{5}{3}}-\frac{70731}{4780}y_{n+\frac{4}{3}}+\frac{227992}{8365}y_{n+1}-\frac{264798}{8365}y_{n+\frac{2}{3}}+\frac{19296}{1195}y_{n+\frac{1}{3}}-\frac{16511}{6692}y_n+\frac{h^3}{376425}\left(60792f_{n+\frac{5}{3}}+133840f_{n+1}-2782f_n\right). \quad (14)$$

$p=6$, $c_{p+3}=9.0\times10^{-6}$.

$$h^2y''_{n+2}=-\frac{29861988}{368060}y_{n+\frac{5}{3}}+\frac{44066883}{368060}y_{n+\frac{4}{3}}-\frac{63973128}{368060}y_{n+1}-\frac{154388262}{368060}y_{n+\frac{2}{3}}+\frac{89899548}{368060}y_{n+\frac{1}{3}}-\frac{13689309}{368060}y_n+h^3\left(\frac{9331249}{8281350}f_{n+\frac{5}{3}}+\frac{135647}{20790}f_{n+1}-\frac{1955008}{1774575}f_n\right). \quad (15)$$

$p=6$, $c_{p+3}=1.35\times10^{-4}$.

The methods eqns. (13), (14) and (15) are of order $p=6$ and error constant, $c_{p+3}=3.107634\times10^{-7}$, $c_{p+3}=9.0\times10^{-6}$ and $c_{p+3}=1.35\times10^{-4}$ respectively.

Other explicit schemes were also generated to evaluate other starting values and Taylor's series was used to evaluate the values for y_{n+i}, $i=\frac{1}{3},\frac{2}{3},1$, as

$$y_{n+i}=y(x_n+ih)=y_n+ihy'_n+\frac{(ih)^2}{2!}y''_n+\frac{(ih)^3}{3!}f_n+\frac{(ih)^4}{4!}f'_n+\frac{(ih)^5}{5!}f''_n+\frac{(ih)^6}{6!}f'''_n.$$

$$y'_{n+i}=y'(x_n+ih)=y'_n+ihy''_n+\frac{(ih)^2}{2!}f_n+\frac{(ih)^3}{3!}f'_n+\frac{(ih)^4}{4!}f''_n+\frac{(ih)^5}{5!}f'''_n.$$

and

$$y''_{n+i}=y''(x_n+ih)=y''_n+ihf_n+\frac{(ih)^2}{2!}f'_n+\frac{(ih)^3}{3!}f''_n+\frac{(ih)^4}{4!}f'''_n. \quad (16)$$

Numerical experiments

Three third order problems out of which one is linear and two are non-linear with exact solutions are solved with our method to test the effectiveness and its accuracy.

Problem 1: $y'''=-e^x$, $y(0)=1$, $y'(0)=-1$, $y''(0)=3$, $h=0.1$

Theoretical solution: $y(x)=2+2x^2-e^x$.

Table 1 shows the maximum absolute error of our predictor-corrector method and that of Olabode [13] block method for Problem 1. It reveals that the new method performed creditably well than that of Olabode [13] of higher order.

Problem 2:

$$y'''=y'(2xy''+y'),\ y(0)=1,\ y'(0)=\frac{1}{2},\ y''(0)=0,\ h=0.1.$$

Theoretical solution: $y(x)=1+\frac{1}{2}\ln\left[\frac{2+x}{2-x}\right]$. In Table 2, y-exact, the y-computed, the errors of the new method and the time(s) of iteration for Problem 2 are shown.

Problem 3: $y'''=-y)^4$, $y(1)=-1$, $y'(1)=-1$, $y''(1)=-2$. $h=0.05$.

Theoretical solution: $y(x)=\frac{1}{x-2}$

Table 3 shows the y-exact, the y-computed, and the errors of the new method and the time of iteration for Problem 3.

Conclusion

This paper has produced 2-step 4-point hybrid method for direct solution of higher order ordinary differential equations. The method developed is symmetric, consistent and convergent which can handle

x	y_{exact}	$y_{computed}$	Error in Olabode, (2009), p=7, k=5	Error in new scheme, p=6, k=2
0.1	0.9148290819243523	0.9148290819245347	7.56477e-11	1.82410e-13
0.2	0.8585972418398302	0.8585972418415010	2.60170e-10	1.67078e-12
0.3	0.8301411924239970	0.8301411924299984	5.76003e-10	6.00142e-12
0.4	0.8281753023587299	0.8281753023735897	8.41270e-10	1.48598e-11
0.5	0.8512787292998718	0.8512787293299923	1.00013e-09	3.01205e-11
0.6	0.8978811996094913	0.8978811996633331	1.09051e-09	5.38418e-11
0.7	0.9662472925295238	0.9662472926178395	1.07048e-09	8.83157e-11
0.8	1.0544590715075328	1.0544590716435931	1.49247e-09	1.36060e-10
0.9	1.1603968888430511	1.1603968890429206	3.15695e-09	1.99870e-10
1.0	1.2817181715409554	1.2817181718237693	4.45905e-09	2.82814e-10

Table 1: The numerical solution of our methods of order 6 compared with the method of Olabode, (2009), of order 7.

x	y_{exact}	$y_{computed}$	Error in new scheme, p=6, k=2	Time(s)
0.1	1.0500417292784914	1.0500418242095606	9.49E-08	0.0027
0.2	1.1003353477310756	1.1003366644736043	1.32E-06	0.0248
0.3	1.1511404359364668	1.1511460842057299	5.65E-06	0.0256
0.4	1.2027325540540821	1.2027483597246535	1.58E-05	0.0261
0.5	1.2554128118829952	1.2554482979429176	3.55E-05	0.0266
0.6	1.3095196042031119	1.3095893044089619	6.97E-05	0.0272
0.7	1.3654437542713962	1.3655690060595540	1.25E-04	0.0278
0.8	1.4236489301936017	1.4238603658481614	2.11E-04	0.0283
0.9	1.4847002785940517	1.4850413496636110	3.41E-04	0.0288
10	1.5493061443340548	1.5498382025385242	5.32E-04	0.0293

Table 2: Numerical solution and errors for problem 2.

x	y_{exact}	$y_{computed}$	Error in new scheme, p=6, k=2	Time(s)
1.05	1.0526315789473684	1.0526315779293796	1.017989e-09	0.0291
1.10	1.1111111111111112	1.1111110853730750	2.573804e-08	0.0313
1.15	1.1764705882352944	1.1764704664689964	1.217663e-07	0.0316
1.20	1.2500000000000002	1.2499996375184190	3.624816e-07	0.0320
1.25	1.3333333333333337	1.3333324687923525	8.645410e-07	0.0323
1.30	1.4285714285714290	1.4285696093065769	1.819265e-06	0.0325
1.35	1.5384615384615392	1.5384579871365800	3.551325e-06	0.0328
1.40	1.6666666666666676	1.6666600337776776	6.632889e-06	0.0332
1.45	1.8181818181818195	1.8181697022691894	1.211591e-05	0.0335
1.50	2.0000000000000018	1.9999779695058428	2.20E-05	0.0338

Table 3: Numerical solution and errors for problem 3.

oscillatory type of problems. The numerical tests results obtained were compared with block method of Olabode [13] which was found to perform favorably than the existing method.

References

1. Jator S (2001) Improvements in Adams-Moulton Methods for the First Order Initial Value Problems. Journal of the Tennessee Academy of Science 76: 57-60.
2. Jator SN, Li J (2009) A Self Starting Linear Multistep Method for the General Second Order Initial Value Problems. Int J Comput Math 86: 817-836.
3. Kayode SJ, Adeyeye O (2011) A 3-step hybrid method for direct solution of second order initial value problems. Aust J Basic Appl Sci 5: 2121-2126.
4. Kayode SJ, Obarhua FO (2013) Continuous y-function Hybrid Methods for Direct Solution of Differential Equations. International Journal of Differential Equations and Applications 6: 191-196.
5. Areo EA, Adeniyi RB (2013) A Self-Starting Linear Multistep Method for Direct Solution of Initial Value Problems of Second Order Ordinary Differential Equations. International Journal of Pure and Applied Mathematics 82: 345-364.
6. Awoyemi DO, Kayode SJ, Adoghe LO (2014) A four-point fully implicit method for numerical integration of third-order ordinary differential equations. Int J Phys Sci 9: 7-12.
7. Kuboye JO, Omar Z (2015) Numerical Solution of Third order Ordinary Differential Equations Using a Seven-Step Block Method. International Journal of Mathematical Analysis 9: 743-754.
8. Yap LK, Ismail F, Senu N (2014) An Accurate Block Hybrid Collocation Method for Third Order Ordinary Differential equations. Journal of Applied Mathematics 2014: 1-9.
9. Kayode SJ, Obarhua FO (2015) 3-step y-function hybrid methods for direct numerical integration of second order IVPs in ODEs. Theoretical Mathematics & Applications 5: 39-51.
10. Majid ZA, Azimi NA, Suleiman M (2009) Solving second order ordinary differential equations using two point four step direct implicit block method. European Journal of Scientific Research 31: 29-36.
11. Akinfenwa OA (2013) Ninth Order Block Piecewise Continuous Hybrid Integrators for Solving Second Order Ordinary Differential Equations. International Journal of Differential Equations and Applications, 12: 49-67.
12. Awoyemi DO, Idowu OM (2005) A Class of Hybrid Collocation Methods for Third Order Ordinary Differential Equations. Int J Computer Math 82: 1287-1293.
13. Olabode BT (2009) An Accurate Scheme by Block Method for Third Order Ordinary Differential Equation. Pacific Journal of Science and Technology 10: 136-142.
14. Olabode BT (2013) Block Multistep Method for the Direct Solution of Third Order of Ordinary Differential Equations. FUTA Journal of Research in Sciences 2: 194-200.
15. Obarhua FO, Kayode SJ (2016) Symmetric Hybrid Linear Multistep Method for General Third Order Differential Equations. Open Access Library Journal 3: 2583.
16. Okunuga SA, Ehijie J (2009) New Derivation of Continuous Multistep Methods using Power Series as Basis Function. J Modern Math Stat 3: 43-50.
17. Majid ZA, Azmi NA, Suleiman M, Ibrahim ZB (2012) Solving Directly General Third Order Ordinary Differential Equations Using Two-Point Four-Step Block Method. Sians Malaysiana 41: 623-632.

Initial and Final Characterized Fuzzy $T_{3\frac{1}{2}}$ and Finer Characterized Fuzzy $R_{2\frac{1}{2}}$-Spaces

Ahmed Saeed Abd-Allah[1]* and A Al-Khedhairi[2]

[1]*Department of Statistics and Operations Research, College of Science, King Saud University, Saudi Arabia*
[2]*Department of Mathematics, College of Science, El-Mansoura University, El-Mansoura, Egypt*

Abstract

Basic notions related to the characterized fuzzy $R_{2\frac{1}{2}}$ and characterized fuzzy $T_{3\frac{1}{2}}$-spaces are introduced and studied. The metrizable characterized fuzzy spaces are classified by the characterized fuzzy $R_{2\frac{1}{2}}$ and the characterized fuzzy T_4-spaces in our sense. The induced characterized fuzzy space is characterized by the characterized fuzzy $T_{3\frac{1}{2}}$ and characterized fuzzy $T_{3\frac{1}{2}}$-space if and only if the related ordinary topological space is $\varphi_{1,2}R_{2\frac{1}{2}}$-space and $\varphi_{1,2}T_{3\frac{1}{2}}$-space, respectively. Moreover, the α-level and the initial characterized spaces are characterized $R_{2\frac{1}{2}}$ and characterized $T_{3\frac{1}{2}}$-spaces if the related characterized fuzzy space is characterized fuzzy $R_{2\frac{1}{2}}$ and characterized fuzzy $T_{3\frac{1}{2}}$, respectively. The categories of all characterized fuzzy $R_{2\frac{1}{2}}$ and of all characterized fuzzy $T_{3\frac{1}{2}}$-spaces will be denoted by CFR-Space and CRF-Tych and they are concrete categories. These categories are full subcategories of the category CF-Space of all characterized fuzzy spaces, which are topological over the category SET of all subsets and hence all the initial and final lifts exist uniquely in CFR-Space and CRF-Tych. That is, all the initial and final characterized fuzzy $R_{2\frac{1}{2}}$ spaces and all the initial and final characterized fuzzy $T_{3\frac{1}{2}}$-spaces exist in CFR-Space and in CRF-Tych. The initial and final characterized fuzzy spaces of a characterized fuzzy $R_{2\frac{1}{2}}$-space and of a characterized fuzzy $T_{3\frac{1}{2}}$-space are characterized fuzzy $R_{2\frac{1}{2}}$ and characterized fuzzy $T_{3\frac{1}{2}}$-spaces, respectively. As special cases, the characterized fuzzy subspace, characterized fuzzy product space, characterized fuzzy quotient space and characterized fuzzy sum space of a characterized fuzzy $R_{2\frac{1}{2}}$-space and of a characterized fuzzy $T_{3\frac{1}{2}}$-space are also characterized fuzzy $R_{2\frac{1}{2}}$ and characterized fuzzy $T_{3\frac{1}{2}}$-spaces, respectively. Finally, three finer characterized fuzzy $R_{2\frac{1}{2}}$-spaces and three finer characterized fuzzy $T_{3\frac{1}{2}}$-spaces are introduced and studied.

Keywords: Fuzzy filter; Fuzzy topological space; Operation; Characterized fuzzy space; Metriz-able characterized fuzzy space; Induced characterized fuzzy space; α-Level characterized space; $\varphi_{1,2}\psi_{1,2}$-fuzzy continuous; Initial and final characterized fuzzy spaces; Characterized fuzzy $T_{3\frac{1}{2}}$-space; Characterized fuzzy $T_{3\frac{1}{2}}$-space; AMS classification; Primary 54E35, 54E52; Secondary 54A4003E72

Introduction

Eklund and Gahler [1] introduced the notion of fuzzy filter and by means of this notion the point-based approach to the fuzzy topology related to usual points has been developed. The more general concept for the fuzzy filter introduced by Gahler [2] and fuzzy filters are classified by types. Because of the specific type of the L-filter however the approach of Eklund and Gahler [1] is related only to the L-topologies which are stratified, that is, all constant L-sets are open. The more specific fuzzy filters considered in the former papers are now called homogeneous. The notion of fuzzy real numbers is introduced by Gahler and Gahler [3], as a convex, normal, compactly supported and upper semi-continuous fuzzy subsets of the set of all real numbers R. The set of all fuzzy real numbers is called the fuzzy real line and will be denoted by R_L, where L is complete chain.

The operation on the ordinary topological space (X,T) has been defined by Kasahara [4] as a mapping φ from T into 2^X such that $A \subseteq A^{\varphi}$, for all $A \in T$. Abd El-Monsef et al. [5], extend Kasahara [4] operation to the power set P (X) of the set X Kandil et al. [6] extended Kasahars's and Abd El-Monsef's operations by introducing operation on the class of all fuzzy sets endowed with an fuzzy topology τ as a mapping φ: $L^X \to L^X$ such that int $\mu \le \mu^{\varphi}$ for all $\mu \in L^X$, where μ^{φ} denotes the value of φ at μ. The notions of fuzzy filters and the operations on the class of all fuzzy sets on X endowed with an fuzzy topology τ are applied in ref. [7] to introduce a more general theory including all the weaker and stronger forms of the fuzzy topology. By means of these notions the notion of $\varphi_{1,2}$-interior of the fuzzy set, $\varphi_{1,2}$-fuzzy convergence and $\varphi_{1,2}$-fuzzy

***Corresponding author:** Ahmed Saeed Abd-Allah, Prince Sattam Bin Abdul-Aziz University, Hotat Bani Tamim, Kingdom of Saudi Arabia
E-mail: asabdallah@hotmail.com

neighborhood filters are defined. The notion of $\varphi_{1,2}$-interior operator for the fuzzy sets is also defined as a mapping $\varphi_{1,2}$.int: $L^X \to L^X$ which fulfill (I1) to (I5). Since there is a one-to-one correspondence between the class of all $\varphi_{1,2}$-open fuzzy subsets of X and these operators, then the class $\varphi_{1,2}$OF (X) of all $\varphi_{1,2}$-open fuzzy subsets of X is characterized by these operators. Hence, the triple (X, $\varphi_{1,2}$.int) as will as the triple (X, $\varphi_{1,2}$ OF (X)) will be called the characterized fuzzy space of $\varphi_{1,2}$-open fuzzy subsets. For each characterized fuzzy space (X, $\varphi_{1,2}$.int) the mapping which assigns to each point x of X the $\varphi_{1,2}$-fuzzy neighborhood filter at x is said to be $\varphi_{1,2}$-fuzzy filter pre topology [7]. It can be identified itself with the characterized fuzzy space (X, $\varphi_{1,2}$.int). The characterized fuzzy spaces are characterized by many of characterizing notions, for example by: $\varphi_{1,2}$-fuzzy neighborhood filters, $\varphi_{1,2}$-fuzzy interior of the fuzzy filters and by the set of all $\varphi_{1,2}$-inner points of the fuzzy filters. Moreover, the notions of closeness and compactness in characterized fuzzy spaces are introduced and studied in ref. [8]. For an fuzzy topological space (X, τ), the operations on (X, τ) and on the fuzzy topological space (I_L, I), where I=[0, 1] is the closed unit interval and I is the fuzzy topology defined on the left unit interval I_L are applied to introduced and studied the notions of characterized fuzzy $R_{2\frac{1}{2}}$-spaces and characterized fuzzy $T_{3\frac{1}{2}}$-spaces or (characterized Tychonoff spaces) [9]. In this paper, Basic notions related to the characterized fuzzy $R_{2\frac{1}{2}}$ and the characterized fuzzy $T_{3\frac{1}{2}}$-spaces are introduced and studied. Some of this the metrizable characterized fuzzy spaces, initial and final characterized fuzzy spaces and three finer characterized fuzzy $R_{2\frac{1}{2}}$-spaces are introduced and classified by the characterized fuzzy $R_{2\frac{1}{2}}$ and characterized fuzzy $T_{3\frac{1}{2}}$-spaces. The metrizable characterized fuzzy space is introduce as a generalization of the weaker and stronger forms of the fuzzy metric space introduced by Gahler and Gahler [3]. For every stratified fuzzy topological space (X, τ_d) generated canonically by an fuzzy metric d on X, the metrizable characterized fuzzy space (X, $\varphi_{1,2}$.int$_{\tau d}$) is characterized fuzzy T_4-space in sense of Abd-Allah [10] and therefore it is characterized fuzzy $R_{2\frac{1}{2}}$ and characterized fuzzy $T_{3\frac{1}{2}}$ L-space. The induced characterized fuzzy space (X, $\varphi_{1,2}$.int$_\omega$) is characterized fuzzy $R_{2\frac{1}{2}}$ and characterized fuzzy $T_{3\frac{1}{2}}$-space if and only if the related ordinary topological space (X, T) is $\varphi_{1,2}$ $T_{3\frac{1}{2}}$-space and $\varphi_{1,}$ $T_{3\frac{1}{2}}$-space, respectively, that is, the notions of characterized fuzzy $R_{2\frac{1}{2}}$-spaces and characterized fuzzy $T_{3\frac{1}{2}}$-spaces are good extension as in sense of Lowen [11]. Moreover, the α-level characterized space (X, $\varphi_{1,2}$.int$_\alpha$) and the initial characterized space (X, $\varphi_{1,2}$.int$_i$) are characterized $R_{2\frac{1}{2}}$-space and characterized $T_{3\frac{1}{2}}$-space if the related characterized fuzzy space (X, $\varphi_{1,2}$.int$_\tau$) is characterized fuzzy $R_{2\frac{1}{2}}$-space and characterized fuzzy $T_{3\frac{1}{2}}$-space, respectively. We show that the finer characterized fuzzy space of the characterized fuzzy $R_{2\frac{1}{2}}$-space and of the characterized fuzzy $T_{3\frac{1}{2}}$-space is also characterized fuzzy $R_{2\frac{1}{2}}$ and characterized fuzzy $T_{3\frac{1}{2}}$-space, respectively. The categories of all characterized fuzzy $R_{2\frac{1}{2}}$ and of all characterized fuzzy $T_{3\frac{1}{2}}$-spaces will be denoted by CFR-Space and CRF-Tych, respectively. We show that these categories are concrete categories and they are full subcategories of the category CF-Space of all characterized fuzzy spaces, which are topological over the category SET of all subsets and hence all the initial and final lifts exist uniquely in CFR-Space and CRF-Tych, respectively. That is, all the initial and final characterized fuzzy $T_{3\frac{1}{2}}$-spaces and all the initial and final characterized fuzzy $T_{3\frac{1}{2}}$-spaces are exist in the categories CFR-Space and CRF-Tych. Moreover, we show that the initial and final characterized fuzzy spaces of the characterized fuzzy $R_{2\frac{1}{2}}$-space and of the characterized fuzzy $T_{3\frac{1}{2}}$-space are characterized fuzzy $R_{2\frac{1}{2}}$ and characterized fuzzy $T_{3\frac{1}{2}}$-spaces, respectively. As an special cases, the characterized fuzzy subspace, characterized fuzzy product space, characterized fuzzy quotient space and characterized fuzzy sum space of the characterized fuzzy $R_{2\frac{1}{2}}$-space and of the characterized fuzzy $T_{3\frac{1}{2}}$-space are also characterized fuzzy $R_{2\frac{1}{2}}$ and characterized fuzzy $T_{3\frac{1}{2}}$-spaces, respectively. Finally, in section 5, we introduce and study three finer characterized fuzzy $R_{2\frac{1}{2}}$ and three finer characterized fuzzy $T_{3\frac{1}{2}}$-spaces as a generalization of the weaker and stronger forms of the completely regular and fuzzy $T_{3\frac{1}{2}}$-spaces introduced [1,12,13]. The relations between such new characterized fuzzy $R_{2\frac{1}{2}}$-spaces and our characterized fuzzy $T_{3\frac{1}{2}}$-spaces are introduced. More general the relations between such new characterized fuzzy $T_{3\frac{1}{2}}$-spaces and our characterized fuzzy $T_{3\frac{1}{2}}$-spaces are also introduced. Meany special cases from these finer characterized fuzzy $R_{2\frac{1}{2}}$-spaces and from finer characterized fuzzy $T_{3\frac{1}{2}}$-spaces are listed in Table 1.

Preliminaries

We begin by recalling some facts on fuzzy sets and fuzzy filters. Let L be a completely distributive complete lattice with different least and last elements 0 and 1, respectively. Consider $L_0=L\backslash\{0\}$ and $L_1=L\backslash\{1\}$. Recall that the complete distributivity of L means that the distributive law $\vee_{i\in I}(\alpha_i \wedge \alpha)=\left(\vee_{i\in I}\alpha_i\right)\wedge\alpha$. Sometimes we will assume more specially that L is a complete chain, that is, L is a complete lattice whose partial ordering is a linear one. The standard example of L is the real closed unit interval I=[0, 1]. For a set X, let L^X be the set of all fuzzy subsets of X, that is, of all mappings μ: $X \to L$. Assume that an order-reversing involution α 7→α′ is fixed. For each fuzzy set μ, let co μ denote the complement of μ defined by: (co μ) (x)=co $\mu(x)$ for all $x \in$ X. For all $x \in$ X and $\alpha \in L_0$. Supμ means the supremum of the set of values of μ. The fuzzy sets on X will be denoted by Greek letters as μ, η, ρ,. . . etc. Denote by $\overline{\alpha}$ the constant fuzzy subset of X with value α ∈ L. The fuzzy singleton x_α is an fuzzy set in X defined by $x_\alpha(x)$=α and $x_\alpha(y)$=0 for all $y \neq x$, α ∈ L_0. The class of all fuzzy singletons in X will be denoted by $S(X)$. For every $x_\alpha \in$ S(X) and $\mu \in L^X$, we write $x_\alpha \leq$ μ if and only if $\alpha \leq \mu(x)$. The fuzzy set μ is said to be quasi-coincident with the fuzzy set ρ and written μ q ρ if and only if there exists $x \in X$ such that $\mu(x)$+ ρ(x)>1. If μ not quasi-coincident with the fuzzy set ρ, then we write $\mu\overline{q}\rho$. The fuzzy filter on X [14] is the mapping M: $L^X \to L$ such that the following conditions are fulfilled:

(F1) M$\left(\overline{\alpha}\right)$ ≤ α for all α ∈ L and $\mathcal{M}$(1)=1.

(F2) $\mathcal{M}(\mu \wedge \eta)=\mathcal{M}(\mu) \wedge \mathcal{M}(\eta)$ for all $\mu, \eta \in L^X$.

The fuzzy filter $\mathcal{M}$ is said to be homogeneous [14] if $M(\overline{\alpha})=\alpha$ for all $\alpha \in L$. For each $x \in X$, the mapping $\dot{x}: L^X \to L$ defined by $\dot{x}(\mu)=\mu(x)$ for all $\mu \in L^X$ is a homogeneous fuzzy filter on X. The homogenous fuzzy filter at the fuzzy set is defined by the same way as follows, for each $\mu \in L^X$, the mapping $\dot{\mu}: L^X \to L$ defined by $\dot{\mu}(\sigma)=\bigwedge_{0<\sigma(x)} \sigma(x)$ for all $\sigma \in L^X$ is also homogenous fuzzy filter on X, called homogenous fuzzy filter at $\mu \in L^X$. Obviously, the relation between homogenous fuzzy filter $\dot{\mu}$ at $\mu \in L^X$ and the homogenous fuzzy filter $\dot{x}$ at $x \in X$ is given by:

$$\dot{\mu}(\eta)=\bigwedge_{\mu(x)\geq 0} \sigma(x) \tag{2.1}$$

for all $\eta \in L^X$. As shown in ref. [15], $\mu \leq \eta$ if and only if $\dot{\mu} \leq \dot{\eta}$ holds for all $\mu, \eta \in L^X$. Let $\mathcal{F}_L X$ and $\mathcal{F}_L X$ denote to the sets of all fuzzy filters and of all homogeneous fuzzy filters on X, respectively. If $\mathcal{M}$ and $\mathcal{N}$ are fuzzy filters on the set X, then $\mathcal{M}$ is said to be finer than $\mathcal{N}$, denoted by $\mathcal{M} \leq \mathcal{N}$, provided $\mathcal{M}(\mu) \geq \mathcal{N}(\mu)$ holds for all $\mu \in L^X$. Noting that if L is a complete chain then M is not finer than N, denoted by $\mathcal{M} / \leq \mathcal{N}$, provided there exists $\mu \in L^X$ such that $\mathcal{M}(\mu) < \mathcal{N}(\mu)$ holds. As shown in ref. [4], if $\mathcal{M}, \mathcal{N}$ and L are three fuzzy filters on a set X, then we have:

$M \neq L \geq N$ implies $M \neq N$ and $M \geq L \neq N$ implies $M \neq N$.

The coarsest fuzzy filter $\mathcal{M}$ on X is the fuzzy filter has the value 1 at 1 and 0 otherwise. Suprema and infimum of sets of fuzzy filters are meant with respect to the finer relation. An fuzzy filter $\mathcal{M}$ on X is said to be ultra [2] fuzzy filter if it does not have a properly finer fuzzy filter. For each fuzzy filter $\mathcal{M} \in \mathcal{F}_L X$ there exists a finer ultra fuzzy filter $U \in \mathcal{F}_L X$ such that $U / \leq \mathcal{M}$. Consider $\mathcal{A}$ is a non-empty set of fuzzy filters on X, then the supremum $\bigvee_{M\in A} M$ exists [2] and given by $(\bigvee_{M\in A} M)(\mu)=\bigwedge_{M\in A} M(\mu)$ for all $\mu = L^X$ but the infimum $\bigwedge_{M\in A} M$ does not exists, in general. As shown in ref. [16], the infimum $\bigwedge_{M\in A} M$ of $\mathcal{A}$ with respect to the finer relation for fuzzy filters exists if and only if $M_1(\mu_1) \wedge \ldots \wedge M_n(\mu_n) \leq \sup(\mu_1 \wedge \ldots \wedge \mu_n)$ holds for all finite subset $\{M_1, \ldots M_n\}$ of $\mathcal{A}$ and $\mu_1, \ldots, \mu_n \in L^X$. In this case the infimum is given by:

$$(\bigwedge_{M\in A} M)(\mu)=\bigvee_{\substack{\mu_1 \wedge \ldots \wedge \mu_n \leq \mu \\ M_1, \ldots M_n \in A}} (M_1(\mu_1) \wedge \ldots \wedge M_n(\mu_n)),$$

for all $\mu \in L^X$.

Fuzzy filter bases. A family $(\mathcal{B}_\alpha)_{\alpha \in L_0}$ of non-empty subsets of L^X is called a valued fuzzy filter base [2] if the following conditions are fulfilled:

(V1) $\mu \in \mathcal{B}_\alpha$ implies $\alpha \leq \sup \mu$.

(V2) For all $\alpha, \beta \in L_0$ with $\alpha \wedge \beta \in L_0$ and all $\mu \in \mathcal{B}_\alpha$ and $\eta \in \mathcal{B}_\beta$ there are $\gamma \geq \alpha \wedge \beta$ and $\sigma \leq \mu \wedge \eta$ such that $\sigma \in \mathcal{B}_\gamma$.

As shown in ref. [2], each valued fuzzy filter base $(\mathcal{B}^\alpha)^{\alpha \in L}$ defines an fuzzy filter $\mathcal{M}$ on X by $M(\mu)=\bigvee_{\eta \in B\alpha, \eta \leq \mu} \alpha$ for all $\mu \in L^X$. Conversely, each fuzzy filter $\mathcal{M}$ can be generated by a valued fuzzy filter base, e.g., by $(\alpha\text{-pr } \mathcal{M})_{\alpha \in L_0}$ with $\alpha\text{-pr } M=\{\mu \in L^X \mid \alpha \leq \mathcal{M}(\mu)\}$. $(\alpha\text{-pr } \mathcal{M})_{\alpha \in L_0}$ is a family of pre filters on X and it is called the large valued filter base of $\mathcal{M}$. Recall that a pre filter on X [17] is a non-empty proper subset of $\mathcal{F}$ of L^X such that (1) $\mu, \eta \in \mathcal{F}_X$ implies $\mu \wedge \eta \in \mathcal{F}$ and (2) from $\mu \in \mathcal{F}$ and $\mu \leq \eta$ it follows $\eta \in \mathcal{F}$. $\mathcal{A}$ subset $\mathcal{B}$ of L^X is said to be superior fuzzy filter base [2] if the following conditions are fulfilled:

(S1) $\overline{\alpha} \in B$ for every $\alpha \in L$.

(S2) For all $\mu, \eta \in \mathcal{B}$ there is a fuzzy set $\sigma \in \mathcal{B}$ such that $\sigma \leq \mu$, $\sigma \leq \eta$ and $\sup \sigma = \sup \mu \wedge \sup \eta$.

Each superior fuzzy filter base $\mathcal{B}$ generated a homogeneous fuzzy filter $\mathcal{M}$ on X by $M(\mu)=\bigvee_{\eta \in B, \eta \leq \mu} \sup \eta$ for all $\mu \in L^X$ and each fuzzy filter $\mathcal{M}$ can be generated by a superior fuzzy filter base, e.g., by base $M=\{\mu \in L^X \mid M(\mu)=\sup \mu\}=\mu \wedge \overline{M\mu} \mid \mu \in L^X\}$, where base M will be called the large superior fuzzy filter base of $\mathcal{M}$. If X is a non-empty set and μ is an fuzzy subset of X, then $B = \{\mu \wedge \overline{\alpha} \mid \alpha \in L\} \cup \{\overline{\alpha} \mid \alpha \in L\}$ is a superior fuzzy filter base of a homogeneous fuzzy filter on X, called superior principal fuzzy filter generated by μ and will be denoted by $[\mu]$. In case L is a complete chain and μ is not constant we have $[\mu](\eta)=\sup \mu$, when $\mu \leq \eta$ and $[\mu](\eta)=\bigwedge_{\eta(x)<\mu(x)} \eta(x)$ otherwise for all $\eta \in L^X$. For each ordinary subset M of X we have that $[\chi_M] = \bigvee_{x \in M} \dot{x}$, where χ_M is the characteristic function of M.

Fuzzy topology

By the fuzzy topology on a set X, we mean a subset of L^X which is closed with respect to all supreme and all finite infimum and contains the constant fuzzy sets $\overline{0}$ and $\overline{1}$ [16,18]. A set X equipped with an fuzzy topology τ on X is called an fuzzy topological space. For each fuzzy topological space (X, τ), the elements of τ are called the open fuzzy subsets of this space. If τ_1 and τ_2 are fuzzy topologies on a set X, then τ_1 is said to be finer than τ_2 and τ_2 is said to be coarser than τ_1, provided $\tau_2 \subseteq \tau_1$ holds. For each fuzzy set $\mu \in L^X$, the strong α-cut and the weak α-cut of μ are the ordinary subsets $S_\alpha(\mu) = \{x \in X \mid \mu(x) > \alpha\}$ and $W_\alpha(\mu) = \{x \in X \mid \mu(x) \geq \alpha\}$ of X respectively. For each complete chain L, the α-level topology and the initial topology [19] of an fuzzy topology τ on the set X are defined as follows:

$$\tau_\alpha = \{S_\alpha(\mu) \in P(X) : \mu \in \tau\} \text{ and } i(\tau) = \inf\{\tau_\alpha : \alpha \in L_1\},$$

respectively, where *inf* is the infimum with respect to the finer relation for topologies. On other hand if (X, T) is an ordinary topological space, then the induced fuzzy topology on X is given by Lowen [17] as the following:

$$\omega(T) = \{\mu \in L^X : S_\alpha(\mu) \in T \text{ for all } \alpha \in L_1\}.$$

The fuzzy topological space(X, τ) and also τ are said to be stratified provided $\alpha \in \tau$ holds for all $\alpha \in L$, that is, all constant fuzzy sets are open [19].

The fuzzy unit interval

The fuzzy unit interval will be denoted by I_L an it is defined in [3] as the fuzzy subset:

$$I_L = \{x \in R^*_L \mid x \leq 1^\sim\},$$

where $I=[0, 1]$ is the real unit interval and $\mathbf{R}^*_L = \{x \in R_L \mid x(0) = 1 \text{ and } 0^\sim \leq x\}$ is the set of all positive fuzzy real numbers. Note that, the binary relation $\leq$ is defined on R_L as follows:

$$x \leq y \Leftrightarrow x_{\alpha 1} \leq y_{\alpha 1} \text{ and } x_{\alpha 2} \leq y_{\alpha 2},$$

for all $x, y \in R_L$, where $x_{\alpha 1} = \inf\{z \in R \mid x(z) \geq \alpha\}$ and $x_{\alpha 2} = \sup\{z \in R \mid x(z) \geq \alpha\}$ for all $\alpha \in L_0$. Note that the family Ω which is defined by:

$$\Im = \{R_\delta | I_L \mid \delta \in I\} \cup \{R^\delta | I_L \mid \delta \in I\} \cup \{0^\sim \mid I_L\}$$

is a base for an fuzzy topology I on I^L, where R_δ and R^δ are the fuzzy subsets of R^L defined by $R_\delta(x) = \bigvee_{\alpha > \delta} x(\alpha)$ and $R^\delta = (\bigvee_{\alpha \geq \delta} x(\alpha))'$ for all x

$\in R_L$ and $\delta \in R$. The restrictions of R_δ and R^δ on I_L are the fuzzy subsets $R_\delta I_L$ and $R^\delta I_L$, respectively. Recall that:

$$R^\delta(x) \wedge R^\gamma(y) \leq R^{\delta+\gamma}(x + y), \tag{2.2}$$

where, $x+y$ is the fuzzy real number defined by $(x + y)(\xi) = \vee_{\gamma,\zeta \in R, \gamma+\zeta=\xi} (x(\gamma) \wedge y(\zeta))$ for all $\xi \in R$.

Operation on fuzzy sets

In the sequel, let a fuzzy topological space (X, τ) be fixed. By the operation [6] on the set X we mean the mapping $\varphi: L^X \to L^X$ such that $\text{int}(\mu) \leq \mu^\varphi$ holds for all $\mu \in L^X$, where, μ^φ denotes the value of φ at μ. The class of all operations on X will be denoted by $O_{(L^X,\tau)}$. By the identity operation on $O_{(L^X,\tau)}$, we mean the operation $1_L^X: L^X \to L^X$ such that $1_L^X(\mu)=\mu$ for all $\mu \in L^X$. The constant operation on $O_{(L^X,\tau)}$ is the operation $c_L^X: L^X \to L^X$ defined by $c_L^X(\mu)=1$ for all $\mu \in L^X$. If $\leq$ is a partially order relation on $O_{(L^X,\tau)}$ defined as follows: $\varphi_1 \leq \varphi_2 \Leftrightarrow \varphi_1(\mu) \leq \varphi_2(\mu)$ for all $\mu \in L^X$, then $(O_{(L^X,\tau)}, \leq)$ is a completely distributive lattice. The operation $\varphi: L^X \to L^X$ is called:

(i) Isotone if $\mu \leq \eta$ implies $\varphi\mu \leq \varphi\eta$, for all $\mu, \eta \in L^X$.

(ii) Weakly finite intersection preerving (wfip, for short) with respect to $A \subseteq L^X$ *if* $\eta \wedge \varphi(\mu) \leq \varphi(\eta \wedge \mu)$ holds, for all $\eta \in \mathcal{A}$ and $\mu \in L^X$.

(iii) Idempotent if $\varphi(\mu) = \varphi(\varphi(\mu))$, for all $\mu \in L^X$.

The operations $\varphi, \psi \in O_{(L^X,\tau)}$ are said to be dual if $\psi(\mu)=co(\varphi(co\mu))$ or equivalently $\varphi(\mu)=co(\psi(co\mu))$ for all $\mu \in L^X$, where $co\mu$ denotes the complement of μ. The dual operation of φ is denoted by $\varphi^\sim$. In the classical case of $L=\{0, 1\}$, by the operation on a set X we mean the mapping φ: P (X) $\to$ P (X) such that int $A \subseteq A^\varphi$ for all $A \in P(X)$ and the identity operation on the class of all ordinary operations $O_{(P(X),T)}$ on X will be denoted by $i_{P(X)}$ and it defined by: $i_{P(X)}(A)=A$ for all $A \in P(X)$.

The φ-open fuzzy sets

Let a fuzzy topological space (X, τ) be fixed and $\varphi \in O_{(L^X,\tau)}$. The fuzzy set $\mu: X \to L$ is said to be φ-open fuzzy set if $\mu \leq \mu^\varphi$ holds. We will denote the class of all φ-open fuzzy sets on X by φ of (X). The fuzzy set μ is called φ-closed if its complement $co\mu$ is φ-open. The operations φ, $\psi \in O_{(L^X,\tau)}$ are equivalent and written $\varphi \sim \psi$ if φ of $(X)=\psi$ of (X).

The $\varphi_{1,2}$-interior fuzzy sets

Let a fuzzy topological space (X, τ) be fixed and

$\varphi_1, \varphi_2 \in O_{(L^X,\tau)}$. Then the $\varphi_{1,2}$-interior of the fuzzy set $\mu: X \to L$ is a mapping $\varphi_{12}.\text{int}\mu: X \to L$ defined by:

$$\varphi_{1,2}.int\ \mu = \vee_{\eta \in \varphi_1 OF(X)\varphi_2 \eta \leq \mu} \eta. \tag{2.3}$$

That is, the $\varphi_{1,2}.\text{int}\mu$ is the greatest φ_1-open fuzzy set η such that $\eta^{\varphi 2}$ less than or equal to μ [19]. The fuzzy set μ is said to be $\varphi_{1,2}$-open if and only if $\mu \leq \varphi_{1,2}.\text{int}\ \mu$. The class of all $\varphi_{1,2}$-open fuzzy sets on X will be denoted by $\varphi_{1,2}OF$ (X). The complement co μ of the $\varphi_{1,2}$-open fuzzy subset μ will be called $\varphi_{1,2}$-closed, the class of all $\varphi_{1,2}$-closed fuzzy subsets of X will be denoted by $\varphi_{1,2}CF$ (X). In the classical case of $L=\{0, 1\}$, the fuzzy topological space (X, τ) is up to an identification by the ordinary topological space (X, T) and $\varphi_{1,2}.\text{int}\ \mu$ is the classical one. Hence in this case the ordinary subset A of X is $\varphi_{1,2}$-open if $A \subseteq \varphi_{1,2}.\text{int}$ A. The complement of a $\varphi_{1,2}$-open subset A of X will be called $\varphi_{1,2}$-closed. The class of all $\varphi_{1,2}$-open and the class of all $\varphi_{1,2}$-closed subsets of X will be denoted by $\varphi_{1,2}O(X)$ and $\varphi_{1,2}C(X)$, respectively. Clearly, F is $\varphi_{1,2}$-closed if and only if $\varphi_{1,2}.cl_T\ F=F$.

Proposition

For each two operations $\varphi_1, \varphi_2 \in O(L^X,\tau)$ and for each $\mu, \eta \in \varphi 1, \varphi 2 \in L^X$, the mapping $\varphi_{1,2}.\text{int}: X \to L$ fulfills the following axioms [7]:

(i) If $\varphi_2 \geq 1_L^X$, then $\varphi_{1,2}.\text{int}\mu \leq \mu$.

(ii) $\varphi_{1,2}.\text{int}$ is isotone, i.e if $\mu \leq \eta$, then $\varphi_{1,2}.\text{int}\mu \leq \varphi_{1,2}.\text{int}\eta$.

$\varphi_{1,2}.int\bar{1} = \bar{1}$.

If $\varphi_2 \geq 1_L X$ is isotone and φ_1 is with respect to $\varphi_1 O\mathcal{F}$ (X), then $\varphi_{1,2}.int(\mu \wedge \eta) = \varphi_{1,2}.int\mu \wedge \varphi_{1,2}.int\eta$.

If φ_2 is isotone and idempotent operation, then $\varphi_{1,2}.int\mu \leq \varphi_{1,2}.int(\varphi_{1,2}.int\mu)$.

$\varphi_{1,2}.int(\vee_{i \in I} \mu_i) = \vee_{i \in I} \varphi_{1,2}.int\mu_i$ *for all* $\mu_i \in \varphi_{1,2}OF(X)$.

Proposition

Let (X, τ) be a fuzzy topological space and $\varphi_1, \varphi_2 \in O_{(L^X,\tau)}$. Then the following are fulfilled:

(i) If $\varphi_2 \geq 1_L X$, then the class $\varphi_{1,2}OF$ (X) of all $\varphi_{1,2}$-open fuzzy sets on X forms an extended fuzzy topology on X [7,21].

If $\varphi_2 \geq 1_{L^X}$ *and* $\varphi_{1,2}.int\bar{1} = \bar{1}$, then the class $\varphi_{1,2}OF$ (X) of all $\varphi_{1,2}$-open fuzzy sets on X forms a supra fuzzy topology on X [21].

If $\varphi_2 \geq 1_L X$ is isotone and φ_1 is with respect to $\varphi_1 OF$ (X), then $\varphi_{1,2}OF$ (X) is an fuzzy pre topology on X [21].

If $\varphi_2 \geq 1_L X$ is isotone and idempotent operation and φ_1 is with respect to $\varphi_1 OF$ (X), then $\varphi_{1,2}OF$ (X) is fuzzy topology on X [16,18].

Because of Propositions 2.1 and 2.2, if the fuzzy topological space(X, τ) be fixed and

$\varphi_1, \varphi_2\ O_{(L^X,\tau)}$. Then the relation between the class $\varphi_{1,2}OF$ (X) of all $\varphi_{1,2}$-open fuzzy sets on X and the mapping $\varphi_{1,2}.\text{int}$ is given by:

$$\varphi_{1,2}OF(X) = \{\mu \in L^X \mid \mu \leq \varphi_{1,2}.int\mu\} \tag{2.4}$$

and the following axioms are fulfilled:

(I1) If $\varphi_2 \geq 1_L^X$, then $\varphi_{1,2}.\text{int}\mu \leq \mu$ holds, for all $\mu \in L^X$.

(I2) If $\mu \leq \eta$, then $\varphi_{1,2}.\text{int}\mu \leq \varphi_{1,2}.\text{int}\eta$ for all $\mu, \eta \in L^X$.

(I3) $\varphi_{1,2}.int\ \bar{1} = \bar{1}$

(I4) If $\varphi_2 \geq 1_{L^X}$ is isotone and φ_1 is with respect to $\varphi_1 OF$ (X), then $\varphi_{1,2}.\text{int}\mu \wedge \wedge \varphi_{1,2}.\text{int}\eta=\varphi_{1,2}.\text{int}(\mu \wedge \eta)$ for all $\mu, \eta \in L^X$. s

(I5) If φ_2 is isotone and idempotent, then $\varphi_{1,2}.\text{int}(\varphi_{1,2}.\text{int}\mu)=\varphi_{1,2}.\text{int}\mu$ for all $\mu \in L^X$.

Characterized Fuzzy Spaces

Independently on the fuzzy topologies, the notion of $\varphi_{1,2}$-interior operator for the fuzzy sets can be defined as a mapping $\varphi_{1,2}.\text{int}: L^X \to L^X$ which fulfill (I1) to (I5). It is well-known that (2.3) and (2.4) give a one-to-one correspondence between the class of all $\varphi_{1,2}$-open fuzzy sets and these operators, that is, $\varphi_{1,2}OF$ (X) can be characterized by the $\varphi_{1,2}$-interior operators. In this case the triple $(X, \varphi_{1,2}.\text{int})$ as well as the triple $(X, \varphi_{1,2}OF(X))$ will be called characterized fuzzy space [7] of the $\varphi_{1,2}$-open fuzzy subsets of X. The characterized fuzzy space $(X, \varphi_{1,2}.\text{int})$ is said to be stratified if and only if $\varphi_{1,2}.int\bar{\alpha} = \bar{\alpha}$ for all $\alpha \in L$. As shown in ref. [7], the characterized fuzzy space $(X, \varphi_{1,2}.\text{int})$ is stratified if the related fuzzy topology is stratified. Moreover, the characterized fuzzy space $(X, \varphi_{1,2}.\text{int})$ is said to have the weak infimum property [21], provided $\varphi_{1,2}.int(\mu \wedge \overline{\alpha}) = \varphi_{1,2}.int\mu \wedge \varphi_{1,2}.int\overline{\alpha}$ *for all* $\mu \in L^X$ *and* $\alpha \in L$. The characterized fuzzy space (X, $\varphi_{1,2}$.int) is said to be strongly stratified [21], provided $\varphi_{1,2}$.int is stratified and have the weak infimum property. If (X, $\varphi_{1,2}$.int) and (X, $\psi_{1,2}$.int) are two characterized fuzzy spaces, then (X, $\varphi_{1,2}$.int) is said to be finer than (X, $\psi_{1,2}$.int) and denoted by $\varphi_{1,2}$.int ≤ $\psi_{1,2}$.int, provided $\varphi_{1,2}$.intμ ≥ $\psi_{1,2}$.intμ holds for all μ ∈ L^X. If τ is a fuzzy topology on the set X and φ_1, φ_2 ∈ $O_{(L}X_{,\tau)}$, then by the initial characterized space of (X, τ) we mean the characterized spaces (X, $(\varphi_{1,2}O(X))_\alpha$) and (X, i($\varphi_{1,2}$O(X))), respectively where $(\varphi_{1,2}O(X))_\alpha$ and i($\varphi_{1,2}$O(X)) are defined as follows:

$$(\varphi_{1,2}O(X))_\alpha = \{S_\alpha\ \mu \in P(X) \mid \mu \in \varphi_{1,2}OF(X)\} \text{ and } i(\varphi_{1,2}O(X)) = \cap\{(\varphi_{1,2}OF(X))_\alpha \mid \alpha \in L_1\}.$$

Sometimes we denoted to the α-level characterized space and the initial characterized space of (X, τ) by $(X,\ \varphi_{1,2}.int_\alpha)$ *and* $(X,\ \varphi_{1,2}.int_i)$, respectively. If T is an ordinary topology on a set X and $\varphi_1, \varphi_2 \in O_{(P(X),T)}$, then by the induced characterized fuzzy space on X we mean the characterized fuzzy space $(X,\ \omega(\varphi_{1,2}OF(X)))$ which is defined by:

$$\omega(\varphi_{1,2}OF(X)) = \{\mu \in L^X \mid S_\alpha\mu \in \varphi_{1,2}O(X) \text{ for all } \alpha \in L_1\}.$$

Sometimes we denoted to the induced characterized fuzzy space for the ordinary topological space (X, *T*) by $(X,\ \varphi_{1,2}.int_\omega)$.

If $\varphi_1 = int_\tau$ *and* $\varphi_2 = 1_L X$, then the class $(\varphi_{1,2}OF(X))$ of all $\varphi_{1,2}$-open fuzzy of X coincide with τ which is defined in [22,23] and hence the characterized fuzzy space $(X,\ \varphi_{1,2}.int)$ coincide with the fuzzy topological space (X, τ).

$\varphi_{1,2}$-fuzzy neighborhood filters

An important notion in the characterized fuzzy space (X, $\varphi_{1,2}$.int) is that of the $\varphi_{1,2}$-fuzzy neighborhood filter at the points and at the ordinary subsets of this space. Let (X, τ) be a fuzzy topological space and $\varphi_1, \varphi_2 \in O_{(L^X,\tau)}$. As follows by (I1) to (I5) for each x ∈ X, the mapping $N_{\varphi 1,2}(x): L^X \to L$ which is defined by:

$$N_{\varphi 1,2}(x)(\mu) = (\varphi_{1,2}.int\ \mu)(x) \tag{2.5}$$

for all $\mu \in L^X$, is a fuzzy filter on X, called $\varphi_{1,2}$-fuzzy neighborhood filter at x [7]. If the related $\varphi_{1,2}$-interior operator fulfill the axioms (I1) and (I2) only, then the mapping $N_{\varphi 1,2}(x): L^X \to L$, defined by (2.5) is fuzzy stack [21], called $\varphi_{1,2}$-fuzzy neighborhood stack at x. Moreover, if the $\varphi_{1,2}$-interior operator fulfill the axioms (I1), (I2) and (I4) such that in (I4) instead of η ∈ L^X we take α^-, then the mapping $N_{\varphi 1,2}(x): L^X \to L$, defined by (2.5) is a fuzzy stack with the cutting property, called $\varphi_{1,2}$-fuzzy neighborhood stack with the cutting property at *x*. The $\varphi_{1,2}$-fuzzy neighborhood filters fulfill the following conditions:

$$(N1)\ x^\cdot \le N_{\varphi 1,2}(x) \text{ holds for all } x \in X$$

$$(N2)\ N_{\varphi 1,2}(x)(\mu) \le N_{\varphi 1,2}(x)(\eta) \text{ holds for all } \mu,\ \eta \in L^X \text{ and } \mu \le \eta.$$

$$N_{\varphi 1,2}(x)(y \mapsto N_{\varphi 1,2}(y)(\mu)) = N_{\varphi 1,2}(y)(\mu), \text{ for all } x \in X \text{ and } \mu \in L^X.$$

Clearly $y \mapsto N_{\varphi 1,2}(y)(\mu)$ is the fuzzy set $\varphi_{1,2}.\text{int}\,\mu$. The characterized fuzzy space $(X, \varphi_{1,2}.\text{int})$ is characterized as the fuzzy filter pre topology [7], that is, as a mapping $N_{\varphi 1,2}: X \to F_L X$ such that (N1) to (N3) are fulfilled.

$\varphi_{1,2}\psi_{1,2}$-Fuzzy continuity

Let now the fuzzy topological spaces (X, τ_1) and (Y, τ_2) are fixed, φ_1, φ_2 ∈ $O_{(L}X_{,\tau 1)}$ and ψ_1, ψ_2 ∈ $O_{(L}Y_{,\tau 2)}$. The mapping f: (X, $\varphi_{1,2}$.int) →(Y, $\psi_{1,2}$.int) is said to be $\varphi_{1,2}\psi_{1,2}$-fuzzy continuous if

$$(\psi_{1,2}.int\ \eta) \circ \mu \le \varphi_{1,2}.int\ (\eta \circ \mu) \tag{2.6}$$

holds for all η ∈ L^Y [7]. If an order reversing involution′ of L is given, we have that f is a $\varphi_{1,2}$ $\psi_{1,2}$-fuzzy continuous if and only if $\varphi_{1,2}.cl\ (\eta \circ \mu) \le (\psi_{1,2}.cl\ \eta) \circ \mu$ holds for all $\eta \in L^Y$. Here $\varphi_{1,2}$.cl and $\psi_{1,2}$.cl, mean the closure operators related to $\varphi_{1,2}$.int and $\psi_{1,2}$.int, respectively which are defined by $\varphi_{1,2}$.cl μ=co ($\varphi_{1,2}$.int coμ) for all μ ∈ L^X. Obviously if f is $\varphi_{1,2}\psi_{1,2}$-fuzzy continuous, then the inverse f^{-1}: (Y, $\psi_{1,2}$.int) → (X, $\varphi_{1,2}$.int) is $\psi_{1,2}\varphi_{1,2}$-fuzzy continuous, that is $(\varphi_{1,2}.int\ h) \circ \mu^{-1} \le \psi_{1,2}.int\ (h \circ f^{-1})$ holds for all h ∈L^X

By means of characterizing $\varphi_{1,2}$-fuzzy neighborhoods $N_{\varphi 1,2}(x)$ *of* $\varphi_{1,2}.int$ *and* $N_{\psi 1,2}(x)$ *of* $\psi_{1,2}.int$, the $\varphi_{1,2}\psi_{1,2}$-fuzzy continuity of f can also be characterized. The mapping f: (X, $\varphi_{1,2}$.int) → (Y, $\psi_{1,2}$.int) is $\varphi_{1,2}\psi_{1,2}$-fuzzy continuous if $N_{\psi 1,2}$ (f(x)) ≥ F_Lf($N_{\varphi 1,2}$ (x)) holds for all x ∈ X. Obviously, in case of L={ 0, 1 }, φ_1=ψ_1=int, φ_2=1_LX and ψ_2=1_LY the $\varphi_{1,2}\psi_{1,2}$-fuzzy continuity coincides with the usual fuzzy continuity.

Initial characterized fuzzy spaces

In the following let X be a set, let *I* be a class and for each i ∈ I, let (X_i, $\delta_{1,2}$.int_i) be a characterized fuzzy space of $\delta_{1,2}$-open fuzzy subsets of X_i and f_i: X → X_i is the mapping from X into X_i. By the initial $\varphi_{1,2}$-fuzzy interior operator of $(\delta_{1,2}.int_i)_i$∈$_I$ with respect to $(f_i)_i$∈$_I$, we mean the coarsest $\varphi_{1,2}$-fuzzy interior operator $\varphi_{1,2}$.int on X for which all mappings f_i: (X, $\varphi_{1,2}$.int) → (X_i, $\delta_{1,2}$.int_i) are $\varphi_{1,2}\delta_{1,2}$-fuzzy continuous. The triple (X, $\varphi_{1,2}$.int) is said to be initial characterized fuzzy space [7] of $((X_i, \delta_{1,2}.int_i))_{i\in I}$ with respect to $(f_i)_{i\in I}$. The initial $\varphi_{1,2}$-fuzzy interior operator $\varphi_{1,2}$.int: L^X → L^X of $(\delta_{1,2}.int_i)_{i\in I}$ with respect to $(f_i)_{i\in I}$ always exists and is given by:

$$\varphi_{1,2}.int\ \mu = \bigvee_{\mu_i f_i \le \mu,\ i \in I} (\delta_{1,2}.int_i\mu_i) \circ f_i \tag{2.7}$$

for all μ ∈ L^X. For each i ∈ I, let $N^i_{\delta 1,2}: X_i \to F_L X_i$ is the representation of $\delta_{1,2}$.int_i as an fuzzy filter pre topology. Then because of (2.5) and (2.7), the mapping $N_{\varphi 1,2}$: X → F_LX which is defined by:

$$N_{\varphi 1,2}(x)(\mu) = \bigvee_{\mu_i \circ f_i \le \mu,\ i \in I} N^i_{\delta 1,2}(f_i(x))(\mu_i)$$

for all x ∈ X and μ ∈ L^X, is the representation of the initial $\varphi_{1,2}$-fuzzy interior operator of $(\psi_{1,2}.int_i)_{i\in I}$ with respect to $(f_i)_{i\in I}$ as the fuzzy filter pre topology.

Characterized Fuzzy Subspaces

Let A be a subset of a characterized fuzzy space (X, $\varphi_{1,2}$.int) and i: A,→ X is the inclusion mapping of A into X. Then the mapping $\varphi_{1,2}$.int_A: L^A → L^A defined by:

$$\varphi_{1,2}.int_A\eta = \bigvee_{\mu \circ i \le \eta} (\varphi_{1,2}.int\ \mu) \circ i$$

for all η ∈ L^A is initial $\varphi_{1,2}$-fuzzy interior operator for $\varphi_{1,2}$.int with respect to the inclusion mapping i: A,→ X. $\varphi_{1,2}$.int_A will be called induced $\varphi_{1,2}$-interior operator of $\varphi_{1,2}$.int on the subset A of X. The triple (A, $\varphi_{1,2}$.int_A) is said to be characterized fuzzy subspace of (X, $\varphi_{1,2}$.int) [7].

Characterized Fuzzy Product Spaces

Assume that (X_i, $\delta_{1,2}$.int_i) is a characterized fuzzy space for each i I, where I is any class. Let X be the cartesian product πX_i of the family $(X_i)_{i\in I}$ and π_i: X → X_i the related projections. The i∈I, mapping $\varphi_{1,2}$.int: L^X → L^X, defined by:

$$\varphi_{1,2}.int\ \mu = \bigvee_{\mu_i \circ \pi_i \le \mu,\ i \in I} (\delta_{1,2}.int_i\mu_i) \circ \pi_i$$

for all μ ∈ L^X, will be called $\varphi_{1,2}$-fuzzy product of the $\delta_{1,2}$L-interior operators $\delta_{1,2}$.int_i. The triple (X, $\varphi_{1,2}$.int) is said to be characterized fuzzy product space [7] of the characterized fuzzy spaces (X_i, $\delta_{1,2}$.int$_i$). The $\varphi_{1,2}$.int will be denoted by $\underset{i\in I}{\pi}\delta_{1,2}.int_i$ and it is initial $\varphi_{1,2}$-fuzzy interior operator of $(\delta_{1,2}.int_i)_{i\in I}$ with respect to the family $(\pi_i)_{i\in I}$ of projections. The characterized fuzzy product space (X, $\varphi_{1,2}$.int) also will be denoted by $\underset{i\in I}{\pi}(X_i, \delta_{1,2}.int_i)$

Final characterized fuzzy spaces

It is well-known (cf. e.g., [11,24]) that in the topological category all final lifts uniquely exist and hence also all final structures exist. They are dually defined. In case of the category CF-Space of all characterized fuzzy spaces the final structures can easily be given, as is shown in the following:

Let I be a class and for each $i \in I$, let (X_i, $\delta_{1,2}$.int$_i$) be an characterized fuzzy space and f_i: $X_i \to X$ is the mapping of X_i into a set X. The final $\varphi_{1,2}$-fuzzy interior operator of $(\delta_{1,2}.int_i)_{i\in I}$ with respect to $(f_i)_{i\in I}$ is the finest $\varphi_{1,2}$.int on X for which all mappings f_i: (X_i, $\delta_{1,2}$.int$_i$) → (X, $\varphi_{1,2}$.int) are $\delta_{1,2}\varphi_{1,2}$-fuzzy continuous [7]. Hence, the triple (X, $\varphi_{1,2}$.int) is the final characterized fuzzy space of $((X_i, \delta_{1,2}.int_i))_{i\in I}$ with respect to $(f_i)_{i\in I}$. The final $\varphi_{1,2}L$-interior operator $\varphi_{1,2}$.int: $L^X \to L^X$ of $(\delta_{1,2}.int_i)_{i\in I}$ with respect to $(f_i)_{i\in I}$ exists and is given by

$$(\varphi_{1,2}.int\ \mu)(x) = \underset{x_i\in f_i^{-1}\{x\},\ i\in I}{\wedge} \delta_{1,2}.int_i(\mu\circ f_i)(x_i)\wedge \mu(x)$$

for all x ∈ X and $\mu \in L^X$.

Characterized Fuzzy Quotient Spaces

Let (X, $\varphi_{1,2}$.int) be a characterized fuzzy space and f: $X \to A$ is an surjective mapping. Then the mapping $\varphi_{1,2}$.int$_f$: $L^A \to L^A$, defined by:

$$(\varphi_{1,2}.int\ \mu)(a) = \underset{x_i\in f_i^{-1}\{a\}}{\wedge} \varphi_{1,2}.int(\mu\circ f_i)(x)$$

for all $a \in A$ and $\mu \in L^A$, is final $\varphi_{1,2}$-fuzzy interior operator of $\varphi_{1,2}$.int with respect to f which is not idempotent. Then the $\varphi_{1,2}$.int$_f$ will be called quotient $\varphi_{1,2}$-fuzzy interior operator and the triple (A, $\varphi_{1,2}$.int$_f$) is said to be characterized fuzzy quotient space [7].

Note that in this case $\varphi_{1,2}$.int is idempotent, $\varphi_{1,2}$.int$_f$ need not be. Even in the classical case of L={0, 1}, φ_1=int and $\varphi_2=1_LX$ we have the following: If $\varphi_{1,2}$.int is up to an identification the usual topology, then $\varphi_{1,2}$.int$_f$ is a pre topology which need not be idempotent. An example is given [25] (p. 234).

Characterized Fuzzy Sum Spaces

Assume that (X_i, $\delta_{1,2}$.int$_i$) is a characterized fuzzy space for each i ∈, where I is any class. Let X be the disjoint union $\underset{i\in I}{\cup}(X_i\times\{i\})$ of the family $(X_i)_{i\in I}$ and for each i ∈ I, let $\varphi_{1,2}$.int: $L^X \to L^X$, defined by:e_i: $X_i \to X$ be the canonical injection from X_i into X given by $e_i(x_i)=(x_i, i)$. Then the mapping $\varphi_{1,2}$.int: $L^X \to L^X$, defined by:

$$(\varphi_{1,2}.int\ \mu)(a,\ i) = \delta_{1,2}.int_i(\mu\circ e_i)(a)$$

for all i ∈ I, of a ∈ X_i and $\mu \in L^X$, is said to be final $\varphi_{1,2}$-fuzzy interior operator with respect to $(e_i)_{i\in I}$.

$(\delta_{1,2}.int_i)_{i\in I}$ $\varphi_{1,2}$.int will be called sum $\varphi_{1,2}$-fuzzy interior operator will be denoted by $^{\Sigma}\delta_{1,2}$.int$_i$. The pair (X, $\varphi_{1,2}$.int) is said to be characterized fuzzy sum space [7] and it will be denoted also by $\sum_{i\in I}(X_i, \delta_{1,2}.int_i)$.

Characterized Fuzzy T_1 And Fuzzy $\Phi_{1,2}T_1$-Spaces

The notions of characterized fuzzy T_s and of characterized fuzzy R_k-spaces are investigated and studied [9,10,26,27] for all $s\in\{0,\ 1,\ 2,\ 2\frac{1}{2},\ 3,\ 3\frac{1}{2},\ 4\}$ and $k\in\{0,\ 1,\ 2,\ 2\frac{1}{2}\}$. These characterized spaces depend only on the usual points and the operation defined on the class of all fuzzy subsets of X endowed with an fuzzy topology τ. Let the fuzzy topological space(X, τ) be fixed and φ_1, $\varphi_2 \in O_{(L}X_{,\tau)}$ then the characterized fuzzy space all fuzzy subsets of X endowed with an fuzzy topology τ. Let the fuzzy topological space (X, τ) be fixed and φ_1, $\varphi_2 \in O_{(L}{}^{X}{}_{,\tau)}$ then the characterized fuzzy space all fuzzy subsets of X endowed with an fuzzy topology τ. Let the fuzzy topological space (X, τ) be fixed and φ_1, $\varphi_2 \in O_{(L}X_{,\tau)}$ then the characterized fuzzy space (X, $\varphi_{1,2}$.int) is said to be characterized fuzzy T_1-space if for all x, y ∈ X such that (X, $\varphi_{1,2}$.int) is said to be characterized fuzzy T_1-space if for all x, y ∈ X such that $x \neq y$ there exist μ, $\eta \in LX$ and α, $\beta \in L0$ such that $\mu(x) < \alpha \le (\varphi_{1,2}.int\mu)(y)$ and $\eta(y) < \beta \le (\varphi_{1,2}.int\eta)(x)$ are hold. The related fuzzy topological space(X, τ) is said to be fuzzy $\varphi_{1,2}$-T_1 if for all $x, y \in X$ such that $x \neq y$, we have $x' /\le N\varphi_{1,2}(y)$ and $y' /\le N\varphi_{1,2}(x)$.

Proposition

Let (X, T) be an ordinary topological space and φ1, φ2 ∈ ∈ $O_{(P(X),T)}$ such that $\varphi_2 \ge i_{P(X)}$ is isotone and idempotent. Then (X, T) is $\varphi_{1,2}T_1$-space if and only if the induced characterized fuzzy space (X, $\varphi_{1,2}$.intω) is characterized fuzzy T_1 [27].

Proposition

Let (X, τ) be an fuzzy $\varphi_{1,2}$-T_1 space and φ_1, $\varphi_2 \in O_{(L}X_{,t)}$ such that φ_2 is isotone and idempotent. Then the α-level characterized space (X, $\varphi_{1,2}$.int$_\alpha$) and the initial characterized space (X, $\varphi_{1,2}$.int$_i$) are T_1-spaces [27].

Proposition

Let X be a set, let I be a class and for each i ∈ I, let the characterized fuzzy space (X_i, $\delta_{1,2}$.int$_i$) is characterized fuzzy T_1 and f_i: X → X_i be an injective mapping for some i ∈ I. Then the initial characterized fuzzy space (X, $\varphi_{1,2}$.int) of $((X_i, \delta_{1,2}.int_i))_{i\in I}$ with respect to $(f_i)_{i\in I}$ is also characterized fuzzy T_1-space [10].

Proposition

Let X be a set, let I be a class and for each $i \in I$, let the characterized fuzzy space (X_i, $\delta_{1,2}$.int$_i$) is characterized fuzzy T_1 and f_i: $X_i \to X$ be an surjective mapping for some i ∈ I. Then the final characterized fuzzy space (X, $\varphi_{1,2}$.int) of $((X_i, \delta_{1,2}.int_i))_{i\in I}$ with respect to $(f_i)_{i\in I}$ (X, $\varphi_{1,2}$.int) is characterized fuzzy T_1-space [27].

Proposition

Let the characterized fuzzy space (X, φ1,2.int) is characterized fuzzy T1 and δ1,2.int is finer than φ1,2.int. Then the characterized fuzzy space (X, δ1,2.int) is also fuzzy T1 [27].

Characterized Fuzzy $R_{2\frac{1}{2}}$ and Characterized Fuzzy R_3-Spaces

Let a fuzzy topological space(X, τ) be fixed and φ_1, $\varphi_2 \in O_{(L}X_{,\tau)}$. Then the characterized fuzzy space (X, $\varphi_{1,2}$.int) is said to be characterized fuzzy $R_{2\frac{1}{2}}$ [9] (resp. fuzzy R_3-space [10] if for all x ∈ X, F ∈ $\varphi_{1,2}$C(X) such that x/ F (resp. F_1, $F_2 \in \varphi_{1,2}C(X)$ such that $F_1 \cap F_2=\emptyset$), there exists an $\varphi_{1,2}\psi_{1,2}$-fuzzy continuous mapping $f:(X,\ \varphi_{1,2}.int)\to(I_L,\ \psi_{1,2}.int_{\Im})$ such that

$$\mu = \{\overline{\alpha}\wedge R^{\delta}|_{R_L}:\ \delta>0\ and\ \alpha\in L\}\cup\{\overline{\alpha}:\alpha\in L\},$$

$$F_1,\ F_2\in\varphi_{1,2}C(X)F_1\cap F_2=\emptyset.$$

for all y ∈ F (resp. the infimum) $\mathcal{N}_{\varphi 1,2}(F_1) \wedge \mathcal{N}_{\varphi 1,2}(F_2)$ does not exist).

Proposition 2.8 [9] Let (X, τ) be a fuzzy topological space, φ_1, φ_2 ∈

$O_{(X,\tau)}$ and Ω is a subbase for the characterized fuzzy space (X, $\varphi_{1,2}$.int$_\tau$). Then, (X, $\varphi_{1,2}$.int$_\tau$) is characterized fuzzy R_2 1_2-space if and only if for all F ∈ Ω′ and x ∈ X such that x ∈/F, there exists a $\varphi_{1,2}\psi_{1,2}$-fuzzy continuous mapping $f:(X,\ \varphi_{1,2}.int)\to(I_L,\ \psi_{1,2}.int_{\Im})$ fuzzy $T_{3\frac{1}{2}}$ and characterized fuzzy T_4-spaces such that $f(x)=\bar{1}$ and $f(y)=\bar{0}$ for all y ∈ F.

Characterized

Let a fuzzy topological space(X, τ) be fixed and $\varphi_1, \varphi_2 \in O_{(L^X,\tau)}$. Then the characterized fuzzy space (X, $\varphi_{1,2}$.int) is said to be characterized fuzzy $R_{2\frac{1}{2}}$ or characterized Tychonoff fuzzy space [9] (resp. fuzzy T_4-space [10] if and only if it is characterized fuzzy $R_{2\frac{1}{2}}$ (resp. characterized fuzzy R_3) and characterized fuzzy T_1-space. The related fuzzy topological space(X, τ) is said to be fuzzy $\varphi_{1,2}$- $T_{3\frac{1}{2}}$ (resp. fuzzy $\varphi_{1,2}$-T_4) if and only if it is fuzzy $\varphi_{1,2}$- $R_{2\frac{1}{2}}$ (resp. fuzzy $\varphi_{1,2}$-R_3) and fuzzy $\varphi_{1,2}$-T_1 space.

Proposition

Every characterized fuzzy T_4-space is characterized fuzzy $T_{3\frac{1}{2}}$ -space [9].

Metrizable Characterized Fuzzy Spaces and Characterized $T_{3\frac{1}{2}}$ -Spaces

By the fuzzy metric on the set X [6], we mean that the mapping d: X × X:→ R^*_L such that the following conditions are fulfilled:

(1) d(x, y)=0~ if and only if x=y.

(2) d(x, y)=d(y, x) for all x, y ∈ X.

(3) d(x, y) ≤ d(x, z)+ d(z, y) holds for all x, y, z ∈ X.

Where 0~ denotes the fuzzy number which has value 1 at 0 and 0 otherwise. The set X equipped with an fuzzy metric on X will be called fuzzy metric space. Each fuzzy metric on a set X generated canonically a stratified fuzzy topology τ_d which has the set B={ξ ∘ d_x: ξ ∈ μ and x ∈ X} as a base, where d_x: X → R^*_L is the mapping defied by: d_x(y)=d(x, y) and

$$\mu = \{\bar{\alpha}\wedge R^\delta|_{R_L^*}: \delta>0 \text{ and } \alpha\in L\}\cup\{\bar{\alpha}:\alpha\in L\},$$

Where $\bar{\alpha}$ has the domain is R_L^* and $R^\delta|_{R_L^*}$ is the restriction of R^δ on R_L^*. Now, consider $\varphi_1, \varphi_2 \in O_{(L^X,\tau_d)}$, then as shown in ref. [20], the characterized fuzzy space (X, $\varphi_{1,2}$.int$_{\tau d}$) is stratified. The stratified characterized fuzzy space (X, $\varphi_{1,2}$.int$_{\tau d}$) is said to be metrizable characterized fuzzy space.

In the following proposition we shall prove that every metrizable characterized fuzzy space is characterized fuzzy T_4-space in sense of Abd-Allah [10].

Proposition

Let (X, τ_d) be an stratified fuzzy topological space generated canonically by an fuzzy metric d on X and $\varphi_1, \varphi_2 \in O_{(L^X,\tau_d)}$, then the metrizable characterized fuzzy space (X, $\varphi_{1,2}$.int$_{\tau d}$) is characterized fuzzy T_4-space.

Proof: Let $F_1, F_2 \in \varphi_{1,2}C(X)$ such that $F_1\cap F_2=\varnothing$. Then for all x ∈ F_1 and y ∈ F2, we get $d(x,y)\neq 0\sim$, that is, there exists δ>0 such that d(x, y)(2δ)>0 and therefore

$$R^{2\delta}|_{R_L^*}\left(d(x,y)\right) = \left(\underset{\alpha\geq 2\delta}{\vee} d(x,y)(\alpha)\right)' < 1,$$

holds. Consider $\mu=R^\circ|_{R_L^*}\circ d_x$ and $\eta=R^\circ|_{R_L^*}\circ d_y$, then

$$\mu(x) = R^\delta|_{RL}{}^*\left(d_x(x)\right) = R^\delta|_{RL}{}^*(0^\sim) = \left(\underset{\alpha\geq\delta}{\vee}(0^\sim)(\alpha)\right)=1 \text{ for all}$$

$$x\in F_1 \text{ and } \eta(y)=R^\delta|_{RL}{}^*\left(d_y(y)\right) = R^\delta|_{RL}{}^*(0^\sim)=\left(\underset{\alpha\geq\delta}{\vee}(0^\sim)(\alpha)\right)=1$$

for all $y\in F_2$. Hence, μ and η are φ1,2-fuzzy neighborhoods in (X, φ1,2.intτd) at all x ∈ F1 and all y ∈ F2, respectively, this means $\underset{x\in F_1}{\wedge} N_{\varphi 1,2}(x)(\mu)\wedge \underset{x\in F_2}{\wedge} N_{\varphi 1,2}(y)(\eta)=1$. Because of the symmetry and triangle inequality of d and (2.2), we get $R^\delta|_{R_L^*}\left(d(x,z)\right)\wedge R^\delta|_{R_L^*}\left(d(y,z)\right)\leq R^{2\delta}|_{R_L^*}\left(d(x,y)\right)<1$ and therefore

$(\mu\wedge\eta)(z) = \left(R^\delta|_{R_L^*}\circ d_x\right)(z)\wedge\left(R^\delta|_{R_L^*} d_y\right)(z) < 1$ holds for all z ∈ X,

that is, sup (μ ∧ η)<1. Hence, the infimum Nφ1,2 (F1) ∧ Nφ1,2 (F2) does exists and therefore (X, $\varphi_{1,2}$.int$_{\tau d}$) is characterized fuzzy R_3-space. Because of Theorem 3.1 [27], it is clear that (X, $\varphi_{1,2}$.int$_{\tau d}$) is characterized fuzzy T_1-space. Consequently, (X, $\varphi_{1,2}$.int$_{\tau d}$) is characterized fuzzy T_4-space.

Example 3.1

From Propositions 2.9 and 3.1, we get that the metrizable fuzzy space in sense of Gahler and Gahler [3] is an example of a metrizable characterized fuzzy T_4-space and $k\in\{0,1,2,2\frac{1}{2},3,3\frac{1}{2}\}$ is also example of a metrizable characterized fuzzy T_k-space for

Characterized $R_{2\frac{1}{2}}$ and characterized $T_{3\frac{1}{2}}$ -spaces

In the following we introduce and study the concepts of characterized $R_{2\frac{1}{2}}$ -space and of characterized $T_{3\frac{1}{2}}$ -spaces in the classical case. Let (X, T) be an ordinary topological space and $\varphi_1, \varphi_2 \in O_{(P(X),T)}$. Then the characterized space (X, $\varphi_{1,2}$.int$_T$) is said to be characterized $R_{2\frac{1}{2}}$ -space if for all x ∈ X, F ∈ $\varphi_{1,2}$C(X) such that x/F, there exists an $\varphi_{1,2}\psi_{1,2}$ continuous mapping f: (X, $\varphi_{1,2}$.int$_T$) → (I, $\psi_{1,2}$.int$_{TI}$) such that f(x)=1 and f(y)=0 for all y ∈ F, where $\psi_{1,2}$.int$_I$ is the usual $\psi_{1,2}$-interior operator on the closed unit interval I and $\psi_1, \psi_2 \in O_{(P(I),TI)}$. Moreover, the ordinary characterized space (X, $\varphi_{1,2}$.int$_T$) is said to be characterized $T_{3\frac{1}{2}}$ -space or classical characterized-Tychonoff space if and only if it is characterized T_1-space and characterized $R_{2\frac{1}{2}}$ -space.

Proposition

Let (X, T) be an ordinary topological space and $\varphi_1, \varphi_2 \in O_{(P(X),T)}$ such that $\varphi_2 \geq i_{P(X)}$ is isotone and idempotent. Then, (X, $\varphi_{1,2}$.int$_T$) is characterized $R_{2\frac{1}{2}}$ -space if and only if the induced characterized fuzzy space (X, $\varphi_{1,2}$.int$_\omega$) is characterized fuzzy $R_{2\frac{1}{2}}$ -space.

Proof: Let (X, $\varphi_{1,2}$.int$_T$) is characterized $R_{2\frac{1}{2}}$ -space, $x\in X$ and $F\in\left(\omega\left(\varphi_{1,2}O(X)\right)\right)'$ such that x /F. Then, there exists $\varphi_{1,2}\delta_{1,2}$-continuous mapping g: $(X,\ \varphi_{1,2}.int_T)\to(I,\ \delta_{1,2}.int_{T_I})$ such that $g(x)=1$ and $g(y)=0$ for all $y\in S_\alpha S = F$ and for all $\alpha\in L_1$, where $\delta_1, \delta_2 \in O_{(P(I),T_I)}$. Hence, the mapping g: $(X,\ \varphi_{1,2}.int_\omega)\to(I,\ \delta_{1,2}.int_{\omega(T_I)})$ is $\varphi_{1,2}\delta_{1,2}$-fuzzy continuous. Consider h: $(I,\ \delta_{1,2}.int_{\omega(T_I)})\to(I_L,\ \psi_{1,2}.int_{\Im})$ is the map-ping defied by $h(z)=\bar{z}$ for all z ∈ I, then h is $\delta_{1,2}\psi_{1,2}$-fuzzy continuous and there-fore there exists an $\varphi_{1,2}\psi_{1,2}$-fuzzy continuous mapping $f=h\circ g:(X,\ \varphi_{1,2}.int_\omega)\to(I_L,\ \psi_{1,2}.int_{\Im})$ such that $f(x) = \bar{1}\, and\, f(y) = \bar{0}$ for all y ∈F. Consequently, (X, $\varphi_{1,2}$.int$_\omega$) is characterized fuzzy $R_{2\frac{1}{2}}$-space.

Conversely, let (X, $\varphi_{1,2}$.int$_\omega$) is characterized fuzzy $R_{2\frac{1}{2}}$-space, x ∈ X and F ∈ $\varphi_{1,2}$C(X) such that $x \notin F$. Then, $x \notin \chi_F$. $\chi_F \in \varphi_{1,2}C(X)$ and $\chi_F \in \left(\omega\left(\varphi_{1,2}O(X)\right)\right)'$. Therefore, there exists an $\varphi_{1,2}\psi_{1,2}$-fuzzy continuous mapping $f:(X, \varphi_{1,2}.int_\omega) \to (I_L, \psi_{1,2}.int_3)$ such that $f(x) = \bar{1}$ and $f(y) = \bar{0}$ for all $y \in \chi_F$. Since $\varphi_{1,2}int_T = (\varphi_{1,2}.int_\omega)_\alpha\, and\, \psi_{1,2}.int_3 = \psi_{1,2}.int_{T_I}$, then there could be found the mapping $f_\alpha = (X, \varphi_{1,2}.int)_T \to (I, \psi_{1,2}.int_{T_I})$ which is $\psi_{1,2}\psi_{1,2}$-continuous with $f_\alpha(x) = 1\, and\, f_\alpha(y) = 0\, for\, all\, y \in F$. Hence, $(X, \varphi_{1,2}.int_T)$ is characterized $R_{2\frac{1}{2}}$-space.

Corollary 3.1

Let (X, T) be an ordinary topological space and φ_1, φ_2 ∈ $O_{(P(X),T)}$ such that $\varphi_2 \geq i_{P(X)}$ is isotone and idempotent. Then, (X, $\varphi_{1,2}$.int$_T$) is characterized $T_{3\frac{1}{2}}$-space if and only if the induced characterized fuzzy space (X, $\varphi_{1,2}$.int$_\omega$) is characterized fuzzy $T_{3\frac{1}{2}}$-space.

Proof: Immediate from Propositions 2.3 and 3.2.

Proposition 3.2 and Corollary 3.1, show that the notions of characterized fuzzy $R_{2\frac{1}{2}}$ and characterized fuzzy $T_{3\frac{1}{2}}$-spaces are good extension as in sense of Lowen [11].

In the following proposition for each fuzzy topological space (X, τ), we show that the α-level characterized space (X, $\varphi_{1,2}$.int$_\alpha$) and the initial characterized space (X, $\varphi_{1,2}$.int$_i$) are characterized $R_{2\frac{1}{2}}$-spaces if the characterized fuzzy space (X, $\varphi_{1,2}$.int$_\tau$) is characterized fuzzy $R_{2\frac{1}{2}}$.

Proposition 3.3

Let (X, τ) be a fuzzy topological space and φ_1, φ_2 ∈ $O_{(L^X,\tau)}$ such that $\varphi_2 \geq 1_L{}^X$ is isotone and idempotent. Then the α-level characterized space (X, $\varphi_{1,2}$.int$_\alpha$) and the initial characterized space (X, $\varphi_{1,2}$.int$_i$) are characterized $R_{2\frac{1}{2}}$-spaces if (X, $\varphi_{1,2}$.int$_\tau$) is characterized fuzzy $R_{2\frac{1}{2}}$-space, there exists

Proof: Consider (X, $\varphi_{1,2}$.int$_\tau$) is characterized fuzzy $R_{2\frac{1}{2}}$-space, x ∈ X and $F \in \left(\left(\varphi_{1,2}O(X)\right)_\alpha\right)'$ such that $x \notin F$. Then $x \notin \chi_F$. and $\chi_F \in \varphi_{1,2}C(X)$. Because of (X, $\varphi_{1,2}$.int$_\tau$) is characterized fuzzy $R_{2\frac{1}{2}}$ Space,-space, there exists an $\varphi_{1,2}\psi_{1,2}$-fuzzy continuous mapping f: (X, $\varphi_{1,2}$.int$_\tau$) → (I_L, $\psi_{1,2}$.int$_I$) and f(y)=0 such that $f(x) = \bar{1}$ and $f(y) = \bar{0}$ for all $y \in X_F$. Since $\varphi_{1,2}$.in t$_\tau$= $\varphi_{1,2}$.int$_\alpha$ and $\psi_{1,2}$.int$_I$=$\psi_{1,2}$.int$_{TI}$, then there could be found the mapping f_α: (X, $\varphi_{1,2}$.int$_\alpha$) → (I, $\psi_{1,2}$.int$_{TI}$) which is $\varphi_{1,2}\psi_{1,2}$-continuous with f_α(x)=1 and f_α(y)=0 for all y ∈ F. Consequently, (X, $\varphi_{1,2}$.int$_\alpha$) is characterized $R_{2\frac{1}{2}}$ space. The second case is similarly, that is, if (X, $\varphi_{1,2}$.int$_\tau$) is characterized fuzzy $R_{2\frac{1}{2}}$-space.

Corollary 3.2

Let (X, τ) be a fuzzy topological space and φ_1, φ_2 ∈ $O_{(L}X_{,\tau)}$ such that $\varphi_2 \geq 1_L X$ is isotone and idempotent. Then the α-level characterized space (X, $\varphi_{1,2}$.int$_\alpha$) and the initial characterized space (X, $\varphi_{1,2}$.int$_i$) are characterized $T_{3\frac{1}{2}}$-spaces if the characterized fuzzy space (X, $\varphi_{1,2}$.int$_\tau$) is characterized fuzzy $T_{3\frac{1}{2}}$.

Proof: Immediate from Propositions 2.4 and 3.3.

In the following it will be shown that the finer characterized fuzzy space of a characterized fuzzy $R_{2\frac{1}{2}}$-space and of a characterized fuzzy $T_{3\frac{1}{2}}$-space is also characterized completely fuzzy $R_{2\frac{1}{2}}$-space and characterized fuzzy $T_{3\frac{1}{2}}$-space, respectively.

Proposition

Let (X, τ) is a fuzzy topological space and φ_1, φ_2∈ O(L^X, τ). If the characterized fuzzy space (X, $\varphi_{1,2}$.int$_\tau$) is characterized fuzzy $R_{2\frac{1}{2}}$ and $\delta_{1,2}$.int$_\tau$ is finer than $\varphi_{1,2}$.int$_\tau$, then (X, $\delta_{1,2}$.int$_\tau$) is also characterized fuzzy and $\delta_{1,2}$.int$_\tau$ $R_{2\frac{1}{2}}$-space.

Proof: Let Ω is a sub base for the characterized fuzzy space (X, $\varphi_{1,2}$.int$_\tau$), x ∈ X and F ∈ Ω′ such that $x \notin F$. Such that $x \notin F$ Then, there is V_1,. . ., V_n ∈ Ω such that x ∈ (V_1 ∩.. . ∩V_n) ⊆ F′ and therefore $x \notin V_i'$, V_i' ∈ Ω′ for all i ∈ {1,. . ., n}. Because of Proposition 2.8, there exists a $\varphi_{1,2}\psi_{1,2}$-fuzzy continuous mappings f_i: (X, $\varphi_{1,2}$.int$_\tau$) → (I_L, $\psi_{1,2}$.int$_I$) such that $f_i(x) = \bar{1}$ and $f_i(y)' = \bar{0}$ is also fulfilled for all $y \in (V_1' \cup \ldots \cup V_n')$. In particular this means that $f_i(x) = \bar{1}$ and $f_i(y) = \bar{0}$ for all y ∈ F and i ∈ {1,. . ., n}. Since $\delta_{1,2}$.int$_\tau$ is finer than $\varphi_{1,2}$.int$_\tau$, then any one of these mappings f_i: X → I_L gives us the required $\delta_{1,2}\psi_{1,2}$-fuzzy continuous mappings g: (X, $\delta_{1,2}$.int$_\tau$) → (I_L, $\psi_{1,2}$.int$_I$) such that $g(x) = \bar{1}$ and $g(y) = \bar{0}$ and f_i(y)=0 for all y ∈ F and i ∈ {1,. . ., n}. Since $\delta_{1,2}$.int$_\tau$ is finer than $\varphi_{1,2}$.int$_\tau$, then any one of these mappings f_i: X → I_L gives us the required $\delta_{1,2}\psi_{1,2}$-fuzzy for all y ∈ F. Consequently, (X, $\delta_{1,2}$.int$_\tau$) is characterized fuzzy $R_{2\frac{1}{2}}$ Space.

Corollary 3.3 Let (X, τ) be a fuzzy topological space and φ_1, φ_2 ∈ $O_{(L}X_{,\tau)}$. If (X, $\varphi_{1,2}$.int$_\tau$) is characterized fuzzy $T_{3\frac{1}{2}}$-space and $\delta_{1,2}$.int$_\tau$ is finer than $\varphi_{1,2}$.int$_\tau$, then (X, $\delta_{1,2}$.int$_\tau$) is also characterized fuzzy $T_{3\frac{1}{2}}$-space.

Proof: Immediate from Propositions 2.7 and 3.4.

Initial and Final Characterized Fuzzy $R_{2\frac{1}{2}}$ and Fuzzy $T_{3\frac{1}{2}}$-Spaces

In this section we are going to introduce and study the notion of initial and final characterized fuzzy $R_{2\frac{1}{2}}$-spaces and the notions of initial and final characterized fuzzy $T_{3\frac{1}{2}}$-spaces. The characterized fuzzy subspace, characterized fuzzy product space, characterized fuzzy quotient space and characterized fuzzy sum space are studied as special case from the initial and final characterized fuzzy $R_{2\frac{1}{2}}$ and fuzzy $T_{3\frac{1}{2}}$-spaces. New additional properties for the initial and final characterized fuzzy $R_{2\frac{1}{2}}$-spaces and for the initial and final characterized fuzzy $T_{3\frac{1}{2}}$-spaces are given. The categories of all characterized fuzzy $R_{2\frac{1}{2}}$ and of all characterized fuzzy $R_{2\frac{1}{2}}$-spaces will be denoted by CFR-Space and CRF-Tych, respectively. Note that the categories CFR-Space and CRF-Tych are concrete categories. The concrete categories CFR-Space and CRF-Tych are full subcategories of the category CF-Space of all characterized fuzzy spaces, which are topological over the category SET of all subsets. Hence, all the initial and final lifts exist uniquely in the categories CFR-Space and CRF-Tych, respectively.

This means that they also topological over the category SET. That is, all the initial and final characterized fuzzy $R_{2\frac{1}{2}}$-spaces and all the initial and final characterized fuzzy $T_{3\frac{1}{2}}$-spaces exist in CFR-Space and CRF-Tych, respectively.

In the following let X be a set, let I be a class and for each $i \in I$, let the characterized fuzzy space $(X_i, \delta_{1,2}.\text{int}_i)$ of all $\delta_{1,2}$-open fuzzy subsets of X_i is characterized fuzzy $R_{2\frac{1}{2}}$-space. For some $i \in I$, let $f_i: X \to X_i$ is $\varphi_{1,2}\delta_{1,2}$-closed injective mapping from X into X_i. Then we show in the following that the initial characterized fuzzy space $(X,\varphi_{1,2}.\text{int})$ of $((X_i, \delta_{1,2}.\text{int}_i))_i \in_I$ with respect to $(f_i)_{i\in I}$ is also characterized fuzzy $R_{2\frac{1}{2}}$-space. More general, we show under the same conditions, that the initial characterized fuzzy space $(X, \varphi_{1,2}.\text{int})$ of $((X_i, \delta_{1,2}.\text{int}_i))_{i\in I}$ with respect to $(f_i)_{i\in I}$ is characterized fuzzy $T_{3\frac{1}{2}}$-space if all the characterized fuzzy spaces $(X_i, \delta_{1,2}.\text{int}_i)$ are characterized fuzzy $T_{3\frac{1}{2}}$-spaces for all $i \in I$. Moreover, as special cases we show that the characterized fuzzy subspace, characterized fuzzy product space and characterized fuzzy filter pre topology of a characterized fuzzy $R_{2\frac{1}{2}}$-space and of a characterized fuzzy $T_{3\frac{1}{2}}$-space are characterized fuzzy $R_{2\frac{1}{2}}$-spaces and characterized fuzzy $T_{3\frac{1}{2}}$-spaces, respectively.

Proposition

Let X be a set and I be a class. For each $i \in I$, let the characterized fuzzy space $(X_i, \delta_{1,2}.\text{int}_i)$ of all $\delta_{1,2}$-open fuzzy subsets of X_i is characterized fuzzy $R_{2\frac{1}{2}}$-space. If $f_i: X \to X_i$ is an $\varphi_{1,2}\delta_{1,2}$-closed injective mapping from X into X_i for some $i \in I$, then the initial characterized fuzzy space $(X, \varphi_{1,2}.\text{int})$ of $((X_i, \delta_{1,2}.\text{int}_i))_{i\in I}$ with respect to $(f_i)_{i\in I}$ is also characterized fuzzy $R_{2\frac{1}{2}}$-space.

Proof: Let $x \in X$ and $F \in \varphi_{1,2}C(X)$ such that x F. Since $f_i: X \to X_i$ is $\varphi_{1,2}\delta_{1,2}$-closed injective for some $i \in I$, then $f_i(F) \in \delta_{1,2}C(X_i)$ and $f_i(x) \notin f_i(F)$. Because of $(X_i, \delta_{1,2}.\text{int}_i)$ is characterized fuzzy $R_{2\frac{1}{2}}$-space for all $i \in I$, then there

exists an $\delta_{1,2}\psi_{1,2}$-fuzzy continuous mapping g: $(X_i, \delta_{1,2}.\text{int}_i) \to (I_L, \psi_{1,2}.\text{int}_i)$ such that $g(fi(x)) = \bar{1}$ and $g(fi(x)) = \bar{0}$ for all $y \in F$. Therefor the composition $h = g\, fi$: $(X, \varphi_{1,2}.\text{int}) \to (I_L, \psi_{1,2}.\text{int}_i)$ is $\varphi_{1,2}\psi_{1,2}$-fuzzy continuous mapping such that $h(x) = (g \circ fi)(x)) = \bar{1}$ and $h(y) = (g \circ fi)(y)) = \bar{0}$ for all y F. Consequently, $(X, \varphi_{1,2}.\text{int})$ is characterized fuzzy $R_{2\frac{1}{2}}$-space.

Corollary 4.1 Let X be a set and I be a class. For each $i \in I$, let the characterized fuzzy space $(X_i, \delta_{1,2}.\text{int}_i)$ of all $\delta_{1,2}$-open fuzzy subsets of X_i is characterized fuzzy $T_{3\frac{1}{2}}$-space. If $f_i: X \to X_i$ is an $\varphi_{1,2}\delta_{1,2}$-closed injective mapping from X into X_i for some $i \in I$, then the initial characterized fuzzy space $(X, \varphi_{1,2}.\text{int})$ $((X_i, \delta_{1,2}.\text{int}_i))_{i\in I}$ of with respect to $(f_i)_{i\in I}$ is also characterized fuzzy $T_{3\frac{1}{2}}$-space.

Proof: Immediate from Propositions 2.5 and 4.1.

Corollary 4.2

The characterized fuzzy subspace $(A, \varphi_{1,2}.\text{int}_A)$ and the characterized fuzzy product space $\prod_{i\in I}(X_i, \psi_{1,2}.\text{int}_i)$ of a characterized fuzzy $R_{2\frac{1}{2}}$-space (resp. characterized fuzzy $T_{3\frac{1}{2}}$-space) are also characterized fuzzy $R_{2\frac{1}{2}}$-space (resp. characterized $T_{3\frac{1}{2}}$-space)

Proof: Follows immediately from Proposition 4.1 and Corollary 4.1. 2

As shown in ref. [7], the characterized fuzzy space $(X, \varphi_{1,2}.\text{int})$ is characterized as a fuzzy filter pre topology, then we have the following result:

Corollary 4.3

For each $i \in I$, let $\mathcal{N}^i_{\delta_{1,2}} X_i \to F_L X_i$ is $\delta_{1,2}.\text{int}_i$ as the fuzzy filter pre topology is characterized fuzzy R_2 fuzzy $T_{3\frac{1}{2}}$). Then, the representation of the initial $\varphi_{1,2}$-interior operator $\mathcal{N}_{\varphi_{1,2}} X \to F_L X$ of the initial characterized fuzzy space $(X, \varphi_{1,2}.\text{int})$ of $((X_i, \delta_{1,2}.\text{int}_i))_{i\in I}$ with respect to $(f_i)_{i\in I}$ as a fuzzy filter pre topology which is defined by:

$$\mathcal{N}_{\varphi 1,2}(x)(\mu) = \bigvee_{\mu_i \circ f_i \le \mu,\, i\in I} \mathcal{N}^i_{\delta 1,2}(f_i(x))(\mu_i)$$

for all $x \in X$ and $\mu \in L^X$ is also characterized fuzzy $R_{2\frac{1}{2}}$ (resp. characterized fuzzy $T_{3\frac{1}{2}}$).

Now, if we consider the case of I being a singleton, then we have the following results as special cases from Proposition 4.1 and Corollary 4.1.

Proposition

Let (X, τ_1) and (Y, τ_2) are two fuzzy topological spaces, $\delta_1, \delta_2 \in O_{(L^Y,\tau_2)}$ and $\delta_1, \delta_2 \in O_{(L^Y,\tau_2)}$. If the mapping f: $X \to Y$ is an $\varphi_{1,2}\delta_{1,2}$-closed injective from X into Y and $(Y, \delta_{1,2}.\text{int})$ is characterized fuzzy $R_{2\frac{1}{2}}$ terized fuzzy $T_{3\frac{1}{2}}$) L-space, then the initial characterized fuzzy space $(X(Y, \delta_{1,2}.\text{int})$ with respect to f is also characterized fuzzy $R_{2\frac{1}{2}}$ (resp. fuzzy $T_{3\frac{1}{2}}$) L-space.

Proof: Straight forward.

Corollary 4.4

Let (Y, τ_2) be an fuzzy topological spaces and $\delta_1, \delta_2 f$: $X \to Y$ is an $\varphi_{1,2}\delta_{1,2}$-closed injective mapping from X into Y fuzzy $\delta_{1,2}T_{3\frac{1}{2}}$-space), then the initial fuzzy topological space $(X, f^{-1}(\tau_2))$ of (Y, τ_2) with respect to f is fuzzy $\varphi_{1,2}$ $R_{2\frac{1}{2}}$ – space (resp. fuzzy $\varphi_{1,2}T_{3\frac{1}{2}}$-space) for all $\varphi_1, \varphi_2 \in O_{(L^X, f^{-1}(\tau_2))}$.

Proof: Follows immediately from Proposition 4.2. 2

In the following let X be a set and I be a class. For each $i \in I$, let the characterized fuzzy space $(X_i, \delta_{1,2}.\text{int}_i)$ of all $\delta_{1,2}$-open fuzzy subsets of X_i is characterized fuzzy $R_{2\frac{1}{2}}$-space. For some $i \in I$, let f_i: $X_i \to X$

is surjective mapping from X_i into X and f_i^{-1} is $\varphi_{1,2}\delta_{1,2}$-closed in the classical sense. Then as in case of the initial characterized fuzzy spaces, we show in the following that the final characterized fuzzy space (X, $\varphi_{1,2}$.int) of $((X_i, \delta_{1,2}.\text{int}_i))_{i\in I}$ with respect to$(f_i)_{i\in I}$ is also characterized fuzzy $R_{2\frac{1}{2}}$-space. More general, we show under the same conditions that, the final characterized fuzzy space (X, $\varphi_{1,2}$.int) of $((X_i, \delta_{1,2}.\text{int}_i))_{i\in I}$ with respect to $(f_i)_{i\in I}$ is characterized fuzzy $T_{3\frac{1}{2}}$ space if each of the characterized fuzzy spaces $(X_i, \delta_{1,2}.\text{int}_i)$ is characterized fuzzy $T_{3\frac{1}{2}}$-spaces for all $i \in I$. Moreover, as special cases we show that the characterized fuzzy quotient space and the characterized fuzzy sum space of the characterized fuzzy $R_{2\frac{1}{2}}$-space and of the characterized fuzzy $T_{3\frac{1}{2}}$-space are characterized fuzzy $R_{2\frac{1}{2}}$-spaces and characterized fuzzy $T_{3\frac{1}{2}}$-spaces, respectively. Proposition 4.3 Let X be a set and I be a class. For each $i \in I$, let the characterized fuzzy space $(X_i, \delta_{1,2}.\text{int}_i)$ of all $\delta_{1,2}$-open fuzzy subsets of X_i is characterized fuzzy $R_{2\frac{1}{2}}$-space. If f_i: $X_i \to X$ is an subjective $\delta_{1,2}\varphi_{1,2}$-fuzzy open mapping from X_i into X and f_i^{-1} is $\varphi_{1,2}\delta_{1,2}$-closed for some $i \in I$, then the final characterized fuzzy space (X, $\varphi_{1,2}$.int) of $((X_i, \delta_{1,2}.\text{int}_i))_{i\in I}$ with respect to $(f_i)_{i\in I}$ is also characterized fuzzy $R_{2\frac{1}{2}}$-space.

Proof: Let $x \in X$ and $F \in \varphi_{1,2}C(X)$ such that x F. Since f_i: $X_i \to X$ is surjective and f_i^{-1} is $\varphi_{1,2}\delta_{1,2}$-closed for some $i \in I$, then there exists $K \in \delta_{1,2}C(X_i)$ and $x_i \in X_i$ for which $x_i=f_i^{-1}(x)$ and $K=f_i^{-1}(F)$ such that $x_i \notin K$. Because of $(X_i, \delta_{1,2}.\text{int}_i)$ is characterized fuzzy $R_{2\frac{1}{2}}$-space for all $i \in I$, then there exists an $\delta_{1,2}\,\psi_{1,2}$ fuzzy continuous mapping g: $(X_i, \delta_{1,2}.\ \text{int}_i) \to (I_L, \psi_{1,2}.\text{int}_j)$ such that $g(x_i)=\bar{1}$ and $g(z)=\bar{0}$ for all $z \in K$, that is, $g(f_i^{-1}(x))=\bar{1}$ and $g(f_i^{-1}(s))=\bar{0}$ for all $s \in F$. Therefore, there exists a mapping $h=g\circ f_i^{-1}:(X,\varphi_{1,2}.\text{int})\to(I_L,\psi_{1,2}.\text{int}_3)$ such that $h(x)=\bar{1}$ and $h(s)=\bar{0}$ for all $s \in F$. Since f_i is $\delta_{1,2}\phi_{1,2}$ fuzzy open, then $\varphi_{1,2}\,\text{int}\,\mu\circ f_i^{-1}=f_i(\varphi_{1,2}.\text{int}\,\mu)\le\delta_{1,2}.\text{int}_i\,f_i(\mu)=\delta_{1,2}.\text{int}_i(\mu\circ f_i^{-1})$ holds for all $\in L^X$ and $i \in I$, which means that $f_i^{-1}:(X,\varphi_{1,2}.\text{int})\to(X_i,\varphi_{1,2}.\text{int}_i)$ is $\varphi_{1,2}$ $\delta_{1,2}$-fuzzy continuous. Hence, the composition $h=g\circ f_i^{-1}:(X,\varphi_{1,2}.\text{int})\to(I_L,\varphi_{1,2}.\text{int}_3)$ is $\varphi_{1,2}\psi_{1,2}$-fuzzy continuous mapping and therefore the final characterized fuzzy space (X, $\varphi_{1,2}$.int) is characterized fuzzy $R_{2\frac{1}{2}}$-space.

Corollary 4.5

Let X be a set and I be a class. For each $i \in I$, let the characterized fuzzy space $(X_i, \delta_{1,2}.\text{int}_i)$ of all $\delta_{1,2}$-open fuzzy subsets of X_i is characterized fuzzy $T_{3\frac{1}{2}}$-space. If f_i: $X_i \to X$ is an surjective $\delta_{1,2}\varphi_{1,2}$-fuzzy open mapping from X_i into X and f_i^{-1} is $\varphi_{1,2}\delta_{1,2}$-closed for some $i \in I$, then the final characterized fuzzy space (X, $\varphi_{1,2}$.int) of $((X_i, \delta_{1,2}.\text{int}_i))_{i\in I}$ with respect to $(f_i)_{i\in I}$ is also characterized fuzzy $T_{3\frac{1}{2}}$-space.

Proof: Immediate from Propositions 2.6 and 4.3. 2

Corollary 4.6

The characterized fuzzy quotient space (A, $\varphi_{1,2}$.int$_f$) and the char characterized fuzzy $T_{3\frac{1}{2}}$-space) are also characterized fuzzy $R_{2\frac{1}{2}}$ (resp. characterized fuzzy $T_{3\frac{1}{2}}$) L-spaces.

Proof: Follows immediately from Proposition 4.3 and Corollary 4.5. 2

Now, if we consider the case of I being a singleton, then we have the following results as special cases from Proposition 4.3 and Corollary 4.5.

Proposition 4.4 Let (X, τ_1) and (Y, τ_2) are two fuzzy topological spaces, $\varphi_1, \varphi_2 \in O_{(L^X, f(\tau_2))}$ and $\delta_1, \delta_2 \in O_{(L^Y, \tau_2)}$. If f: Y → X is an subjective $\delta_{1,2}\varphi_{1,2}$-fuzzy open mapping from X into Y and f^{-1} is $\varphi_{1,2}\delta_{1,2}$-closed, then the final characterized fuzzy space (X, $\varphi_{1,2}$.int) of (Y, $\delta_{1,2}$.int) with respect to f is characterized fuzzy $R_{2\frac{1}{2}}$ (resp. characterized fuzzy $T_{3\frac{1}{2}}$)L-space if (Y, $\delta_{1,2}$.int) is characterized fuzzy $R_{2\frac{1}{2}}$ (resp. characterized fuzzy $T_{3\frac{1}{2}}$) L-spaces.

Proof: Straight forward.

Corollary 4.7

Let (Y, τ_2) be an fuzzy topological spaces and $\delta_1, \delta_2 \in O_{(L^Y, \tau_2)}$, f: Y → X is an $\delta_{1,2}\varphi_{1,2}$-fuzzy open surjective mapping from Y into X and f^{-1} $\varphi_{1,2}\delta_{1,2}$-closed, then the final fuzzy topological space(X, $f(\tau_2)$) of (Y, τ_2) with respect to f is fuzzy $\varphi_{1,2}$ $R_{2\frac{1}{2}}$-space (resp. fuzzy $\varphi_{1,2}T_{3\frac{1}{2}}$)-space if (Y, τ_2) is fuzzy $\delta_{1,2}$ $R_{2\frac{1}{2}}$-space (resp. fuzzy $\delta_{1,2}T_{3\frac{1}{2}}$)-space for all $\varphi_1, \varphi_2 \in O_{(L^X, f(\tau_2))}$.

Proof: Follows immediately from Proposition 4.4. 2.

Finer Characterized Fuzzy $R_{2\frac{1}{2}}$ and Finer Characterized Fuzzy $T_{3\frac{1}{2}}$-Spaces

In this section we are going to introduce and study some finer characterized fuzzy $R_{2\frac{1}{2}}$ and finer characterized fuzzy $T_{3\frac{1}{2}}$-paces as a generalization of the weaker and stronger forms of the completely fuzzy regular and fuzzy $T_{3\frac{1}{2}}$-spaces introduced [28,12,13]. The relations between such characterized fuzzy $R_{2\frac{1}{2}}$-spaces and our characterized fuzzy $R_{2\frac{1}{2}}$-spaces which presented [9] are introduced. More generally, the relations between such characterized fuzzy $T_{3\frac{1}{2}}$-spaces and our characterized fuzzy $T_{3\frac{1}{2}}$-spaces are also introduced.

Characterized fuzzy $R_{2\frac{1}{2}}$ H and characterized fuzzy $T_{3\frac{1}{2}}$ H-spaces. In the following we introduce and study the concept of characterized completely fuzzy regular Hutton and characterized fuzzy $T_{3\frac{1}{2}}$ Hutton-spaces as a generalization of the weaker and stronger forms of the completely fuzzy regular and fuzzy $T_{3\frac{1}{2}}$-spaces in sense of Hutton [28], respectively. The relation between characterized completely fuzzy regular Hutton-spaces and the characterized fuzzy $R_{2\frac{1}{2}}$-spaces in our sense is introduced. More generally, the relations between characterized fuzzy $T_{3\frac{1}{2}}$ Hutton-spaces and the characterized fuzzy $T_{3\frac{1}{2}}$-spaces in our sense is also introduced. Let (X, τ) be a fuzzy topological space and $\varphi_1, \varphi_2 \in O_{(L^X, \tau)}$. Then the characterized fuzzy space (X, $\varphi_{1,2}$.int) is said to be characterized completely fuzzy regular Hutton-space or (characterized fuzzy $R_{2\frac{1}{2}}$ H-space, for short) if for an

$\mu \in \varphi_{1,2}OF(X)$, there exists a collection$(\eta_\alpha)_{\alpha\in L}$ in L^X and an $\varphi_{1,2}\psi_{1,2}$-fuzzy continuous mapping g: $(X, \varphi_{1,2}.\text{int}) \to (I_L, \psi_{1,2}.\text{int}_I)$ such that $\mu = \vee_{\alpha\in L} \eta_\alpha$ and $\eta_\alpha(y) \le g(y)(1-) = \wedge_{t<1} g(y)(t) \le g(y)(0_+) = \vee_{s>0} g(y)(s) \le \mu(y)$ holds for all $y \in X$. Then characterized fuzzy space $(X, \varphi_{1,2}.\text{int})$ is said to be characterized fuzzy $T_{3\frac{1}{2}}$ Hutton-space or (characterized fuzzy $T_{3\frac{1}{2}}$ H-space, for short) if and only if it is characterized fuzzy $R_{2\frac{1}{2}}$ H and characterized fuzzy $T_{3\frac{1}{2}}$-spaces.

In the classical case of $L=\{0, 1\}$, $\varphi_1=\text{int}_\tau$, $\psi_1=\text{int}_I$, 1 and $\psi_2 = 1_{L^I}$, the $\varphi_{1,2}\psi_{1,2}$-fuzzy continuity of f is up to an identification the usual fuzzy continuity of f. Then in this case the notions of characterized fuzzy $R_{2\frac{1}{2}}$ H-spaces and of characterized fuzzy $T_{3\frac{1}{2}}$ H-spaces are coincide with the notion of fuzzy completely regular spaces and the notion fuzzy $T_{3\frac{1}{2}}$-spaces defined by Hutton [28], respectively. Another special choices for the operations φ_1, φ_2, ψ_1 and ψ_2 are obtained (Table 1).

In the following proposition, we show that the characterized fuzzy $R_{2\frac{1}{2}}$-spaces which are presented [9] are more general than the characterized fuzzy $R_{2\frac{1}{2}}$ H-spaces.

Proposition 5.1

Let (X, τ) be an fuzzy topological space and $\varphi_1, \varphi_2 \in O_{(L^X,\tau)}$.

Then every characterized fuzzy $R_{2\frac{1}{2}}$ H-space $(X, \varphi_{1,2}.\text{int})$ is characterized fuzzy $R_{2\frac{1}{2}}$-space.

Proof: Let $(X, \varphi_{1,2}.\text{int})$ is characterized fuzzy $R_{2\frac{1}{2}}$ H-space, $x \in X$ and $F \in \varphi_{1,2}C(X)$ such that $x \notin F$. Then, $\chi_{F'} \in \varphi_{1,2} \in OF(X)$ and $\chi_{F'}(x) = 1$, therefore $\chi_{F'}(x) \ge \alpha$ holds for all $\alpha \in L$. Hence, $\chi_{F'} = \vee_{\alpha\in F', \alpha\in L} x_\alpha$ and therefore for all $x \in F'$, there exists a family $(x_\alpha)_{\alpha\in L}$ in L^X such that $\chi_{F'} = \vee_{\alpha\in L} x_\alpha$ and $x_\alpha(y) < g(y)(1-) < g(y)(0+) < \chi_{F'}(y)$ holds for all $y \in X$. In case of $y \in F$, we get $0 \le g(y)(1-) \le g(y)(0+) \le 0$ holds for all $y \in F$ and therefore $g(y) = \overline{0}$ for all $y \in F$. In case of $y=x$, we get $x_\alpha(x)=\alpha$ $g(x)(1-) \le g(x)(0+) \le 1$ holds for all $\alpha \in L$ and this means that $g(x)(s)=1$ for all $s < 1$ and therefore $g(y) = \overline{1}$ Consequently, $(X, \varphi_{1,2}.\text{int})$ is characterized fuzzy $R_{2\frac{1}{2}}$-space in sense [9].

Corollary 5.1 Let (X, τ) be an fuzzy topological space and $\varphi_1, \varphi_2 \in O_{(L^X,\tau)}$. Then every characterized fuzzy $T_{3\frac{1}{2}}$ H-space is characterized fuzzy $T_{3\frac{1}{2}}$-space.

Proof: Follows immediately from Proposition 5.1.

The following example shows that the inverse of Proposition 5.1 and of Corollary 5.1 is not true in general.

Example 5.1.

Let $X=\{x, y\}$ with $x \ne y$ and $\tau = \{\overline{0}, \overline{1}, x_{\frac{1}{2}}, x_1 \vee y_{\frac{1}{2}}, x_{\frac{1}{2}} \vee y_1, x_{\frac{1}{2}} \vee y_{\frac{1}{2}}\}$ is an fuzzy topology on X. Choose $\varphi_1=\text{int}_\tau$, $\varphi_2=\text{cl}_\tau$, $\psi_1=\text{int}_I$ and $\psi_2=\text{cl}_I$. Hence, $\varphi_{1,2}CF(X) = \{\overline{0}, \overline{1}, y_1, x_{\frac{1}{2}}, y_{\frac{1}{2}}, x_{\frac{1}{2}} \vee y_{\frac{1}{2}}, x_{\frac{1}{2}} \vee y_1\}$ and there is the only case of $x \in X$, $F=\{y\} \in \varphi_{1,2}C(X)$ such that $x \notin F$. Since the mapping f: $(X, \varphi_{1,2}.\text{int}_\tau) \to (I_L, \psi_{1,2}.\text{int}_I)$ which is defined by $f(x) = \overline{1}$ and $f(y) = \overline{0}$ for all $y \ne x$ is $\varphi_{1,2}\psi_{1,2}$-fuzzy continuous, then $(X, \varphi_{1,2}.\text{int}_\tau)$ is characterized fuzzy $R_{2\frac{1}{2}}$-space in sense [9]. Obviously, $(X, \varphi_{1,2}.\text{int}_\tau)$ is characterized fuzzy T1-space, therefore $(X, \varphi_{1,2}.\text{int}_\tau)$ is characterized fuzzy $T_{3\frac{1}{2}}$-space.

On other hand, let $(X, \varphi_{1,2}.\text{int}\tau)$ is characterized fuzzy $T_{3\frac{1}{2}}$ H-space, then$(X, \varphi_{1,2}.\text{int}_\tau)$ is characterized fuzzy $R_{2\frac{1}{2}}$ H and characterized fuzzy T_1-space. Since $x_{\frac{1}{2}} \in \tau = \varphi_{1,2}OF(X)$ and $x_{\frac{1}{2}} = \vee_{\alpha\in L}\left(\frac{1}{2} \wedge x\alpha\right)$ then there exists a collection $(\eta_\alpha)_{\alpha\in L} = \vee_{\alpha\in L}\left(\frac{1}{2} \wedge x_\alpha\right)_{\alpha\in L}$ such that $x_{\frac{1}{2}} = \vee_{\alpha\in L} \eta_\alpha$. Moreover, for an $\varphi_{1,2}\psi_{1,2}$-fuzzy continuous mapping f: $(X, \varphi_{1,2}.\text{int}\tau) \to (I_L, \psi_{1,2}.\text{int}_I)$ such that $f(x) = \overline{1}$ and $f(y) = \overline{0}$ for all $y \ne x$, we get the inequality

$$\eta_\alpha(z) \le f(z)(1-) \le f(z)(0+) \le x_{\frac{1}{2}}(z)$$

holds only when $z=y$, but it is not holds when $z=x$, because $(\frac{1}{2} \wedge \alpha) \le 1 \le \frac{1}{2}$ and this is a contradiction. Hence, $(X, \varphi_{1,2}.\text{int}_\tau)$ is not characterized fuzzy $R_{2\frac{1}{2}}$ H-space and therefore it is not characterized fuzzy $T_{3\frac{1}{2}}$ H-space.

Characterized fuzzy $R_{2\frac{1}{2}}$ K and characterized fuzzy $T_{3\frac{1}{2}}$ K-spaces.

In the following we introduce and study the concept of characterized completely fuzzy regular Katasars spaces and characterized fuzzy $T_{3\frac{1}{2}}$ Katasars spaces as a generalization of the weaker and stronger forms of the completely fuzzy regular and fuzzy $T_{3\frac{1}{2}}$-spaces introduced by Katasars [13], respectively. The relation between characterized fuzzy completely regular Katasars spaces and the characterized fuzzy $R_{2\frac{1}{2}}$-spaces in sense Abd-Allah and Khedhairi [9] is introduced. More generally, the relations between characterized fuzzy $T_{3\frac{1}{2}}$ Katasars spaces and the characterized fuzzy $T_{3\frac{1}{2}}$-spaces in sense of [9] is also introduced.

Let (X, τ) be an fuzzy topological space and $\varphi_1, \varphi_2 \in O_{(L^X,\tau)}$. Then the characterized fuzzy space $(X, \varphi_{1,2}.\text{int})$ is said to be characterized completely fuzzy regular Katasars-space or (characterized fuzzy $R_{2\frac{1}{2}}$ K-space, for short) if for every $x \in X$ and $\mu \in L^X$ such that $\mu(x)>\alpha$, $\alpha \in L_0$, there exists an $\varphi_{1,2}\psi_{1,2}$-fuzzy continuous mapping g: $(X, \varphi_{1,2}.\text{int}) \to (I_L, \psi_{1,2}.\text{int}_I)$ such that $g(y)(0+) \le \mu(y)$ and $g(y)(1-)>\alpha$ are holds for all $y \in X$ and $\alpha \in L_0$. The characterized fuzzy space $(X, \varphi_{1,2}.\text{int})$ is said to be characterized fuzzy $T_{3\frac{1}{2}}$ Katasars-space or (characterized fuzzy $T_{3\frac{1}{2}}$ K-space, for short) if and only if it is characterized fuzzy $R_{2\frac{1}{2}}$ K-space and characterized fuzzy T_1-space.

In the classical case of $L=\{0, 1\}$, $\varphi_1=\text{int}_\tau$, $\psi_1=\text{int}_I$, $\psi_2 = 1_{L^I}$ and $\psi_2 = 1_{L^I}$, the $\varphi_{1,2}\psi_{1,2}$-fuzzy continuity of f is up to an identification the usual fuzzy continuity of f. Then in this case the notions of characterized fuzzy $R_{2\frac{1}{2}}$ K-space and of characterized fuzzy $T_{3\frac{1}{2}}$ K-spaces are coincide with the notion of completely fuzzy regular spaces and the notion of fuzzy $T_{3\frac{1}{2}}$ -spaces presented by Katasars [13], respectively. Another special choices for the operations φ_1, φ_2, ψ_1 and ψ_2 are obtained in Table 1. In the following proposition we show that the notion of characterized fuzzy $R_{2\frac{1}{2}}$ -spaces which are presented [9] are more general than the characterized fuzzy $R_{2\frac{1}{2}}$ K-spaces.

Proposition

Let (X, τ) be an fuzzy topological space and $\varphi_1, \varphi_2 \in O_{(L^X,\tau)}$. Then every characterized fuzzy $R_{2\frac{1}{2}}$ *K*-space $(X, \varphi_{1,2}.\text{int})$ is characterized fuzzy $R_{2\frac{1}{2}}$ space.

Proof: Let $(X, \varphi_{1,2}.\text{int})$ is a characterized fuzzy $R_{2\frac{1}{2}}$ *K*-space, *x X* and $F \in \varphi_{1,2}C(X)$ such that $x \notin F$. Then, $\chi_{F'}(x) = 1$ and $\chi_{F'}(x) = 1$, therefore $\chi_{F'}(x) \geq \alpha$ holds for all $\alpha \in L$. Because of $(X, \varphi_{1,2}.\text{int})$ is characterized fuzzy $R_{2\frac{1}{2}}$ *K*-space, then there exists a $\varphi_{1,2}\psi_{1,2}$-fuzzy continuous mapping g: $(X, \varphi_{1,2}.\text{int}) \to (I_L, \psi_{1,2}.\text{int}_1)$ such that $\bigvee_{t>0} g(y)(t) \leq \chi_{F'}(y)$ and $\bigvee_{s>1} g(y)(s) > \alpha$ are hold for all $y \in X$ and $\alpha \in L$. In case of $y \in F$, we have $\bigvee_{t>0} g(y)(t) \leq 0$, that is, $g(y)(t)=0$ for all $t>0$, $y \in F$ and therefore $g(y)\ \ \bar{0}$ for all $y \in F$. In case of y=x, we have $\bigvee_{s>1} g(x)(s)| > \alpha$ holds for all $\alpha \in L$, and therefore $g(x) = \bar{1}$. Hence, there exists a $\varphi_{1,2}\psi_{1,2}$-fuzzy continuous mapping g: $(X, \varphi_{1,2}.\text{int}) \to (I_L, \psi_{1,2}.\text{int}_1)$ such that $g(y) = \bar{0}$ and $g(y) = \bar{0}$ for all $y \in F$. Consequently, $(X, \varphi_{1,2}.\text{int})$ is characterized fuzzy $R_{2\frac{1}{2}}$ -space in sense [9].

Corollary 5.2 Let (X, τ) be an fuzzy topological space and $\varphi_1, \varphi_2 \in O_{(L^X,\tau)}$. Then every characterized fuzzy $T_{3\frac{1}{2}}$ K-space is characterized fuzzy $T_{3\frac{1}{2}}$ -space.

Proof: Follows immediately from Proposition 5.2.

The following example shows that the inverse of Proposition 5.2 and of Corollary 5.2 is not true in general.

Example 5.2.

Consider the characterized fuzzy space $(X, \varphi_{1,2}.\text{int}_\tau)$ which is defined in Example 5.1, then as shown in Example 5.1, $(X, \varphi_{1,2}.\text{int}_\tau)$ is characterized fuzzy $R_{2\frac{1}{2}}$ -space in sense [9] and characterized fuzzy T_1-space, therefore $(X, \varphi_{1,2}.\text{int}_\tau)$ is characterized fuzzy $T_{3\frac{1}{2}}$ -space in sense [9].

On other hand, for any $\varphi_{1,2}\psi_{1,2}$-fuzzy continuous mapping *f*: $(X, \varphi_{1,2}.\text{int}_\tau) \to (I_L, \psi_{1,2}.\text{int}_1)$ such that $f(y) = \bar{1}$ and $f(y) = \bar{0}$ for all $y \neq x$, we shall consider $x_{\frac{1}{2}} \in \varphi_{1,2}.OF(X)$ with $x_{\frac{1}{2}}(x) = \frac{1}{2} > 0$, that is, there exists some $\alpha = \frac{1}{2} \in L$ such that $x_{\frac{1}{2}}(x) = \alpha$. Therefore, $f(z)(1-) = \bigwedge_{t<1} f(z)(t) > \frac{1}{2}$ holds only when $z=x$ and it is not fulfilled when $z=y$. Moreover, $f(z)(0+) = \bigwedge_{s>0} f(z)(s) \leq x_{\frac{1}{2}}(z)$ holds only when $z=y$ and it is not fulfilled when $z=x$. Hence, $(X, \varphi_{1,2}.\text{int}_\tau)$ is not characterized fuzzy $R_{2\frac{1}{2}}$ *K*-space and therefore it is not characterized fuzzy $T_{3\frac{1}{2}}$ *K*-space.

Characterized Fuzzy $R_{2\frac{1}{2}}$ KE and Characterized Fuzzy $T_{3\frac{1}{2}}$ KE-Spaces

In the following we introduce and study the concepts of characterized completely fuzzy regular Kandil and Shafee spaces and of characterized fuzzy $T_{3\frac{1}{2}}$ Kandil and Shafee spaces as a generalization of the weaker and stronger forms of the completely fuzzy regular and fuzzy $R_{2\frac{1}{2}}$-spaces presented by Kandil and Shafee [12], respectively. The relation between characterized completely fuzzy regular Kandil and Shafee spaces and the characterized fuzzy $R_{2\frac{1}{2}}$ -spaces which are presented [6]. More generally, the relations between characterized fuzzy $T_{3\frac{1}{2}}$ Kandil El-Shafee-spaces and the characterized fuzzy $T_{3\frac{1}{2}}$ -spaces in sense [9] is also introduced.

Let (X, τ) be an fuzzy topological space and $\varphi_1, \varphi_2 \in O_{(L^X,\tau)}$. Then the characterized fuzzy space $(X, \varphi_{1,2}.\text{int})$ is said to be characterized completely fuzzy regular Kandil and Shafee space or (characterized fuzzy $R_{2\frac{1}{2}}$ KE-space, for short) if for every $x_\alpha \in S(X)$ and $\mu \in \varphi_{1,2}CF(X)$ such that $x_\alpha \bar{q} \mu$, there exists an $\varphi_{1,2}\psi_{1,2}$-fuzzy continuous mapping *f*: $(X, \varphi_{1,2}.\text{int}) \to (I_L, \psi_{1,2}.\text{int}_1)$ such that $f(y)(0+) \leq \mu'(y)$ and $f(y)(1-) \geq x_\alpha(y)$ are hold for all $y \in X$ and $\alpha \in L$. The characterized fuzzy space $(X, \varphi_{1,2}.\text{int})$ is said to characterized quasi fuzzy T_1-space or (characterized QFT_1-space, for short) if for all $x, y \in X$ such that $x \neq y$ we have $x_\alpha \bar{q} \varphi_{1,2}.\text{cl}y_\beta$ and $\varphi_{1,2}.\text{cl}x_\alpha \bar{q} y_\beta$ for all $\alpha, \beta \in L$. As easily seen that every characterized QFT_1-space is characterized fuzzy T_1-space. The characterized fuzzy space $(X, \varphi_{1,2}.\text{int})$ is said to be characterized fuzzy $T_{3\frac{1}{2}}$ Kandil El-Shafee-space or (characterized fuzzy $T_{3\frac{1}{2}}$ KE-space, for short) if and only if it is characterized fuzzy $T_{3\frac{1}{2}}$ KE and characterized QFT_1-spaces. Obviously, every characterized fuzzy $T_{3\frac{1}{2}}$ KE-space is characterized fuzzy $T_{3\frac{1}{2}}$ K-space. In the classical case of $L=\{0, 1\}$, $\varphi_1=\text{int}_\tau$, $\psi_1=\text{int}_I$, $\varphi_2 = 1_{L^X}$ and $\psi_2 = 1_{L^I}$, the $\varphi_{1,2}\psi_{1,2}$-fuzzy continuity of f is up to an identification the usual fuzzy continuity of *f*. Hence, the notions of characterized fuzzy $R_{2\frac{1}{2}}$ KE-spaces and of characterized fuzzy $T_{3\frac{1}{2}}$ KE-spaces are coincide with the notion of completely fuzzy regular spaces and the notion fuzzy fuzzy $T_{3\frac{1}{2}}$ -spaces presented by Kandil and Shafee [12], respectively. Another special choices for the operations φ_1, φ_2, ψ_1 and ψ_2 are obtained in Table 1.

In the following proposition we show that the characterized fuzzy $R_{2\frac{1}{2}}$ -spaces which are presented [9] are more general than the characterized fuzzy $R_{2\frac{1}{2}}$ KE-spaces.

Proposition 5.3

Let (X, τ) be an fuzzy topological space and $\varphi_1, \varphi_2 \in O_{(L^X,\tau)}$ Then every characterized fuzzy $R_{2\frac{1}{2}}$ KE-space $(X, \varphi_{1,2}.\text{int})$ is characterized fuzzy $R_{2\frac{1}{2}}$ -space.

Proof: Let $(X, \varphi_{1,2}.\text{int})$ is a characterized fuzzy $R_{2\frac{1}{2}}$ KE-space, $x \in X$ and $F \in \varphi_{1,2}C(X)$ such that $x \notin F$. Then, $\chi_{F'} \in \varphi_{1,2}OF(X)$ and $\chi_{F'}(x) = 1$, therefore $x_1 \bar{q} \chi_F$. Because of $(X, \varphi_{1,2}.\text{int})$ is characterized fuzzy $R_{2\frac{1}{2}}$ KE-space, then there exists a $\varphi_{1,2}\psi_{1,2}$-fuzzy continuous mapping f: $(X, \varphi_{1,2}.\text{int}) \to (IL, \psi_{1,2}.\text{int}_I)$ such that $f(y)(0+) \leq \chi_{F'}(y)$ and $f(y)(1-) \geq x_1(y)$ are hold for all $y \in X$. In case of $y \in F$, we have $0 \leq f(y)(1-) \leq f(y)(0+) \leq 0$, that is, $f(y)(s)=0$ for all $s>0$and therefore $f(y)=\bar{0}$ for all $y \in F$. In case of y=x, we have $1 \leq f(x)(1-) \leq f(x)(0+) \leq 1$ holds and then $f(x)(s)=1$ for all $s < 1$, therefore $f(x)=\bar{1}$. Hence, there exists a $\varphi_{1,2}\psi_{1,2}$-fuzzy continuous mapping f: $(X, \varphi_{1,2}.\text{int}) \to (I_L, \psi_{1,2}.\text{int}_I)$ such that $f(x)=\bar{1}$ and $f(y)=\bar{0}$ for all $y \in F$. Consequently, $(X, \varphi_{1,2}.\text{int})$ is characterized fuzzy $R_{2\frac{1}{2}}$-space in sense [9].

Corollary 5.3

Let (X, τ) be an fuzzy topological space and $\varphi_1, \varphi_2 \in O_{(L^X,\tau)}$. Then every characterized fuzzy $T_{3\frac{1}{2}}$ KE-space is characterized fuzzy $T_{3\frac{1}{2}}$-space.

Proof: Follows immediately from Proposition 5.3 and the fact that every characterizedQFT$_1$-space is characterized fuzzy T_1-space.

The following example shows that the inverse of Proposition 5.3 and Corollary 5.3 are not true in general.

Example 5.3.

Consider the characterized fuzzy space $(X, \varphi_{1,2}.\text{int}_\tau)$ which is defined in Example 5.1, then as shown in Example 5.1, $(X, \varphi_{1,2}.\text{int}_\tau)$ is characterized fuzzy $R_{2\frac{1}{2}}$-space in sense [9] and characterized fuzzy T_1-space, therefore $(X, \varphi_{1,2}.\text{int}_\tau)$ is characterized fuzzy $T_{3\frac{1}{2}}$-space in sense [9].

Now, choose $x_{\frac{1}{2}} \in S(X)$ and $\mu = x_{\frac{1}{2}} \in \varphi_{1,2}CF(X)$ then $\mu' = x_{\frac{1}{2}} \vee y_1 \in \varphi_{1,2}OF(X)$ such that $x_{\frac{1}{2}}\bar{q}\mu$ .Hence, for any $\varphi_{1,2}\psi_{1,2}$-fuzzy continuous mapping f: $(X, \varphi_{1,2}.\text{int}_\tau) \to (I_L, \psi_{1,2}.\text{int}_I)$ such that $f(x)=\bar{1}$ and $f(y)=\bar{0}$ for all $y \neq x$, we get $x_{\frac{1}{2}}(z) \leq f(z)(1-) = \bigwedge_{\tau<1} f(z)(t)$ holds for all $z \in X$. But $\mu'(z) = \left(x_{\frac{1}{2}} \vee y_1\right)(z) \geq f(z)(0+) \bigwedge_{\sigma>0} f(z)(s)$ holds only for $z=y$ and it is not fulfilled for $z=x$. Consequently, $(X, \varphi_{1,2}.\text{int}_\tau)$ is not characterized fuzzy $R_{2\frac{1}{2}}$ KE-space and therefore it is not characterized fuzzy $T_{3\frac{1}{2}}$ KE-space.

Conclusion

In this paper, basic notions related to the characterized fuzzy $R_{2\frac{1}{2}}$ and the characterized fuzzy $T_{3\frac{1}{2}}$-spaces which are presented [9] are introduced and studied. These notions are named metrizable characterized fuzzy spaces, initial and final characterized fuzzy spaces, some finer characterized fuzzy $R_{2\frac{1}{2}}$ and characterized fuzzy $T_{3\frac{1}{2}}$-spaces. The metrizable characterized fuzzy space is introduced as a generalization of the weaker and stronger forms of the fuzzy metric space introduced by Gahler and Gahler [3]. For every stratified fuzzy topological space generated canonically by an fuzzy metric we proved that, the metrizable characterized fuzzy space is characterized fuzzy T_4-space in sense of Abd-Allah [10] and therefore, it is characterized fuzzy $R_{2\frac{1}{2}}$ and characterized fuzzy $T_{3\frac{1}{2}}$-space. The induced characterized fuzzy space is characterized fuzzy $R_{2\frac{1}{2}}$ and characterized fuzzy $T_{3\frac{1}{2}}$-space if and only if the related ordinary topological space is $\varphi_{1,2}$ $R_{2\frac{1}{2}}$-space and $\varphi_{1,2}$ $T_{3\frac{1}{2}}$-space, respectively. Hence, the notions of characterized fuzzy $R_{2\frac{1}{2}}$ and of characterized fuzzy $T_{3\frac{1}{2}}$ are good extension in sense of Lowen [11]. Moreover, the α-level characterized space and the initial characterized space are characterized -space and characterized $T_{3\frac{1}{2}}$-space if the related characterized fuzzy space is characterized fuzzy $R_{2\frac{1}{2}}$-space and characterized fuzzy $T_{3\frac{1}{2}}$-space, respectively. We shown that the finer characterized fuzzy space of a characterized fuzzy $R_{2\frac{1}{2}}$-space and of a characterized fuzzy $T_{3\frac{1}{2}}$-space is also characterized fuzzy $R_{2\frac{1}{2}}$ and characterized fuzzy $T_{3\frac{1}{2}}$-space, respectively. The categories of all characterized fuzzy $R_{2\frac{1}{2}}$ and of all characterized fuzzy $T_{3\frac{1}{2}}$-spaces will be denoted by CFR-Space and CRF-Tych and they are concrete categories. These categories are full subcategories of the category CF-Space of all characterized fuzzy spaces, which are topological over the category SET of all subsets and hence all the initial and final lifts exist uniquely in CFR-Space and CRF-Tych, respectively. That is, all the initial and final characterized fuzzy $R_{2\frac{1}{2}}$-spaces exist in CFR-Space and also all the initial and final characterized fuzzy $T_{3\frac{1}{2}}$-spaces exist in CRF-Tych. We shown that the initial and final characterized fuzzy spaces of a characterized fuzzy $R_{2\frac{1}{2}}$-space and of characterized fuzzy $T_{3\frac{1}{2}}$-space are characterized fuzzy $R_{2\frac{1}{2}}$ and characterized fuzzy $T_{3\frac{1}{2}}$-spaces, respectively. As special cases, the characterized fuzzy subspace, characterized fuzzy product space, characterized fuzzy quotient space and characterized fuzzy sum space of a characterized fuzzy $R_{2\frac{1}{2}}$-space and of a characterized fuzzy $T_{3\frac{1}{2}}$-space are also characterized fuzzy $R_{2\frac{1}{2}}$ and characterized fuzzy $T_{3\frac{1}{2}}$-spaces, respectively. Finally, we introduced and studied three finer characterized fuzzy $R_{2\frac{1}{2}}$ and three finer characterized fuzzy $T_{3\frac{1}{2}}$ L-spaces as a generalization of the weaker and stronger forms of the completely regular and the fuzzy $T_{3\frac{1}{2}}$-spaces introduced [28,12,13]. These fuzzy spaces are named characterized fuzzy $R_{2\frac{1}{2}}$ H, characterized fuzzy $R_{2\frac{1}{2}}$ K, characterized fuzzy $R_{2\frac{1}{2}}$ KE, characterized fuzzy $T_{3\frac{1}{2}}$ H, characterized fuzzy $T_{3\frac{1}{2}}$ K and characterized fuzzy $T_{3\frac{1}{2}}$ KE-spaces. The relations between characterized fuzzy $R_{2\frac{1}{2}}$ H, characterized fuzzy $R_{2\frac{1}{2}}$ K, characterized fuzzy $R_{2\frac{1}{2}}$ KE-spaces and the characterized fuzzy $R_{2\frac{1}{2}}$-space which are presented [9] are introduced. More generally, the relations between characterized fuzzy $T_{3\frac{1}{2}}$ H, characterized fuzzy $T_{3\frac{1}{2}}$ K, characterized fuzzy $T_{3\frac{1}{2}}$ KE-spaces and the characterized fuzzy $T_{3\frac{1}{2}}$-spaces are also introduced. Meany special cases from these finer characterized fuzzy $R_{2\frac{1}{2}}$ and finer characterized fuzzy $T_{3\frac{1}{2}}$-spaces are listed in Table 1.

	Operations	Char.fuzzy $R_{2\frac{1}{2}}$ H-space	Char.fuzzy $R_{2\frac{1}{2}}$ K-space	Char.fuzzy $R_{2\frac{1}{2}}$ KE-space	Char.fuzzy $T_{3\frac{1}{2}}$ H-space	Char.fuzzy $T_{3\frac{1}{2}}$ K-space	Char.fuzzy $T_{3\frac{1}{2}}$ KE-space
1	$\varphi1=int\tau$, $\varphi2=1_LX$ $\psi1=intl$, $\psi2=1_LI$	Fuz. $R_{2\frac{1}{2}}$ H space [10,21]	Fuz. $R_{2\frac{1}{2}}$ K space [10,25]	Fuz. $R_{2\frac{1}{2}}$ KE space [10,23]	Fuz. $T_{3\frac{1}{2}}$ H space [10,21]	Fuz. $T_{3\frac{1}{2}}$ K space [10,25]	Fuz. $T_{3\frac{1}{2}}$ KE space [10,23]
2	$\varphi_1=int_\tau$, $\varphi_2=cl_\tau$ $\psi_1=int_l, \psi_2=cl_l$	Fuz.θ $R_{2\frac{1}{2}}$ H-space	Fuz.θ $R_{2\frac{1}{2}}$ K-space	Fuz.θ $R_{2\frac{1}{2}}$ KE-space	Fuz.θ $T_{3\frac{1}{2}}$ H-space	Fuz.θ $T_{3\frac{1}{2}}$ K-space	Fuz.θ $T_{3\frac{1}{2}}$ KE-space
3	$\varphi_1=int_\tau$, $\varphi_2=int_\tau \circ cl_\tau$ $\psi_1=int_l, \psi_2=int_l \circ cl_l$	Fuz.δ $R_{2\frac{1}{2}}$ H-space	Fuz.δ $R_{2\frac{1}{2}}$ K-space	Fuz.δ $R_{2\frac{1}{2}}$ KE-space	Fuz.δ $T_{3\frac{1}{2}}$ H-space	Fuz.δ $T_{3\frac{1}{2}}$ K-space	Fuz.δ $T_{3\frac{1}{2}}$ KE-space
4	$\varphi1=int\tau$, $\varphi2=1_LX$ $\psi_1=int_l, \psi_2=cl_l$	Fuz.W $R_{2\frac{1}{2}}$ H-space	Fuz.W $R_{2\frac{1}{2}}$ K-space	Fuz.W $R_{2\frac{1}{2}}$ KE-space	Fuz.W $T_{3\frac{1}{2}}$ H-space	Fuz.W $T_{3\frac{1}{2}}$ K-space	Fuz.W $T_{3\frac{1}{2}}$ KE-space
5	$\varphi_1=int_\tau$, $\varphi_2=cl_\tau$ $\psi1=intl$, $\psi2=1_LI$	Fuz.S.θ $R_{2\frac{1}{2}}$ H-space	Fuz.S.θ – K-space	Fuz.S.θ $R_{2\frac{1}{2}}$ KE-space	Fuz.S.θ $T_{3\frac{1}{2}}$ H-space	Fuz.S.θ $T_{3\frac{1}{2}}$ K-space	Fuz.S.θ $T_{3\frac{1}{2}}$ KE-space
6	$\varphi1=int\tau$, $\varphi2=1_LX$ $\psi_1=int_l, \psi_2=int_l \circ cl_l$	Fuz.A $R_{2\frac{1}{2}}$ H-space	Fuz.A $R_{2\frac{1}{2}}$ K-space	Fuz.A $R_{2\frac{1}{2}}$ KE-space	Fuz.A $T_{3\frac{1}{2}}$ H-space	Fuz.A $T_{3\frac{1}{2}}$ K-space	Fuz.A $T_{3\frac{1}{2}}$ KE-space
7	$\varphi_1=int_\tau$, $\varphi_2=cl_{\tau\ \psi1}$ $\psi_1=int_l, \psi_2=int_l \circ cl_l$	Fuz.A.S.θ $R_{2\frac{1}{2}}$ H-space	Fuz.A.S.θ $R_{2\frac{1}{2}}$ K-space	Fuz.A.S.θ $R_{2\frac{1}{2}}$ KE-space	Fuz.A.S.θ $T_{3\frac{1}{2}}$ H-space	Fuz.A.S.θ $T_{3\frac{1}{2}}$ K-space	Fuz.A.S.θ $T_{3\frac{1}{2}}$ KE-space
8	$\varphi_1=int_\tau$, $\varphi_2=int_\tau \circ cl_\tau$	Fuz. super $R_{2\frac{1}{2}}$ H-space	Fuz. super $R_{2\frac{1}{2}}$ K-space	Fuz. super $R_{2\frac{1}{2}}$ KE-space	Fuz. super $T_{3\frac{1}{2}}$ H-space	Fuz. super $T_{3\frac{1}{2}}$ K-space	Fuz. super $T_{3\frac{1}{2}}$ KE-space
	$\varphi_1=int_\tau$, $\varphi_2=int_\tau \circ cl_\tau$ $\psi_1=int_l, \psi_2=cl_l$	Fuz.W.θ $R_{2\frac{1}{2}}$ H-space	Fuz.W.θ $R_{2\frac{1}{2}}$ K-space	Fuz.W.θ $R_{2\frac{1}{2}}$ KE-space	Fuz.W.θ $T_{3\frac{1}{2}}$ H-space	Fuz.W.θ $T_{3\frac{1}{2}}$ K-space	Fuz.W.θ $T_{3\frac{1}{2}}$ KE-space
10	$\varphi1=cl\tau \circ int\tau$, $\varphi2=1_LX$ $\psi1=intl$, $\psi2=1_LI$	Fuz.semi $R_{2\frac{1}{2}}$ H-space	Fuz.semi $R_{2\frac{1}{2}}$ K-space	Fuz.semi $R_{2\frac{1}{2}}$ KE-space	Fuz.semi $T_{3\frac{1}{2}}$ H-space	Fuz.semi $T_{3\frac{1}{2}}$ K-space	Fuz.semi – KE-space
11	$\varphi1=cl\tau \circ int\tau$, $\varphi2=1_LX$ $\psi1=cll \circ intl$, $\psi2=1_LI$	Fuz.irr. $R_{2\frac{1}{2}}$ H-space	Fuz.irr. $R_{2\frac{1}{2}}$ K-space	Fuz.irr. $R_{2\frac{1}{2}}$ KE-space	Fuz.irr. $T_{3\frac{1}{2}}$ H-space	Fuz.irr. $T_{3\frac{1}{2}}$ K-space	Fuz.irr. $T_{3\frac{1}{2}}$ KE-space
12	$\varphi1=cl\tau \circ int\tau$, $\varphi2=1_LX$ $\psi_1=cl_l \circ int_l$, $\psi_2=Scl_l$	Fuz. semi -irr. $R_{2\frac{1}{2}}$ H-space	Fuz. semi –irr. $R_{2\frac{1}{2}}$ K-space	Fuz. semi -irr. $R_{2\frac{1}{2}}$ KE-space	Fuz. semi-irr. $T_{3\frac{1}{2}}$ H-space	Fuz. semi -irr. $T_{3\frac{1}{2}}$ K-space	Fuz. semi -irr. $T_{3\frac{1}{2}}$ KE-space
13	$\varphi_1=cl_\tau \circ int_\tau$, $\varphi_2=Scl_\tau$ $\psi1=cll \circ intl$, $\psi2=1_LI$	Fuz.S-irr. $R_{2\frac{1}{2}}$ H-space	Fuz.S-irr. $R_{2\frac{1}{2}}$ K-space	Fuz.S-irr. $R_{2\frac{1}{2}}$ KE-space	Fuz.S-irr. $T_{3\frac{1}{2}}$ H-space	Fuz.S-irr. $T_{3\frac{1}{2}}$ K-space	Fuz.S-irr. $T_{3\frac{1}{2}}$ KE-space
14	$\varphi1=int\tau \circ cl\tau \circ int\tau$, $\varphi2=1_LX$ $\psi1=intl$, $\psi2=1_LI$	Fuz.λ $R_{2\frac{1}{2}}$ H-space	Fuz.λ $R_{2\frac{1}{2}}$ K-space	Fuz.λ $R_{2\frac{1}{2}}$ KE-space	Fuz.λ $T_{3\frac{1}{2}}$ H-space	Fuz.λ $T_{3\frac{1}{2}}$ K-space	Fuz.λ $T_{3\frac{1}{2}}$ KE-space
15	$\varphi1=int\tau \circ cl\tau$, $\varphi2=1_LX$ $\psi1=intl$, $\psi2=1_LI$	Fuz.pre $R_{2\frac{1}{2}}$ H-space	Fuz.pre $R_{2\frac{1}{2}}$ K-space	Fuz.pre $R_{2\frac{1}{2}}$ KE-space	Fuz.pre $T_{3\frac{1}{2}}$ H-space	Fuz.pre $T_{3\frac{1}{2}}$ K-space	Fuz.pre $T_{3\frac{1}{2}}$ KE-space
16	$\varphi1=cl\tau \circ int\tau \circ cl\tau$, $\varphi2=1_LX$ $\psi1=intl$, $\psi2=1_LI$	Fuz.β $R_{2\frac{1}{2}}$ H-space	Fuz.β $R_{2\frac{1}{2}}$ K-space	Fuz.β $R_{2\frac{1}{2}}$ KE-space	Fuz.β $T_{3\frac{1}{2}}$ H-space	Fuz.β $T_{3\frac{1}{2}}$ K-space	Fuz.β $T_{3\frac{1}{2}}$ KE-space
17	$\varphi1=cl\tau \circ int\tau$, $\varphi2=1_LX$ $\psi_1=int_l, \psi_2=cl_l$	Fuz. W semi $R_{2\frac{1}{2}}$ H-space	Fuz. W semi $R_{2\frac{1}{2}}$ K-space	Fuz. W semi $R_{2\frac{1}{2}}$ KE-space	Fuz. W semi – H-space	Fuz. W semi $T_{3\frac{1}{2}}$ K-space	Fuz. W semi $T_{3\frac{1}{2}}$ KE-space
18	$\varphi1=int\tau \circ cl\tau$, $\varphi2=1_LX$ $\psi_1=int_l, \psi_2=cl_l$	Fuz. W pre $R_{2\frac{1}{2}}$ H-space	Fuz. W pre $R_{2\frac{1}{2}}$ K-space	Fuz. W pre $R_{2\frac{1}{2}}$ KE-space	Fuz. W pre $T_{3\frac{1}{2}}$ KE-space	Fuz. W pre $T_{3\frac{1}{2}}$ K-space	Fuz. W pre $T_{3\frac{1}{2}}$ KE-space
19	$\varphi1=int\tau \circ cl\tau \circ int\tau$, $\varphi2=1_LX$ $\psi_1=int_l, \psi_2=cl_l$	Fuz.W λ $R_{2\frac{1}{2}}$ 2 H-space	Fuz.W λ $R_{2\frac{1}{2}}$ K-space	Fuz.W λ $R_{2\frac{1}{2}}$ KE-space	Fuz.W λ $T_{3\frac{1}{2}}$ H-space	Fuz.W λ $T_{3\frac{1}{2}}$ K-space	Fuz.W λ $T_{3\frac{1}{2}}$ KE-space
20	$\varphi1=cl\tau \circ int\tau \circ cl\tau$, $\varphi2=1_LX$ $\psi_1=int_l, \psi_2=cl_l$	Fuz. W β v H-space	Fuz. W β $R_{2\frac{1}{2}}$ K-space	Fuz. W β $R_{2\frac{1}{2}}$ KE-space	Fuz. W β $T_{3\frac{1}{2}}$ H-space	Fuz. W β $T_{3\frac{1}{2}}$ K-space	Fuz. W β $T_{3\frac{1}{2}}$ KE-space

21	$\varphi 1=cl\tau \circ int\tau$, $\varphi 2=1_L X$ $\psi_1=int_I, \psi_2=int_I \circ cl_I$	Fuz. A semi $R_{2\frac{1}{2}}$ H-space	Fuz. A semi $R_{2\frac{1}{2}}$ K-space	Fuz. A semi $R_{2\frac{1}{2}}$ KE-space	Fuz. A semi $T_{3\frac{1}{2}}$ H-space	Fuz. A semi $T_{3\frac{1}{2}}$ K-space	Fuz. A semi $T_{3\frac{1}{2}}$ KE-space
22	$\varphi 1=int\tau \circ cl\tau \circ int\tau$, $\varphi 2=1_L X$ $\psi_1=int_I, \psi_2=int_I \circ cl_I$	Fuz.A λ $R_{2\frac{1}{2}}$ H-space	Fuz.A λ $R_{2\frac{1}{2}}$ K-space	Fuz.A λ $R_{2\frac{1}{2}}$ KE-space	Fuz.A λ $T_{3\frac{1}{2}}$ H-space	Fuz.A λ $T_{3\frac{1}{2}}$ K-space	Fuz.A λ $T_{3\frac{1}{2}}$ KE-space
23	$\varphi 1=cl\tau \circ int\tau \circ cl\tau$, $\varphi 2=1_L X$ $\psi_1=int_I, \psi_2=int_I \circ cl_I$	Fuz.A β $R_{2\frac{1}{2}}$ H-space	Fuz.A β $R_{2\frac{1}{2}}$ K-space	Fuz.A β $R_{2\frac{1}{2}}$ KE-space	Fuz.A β $T_{3\frac{1}{2}}$ H-space	Fuz.A β $T_{3\frac{1}{2}}$ K-space	Fuz.A β $T_{3\frac{1}{2}}$ KE-space
24	$\varphi 1=cl\tau \circ int\tau \circ cl\tau$, $\varphi 2=1_L X$ $\psi_1=int_I, \psi_2=int_I \circ cl_I$	Fuz.θ semi $R_{2\frac{1}{2}}$ H-space	Fuz.θ semi $R_{2\frac{1}{2}}$ K-space	Fuz.θ semi $R_{2\frac{1}{2}}$ KE-space	Fuz.θ semi. T $T_{3\frac{1}{2}}$ H-space	Fuz.θ semi _ K-space	Fuz.θ semi $R_{2\frac{1}{2}}$ KE-space
25	$\varphi 1=cl\tau \circ int\tau$, $\varphi 2=1_L X$ $\psi_1=int_I, \psi_2=Scl_I$	Fuz. semi. W. $R_{2\frac{1}{2}}$ H-space	Fuz. semi. W. _ K-space	Fuz. semi. W. $R_{2\frac{1}{2}}$ KE-space	Fuz. semi. W. $T_{3\frac{1}{2}}$ H-space	Fuz. semi. W. $T_{3\frac{1}{2}}$ K-space	Fuz. semi. W. $T_{3\frac{1}{2}}$ KE-space
26	$\varphi 1=int\tau \circ cl\tau \circ int\tau$, $\varphi 2=1_L X$ $\psi 1=intI \circ clI \circ intI$, $\psi 2=1_L I$	Fuz.λ. irr. $R_{2\frac{1}{2}}$ H-space	Fuz.λ. irr. $R_{2\frac{1}{2}}$ K-space	Fuz.λ. irr. $R_{2\frac{1}{2}}$ KE-space	Fuz.λ. irr. $T_{3\frac{1}{2}}$ H-space	Fuz.λ. irr. $T_{3\frac{1}{2}}$ K-space	Fuz.λ. irr. $T_{3\frac{1}{2}}$ KE-space
27	$\varphi 1=int\tau \circ cl\tau$, $\varphi 2=1_L X$ $\psi 1=intI \circ clI$, $\psi 2=1_L I$	Fuz. pre-irr. R $R_{2\frac{1}{2}}$ H-space	Fuz. pre-irr. $R_{2\frac{1}{2}}$ K-space	Fuz. pre-irr. $R_{2\frac{1}{2}}$ KE-space	Fuz. pre-irr. $T_{3\frac{1}{2}}$ H-space	Fuz. pre-irr. $T_{3\frac{1}{2}}$ K-space	Fuz. pre-irr. $T_{3\frac{1}{2}}$ KE-space
28	$\varphi 1=cl\tau \circ int\tau \circ cl\tau$, $\varphi 2=1_L X$ $\psi 1=clI \circ intI\ clI$, $\psi 2=1_L I$	Fuz.β. irr. $R_{2\frac{1}{2}}$ H-space	Fuz.β. irr. $R_{2\frac{1}{2}}$ K-space	Fuz.β. irr. $R_{2\frac{1}{2}}$ KE-space	Fuz.β. irr. $T_{3\frac{1}{2}}$ H-space	Fuz.β. irr. $T_{3\frac{1}{2}}$ K-space	Fuz.β. irr. $T_{3\frac{1}{2}}$ KE-space
29	$\varphi 1=int\tau$, $\varphi 2=1_L X$ $\psi_1=int_I \circ cl_I, \psi_2=cl_I$	Fuz.(θ, S) $R_{2\frac{1}{2}}$ H-space	Fuz.(θ, S) $R_{2\frac{1}{2}}$ K-space	Fuz.(θ, S) $R_{2\frac{1}{2}}$ KE-space	Fuz.(θ, S) $T_{3\frac{1}{2}}$ H-space	Fuz.(θ, S) $T_{3\frac{1}{2}}$ K-space	Fuz.(θ, S) $T_{3\frac{1}{2}}$ KE-space

Table 1: Some special classes of Char.fuzzy $R_{2\frac{1}{2}}$ H-spaces, Char.fuzzy $R_{2\frac{1}{2}}$ K-spaces, Char.fuzzy $R_{2\frac{1}{2}}$ KE-spaces, Char.fuzzy $T_{3\frac{1}{2}}$ H-spaces, Char. fuzzy $T_{3\frac{1}{2}}$ K-spaces, Char.fuzzy $T_{3\frac{1}{2}}$ KE-spaces.

References

1. Eklund P, Gahler W (1992) Fuzzy filter functors and convergence. In: Applications of Cat-egory Theory Fuzzy Subsets, Kluwer Academic Publishers, Dordrecht, pp: 109-136.
2. Gahler W (1995) The general fuzzy filter approach to fuzzy topology. I Fuzzy Sets and Systems 76: 205-224.
3. Gahler S, Gahler W (1994) Fuzzy real numbers. Fuz Se Syst 66: 137-158.
4. Kasahara S (1979) Operation-compact spaces. Math Japon 24: 97-105.
5. Abd El-Monsef ME, Zeyada FM, Mashour AS, El-Deeb SN (1983) Operations on the power set P (X) of a topological space (X, T), Colloquium on topology. Janos Bolyai Math Soc Eger, Hungry.
6. Kandil A, Abd-Allah AS, Nouh AA (1999), Operations and its applications on L-fuzzy bitopological spaces. Part I, Fuzzy Sets and Systems 106: 255-274.
7. Abd-Allah AS (2002) General notions related to fuzzy filters. J Fuz Math 10: 321-358.
8. Abd-Allah S, El-Essawy M (2003) On characterizing notions of characterized spaces Fuzzy. J Math 11: 853-857.
9. Abd-Allah AS and Al-Khedhairi AA Characterized fuzzy $R_{2\frac{1}{2}}$ and characterized fuzzy $T_{3\frac{1}{2}}$ spaces, Indian Univ. Math. Journal, (submitted).
10. AbdAllah A.S (2010) On characterized fuzzy regular and fuzzy normal spaces. J Egypt Math Soc 18: 247-276.
11. Lowen R (1978) A comparison of different compactness notions in fuzzy topological spaces. J Math. Anal Appl 64: 446-454.
12. Kandil A, El-Shafee ME (1988) Regularity axioms in fuzzy topological spaces and F R_i-Proximities. Fuzzy Sets and Systems 27: 217-231.
13. Katsaras A (1980) Fuzzy proximity and fuzzy completely regular spaces, I. An Stinto Univ Issi Sect Ia Math 26: 31-41.
14. Gahler W (1995) The general fuzzy filter approach to fuzzy topology. I Fuzzy Sets and Systems 205-224.
15. Bayoumi F and Ibedou I (2004) The relation between G i-spaces and fuzzy proximity spaces, fuzzy uniform spaces. Chaos Solitons and Fractals 20: 955-966.
16. Chang CL (1968) N Fuzzy topological spaces. J Math Anal Appl 24: 182-190.
17. Lowen R(1979) Convergence in fuzzy topological spaces. General Topology and Appl 10: 147-160.
18. Goguen JA(1967) L-fuzzy sets. J Math Anal Appl 18: 145-174.
19. Lowen R (1976) Fuzzy topological spaces and fuzzy compactness. J Math.Anal Appl 56: 621-633.
20. Abd-Allah S (2013) Characterized proximity fuzzy spaces, Characterized compact fuzzy spaces and Characterized uniform fuzzy spaces. Mathematical Theory and Modelings 3: 98-119.
21. Gahler W, Abd-Allah AS and Kandil A (2000) On Extended fuzzy topologies, Fuzzy Sets and Systems 109: 149-172.
22. Bayoumi F and Ibedou I (2006) G $T_{3\frac{1}{2}}$ spaces, I, J. Egypt. Math Soc. 14: 243-264.
23. Gahler W, Bayoumi F, Kandil A, Nouh AA (1998) The theory of global fuzzy neighborhood structures, II, fuzzy topogenous orders. Fuzzy Sets and Systems 98: 153-173.
24. Adamek J, Herrlich H, Strecker G (1990) Abstract and Concrete Categories, John Willy and Sons Inc, pp: 1-524.
25. Gahler W (1977) Grundstrukturen der Analysis, Akademie Verlag Berlin, Birkhauser Verlag Basel-Stuttgart.
26. Abd-Allah SA (2008) Fuzzy regularity axioms and fuzzy $T_{3\frac{1}{2}}$ axioms in characterized spaces. J Egypt Math Soc 16: 225-253.
27. Abd-Allah S (2007) Some separation axioms in characterized fuzzy spaces. J Fuzzy Math 5: 291-329.
28. Hutton (1977) Uniformities on fuzzy topological spaces. J Math Ana Appl 58: 559-571.

Permissions

All chapters in this book were first published in JACM, by OMICS International; hereby published with permission under the Creative Commons Attribution License or equivalent. Every chapter published in this book has been scrutinized by our experts. Their significance has been extensively debated. The topics covered herein carry significant findings which will fuel the growth of the discipline. They may even be implemented as practical applications or may be referred to as a beginning point for another development.

The contributors of this book come from diverse backgrounds, making this book a truly international effort. This book will bring forth new frontiers with its revolutionizing research information and detailed analysis of the nascent developments around the world.

We would like to thank all the contributing authors for lending their expertise to make the book truly unique. They have played a crucial role in the development of this book. Without their invaluable contributions this book wouldn't have been possible. They have made vital efforts to compile up to date information on the varied aspects of this subject to make this book a valuable addition to the collection of many professionals and students.

This book was conceptualized with the vision of imparting up-to-date information and advanced data in this field. To ensure the same, a matchless editorial board was set up. Every individual on the board went through rigorous rounds of assessment to prove their worth. After which they invested a large part of their time researching and compiling the most relevant data for our readers.

The editorial board has been involved in producing this book since its inception. They have spent rigorous hours researching and exploring the diverse topics which have resulted in the successful publishing of this book. They have passed on their knowledge of decades through this book. To expedite this challenging task, the publisher supported the team at every step. A small team of assistant editors was also appointed to further simplify the editing procedure and attain best results for the readers.

Apart from the editorial board, the designing team has also invested a significant amount of their time in understanding the subject and creating the most relevant covers. They scrutinized every image to scout for the most suitable representation of the subject and create an appropriate cover for the book.

The publishing team has been an ardent support to the editorial, designing and production team. Their endless efforts to recruit the best for this project, has resulted in the accomplishment of this book. They are a veteran in the field of academics and their pool of knowledge is as vast as their experience in printing. Their expertise and guidance has proved useful at every step. Their uncompromising quality standards have made this book an exceptional effort. Their encouragement from time to time has been an inspiration for everyone.

The publisher and the editorial board hope that this book will prove to be a valuable piece of knowledge for researchers, students, practitioners and scholars across the globe.

List of Contributors

Selvarani S and Beulah RD
Department of Mathematics, VLB Janakiammal College of Arts and Science, Coimbatore-641 042, Tamilnadu, India

Salahuddin
Department of Mathematics, Jazan University, Jazan, Kingdom of Saudi Arabia

Ferreira Cordeiro SG and Leonel ED
Department of Structural Engineering, School of Engineering of Sao Carlos, University of Sao Paulo, Brazil

Jameel AF
School of Mathematical Sciences, 11800 USM, University Science Malaysia, Penang, Malaysia

Abdel-Baset H Mekky and Abdulaziz S Al-Aboodi
Buraydah Colleges, Al-Qassim, King of Saudi Arabia

Hanan G Elhaes
Faculty of Women for Arts, Science, and Education, Ain Shams University, Egypt

Mohamed M. El-Okr
Faculty Science, El Azhar University, Egypt

Medhat A. Ibrahim
Spectroscopy Department, National Research Centre, Egypt

Adel Ouannas
LAMIS Laboratory, Department of Mathematics and Computer Science, University of Tebessa, Algeria

Fuhua Chen and Katherine Hastings
West Liberty University, West Virginia, USA

Debiprasad Acharya
Department of Mathematics, N.V.College, Nabadwip, Nadia-741302, W.B, India

Manjusri Basu and Atanu Das
Department of Mathematics, University of Kalyani, Kalyani-741235, India

Don Liu and Yifan Wang
Mathematics and Statistics, Louisiana Tech University Ruston, LA 71272, USA

James Tesiero
University of Maine, Orono, USA

Abd-Allah AS
Department of Mathematics, College of Science, El-Mansoura University, El-Mansoura, Egypt

Al-Khedhairi A
Department of Statistics and Operations Research, College of Science, King Saud University, PO Box 2455, Riyadh 11451, Saudi Arabia

Kathuria L and Raka M
Centre for Advanced Study in Mathematics, Panjab University, Chandigarh-160014, India

Mohammed El Mokhtar Ould El Mokhtar
Qassim University, Department of Mathematics, Qassim University, Buraidah, Kingdom of Saudi Arabia

Abhijit Baidya, Uttam Kumar Bera and Manoranjan Maiti
Department of Mathematics, National Institute of Technology, Agartala, Jirania 799055, West Tripura, India

Jehad R Kider
Department of Applied Science, University of Technology, Iraq

Fabien Kenmogne
Laboratory of Modelling and Simulation in Engineering and Biological Physics, Faculty of Science, University of Yaound´e I, Cameroon

Ghosh S
Department of Computer Science, BIT Mesra, Kolkata Extension Centre, 1582 Rajdanga Main Road, Kolkata-700107, India

Sohalya MA, Yassena MT and Elbaza IM
Faculty of Science, Department of Mathematics, Mansoura University, Egypt

Ahmed S and Bano Z
Department of Mathematics, Sukkur Institute of Business Administration, Sukkur, 62500, Pakistan

El-Damcese MA and Shama MS
Department of Mathematics, Faculty of Science, Tanta University, Tanta, Egypt

Shawky AI
Department of Statistics, Faculty of Science, King Abdulaziz University, Saudi Arabia

Badr MM
Department of Statistics, Faculty of Science for Girls, King Abdulaziz University, Saudi Arabia

Essa KSM, Marrouf AA and El-Otaify MS
Mathematics and Theoretical Physics, NRC, Atomic Energy Authority, Cairo, Egypt

Mohamed AS and Ismail G
Department of Mathematics, Faculty of Science, Zagazig University, Egypt

Eghbal Hosseini
Philosophy Doctor of Operational Research and Optimization, Department of Mathematics, UAE

Raslan KR and Ali KK
Mathematics Department, Faculty of Science, Al-Azhar University, Nasr-City, Cairo, Egypt

EL-Danaf TS
Mathematics Department, Faculty of Science, Menoufia University, Shebein El-Koom, Egypt

Mohedul Hasan Md and Abdul Matin Md
Department of Mathematics, Dhaka University, Dhaka, Bangladesh

Hazaimeh MH
Department of Mathematics, Zayed University, Dubai, UAE

Mohammadi B and Alizade E
Department of Mathematics, Marand Branch, Islamic Azad University, Tehran, Iran

Mansour Alyazidi-Asiry
Department of Mathematics, College of Sciences, King Saud University, Riyadh, Saudi Arabia

Jo Boaler, Lang Chen, Cathy Williams and Montserrat Cordero
Stanford Graduate School of Education & Stanford Cognitive & Neuroscience Lab, Stanford University, USA

GuagSan Y
Harbin Macro Dynamics Institute, P. R. China

Joseph KA
Federal University of Technology Akure, Nigeria

Mishra VN
Applied Mathematics and Humanities Department, Sardar Vallabhbhai National Institute of Technology, 395 007, Surat, Gujarat, India

Deepmala
SQC and OR Unit, Indian Statistical Institute, 203 B. T. Road, 700 108, Kolkata, West Bengal, India

Subramanian N
Department of Mathematics, SASTRA University, 613 401, Thanjavur, India

Mishra LN
Department of Mathematics, National Institute of Technology, 788 010, Silchar, District Cachar, Assam, India

Kayode SJ and Obarhua FO
Department of Mathematical Sciences, Federal University of Technology, Nigeria

Ahmed Saeed Abd-Allah
Department of Statistics and Operations Research, College of Science, King Saud University, Saudi Arabia

A Al-Khedhairi
Department of Mathematics, College of Science, El-Mansoura University, El-Mansoura, Egypt

Index

www.ingramcontent.com/pod-product-compliance
Lightning Source LLC
Chambersburg PA
CBHW080628200326
41458CB00013B/4553